HANDBOOK OF WEAVING

Sabit Adanur, B.S., M.S., Ph.D.
Professor, Department of Textile Engineering
Auburn University, Alabama, U.S.A.

SULZER

A CRC title, part of the Taylor & Francis imprint, a member of the Taylor & Francis Group, the academic division of T&F Informa plc.

Published in 2001 by
CRC Press
Taylor & Francis Group
6000 Broken Sound Parkway NW, Suite 300
Boca Raton, FL 33487-2742

CRC Press is an imprint of Taylor & Francis Group

No claim to original U.S. Government works
Printed in the United States of America on acid-free paper
10 9 8 7 6 5 4 3 2

International Standard Book Number-10: 1-58716-013-7 (Hardcover)
International Standard Book Number-13: 978-1-58716-013-4 (Hardcover)
Library of Congress Card Number 00-107625

Library of Congress Cataloging-in-Publication Data

Main entry under title:
Handbook of Weaving

Taylor & Francis Group
is the Academic Division of T&F Informa plc.

Visit the Taylor & Francis Web site at
http://www.taylorandfrancis.com

and the CRC Press Web site at
http://www.crcpress.com

Table of Contents

Introduction

Clothing is one of the three basic needs of human kind, the other two being food and shelter. Therefore, fabric formation in general and weaving in particular are probably as old as human history. One of the first necessities of early humans was a piece of cloth to cover their body for decency and to protect themselves from the adverse effects of the environment.

Weaving is a mixture of science and art. Despite all the technological advances, weaving is still not a positively controlled process. That is, it is hardly possible to control the individual fiber which is the smallest meaningful building block in a woven structure. This fact makes weaving an interesting technology.

The flow of the book follows the flow of the woven fabric manufacturing process. Chapter 1 is a brief introduction to fabric formation by weaving. Major steps in various fabric manufacturing and technology are summarized. A brief history of fabric formation by weaving is given. Chapter 2 gives an overview of polymers, fibers and yarns used in woven fabrics, for the reader's convenience. These topics are not the subject of the book and therefore their summaries are included only as a convenience to the reader. There are some excellent books in these areas and the reader is advised to refer to them for in-depth information. Woven fabric design and construction is included in Chapter 3. Basic woven fabric designs are explained. There are an endless number of woven designs that would be impossible to cover in this book. However, most of those designs are based on the six basic designs that are explained in the chapter. Chapter 4 is about weaving preparation which is an essential part of woven fabric manufacturing. Winding, warping, slashing (sizing), drawing-in and tying-in are included in this chapter. After these processes, the warp and filling yarns are ready to be converted to a woven fabric structure. Basic weaving fundamentals are included in Chapter 5. Warp let-off, shedding, filling insertion, beat-up, take-up and fabric control are introduced. Then, in the following several chapters basic weave motions are covered in detail and machine specific. Chapter 6 deals with major shedding systems: cam, dobby and jacquard. A summary of the shuttle loom, which is becoming obsolete, is given in Chapter 7. Air-jet weaving is the subject of Chapter 8. This chapter also includes the other fluid insertion system, water-jet. Projectile weaving and rapier weaving are the subjects of Chapters 9 and 10, respectively. For the first time in the history of weaving, there is a commercial multiphase weaving machine on the market. This new machine is included in Chapter 11, along with a brief history of other multiphase developments. Certain fabric types such as denim, terry fabrics and industrial fabrics, require special attention during weaving. Chapter 12 is devoted to the manufacturing of these special fabrics. Since the end product of weaving is the fabric, a book on weaving would be incomplete without some discussion of fabric structure and properties. This is the subject of Chapter 13. Weaving plant operations are included in Chapter 14. Computers and automation in the weave room are also included. A brief discussion of ISO quality and environmental standards is also given. Finally, an attempt is made in Chapter 15 to review the trends for the future of weaving.

Units present a challenge in writing a book in this field. Both metric and standard units had to be used in the book. Quite often, the corresponding value of a quantity in the other unit system is given in parenthesis. It should be noted that, sometimes, rather than giving the exact conversion, the value that is used most in practice in that unit system is included in the parenthesis. A unit conversion table, to convert units from one system to the other, is given in Appendix 5.

This book could not have been a reality without help. I thank God for everything that made this book possible. I would like to thank Mr. Pete Egloff who made his company's human and technical resources freely available for the successful outcome of the book. Mr. Louie Dejonckheere and Mr. Rene Koenig provided some of the technical information for the book and offered editorial comments. Special thanks are extended to Prof. Dr. William K. Walsh, Textile Engineering Department Head, Prof. Dr. Larry Benefield, Dean of the College of Engineering and other officials of Auburn University for fostering an environment that allows and encourages productive work. Last, but not least, this book would not have been possible without my wife Nebiye's continuous help and support.

I would like to acknowledge many people, institutions, uiniversities and companies who have contributed to the book by providing information, pictures, graphs and data. Each chapter has been reviewed and edited by several professionals for technical content, and this is greatly appreciated. I would like to thank the following individuals for their help in preparation of this book: Uwe Nick, Urs Nef, Karl Peter, Bruce Luther, Don Cotney, Steve Sartain, Bernhard Egli, Ernst Frick, Peter Frei, Werner Mendler and Hanspeter Trumpi. My sincere apologies to those whose names I might have failed to mention for their valuable time and contributions.

There has been a direct correlation between a nation's standard of living and the amount of textiles used. This makes the future of the textile industry and weaving process bright. My hope is that this book will be useful for industry and academic professionals as well as students in shaping that future.

SABIT ADANUR

1

Fabric Formation by Weaving

1.1 FABRIC FORMATION METHODS

In general terms, a textile fabric may be defined as an assembly of fibers, yarns or combinations of these. There are several ways to manufacture a fabric. Each manufacturing method is capable of producing a wide variety of fabric structures that depend on the raw materials used, equipment and machinery employed and the set up of control elements within the processes involved. Fabric selection for a given application depends on the performance requirements imposed by the end use and/or the desired aesthetic characteristics of the end user, with consideration for cost and price. Fabrics are used for many applications such as apparel, home furnishings, and industrial.

Figure 1.1 shows the major steps in fabric manufacturing. Polymers are the resource for man-made fibers. Polymers are derived mostly from oil. Plant fibers and animal fibers constitute the natural fibers. After the fabric is formed, it is generally subjected to a finishing and/or dyeing process, in which the raw fabric properties are modified for the end use.

The most commonly used fabric forming methods are weaving, braiding, knitting, tufting, and nonwoven manufacturing. Figure 1.2 shows the schematics of fabrics produced by these methods.

Weaving is the interlacing of warp and filling yarns perpendicular to each other. There are practically an endless number of ways of interlacing warp and filling yarns. Each different way results in a different fabric structure.

Braiding is probably the simplest way of fabric formation. A braided fabric is formed by diagonal interlacing of yarns. Although there are two sets of yarns involved in the process, these are not called warps and fillings as in the case of woven fabrics. Each set of yarns moves in an opposite direction. Braiding does not require shedding, filling insertion, and beat up. The yarns do not have to go through harnesses and reed. Braiding is generally classified as two dimensional and three dimensional braiding. Two dimensional braiding includes circular and flat braids. The application of two dimensional braiding is very limited in apparel manufacturing. Three dimensional braiding is a relatively new topic, and mainly developed for industrial composite materials.

Knitting is interlooping of one yarn system into vertical columns and horizontal rows of loops called wales and courses, respectively. There are two main types of knitting: weft knitting and warp knitting. In weft knitting, the yarns flow along the horizontal direction in the structure (filling or course direction); in warp knitting, they flow along the vertical direction

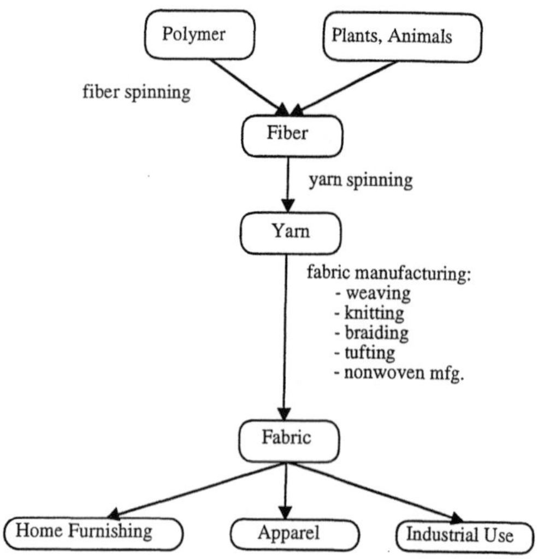

FIGURE 1.1 Major steps in fabric manufacturing.

(warp or wale direction). Special needles are used to form the yarn loops. The basis of knit fabric structure is the continuing intersection of loops. This feature provides unique characteristics to the knit fabrics compared to woven and braided fabrics. Practically, there are an endless number of knit fabric structures. Knit fabrics are widely used in apparel and home furnishings. They are also used in technical textiles, such as artificial arteries, bandages, casts, composites, sporting equipment, etc.

Tufting is the process of manufacturing some carpets and similar structures. A surface yarn system of loops is "sewn" or "stitched" through a primary backing fabric, usually a woven or nonwoven fabric. The loops are arranged in vertical columns (rows) and horizontal lines (stitches). Loops can be in the form of cut or uncut loops (piles) or a combination of thereof. The fabric is usually back-coated in a later process to secure tufted loops.

Bonding is the method of manufacturing nonwovens using either textile, paper, extrusion, or some combination of these technologies, to form and bond polymers, fibers, filaments, yarns or combination sheets into a flexible, porous structure. The resulting structure is quite different than the structures mentioned above. In fact, some nonwoven products are claimed by both the textile industry and paper industry.

Each of these major fabric manufacturing processes produces unique structures. Big industries have evolved around each method of fabric manufacturing in almost every country in the world. Around 40 countries have major textile industries and another 50 countries produce woven fabrics in various amounts. Approximately 70% of the fabrics made in the world are woven fabrics. Figure 1.3 shows the major steps in manufacturing of finished woven fabrics. This book is concerned only with the weaving process and woven fabrics.

Weaving of yarns into a fabric is performed on a "weaving machine" which has also been called a "loom". However, "loom" is more descriptive of a "shuttle weaving machine". For shuttleless "looms", the term "weaving machine" is more appropriate since these machines are as sophisticated as any other machine that exists today. A typical modern weaving machine consists of up to 5000 parts working together in a very precise manner. Although shuttle looms have been obsolete, reference is still made to them, for comparison purposes, when describing shuttleless weaving machines.

1.2 WEAVING AND WOVEN FABRICS

1.2.1 History of Weaving

Weaving is probably as old as human civilization. One of the basic necessities of humans is to cover their bodies to protect themselves from outside effects (hot, cold) and look more "civilized" to the eye. Other reasons for development of different clothing throughout the history are social status, religious requirements, etc. Clothing trends depend on location as well.

Historical findings suggest that Egyptians made woven fabrics some 6000 years ago. Chinese made fine fabrics from silk over 4000 years ago. It is believed that the hand loom has been invented many times in different civilizations [2].

Weaving started as a domestic art and stayed as a cottage industry until the invention of the fly shuttle. The fly shuttle, invented in 1733 by Kay, was hand operated. In 1745, de Vaucanson

a. woven

b. braided

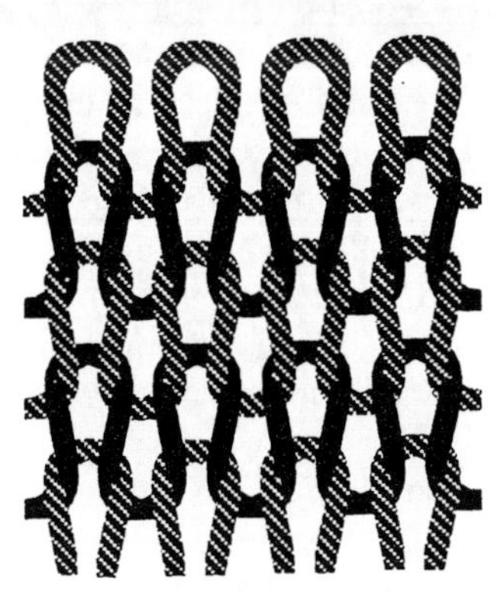

c. weft knit

d. warp knit

e. tufted

f. nonwoven

FIGURE 1.2 Schematics of major fabric types [1].

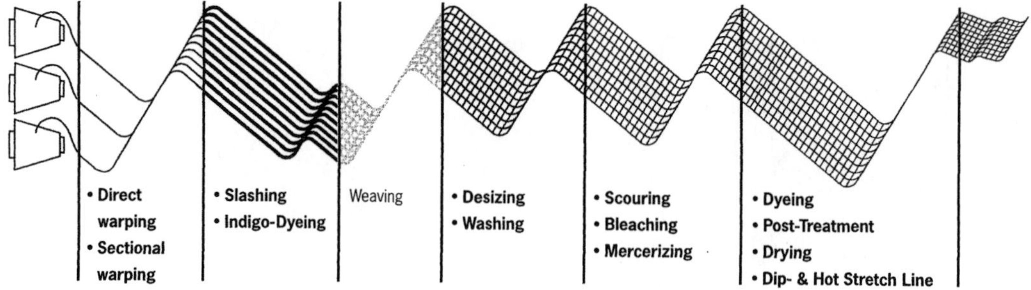

FIGURE 1.3 Major manufacturing steps of woven fabrics (courtesy of Benninger).

made a loom, further developed by Jacquard, to control each warp end separately. In 1785, Cartwright invented the power loom which could be operated from a single point. In the early 1800s, looms made of cast iron, were operated by steam power. Power loom required stronger warp yarn, resulting in development of the first sizing machine in 1803. In the 1830s, there were some 100,000 shuttle looms operating in England. The processing principles of these looms were pretty much the same as today's shuttle looms.

In the early 20th century, improvements were made in the winding and warping of yarns. The looms were improved further, including warp tying machines and warp drawing-in machines. After the end of the World War II, the modern textile industry began to emerge. Invention of synthetic fibers changed the scope of textile industry drastically. In 1930 an engineer named Rossmann developed the first prototype of projectile weaving machines. In 1953 the first commercial projectile weaving machines were shipped. Production of rapier and airjet weaving machines started in 1972 and 1975, respectively.

The fundamental principle of weaving has remained unchanged for centuries. Today, as in the past, woven fabrics are made by crossing yarns over and under at right angles to one another. This method of producing textiles has many advantages, e.g., stability and resistance to deformation by compression and tensile stress. These are the features that distinguish woven fabrics from the cheaper nonwovens and knitted goods. Until recently, all the woven fabrics in the world have been produced on single phase weaving machines, and the focus in technological progress has been on accelerating the conventional weaving process. Over a period measured in centuries the filling insertion rate, for example, has increased from a few meters per minute to over 2000 m/min. Major increases have been revealed at textile machinery shows. There are three major textile machinery shows in the world: ITMA which is held every four years in Europe, ATME-I held every four years in the US, and OTEMAS held in Japan. These machinery shows usually alternate with each other. The filling insertion rates of today's conventional single phase weaving machines are approaching physical limits. Further major increases in performance can only be achieved with new technologies such as multi phase weaving machines (Chapter 11).

1.2.2 Weaving Machines

In practice, the weaving machines are named after their filling insertion systems. Schematics of the filling insertion systems that are used in the market are shown in Figure 1.4. Based on the filling insertion systems, the weaving machines can be classified as shuttle and shuttleless weaving machines. Shuttle looms have been used for centuries to make woven fabrics. In this type of loom, a shuttle, which carries the filling yarn wound on a quill, is transported from one side to the other and back. In the mid 20th century, other weaving machines started to emerge that used other forms of filling insertion mechanisms such as air, projectile, rapier and water. In reference to shuttle looms, these machines are called shuttleless looms or shuttleless weaving machines. Today, the shuttle looms have become

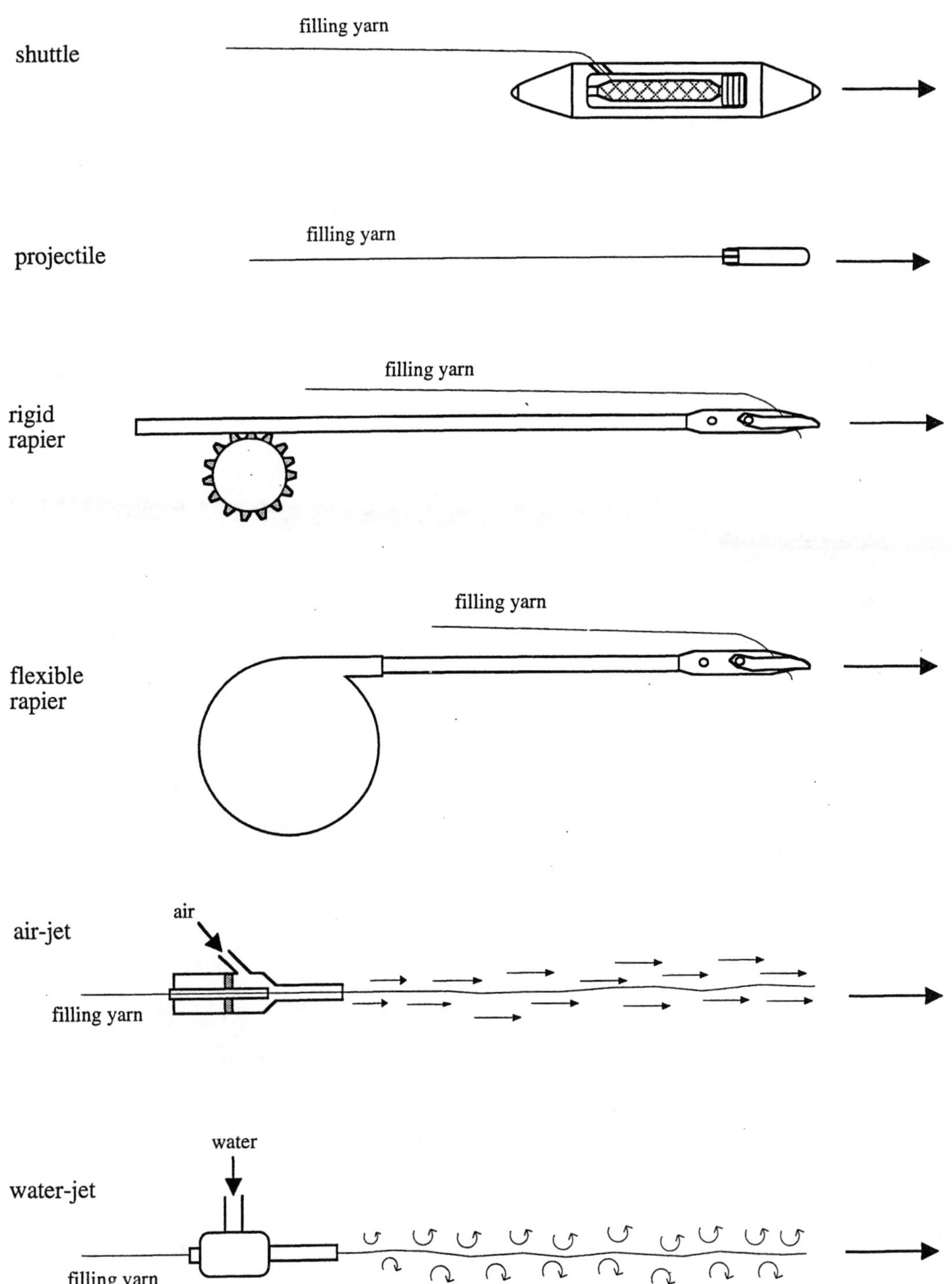

FIGURE 1.4 Principles of filling insertion systems.

obsolete and are not manufactured anymore except for some very special niche markets. The existing shuttle looms have been replaced by the shuttleless weaving machines in industrialized countries. Nevertheless, approximately 2.6 million of the 3.2 million looms in existence throughout the world in 1998 were still shuttle looms [3]. However, they are being replaced by the shuttleless weaving machines at a fast rate. Today, the three most popular weaving machines are air-jet, rapier and projectile machines (Figure 1.5). Table 1.1 gives a comparison of these machines, which are all single phase machines.

The total number of weaving machines is decreasing in the textile industry. However, it should be noted that the total productivity is either increasing or staying constant. Since the early beginnings of mechanical weaving, productivity in the weaving mill has practically doubled every 25 years. For example, square yards of woven fabric per loom hour have increased 355% in the US since 1974 [4]:

1974	8.3
1984	10.3
1994	29.5

Mill consumption of raw materials has increased 50% since 1974 [4]:

1974	10.8 billion pounds
1984	10.8 billion pounds
1994	16.2 billion pounds

Appendix 1 shows the latest weaving machines and their producers in the world.

1.2.3 Woven Fabrics

Woven fabrics can be classified in many ways:

(1) Classification by weave type, i.e., plain, twill, satin, leno, etc.
(2) Classification by common names, i.e., denim, cheesecloth, percale, etc.
(3) Classification by weight, i.e., heavy fabrics and light fabrics
(4) Classification by coloration method, i.e., solution dyed, stock dyed, yarn or piece dyed
(5) Classification by the end-use, i.e., apparel fabrics, home furnishings and industrial fabrics

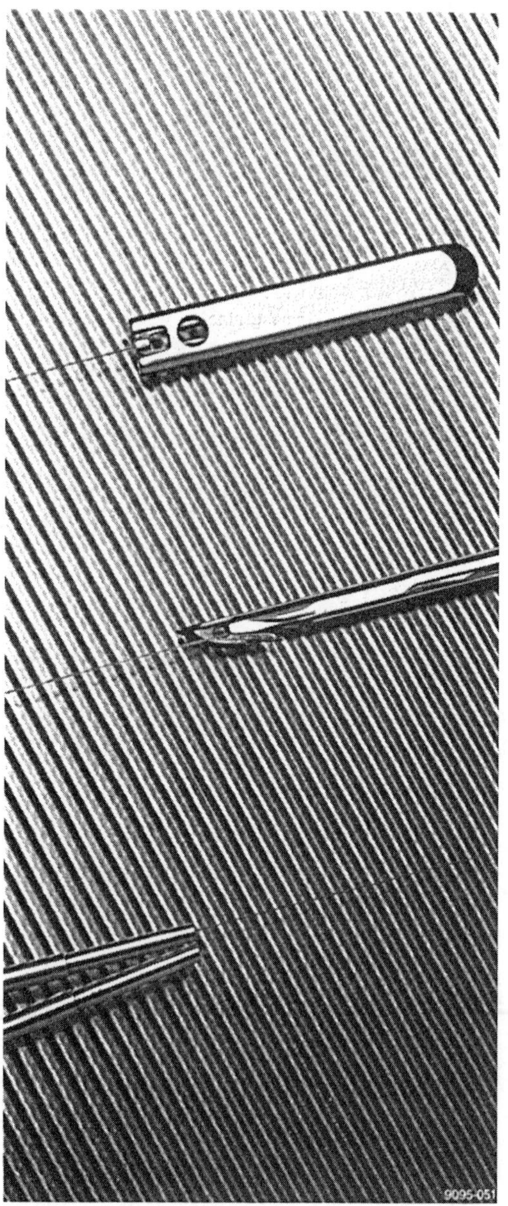

FIGURE 1.5 The three most popular filling insertion devices: projectile (top), rapier (middle) and air-jet.

It is reported that after the period of survival, a typical family in a developed country may spend up to 20% of their income on clothing. Industrialized nations spend more on clothing. Fashion changes garments every few months. Although throwing away textiles before the end of their usable lives is wasteful, it stimulates new developments and business. Not fully used garments are either re-used by somebody else

TABLE 1.1 Typical characteristics of single phase projectile, rapier and air-jet weaving machines.

	Projectile	Rapier	Air-jet
Weft insertion rate (maximum, m/min)	1400	1260	2000
Nominal width (cm)	190–540	140–360	140–430
Weft yarn color	1–6	2–12	2–4
Shed formation	cam	cam	cam
	dobby	dobby	dobby
	jacquard	jacquard	inside treadle motion
			jacquard

or recycled as secondary raw materials for the textile industry.

REFERENCES

1. Adanur, S., Wellington Sears Handbook of Industrial Textiles, Technomic Publishing Co., Inc., 1995.
2. Lord, P. R., and Mohamed, M. H., "Weaving: Conversion of Yarn to Fabric", Merrow Technical Library, 1982.
3. Seidl, R., "Current Trends in Weaving Machine Construction", Melliand International (1), 1998.
4. American Textile Manufacturers Institute (ATMI), April 1995.

SUGGESTED READING

- Rozelle, W. N., "Slashing Receives New Demands in Yarns, Processes", Textile World, December 1998.
- Slashing Short Course, Auburn University, Auburn, AL, September 1998.
- Isaacs III, M., "Loom Makers Get Ready for ITMA 99", Textile World, May 98.
- Ishida, T., "Historical Developments in Weaving Machinery", JTN, No. 454, September 1992.
- Ishida, T., "Historical Developments in Weaving Machinery", JTN, No. 460, March 1993.

REVIEW QUESTIONS

1. Compare the major fabric types shown in Figure 1.2 for the following:
   - raw materials
   - manufacturing processes
   - repeat unit
   - major properties
   - cost
2. Does a woven or knit fabric have better insulation properties? Explain.
3. In how many ways can you classify the weaving machines?
4. Give your own definition of woven fabrics.

2

Overview of Polymers, Fibers and Yarns Used in Woven Fabrics

The purpose of this chapter is to give an overview of polymers, fibers and yarns that are used in woven fabrics, as a convenience to the reader who may not be familiar with these areas. It is not the intent of this chapter to include details, but rather to give a brief introduction to these subjects. There are excellent sources in these areas and the reader is referred to those for in-depth understanding of these topics.

Woven fabrics are made of yarns, yarns are made of fiber(s), and fibers are made of polymer molecules. Therefore, a polymer is the very first step in making a woven fabric.

2.1 POLYMERS

"Poly" means "many" and "mers" means "parts". A polymer is a molecular chain-like structure from which manmade fibers are derived. A polymer is produced by linking together molecular units called monomers. Polymerization is a chemical reaction in which small molecules are combined to form much larger, long chain polymer molecules. There are two types of polymerization: addition polymerization and condensation polymerization. In addition polymerization, the molecular formula of the repeating unit is identical with the monomer. The molecular weight of the newly formed polymer is the sum of the molecular weight of the combined monomer units. Polymerization takes place by rearranging the chemical bonds. In condensation polymerization, the repeating unit of the polymer has less atoms than the monomer or monomers. Synthetic polymers are derived from oil [1].

All natural and manmade textile fibers are made of polymers. Polymers can be grouped under two categories: homopolymer and copolymer. In a homopolymer, the repeating units of the polymer are the same. Homopolymers can be periodic. A copolymer has at least two types of monomer precursors. Copolymers can be alternating, random, block, graft and branched [2,3].

The polymer names are given by the International Union of Pure and Applied Chemistry (IUPAC). However, the common name of the polymer that is used in practice may be different than the IUPAC name. Table 2.1 shows some synthetic polymers that are used to produce textile fibers. There are thousands of polymers that are patented in the world.

2.2 TEXTILE FIBERS

2.2.1 Fiber Classification and Spinning

Fibers that are used in textiles can be classified under two main categories: natural fibers and

TABLE 2.1 Examples of polymers used to make textile fibers [3].

Molecular Formula	IUPAC Name	Common Name
$—[CH_2CH_2]—$	Poly(ethene)	Polyethylene
$—[CH_2CH(CH_3)]—$	Poly(propene)	Polypropylene
$—[CH_2CH_2OCOC_6H_4COO]—$	Poly(ethylene terephthalate)	Polyester
$—[NH(CH_2)_6NHCO(CH_2)_4CO]—$	Poly(hexamethylene adipamide)	Nylon 6,6
$—[NH(CH_2)_5CO]—$	Poly(hexanolactam)	Nylon 6

man-made (synthetic) fibers. Table 2.2 shows the most commonly used fibers in textile industry. Natural fibers are found in animals or plants. Man-made fibers are produced from polymers.

Man-made fibers are manufactured by spinning the polymer. There are three major types of spinning processes: melt, dry and wet spinning (Figure 2.1). In melt spinning, the polymer is melted by heating. The molten polymer is pumped through the tiny holes of a spinnerette; thus the fiber is formed. The fiber is then cooled and solidified. In dry spinning, polymer is dissolved in a solvent. After extrusion through a spinnerette, the solvent is evaporated and the fiber is solidified. In wet spinning, the polymer is dissolved in a solvent similar to dry spinning. After extrusion, the solvent is removed in a liquid coagulating medium. Among the three methods, melt spinning is the most common. If a polymer can be melted, then melt spinning is the choice for fiber production. A typical fiber extrusion machine includes hopper, extruder, metering pump, spinneret, quench tank, finish applicators, godets and winder. Spinneret hole shapes vary depending on the end use of the fiber. Figure 2.2 shows various spinneret hole shapes. It should be noted that in addition to melt, dry and wet spinning, there are other spinning methods, as well.

Important fiber characteristics that have an impact on fabric appearance, properties, and performance are as follows:

- length
- luster
- crimp level
- color
- bulk

TABLE 2.2 Classification of commonly used fibers in textiles.

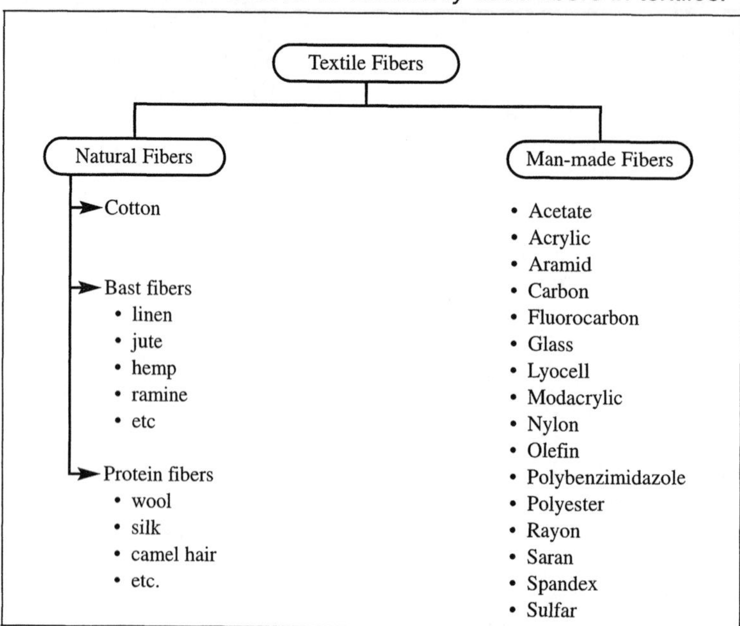

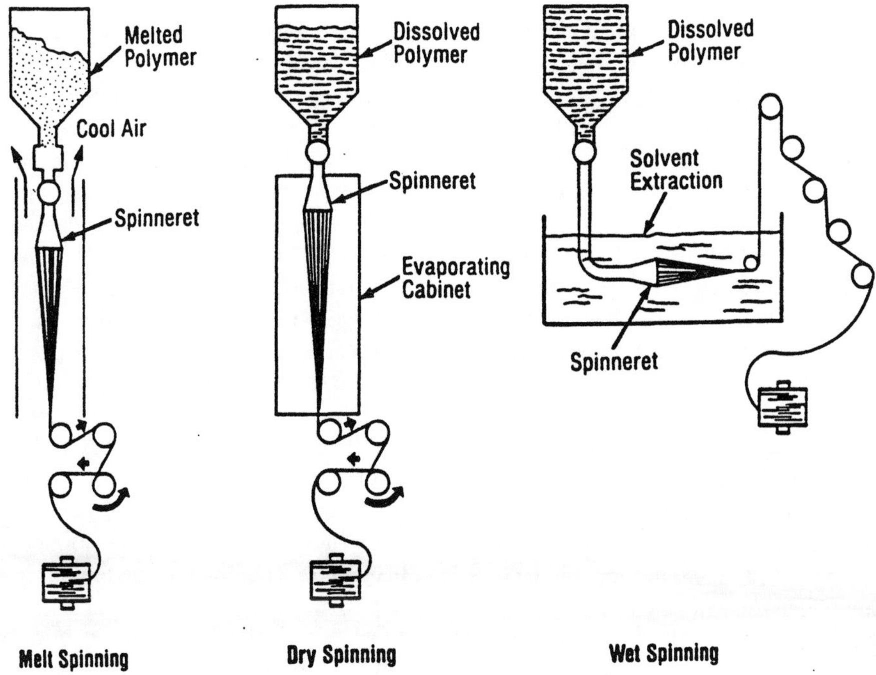

FIGURE 2.1 Major types of spinning processes (courtesy of Hoechst Celanese).

- type (staple, filament, etc.)
- stiffness (flexibility)
- grade
- cross-sectional shape
- maturity (natural fibers)
- finish presence
- surface characteristics
- density (fineness)
- strength
- melting point
- moisture content
- dye affinity
- shrinkage
- generia
- elongation
- solubility
- refractive index
- resiliency
- flame resistance
- abrasion resistance
- chemical resistance
- absorbency
- glass transition temperature

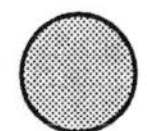

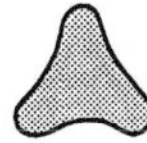

FIGURE 2.2 Examples of spinneret hole shapes.

2.2.2 Microfilaments and Microfibers

There has been a recent popularity of extremely fine fibers in the textile industry. The term "micro" is used for filaments with single titres of less than 1.0 dtex, while filaments with somewhat coarser single titres, such as 1.25 or 1.30 dtex, are referred to as fine-fibril filaments or finest filaments (Table 2.3).

TABLE 2.3 Comparison of various fiber counts.

	dtex
polyester (typical)	2.2
wool	1.7
cotton	1.4
silk	1.0
micro	0.55

In the range of staple fiber yarns, microfibers are either blends with 1.1 dtex cotton or blends with 1.3 dtex wool. This adjustment is necessary, even though finer staple fibers, e.g., 0.8 dtex, can be spun. Otherwise, microfibers would be too delicate in comparison with the natural fibers blended with them.

For this reason, problems such as the formation of slubs, would arise when spinning staple fibers. Polyester staple fibers of count 0.8 dtex are also used for ring spinning pure polyester yarns. The great quantity of fibers in the yarn cross section and the resulting increased fiber surface lead to an increase in drafting forces, and, thus, to drafting disturbances.

Spinning problems with microfibers that, in many cases, can be remedied only by lowering the processing speed, have, to date, obviated a real market penetration of staple fiber yarns.

Today's high, state of the art spinning technology enables the production of synthetic microfilaments that are a good deal finer than natural fibers. The spinning of microfilaments with fine and finest single titres is easier with polyester than with polyamide. For this reason, polyester microfilaments, such as Trevira®, Finesse®, Diolen Mikro®, Micrell® and Setila®, were first in the course of this development. Today, microfilaments (Tactel®, Meryl®) of perfect quality are made of polyamide as well. Modern spinning technology has virtually no restrictions. Technically, it should be possible to spin microfilaments with single titres of 0.1 dtex. Because of the textile properties of the fabrics, the manufacturers of man-made fibers have decided to produce polyester fibers with single titres of 0.55 to 0.65, and polyamide fibers with single titres of 0.80 to 0.85 dtex. Fabrics of microfilaments with still finer single titres are losing fastness, have poor appearance, and too soft a hand. Fabrics of this sort could be used for ready-to-wear clothes only with restrictions.

The different counts of the single polyester and polyamide titres are caused by the fact that the softer polyamide with single titres of 0.8 dtex has rather the same textile properties as polyester with single titres of 0.5 dtex.

Fabric constructions of microfibers enable a great variety of applications. Fabric finishing, therefore, is of great importance and makes considerable demands on the ingenuity of the finisher. Application areas of microfibers include functional garments (rain clothes, sportswear, mountain wear), and aesthetics and fashion (ladies' and men's outer wear, blouses and shirts).

Within the area of functional garments, the requirements focus on waterproofness but permeability to steam (Figure 2.3), wind tightness and abrasion resistance. These demands are, as a rule, fulfilled by densely woven microfilament fabrics. Whenever necessary, the desired properties are reinforced with the aid of hot calendering and waterproofing. The latter is done with water repellent fluorocarbon products. Waterproofness of up to 800 mm water column can be obtained by coating.

Fashionable garments are somewhat lighter in weight. The shaping of the surface is very important and hence varied. Presently, peachskin treatment is quite significant and is mostly carried out prior to dyeing. Another important topic is the vivid surface, e.g., crinkle, crash and embossed effects. Finishing technologies are currently seeing a real development boom, since the surface treatment of microfibers allows for the most varied combinations. Further, it is interesting to note that fashionable finishings show an ecologically positive trend.

Today, all of the leading manufacturers of man-made fibers carry microfilaments and fibers of polyester and polyamide in their range of commodities. The offered yarns are either flat or texturized. The entrenched brand names are supplemented by additional designations, such as Diolen Mikro®, Trevira Finesse®, ICI Tacel®, etc. Presently, the titre program comprises the yarn count range of dtex 50 f 72 up to dtex 334 f 574, with all current intermediate stages.

2.2.3 Processing Microfilaments

Fabrics woven of microfilaments are highly dense structures. This is a must with functional garments, in order to obtain the desired impermeability to water. With aesthetic clothes, fabric density affords appealing fabric volume and a sufficient degree of antislip properties. For filling yarns, finer titres like 50 or 76 dtex are preferred. With fine microfilaments, high filling densities call for high quality warps.

FIGURE 2.3 Waterproofness with permeability to steam.

The makers of microfilaments offer sized warp beams, preferably with different standard numbers of yarns, but also with yarn quantities demanded by the customer. Because of the high warp densities, half warp beams must first be sized and assembled afterwards. All microfilaments can, indiscriminately and without any restrictions, be processed on projectile, rapier and air-jet weaving machines. For filling insertion, prewinders and yarn brakes must be used that are suited to processing filament yarns.

In order to obtain satisfactory running behavior with texturized microfilaments, they must be more thoroughly intermingled. The setting parameters of the weaving machine must be most carefully observed so as to obtain optimum fabric appearance with these dense fabrics. Microfilaments may be used for fabric construction in the combinations listed in Table 2.4. This applies to polyester as well as to polyamide yarns.

At ITMA-99 in Paris, Corterra Fibers introduced a new fiber, Polythrimethylene Tereph-

TABLE 2.4 Warp and filling combinations for microfiber fabrics.

Usage	Warp	Filling
Functional and fashionable garments	micro flat	micro flat
	micro flat	micro texturized
Fashionable garments	micro flat	staple fiber yarns of various materials
	micro flat	PE/PA fine fibrils
	cotton	micro flat + texturized

thalate (PTT). This is the first commercial fiber introduced in the last 25 years for apparel markets. Its chemical properties are similar to PET but its physical properties are different due to micromorphology. The glass transition temperature is lower due to greater chain mobility, which allows atmospheric dyeability.

2.3 YARNS

2.3.1 Yarn Manufacturing and Classification

Yarns are the basic building blocks of woven fabrics. After the fibers are formed or obtained, they are made into yarns. There are various types of yarns, and the classification can be made based on different characteristics of yarns.

Based on the manufacturing method, yarns are classified as continuous filament yarns and staple yarns. Manufacturing of continuous filaments is relatively easy. Continuous filament yarns are made by spinning the polymer. A number of spun fibers (also called filaments) are collected together to produce the desired continuous filament yarn size. These yarns are called the producer's yarns. They contain the minimum required twist to maintain the yarn's integrity, ranging between 0–2.5 turns per inch. Continuous filaments are smooth and lustrous. Most continuous filament yarns have a finish on them to protect the individual filaments from abrasion and snagging.

Flat continuous filament yarns, as produced, are not suitable for many apparel fabrics. Therefore, these yarns are subjected to texturing process. Texturing makes the continuous filament yarns bulkier and stretchable. Different texturing methods are used such as false twist, crimping, knit-de-knit and air-jet.

By the nature of the fiber spinning process, the fiber lengths are very long. Therefore, the man-made fibers are cut into short staples to make the staple yarn. Of course, natural fibers already come in short staple form. Staple (spun) yarns are made of twisting and entangling short fibers together. Both natural and man-made fibers can be used to make staple yarns. The length of the staple typically varies between 2.5–4.00 cm.

Manufacturing of staple yarns is called yarn spinning. There are several well-established spinning methods. The major methods are ring spinning, open-end (rotor) spinning, air-jet spinning and friction spinning. Prior to the actual spinning process, spinning preparation takes place. This may include blending, opening, cleaning, carding, drawing, combing and roving, depending on the requirements on the yarn. Not every step is mandatory for every staple yarn.

In ring spinning, fibers in the roving are twisted by a traveler rotating on a ring (Figure 2.4), thus forming the yarn. Simultaneously, the yarn is wound on the rotating spinner's package mounted on a spindle.

In open-end spinning (also called rotor spinning), fibers are twisted together inside a rotor that is rotating at high speeds (Figure 2.5). The fibers are fed into the rotor from one side in the

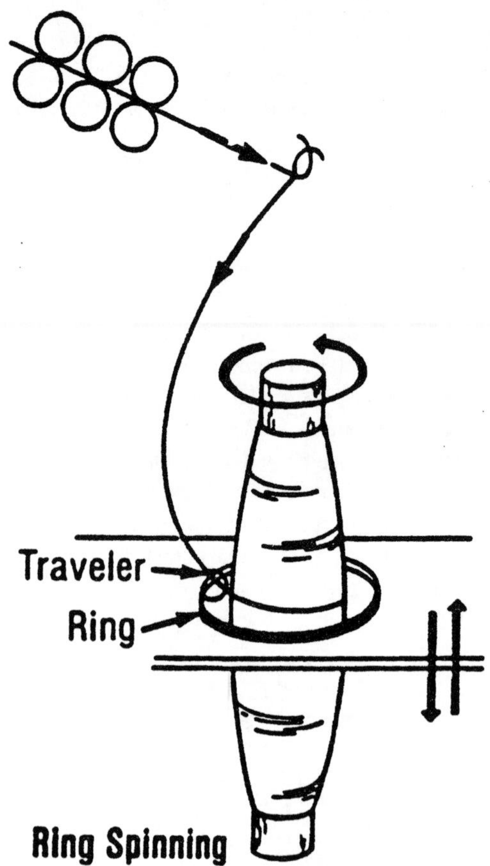

FIGURE 2.4 Schematic of ring spinning (courtesy of Hoechst Celanese).

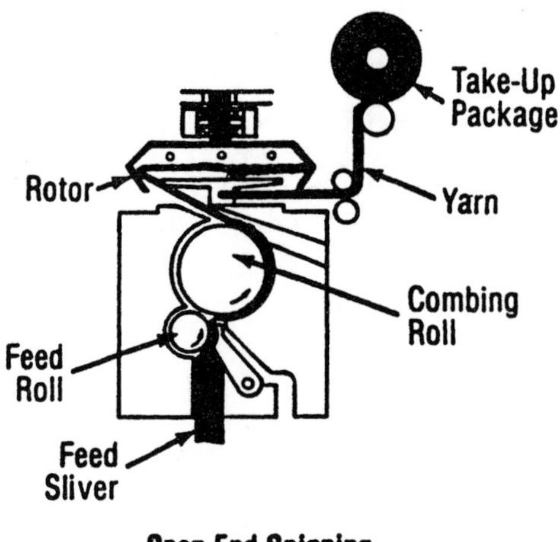

FIGURE 2.5 Schematic of open-end spinning (courtesy of Hoechst Celanese).

form of sliver, twisted by the rotation of the rotor producing the yarn and the yarn is wound on a package.

There are two methods of air jet spinning: Murata jet spinning (MJS) and Murata vortex spinning (MVS). In MJS, two air jet nozzles are used to twist and entangle the fibers in the sliver. The air vortexes inside the nozzles are in opposite directions. Therefore, the first nozzle twists the fibers in one direction and the second nozzle twists the fibers in the other direction (Figure 2.6). However, there is no positively controlled twist given to the yarn. Murata vortex spinning, which is relatively new, was developed for 100% cotton yarns. Yarn does have real twist in its structure, which is similar to twist in ring spun yarns.

In friction spinning, the fibers are dropped in the nip between two revolving perforated drums (Figure 2.7). The rotation of the drums gives twist and entanglement to the fibers.

The yarns that are produced with each spinning method have quite different structures and properties as far as sizing and weaving are concerned. The ring spun yarns are characterized by high level and relatively uniform twist. In open end spun yarns, there is a distinct core of fibers with relatively low twist; other fibers are wrapped around the core. The structure of air-jet yarns is in between the open end and ring spun yarns. On average, ring spun yarns are 20–25% stronger than air-jet spun yarns and 30% stronger than open end yarns. The strength of MVS yarns is closer to ring spun yarns than the other methods. Ring spun yarns also have the highest elongation followed by jet spun and open end yarns. The evenness of jet spun yarns is more than open end yarns that are, in turn, more consistent than ring spun yarns. As a result, the jet and open end spun yarns have fewer slubs, thin and thick places which result in less warp stops at the loom. Ring spun yarn has the highest hairiness due to a high twist level that causes the fibers to protrude from the yarn structure. Open end and jet spun yarns are more susceptible to handling damage than ring spun yarns. Ring spun yarns are costlier than open end yarns which in turn are costlier than Murata vortex spun yarns.

Two or more single yarns can be twisted together to obtain ply yarns. Several plied yarns

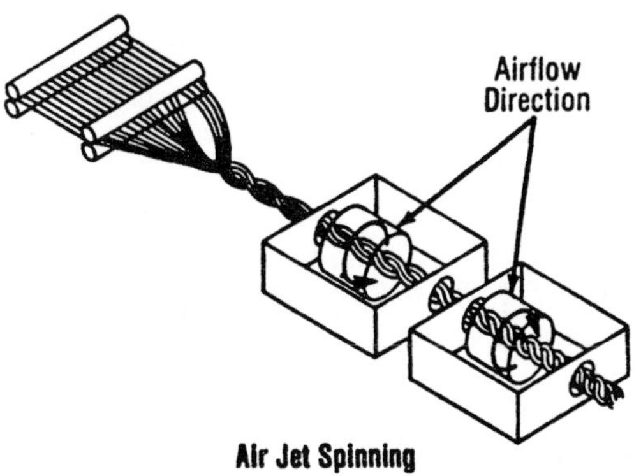

FIGURE 2.6 Schematic of Murata jet spinning (courtesy of Hoechst Celanese).

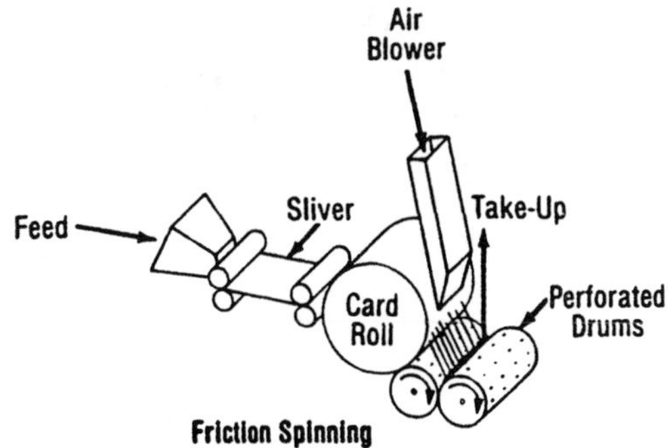

FIGURE 2.7 Schematic of friction spinning (courtesy of Hoechst Celanese).

can be further plied and twisted to form increasingly heavier yarns called cords.

Important yarn characteristics that have an impact on fabric appearance, properties, and performance are as follows:

- structure (single, ply, spun, filament)
- twist level and direction
- coloration
- bulk
- hairiness
- presence of size
- fiber characteristics
- grade (appearance)
- structure due to spinning method (ring, open end, air jet, friction)
- surface properties
- count
- strength
- elongation
- abrasion resistance
- moisture absorbency
- stiffness
- evenness
- diameter

2.3.2 Yarn Count Systems

There are two systems that are used to indicate the "number" or "count" of yarns: direct and indirect systems. In the direct yarn number system, weight per unit length is specified. Heavier yarn has greater weight per unit length. In the indirect yarn number system, length per unit weight is specified. Less heavy yarn has greater length per unit weight. Traditionally, the direct system is used in the metric system and indirect system is used in the English system. However, there are exceptions to this. For example, N_m is a metric system, yet it is used in the indirect system (m/g).

Examples of direct yarn number system units are tex and denier. The tex is the weight (in grams) of 1000 meters of yarn. For example, a 400 tex yarn weighs 400 grams per kilometer. Denier is the weight (in grams) of 9000 meters of yarn (denier = 9 tex).

The following formula is used to calculate the yarn length in the direct system:

$$n = (s \cdot w)/L \tag{2.1}$$

where n: yarn linear density (tex or denier)
L: yarn length (m)
w: yarn weight (g)
s: standard length (1000 m for tex, 9000 m for denier)

The higher the number (n), the heavier the yarn; that is why this system is called the direct system.

In the indirect yarn number system, the number of hanks in one pound of yarn is indicated. If there is one hank in one pound of yarn, then that yarn is called 1s (ones) single count yarn; if there are 7 hanks, it is called 7s (sevens) single count, etc. The length of a hank is different for different kinds of yarns. A cotton hank is 840 yards. For worsted yarn, one hank is 560 yards and for linen yarn one hank is 300 yards. Thus, a two hank worsted yarn contains 2 × 560 = 1120 yards per lb and called 2s (twos) count yarn. In the indirect system:

$$N_e = L/(s \cdot w) \tag{2.2}$$

where Ne: cotton count
L: length of the yarn (yards)
w: yarn weight (lbs)
s: standard hank (e.g., 840 yards for cotton)

The higher the number (N_e), the finer the yarn. Due to this inverse relationship, this system is called the indirect system. The relation between metric count and cotton count is:

$$N_m = 1.693\ N_e \tag{2.3}$$

2.3.3 Yarn Twist

There are two twist directions of fibers in a yarn: "right hand" twist and "left hand" twist. The letters S and Z are used to designate left and right twist, respectively (Figure 2.8).

Twist Multiplier

In practice, twist multiplier is used to calculate the turns per inch necessary for a given size

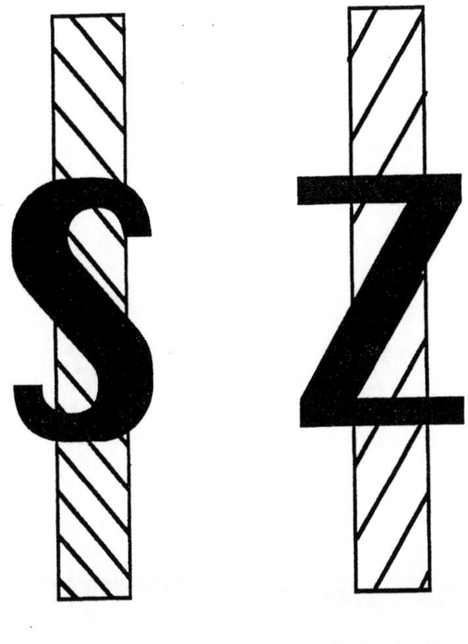

FIGURE 2.8 Yarn twist directions.

spun yarn. Twist multiplier is determined from the turns per inch and the cotton count:

$$TM = \frac{T}{\sqrt{N_e}} \tag{2.4}$$

where T: turns per inch

Typical twist multiplier range for cotton yarns is 2.5–5.

REFERENCES

1. Dictionary of Fiber & Textile Technology, Hoechst Celanese, 1990.
2. Brandrup, J. and Immergut, E. H., Polymer Handbook, Third Edition, John Wiley & Sons, 1989.
3. Broughton, R. M. and Brady, P. H., Fiber Forming Polymers in Wellington Sears Handbook of Industrial Textiles, Technomic Publishing Co., Inc., 1995.

SUGGESTED READING

- Adanur, S., Wellington Sears Handbook of Industrial Textiles, Technomic Publishing Co., Inc., 1995.
- Slashing Short Course, Auburn University, Auburn, AL, September 1998.

REVIEW QUESTIONS

1. Find out what a generic name and a trade name for a fiber is.
2. Compare the ring spun, open-end, air-jet spun and friction spun yarns for the following:
   - tensile strength
   - hairiness
   - structure
3. What is fiber spinning versus yarn spinning? Explain.
4. Why is the yarn linear density expressed in different units rather than in engineering units of mass/volume?
5. How do the crystallinity and molecular orientation affect the fiber properties?
6. If you were to design a battledress uniform fabric, how would you choose the fiber, yarn and fabric structures?

3

Woven Fabric Design and Construction

Weaving is the method of interlacing the warp and filling yarns to form a fabric. The structure of the fabric and its appearance are affected by the pattern of interlacing to a large extent. As a result, fabrics made of the same yarns may differ greatly in appearance and properties if the interlacing pattern is different.

There is practically an unlimited number of weaves that can be developed. This gives the designer endless possibilities to develop a fabric for any purpose. The possibilities are only limited by the imagination of the designer. This is an obvious advantage that textile technology offers.

3.1 WOVEN FABRIC CONSTRUCTION

Woven fabrics are made of two sets of yarns: warp and filling. These yarns are interlaced at 90° to each other (Figure 3.1). The warp yarns are parallel to each other and run lengthwise through the fabric or along the weaving machine direction. In general, there are thousands of warp ends on a typical weaving machine making a fabric. A single warp yarn is called a "warp end" or an "end". Sometimes, a warp yarn is also called the machine direction yarn, especially in industrial fabric manufacturing. Filling yarns run perpendicular to the warp yarns. A single yarn of filling is called a "pick". Other names that are used for filling yarns are "weft", "shute", and "cross-machine direction yarn". The name usually depends on the industry.

Figure 3.2 shows the common reference points on a weaving machine. The front of the machine is where the fabric beam is, and is also called "weaver's side". The back of the machine, where the warp beam is, is called "warp side". Facing the machine from front, the right of the observer indicates the right side of the weaving machine. This is the side where the pick is received (receiving side). The left side, where the pick is inserted from, is called the picking side. Although most modern weaving machines use the left side as the picking side, in some machines the right side is the picking side. The warp yarns are numbered starting from the left side of the weaving machine. The harness numbering starts from the front side of the loom. These reference points are important to avoid confusion among the professionals.

3.1.1 Symbolic Representation of a Weave

The pattern in which the warp and filling yarns are interlaced is called order of interlacing. Order of interlacing is a result of order of entering the warp yarns through the heddles and order of lifting the harnesses.

Order of interlacing of a fabric is called the weave. The weave is symbolically shown using

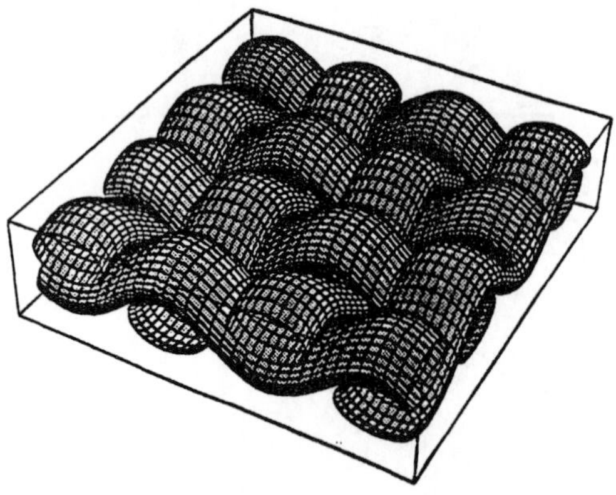

FIGURE 3.1 Interlacing of warp and filling yarns in plain weave.

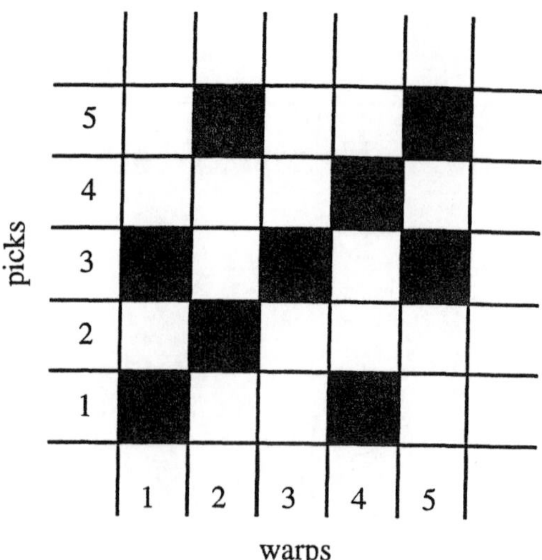

FIGURE 3.3 Weave diagram of a 5 harness design.

a weave diagram as shown in Figure 3.3. In the weave diagram, the columns represent the warp yarns (ends) and the rows represent the filling yarns (picks). The ends are numbered from left to right, and the picks are numbered from bottom to top. A square in the diagram represents the intersection of one end and one pick. If the warp yarn is over the filling yarn in that intersection, then the square is filled or marked with an X (that is why a weave diagram is also called an X-diagram), or any other symbol. Leaving a square blank means that the warp yarn is under the filling yarn. For example, in the figure the first warp yarn is over the first filling yarn, the second warp yarn is under the first filling yarn, and so on.

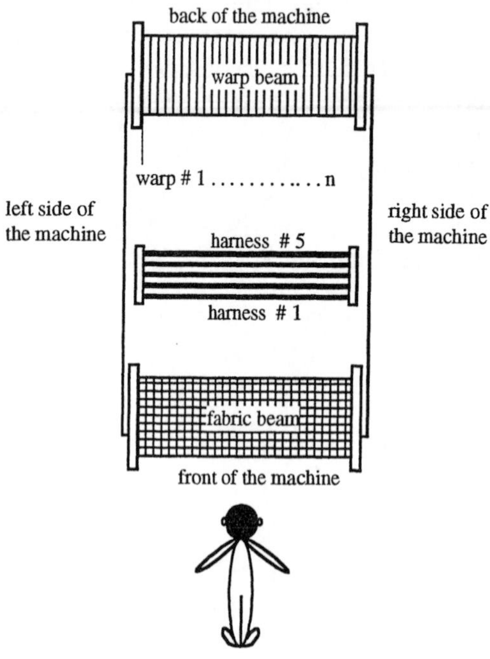

FIGURE 3.2 Major reference points on a weaving machine.

The weave diagram should show at least the minimum number of warps and fillings needed to identify the woven structure completely. This is called a "repeat unit" or "unit cell". Describing the repeat unit is usually enough to identify the whole fabric structure, since the rest of the fabric is formed by extending the repeat unit in the warp and filling directions as shown in Figure 3.4.

3.1.2 Drawing-in-Draft (DID)

After a drop wire, each warp yarn is threaded through a heddle that is attached to the harness frame (see Section 4.4 in Chapter 4). Starting from the left side of the weaving machine the warp yarns are selected one by one, the first, the second, the third, etc., across the machine in straight order. Drawing-in-Draft (DID) diagram indicates which warp end is attached to which harness as shown in Figure 3.5. The DID is also called "entering plan". The vertical columns in the DID represent the warp yarns and the horizontal rows represent the harnesses which are numbered sequentially from bottom to top. If a

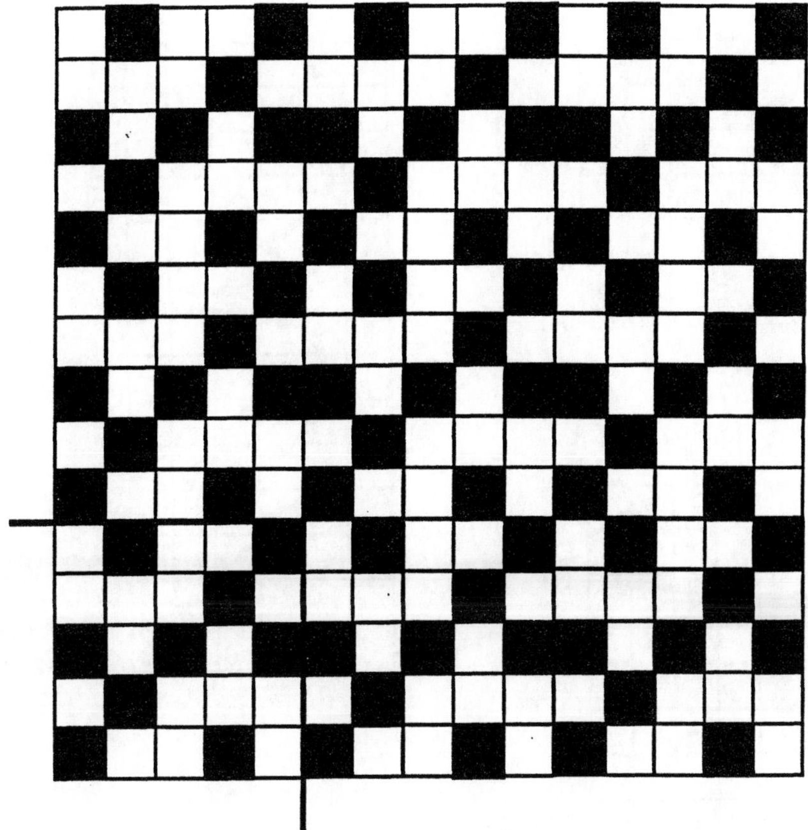

FIGURE 3.4 Repeat of the unit cell.

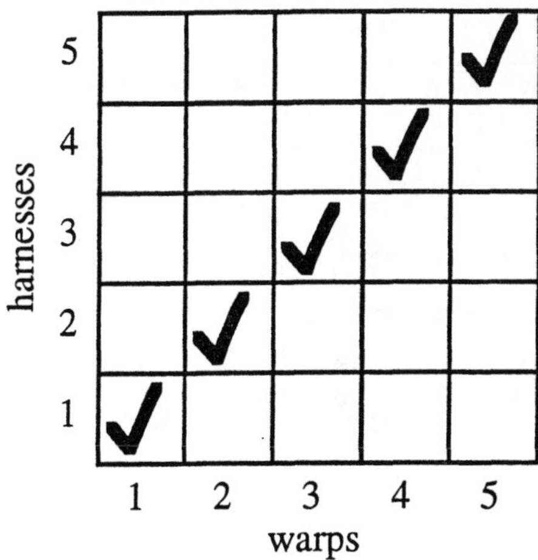

FIGURE 3.5 Drawing-in-draft (DID) diagram for the design in Figure 3.3.

warp yarn is controlled by a particular harness, a mark is placed at the intersection of the respective column and raw. The mark can be a check mark (✓), an "X" or a fully painted square. For example in Figure 3.5, the warp #1 is controlled by harness number #1, the warp #2 is controlled by harness #2, and so on, which is called a straight draw. The DID diagram should show the configuration for the whole repeat unit (unit cell) of the fabric.

Straight draw is the simplest and therefore the most widely used drawing pattern. However, there also are drawing plans other than straight draw. Some of these plans are shown in Figure 3.6. A particular drawing plan is chosen based on the warp density and/or the requirement of the weave.

If there are warp yarns in the unit cell that have the same interlacing pattern, then these warp yarns can be attached to the same harness,

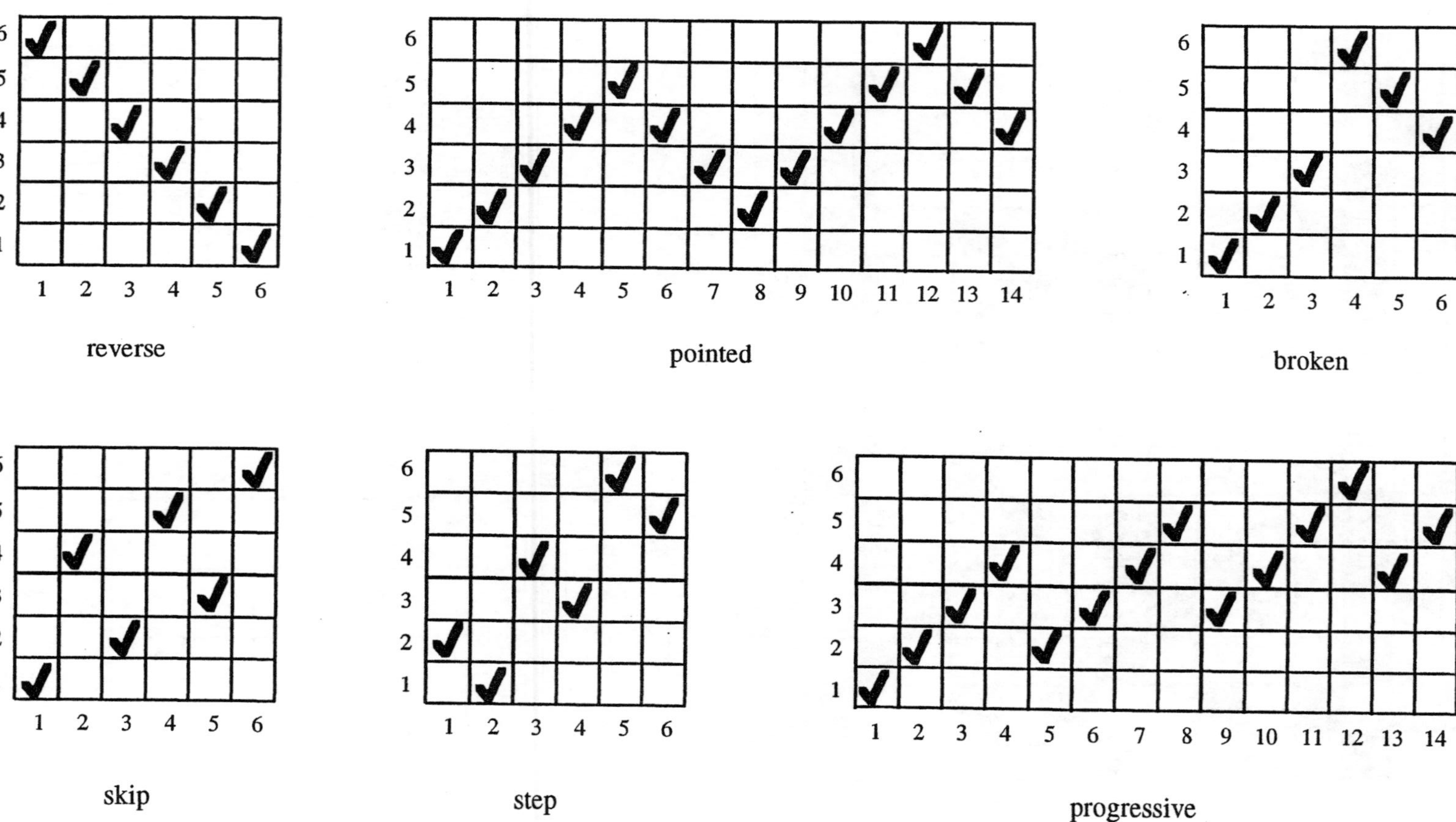

FIGURE 3.6 Different drawing plans.

thus minimizing the number of harnesses needed. This is called least harnesses draw. Of course a straight draw can also be used; however, in that case, the number of harnesses required will be equal to the number of warp yarns in the unit cell as shown in Figure 3.7. The advantage of straight draw is the simplicity of drawing. The disadvantage is the cost of extra harnesses that may not be needed.

3.1.3 The Reed Plan

The warp yarns are drawn through the reed dents sequentially. Naturally, each warp end would correspond to one dent. However, it is not practical to draw only one yarn through a dent since the number of warp yarns is generally more than the number of dents in the reed. Therefore, in practice more than one yarn is placed

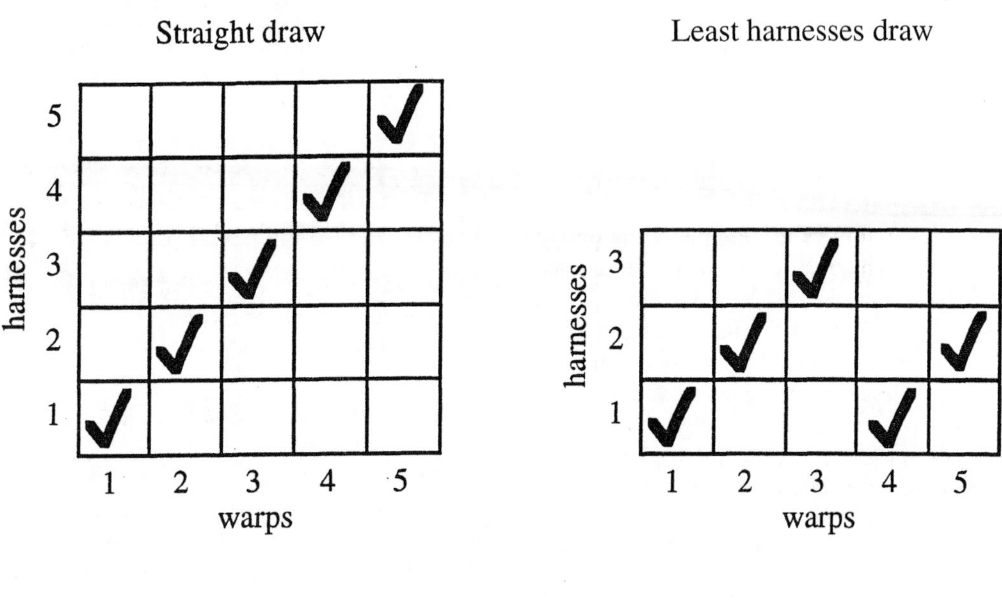

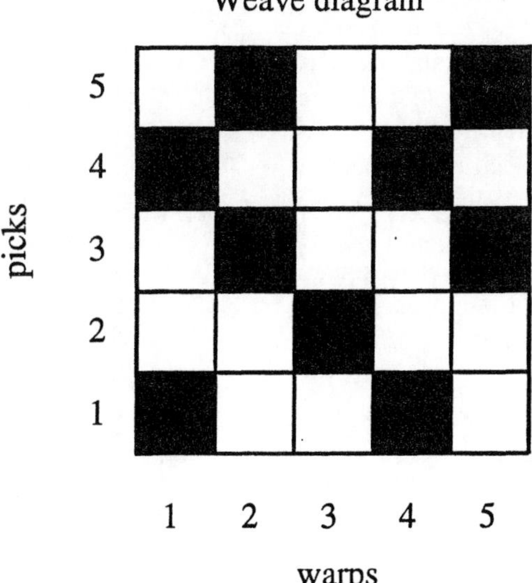

FIGURE 3.7 Straight draw and least harnesses draw.

in a dent. The reed plan shows how many warp yarns are drawn through each dent. The practical number of warp yarns per dent can be between 2 and 4. The upper limit is determined by the warp diameters and the width of the dent. The warp yarns should be able to move up and down freely in the dent during shed change in order to have interlacing with the filling yarns.

Figure 3.8 shows examples of reed plans. If the number of yarns per dent is constant throughout the reed, the reed plan is called "simple reed plan". If the number of warp yarns per dent varies, it is called "complicated reed plan". If a simple reed plan is to be used, it may not be necessary to draw a reed plan but simply state the number of warp yarns per dent.

simple reed plan; 2 yarns/dent

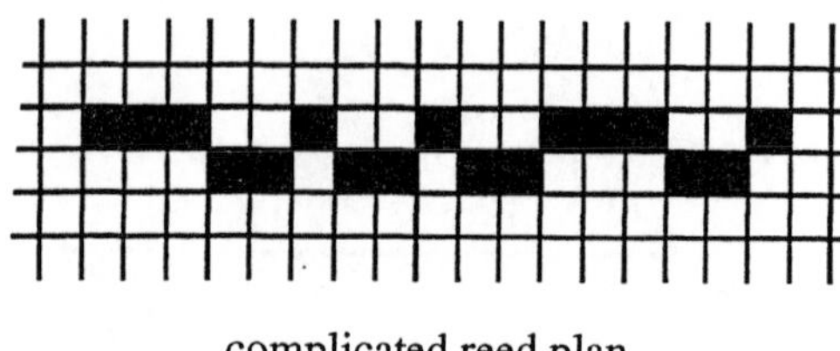

complicated reed plan

FIGURE 3.8 Examples of reed plans.

3.1.4 Cam Draft or Chain Draft (CD)

The next step in describing the fabric design is to specify which harnesses will be lifted and which harnesses will be lowered during each pick in the unit cell of the fabric. This is done using cam draft or chain draft (CD) for cam and dobby shedding. In jacquard shedding, every warp end is controlled individually. CD diagram shows the order of lifting the harnesses and therefore the warp yarns since each warp end is attached to a harness.

In CD diagram, the columns represent the harnesses and the rows represent the picks. The picks are numbered from bottom to top as in the case of unit cell; therefore, the heights of the unit cell and the CD are equal. Similarly, the width of CD and the height of DID are equal since they both show the harnesses. To indicate that a harness is lifted during insertion of a pick, the corresponding square in the CD matrix is filled or marked with an "X" as shown in Figure 3.9. Conversely, a blank square means that the harness is lowered during the insertion of that particular weft. In the figure, harness 1 and harness 4 are lifted during the insertion of the first pick.

In practice, Figures 3.3, 3.5, 3.8 and 3.9 are combined together as shown in Figure 3.10. Such an organization of weave diagram, DID, reed plan and CD gives most of the information for the design and manufacturing of the fabric. The only information missing from this diagram is the selvages. When specifying a fabric, the specifications for the entire warp should be given, including selvages. The DID diagram and reed plan for selvages are constructed similar to the body of the fabric.

3.1.5 Warp and Filling Profiles

Sometimes, it may be useful to show warp and filling profiles of a woven fabric for easier understanding of the structure. For single layer fabrics, drawing of these profiles is relatively easy. Figure 3.11 shows the warp and filling profiles of a five harness woven fabric structure.

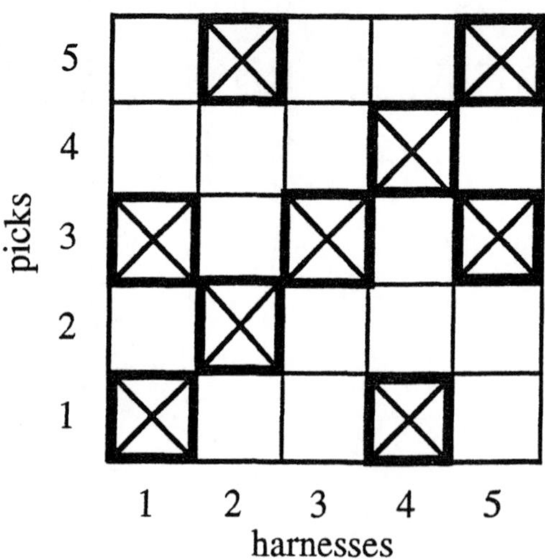

FIGURE 3.9 Cam draft or chain draft (CD) of the design in Figure 3.3.

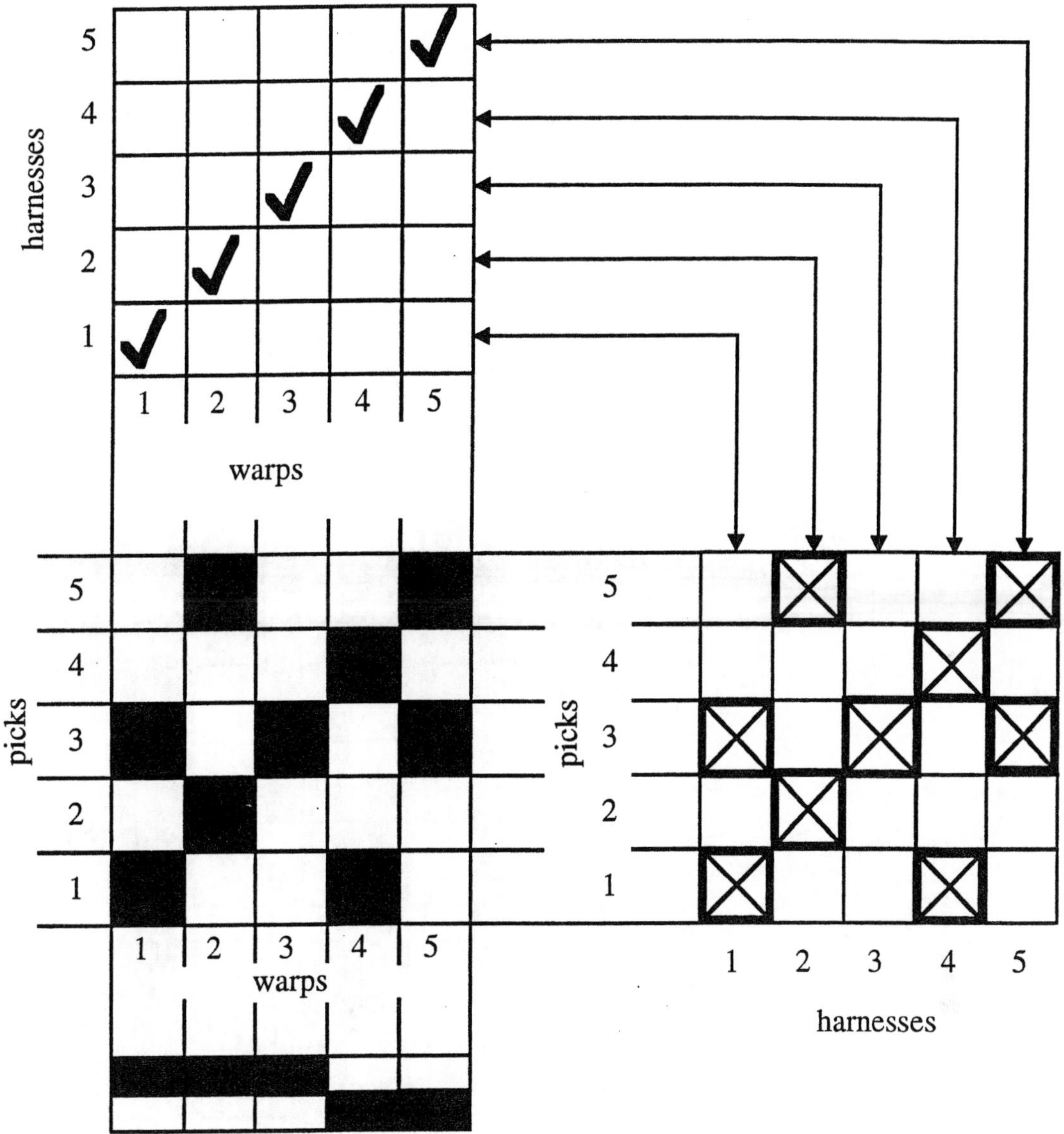

FIGURE 3.10 Combination of weave diagram, DID, reed plan and CD.

Warp and filling profiles are especially helpful to better visualize complex fabric structures such as multilayer fabrics. Figure 3.12 shows the warp and filling profiles of a three-layer forming fabric design that is used in papermaking.

3.2 BASIC WEAVE DESIGNS

It is generally accepted that there are three basic weaves:

- Plain weave
- Twill weave
- Satin weave

Each of these three fabrics has a different texture. Assuming that they have similar yarns and the same number of warp and filling yarns per unit length, they have different properties, e.g., tensile strength.

Although there are some weaves that are difficult to structurally connect to these three basic structures, most of the others are derived from

warp profile:

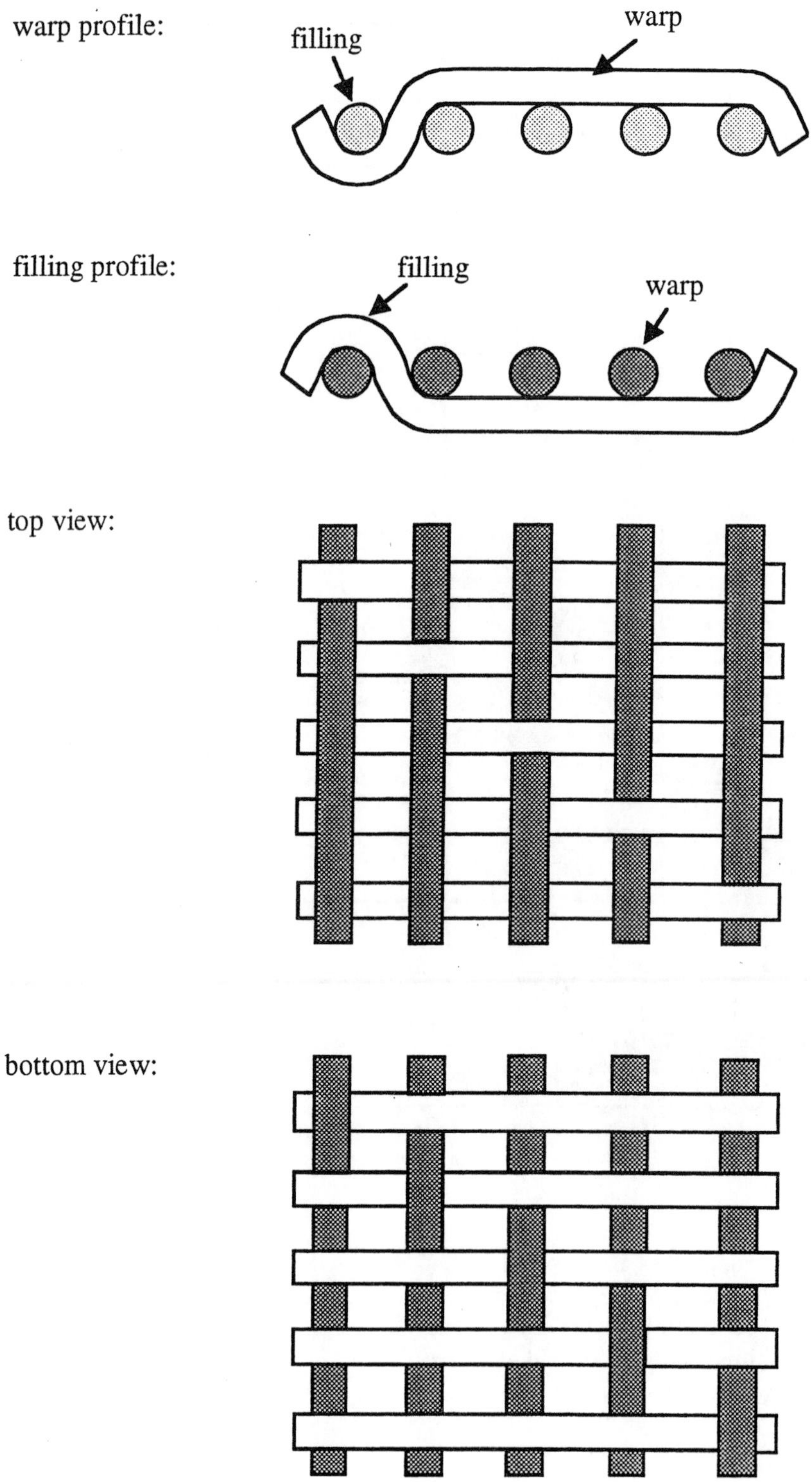

filling

warp

filling profile:

filling

warp

top view:

bottom view:

FIGURE 3.11 Schematics of five harness single layer fabric design (courtesy of AstenJohnson).

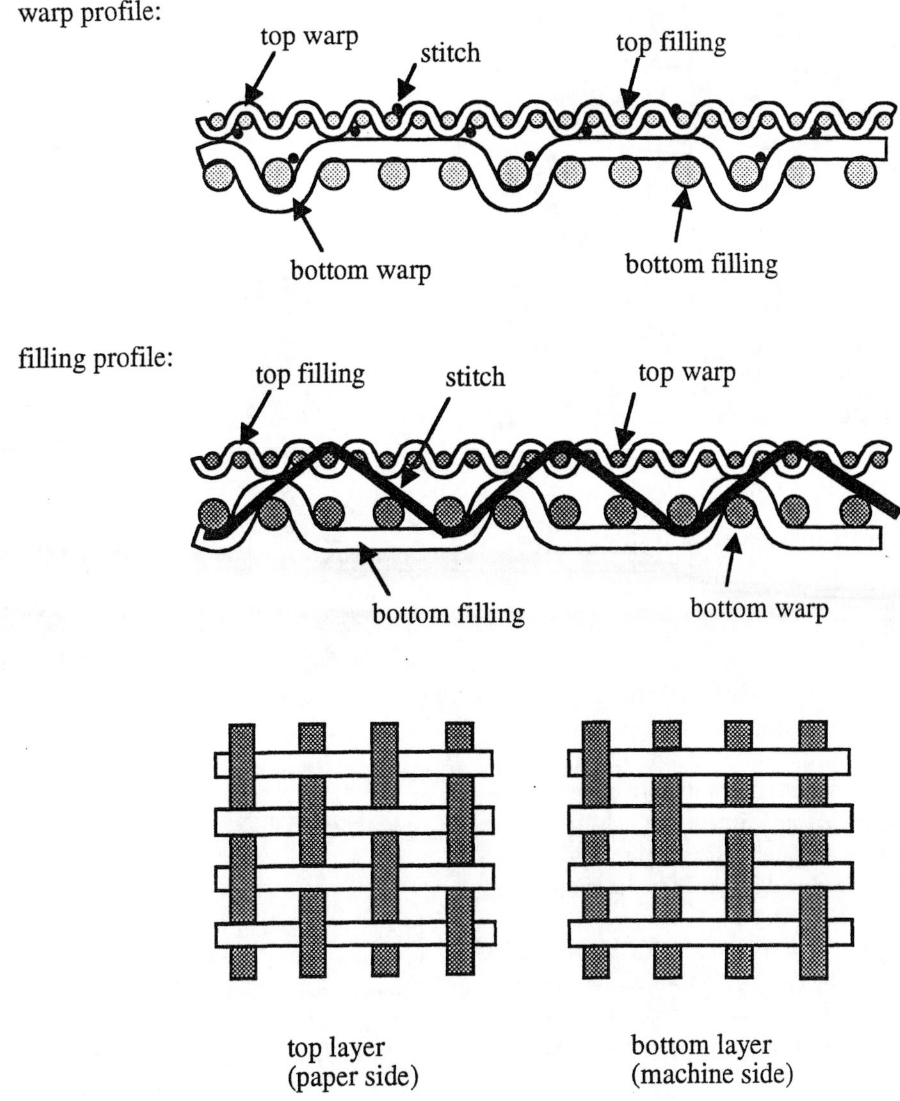

FIGURE 3.12 Schematics of a three layer forming fabric design (courtesy of AstenJohnson).

these three basic weaves. The immediate derivatives of these three structures are warp rib, filling rib, and basket weave. These six designs are explained below. It should be noted that some consider the twill weave as the only basic weave from which all the other weaves are derived.

3.2.1 Plain Weave

Plain weave is the simplest of all weaves. It has one-over one-under interlacing for both warp and filling yarns as shown in Figure 3.13. As a result, the plain weave formula repeats on two warp and two filling yarns and the fabric has the same texture on both top and bottom sides.

Plain weave requires only two harnesses. However, it can be woven on more than two harnesses especially if the warp density is more than 50 ends per inch (epi). Quite often, it is woven on four harnesses.

Due to 1/1 interlacing, the plain weave has the maximum level of yarn crimp in its structure. As a result of this, the plain weave has low modulus compared to other designs that have less crimp in their structure.

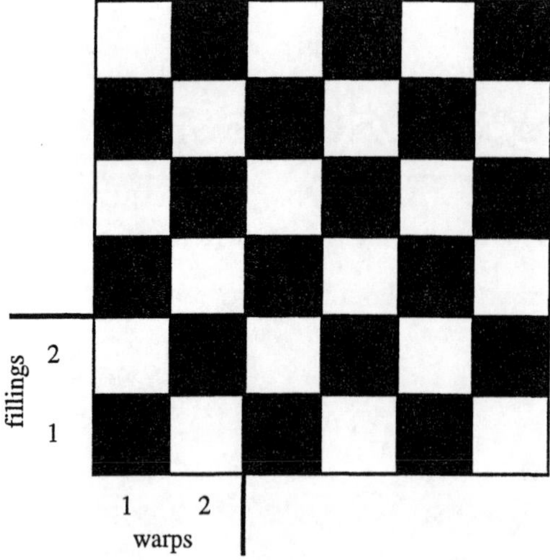

FIGURE 3.13 Symbolic notation of plain weave.

3.2.2 Warp Rib

Warp rib formula has 1/1 interlacing in the filling direction and an interlacing pattern different than 1/1 in the warp direction. This results in a design that has ribs or texture ridges across the fabric in the warp direction which are caused by grouping of filling yarns. Figure 3.14 shows a 2/2 warp rib structure which is the simplest of all warp rib designs.

The repeat units of all warp ribs have two warp yarns. The first warp follows the formula and the second warp does the opposite. Therefore, any warp rib design requires a minimum of two harnesses. The number of filling yarns in the repeat unit is the sum of the digits in the warp rib formula.

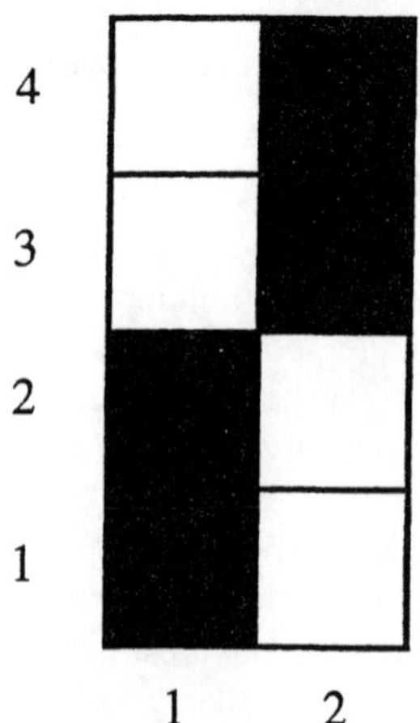

FIGURE 3.14 2/2 Warp rib design.

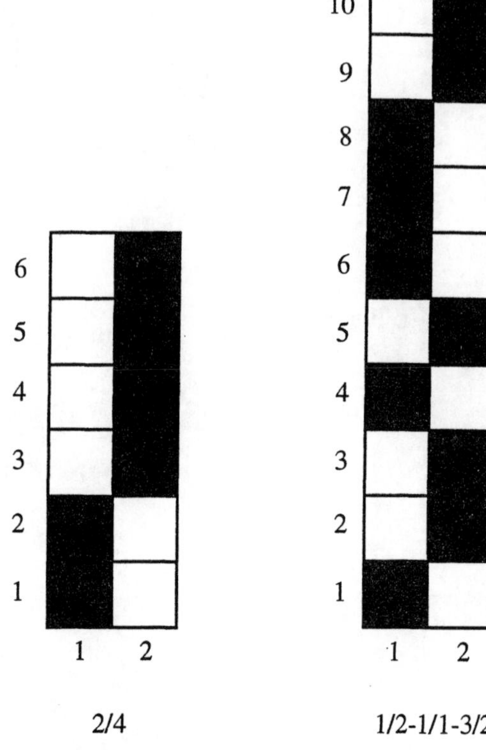

FIGURE 3.15 Irregular warp rib designs.

Warp rib formulae are classified as regular (balanced) or irregular (unbalanced). The numerator and denominator of a regular or balanced warp rib formula is the same number, e.g. 2/2, 3/3, etc. In irregular or unbalanced formula, the digits are different numbers, e.g, 2/3, 4/2, 2/3-3/2, etc. Figure 3.15 shows irregular warp rib designs. If only a portion of the formula has the same number as numerator and denominator, the design is still considered to be an irregular rib.

3.2.3 Filling Rib

Filling rib formula has 1/1 interlacing in the warp direction and an interlacing pattern different than 1/1 in the filling direction. This results in a design that has ribs or texture ridges across the fabric in the filling direction. These ribs are

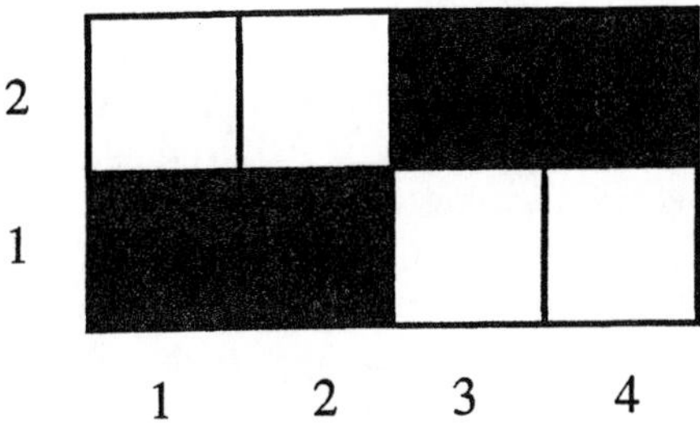

FIGURE 3.16 2/2 Filling rib design.

caused by grouping of warp yarns. Figure 3.16 shows a 2/2 filling rib structure which is the simplest of all filling rib designs.

Analogous to the warp ribs, the repeat units of all filling ribs have two filling yarns. The first filling follows the formula and the second filling does the opposite. Therefore, any filling rib design requires a minimum of two harnesses. The number of warp yarns in the repeat unit is the sum of the digits in the filling rib formula. The regular (balanced) or irregular (unbalanced) formulae apply to filling ribs as well. Figure 3.17 shows examples of irregular filling ribs.

3.2.4 Basket Weaves

Basket weaves are produced by combining warp and filling ribs. In basket weaves, warp and filling yarns are grouped and they interlace together. Figure 3.18 shows a 2/2 basket weave. The number of warp and filling yarns in the unit cell is equal to the sum of the digits in the formula. The basket weaves require a minimum of two harnesses.

Basket weaves can be classified as common formula or uncommon formula (Figure 3.19). In a common formula basket weave, the first warp yarn and the first filling yarn follow the same formula. In an uncommon formula basket weave, the first warp and the first filling follow different formulae.

3.2.5 Twill Weave and Its Derivatives

Twill weave is produced in a stepwise progression of the warp yarn interlacing pattern. The interlacing pattern of each warp yarn starts on a different filling yarn and follows the same formula. This results in the appearance

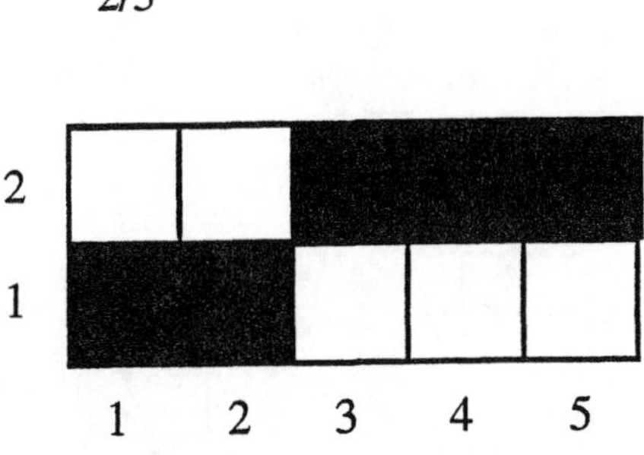

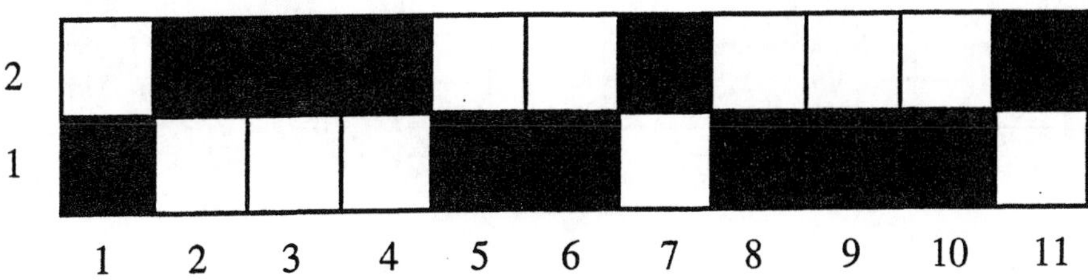

FIGURE 3.17 Irregular filling rib designs.

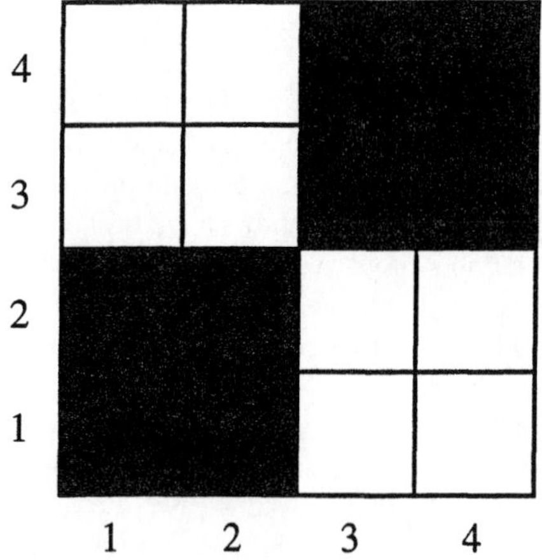

FIGURE 3.18 2/2 Basket weave.

of a diagonal line called twill line in the fabric which is the characteristic of this design. The twill line is not a physical line but an impression caused by the stepwise progression of the interlacing of the design. Depending on the direction of the twill line, the twill weaves are called right-hand or left-hand twills. Figure 3.20 shows 3/3 right- and left-hand twill designs. In right-hand twill, the twill line runs from lower-left to upper-right. In left-hand twill, the twill line runs from lower-right to upper-left. A fabric with a right-hand twill on the surface has a left-hand twill on the back.

Twill weave formulae are classified as regular (balanced) or irregular (unbalanced). The digits of a regular or balanced twill formula are the same number, e.g. 2/2, 3/3, 4/4, etc. (Figure 3.20). In irregular or unbalanced formula, the digits are different numbers, e.g, 1/3, 4/2, 2/3-3/2-1/1, etc. Examples of irregular right- and left-hand twills are shown in Figure 3.21. The sum of the digits in the formula determines the unit cell of the design which also gives the minimum number of harnesses required to weave the design; at least three harnesses are required for a twill weave.

There are an unlimited number of twill weave variations. If the twill line angle is 45°, the design is called common twill. The designs shown in Figures 3.20 and 3.21 are common twills. In a common twill the starting point of each warp interlacing pattern is on the adjacent pick. The twill line angle can be changed by design between 15° and 75° as shown in Figure 3.22. If the twill angle is greater than 45°, it is called steep twill; if it is less than 45°, it is called reclining twill. Twill angle also depends on the warp and filling density. Sometimes, especially in denim manufacturing, the fabric is sheared after weaving which also changes the twill angle.

A variation of twill weave is called broken twill. In this type of design, the start-up point of the pattern is random, which distorts the twill line. Figure 3.23 shows a 3/2 broken twill.

4/2-3/2 common

2/3-3/3 and 3/2-2/4 uncommon

FIGURE 3.19 Examples of common and uncommon formula basket weaves.

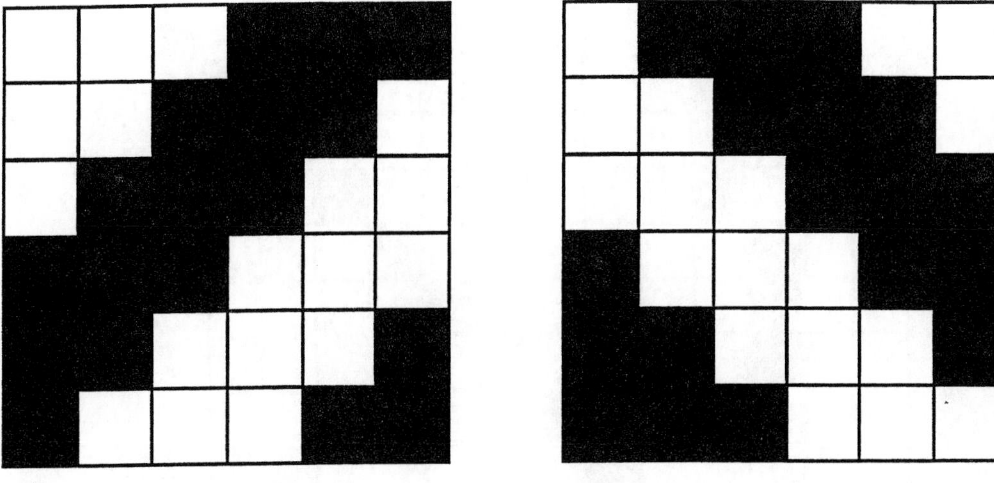

FIGURE 3.20 3/3 Right-hand and left-hand twill.

3.2.6 Satin Weave and Its Derivatives

In satin weave, one yarn has a long float over several of the other yarns on one side of the fabric as shown in Figure 3.24. A yarn is considered to have a float (knuckle) if it stays over or under more than one other yarn. If a yarn is brought to the top surface of the fabric, it is

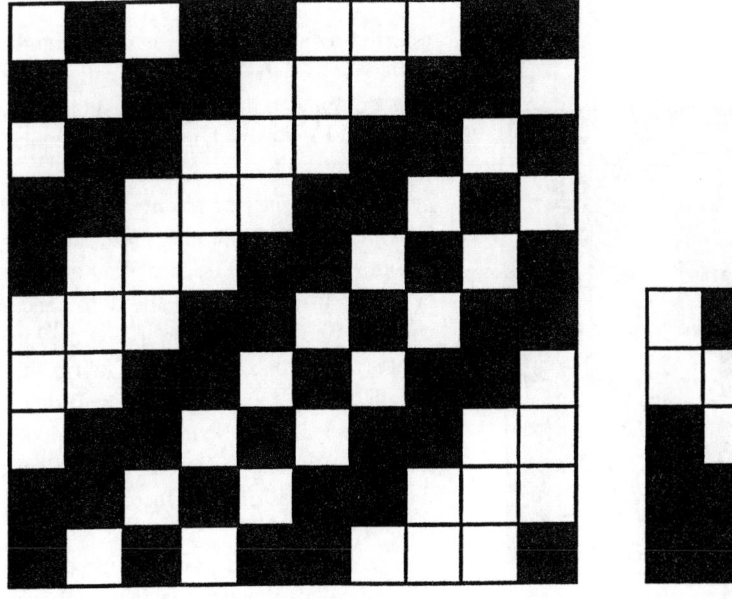

FIGURE 3.21 Irregular right- and left-hand twills.

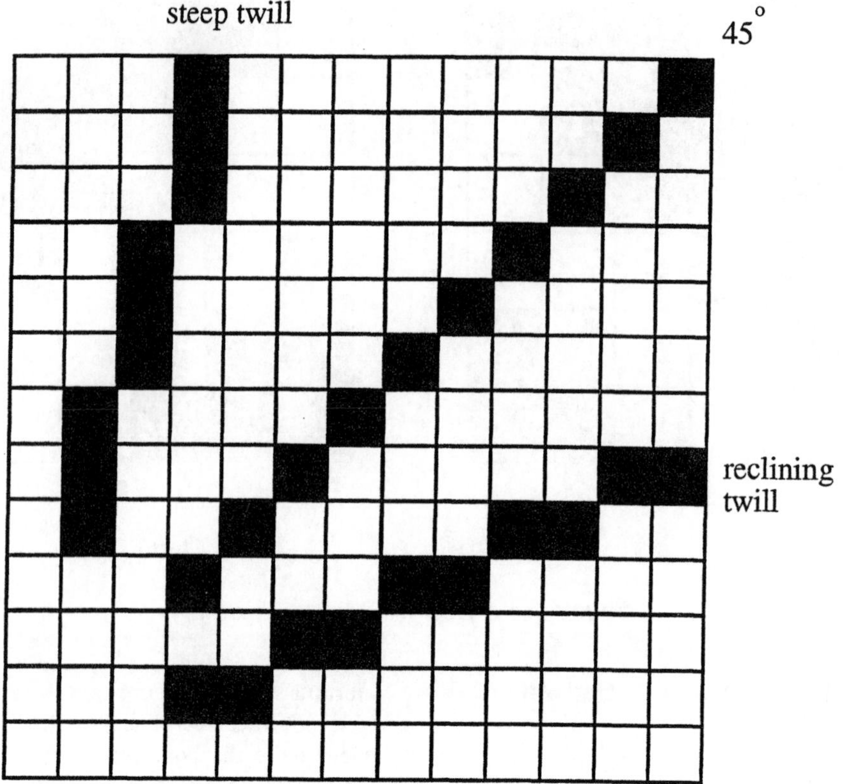

FIGURE 3.22 Variation of twill angle.

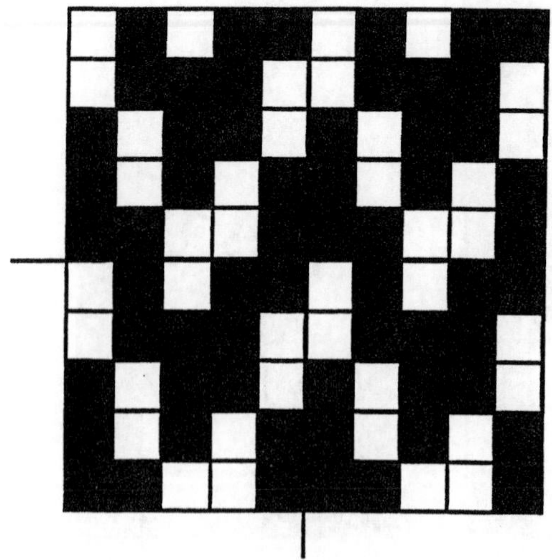

FIGURE 3.23 3/2 Broken twill.

referred to as a "raiser", if it is brought down to the back of the fabric, it is called a "sinker". This dominance of one yarn results in a smooth texture. At least five harnesses are required for a satin weave, which is named after the minimum number of harnesses required to make it, e.g., 5-harness satin, 7-harness satin, etc.

Satin weaves can be classified as warp faced or filling faced based on the dominance of the yarns on one side of the fabric. If the long warp float is on the top surface of the fabric, the design is called warp faced. If the long filling float is on the top, it is called filling faced. A filling faced satin is also called a sateen. Figure 3.25 shows examples of warp faced and filling faced satin weaves.

A counter is used to determine the layout of the unit cell of the satin weave. Each warp yarn has the same interlacing pattern in the weave with a different starting point. Once the first warp yarn interlacing is placed freely on the cell, the counter determines the starting points of the

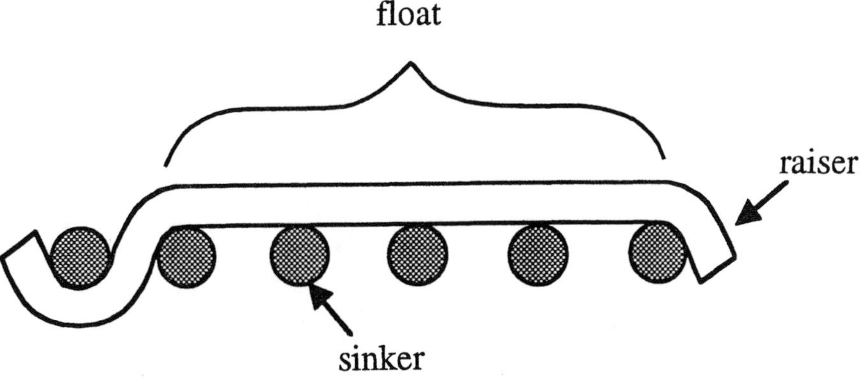

FIGURE 3.24 Long yarn float.

next warp yarns' interlacings. In general, the number of interlacings is kept to a minimum such that the design does not resemble a twill weave. A number cannot be selected freely as a counter; there are rules in selecting counters. Usually a satin has a pair of numbers as useful counters. Table 3.1 shows the useful counters for several satin designs. The numbers in the pairs should not have a whole number relationship to each other and the sum of the pairs must be equal to the minimum number of harnesses required to make the weave.

3.2.7 Other Fabric Designs

There are practically endless numbers of woven fabric designs. It is possible to develop weaves with various patterns such as honeycomb, stripes, check patterns, spot patterns, etc.; but covering all of the possible designs is out of the scope of this book. Some of these special fabrics are discussed in Chapter 12.

There are several sources that may necessitate the development of a new fabric style. A new fabric design may be suggested or required by

7 harness warp faced
counter 5

5 harness filling faced
counter 3

FIGURE 3.25 Warp faced and filling faced satin weaves.

TABLE 3.1 Counters for some satin weaves.

No. of harnesses	Counter pairs
5	2 or 3
7	2 or 5; 3 or 4
8	3 or 5
9	2 or 7; 4 or 5

the customers. Complaints are also a good reason to develop new fabric designs or to modify existing ones. Sometimes, a new fabric idea may come from the associates in the manufacturing plant or competition from outside the plant.

When developing fabric specifications, the end use requirements must be considered. Other considerations are raw material, yarn structure, fabric structure and finishing.

There are several ways and methods to develop fabrics. Developing a completely new design is a good way to avoid any patent infringements, if there is any. Another way is to modify an existing design to obtain a new design; once the weaving machine is set up, changing the DID, CD or both could result in a new design.

REVIEW QUESTIONS

1. Compare the six basic weave designs (plain, warp rib, filling rib, basket, twill and satin) for the following properties. Assume that the yarn counts and densities are the same.
   - tensile strength
   - modulus
   - hand
   - tear resistance
2. What is a three dimensional (3D) woven fabric? Explain.
3. Draw the DID and CD diagrams of a 9 harness sateen weave of your choice.
4. How do you determine the number of minimum harnesses required to produce a woven fabric design? Explain.
5. When does a CD diagram become identical to the X-diagram?

4

Weaving Preparation

Yarn is the basic building block in weaving. Therefore, after yarn manufacturing, the next successive step would be to weave the yarn into a fabric. However, in practice, the condition of yarn produced on the spinning machine is not always good enough to be used directly for fabric formation. Package size, yarn surface characteristics, and other factors make it necessary for both filling yarn and warp yarn to be further processed for efficient fabric formation. These preparatory processes are called weaving preparation, which is the subject of this chapter.

Warp and filling yarns are subjected to different conditions and requirements during weaving. Therefore, the preparation of warp and filling yarns is different. Warp yarn is subjected to higher stresses which requires extra preparation. The filling yarns are not subjected to the same type of stresses as the warp yarns and thus are easily prepared for the weaving process. Depending on the spinning method, the filling yarns may not be prepared at all, but rather taken straight off the spinning process and transported to the weaving process. This is the case with open-end (rotor), air-jet and friction spinning systems which provide a large single-end package suitable for insertion during weaving. However, ring spun yarns need to go through a winding process for several reasons that are explained below. The processes used to prepare yarns for weaving depend on yarn type as well.

Winding is the major preparation process for filling yarn. Warp preparation includes winding, warping, slashing and drawing-in or tying-in. Figure 4.1 shows the major preparation processes for filling and warp yarns.

Spun yarn quality characteristics that are most important for good weaving performance include short- and long-term weight uniformity, imperfections, tensile properties and hairiness. It should be noted that variation in a property is almost always more important than the average value of that property. Regardless of the processes employed, a second concept of quality has to be embraced. Not only must the quality of the yarn itself be maintained and enhanced, but also the quality of yarn packages is extremely important to further processing.

The cost to repair a yarn failure is much less if it occurs prior to the weaving process. In addition, a yarn failure during weaving also increases the chances for off quality fabric. Many if not most of the quality problems encountered during fabric forming are directly related to mistakes made during yarn manufacturing or yarn preparation for weaving.

Since winding is common for both filling and warp preparation, it will be discussed first for both yarn systems. The weaving process is particularly abusive to lengthwise yarns in a woven fabric; therefore, the technology surrounding the

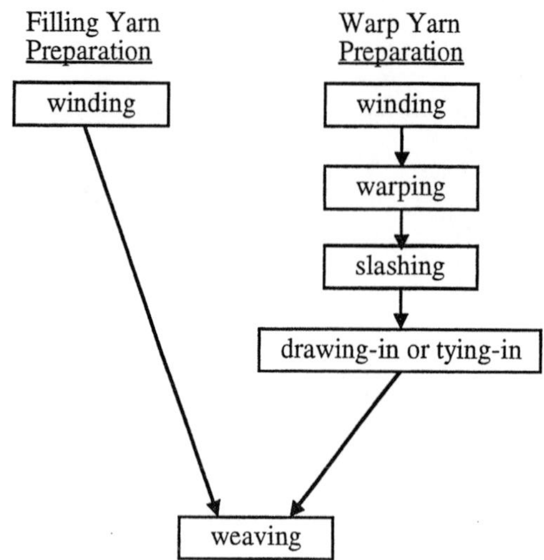

FIGURE 4.1 Yarn preparation steps for weaving.

preparation of warp yarn for weaving is given special attention.

4.1 WINDING

Winding is basically transferring a yarn from one type of package to another (Figure 4.2). This simple definition may make the winding sound like a trivial process; however, it is an important and necessary process that performs the following functions especially for ring spun yarns.

a) Winding produces a yarn package that is suitable for further processing. Ring spinning produces small packages of yarn (called spinner's packages or bobbins) which would be depleted relatively quick during filling insertion or warping. Therefore, the amount of yarn on several small packages is combined by splicing or knotting onto a single package (Figure 4.3). Knotting has been replaced by splicing in modern winding machines.

b) The winding process provides an opportunity to clear yarn defects. Thin and thick places, slubs, neps or loose fibers on the yarn are cleared during winding and, thus, the overall quality of the yarn is improved (Figure 4.4). Staple yarns require this clearing operation most because they may have these kinds of faults more often.

The increasing use of newer spinning technologies resulted in a situation where the old concept of yarn clearing and package quality now has become a part of the spinning process rather than part of a separate winding process. Properly formed packages of defect-free spun yarn are an even more critical factor. Package considerations include condition of the package core, the proper provision of yarn transfer tails; properly formed splices or knots; elimination of internal defects such as slubs, sloughs, tangles, wild yarn, scuffs and ribbon wind; and elimination of external defects such as over-end winding, cobwebs, abrasion scuffs, poor package shape or build, proper density (hardness) and unwindability.

4.1.1 Winding Process

There are three main regions in winding (Figure 4.2).

(1) Region 1. Unwinding of yarn from the spinning package—The yarn package is held in the creel in an optimum position for unwinding. Yarn withdrawal can be done in two ways (Figure 4.5):

a) Side withdrawal. In this method the spool is rotated and therefore the yarn does not rotate during withdrawal. As a result, the yarn twist does not change, which is an advantage.

Since the yarn does not rotate, the spool must rotate for side withdrawal. This requires additional energy and equipment, which is a disadvantage. At high winding speeds, due to inertia, the rotation of the spool can cause yarn tension variations. Upon start-up, higher tensions may be developed because the winder must overcome spool inertia.

b) Over-end withdrawal. In this system, the spool does not rotate. Therefore, the problems associated with rotating a spool are avoided. The method is simple and does not require driving the spool.

The disadvantage of this system is ballooning which is due to the way the yarn is withdrawn and unwound from the package at high speeds. Centrifugal force causes the yarn to follow a curved path leading to ballooning upon rotation of the yarn (Chapter 8, Jet Weaving). Ballooning leads to uneven tensions in the yarn. Each time one complete wrap of yarn is removed from the supply package, the twist in that length changes by one turn. This change may be insignificant for regular round yarns, but in cases where flat

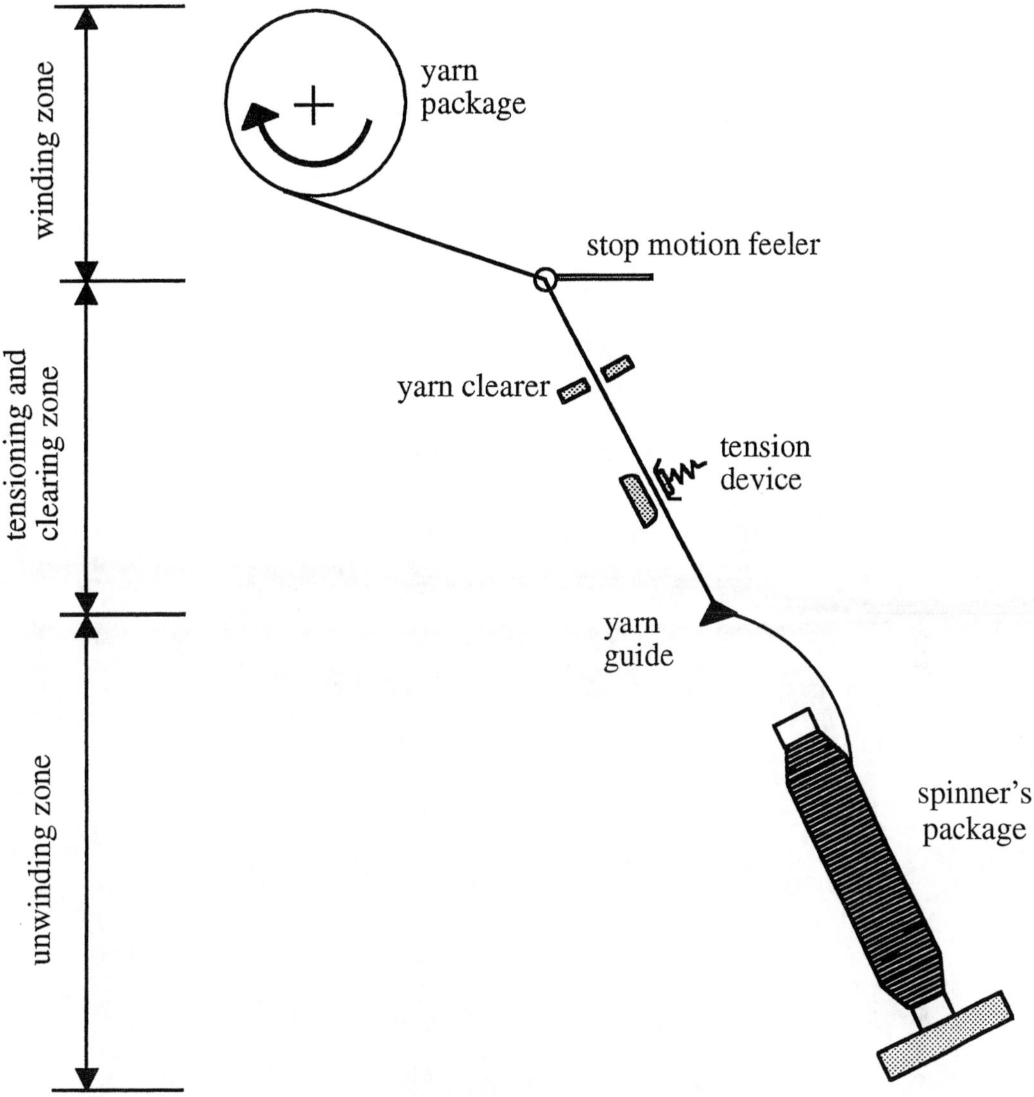

FIGURE 4.2 Schematic of winding process.

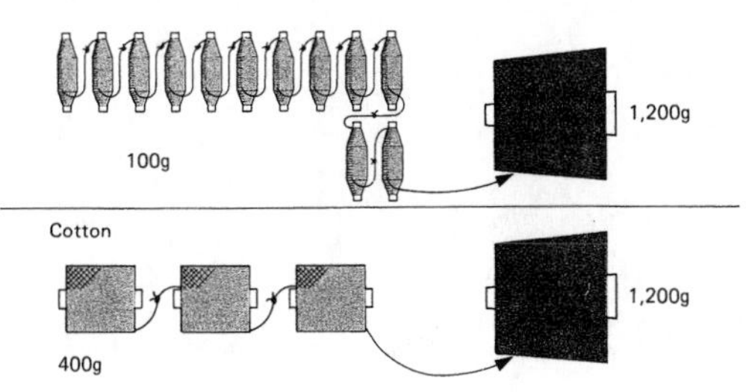

FIGURE 4.3 Building large packages (courtesy of Murata).

yarns of metal, polymer or rubber are used, even one twist is not allowed since yarns must remain flat. These yarns cannot be unwound using the over-end method; therefore, the side withdrawal method must be used.

In fiber optics guided (FOG) missiles, over-end unwinding is used to send the missile to the target while observing the target from a ground station. Variation in tension due to ballooning, as well as twisting, may cause yarn or fiber optic breakage (Figure 4.6).

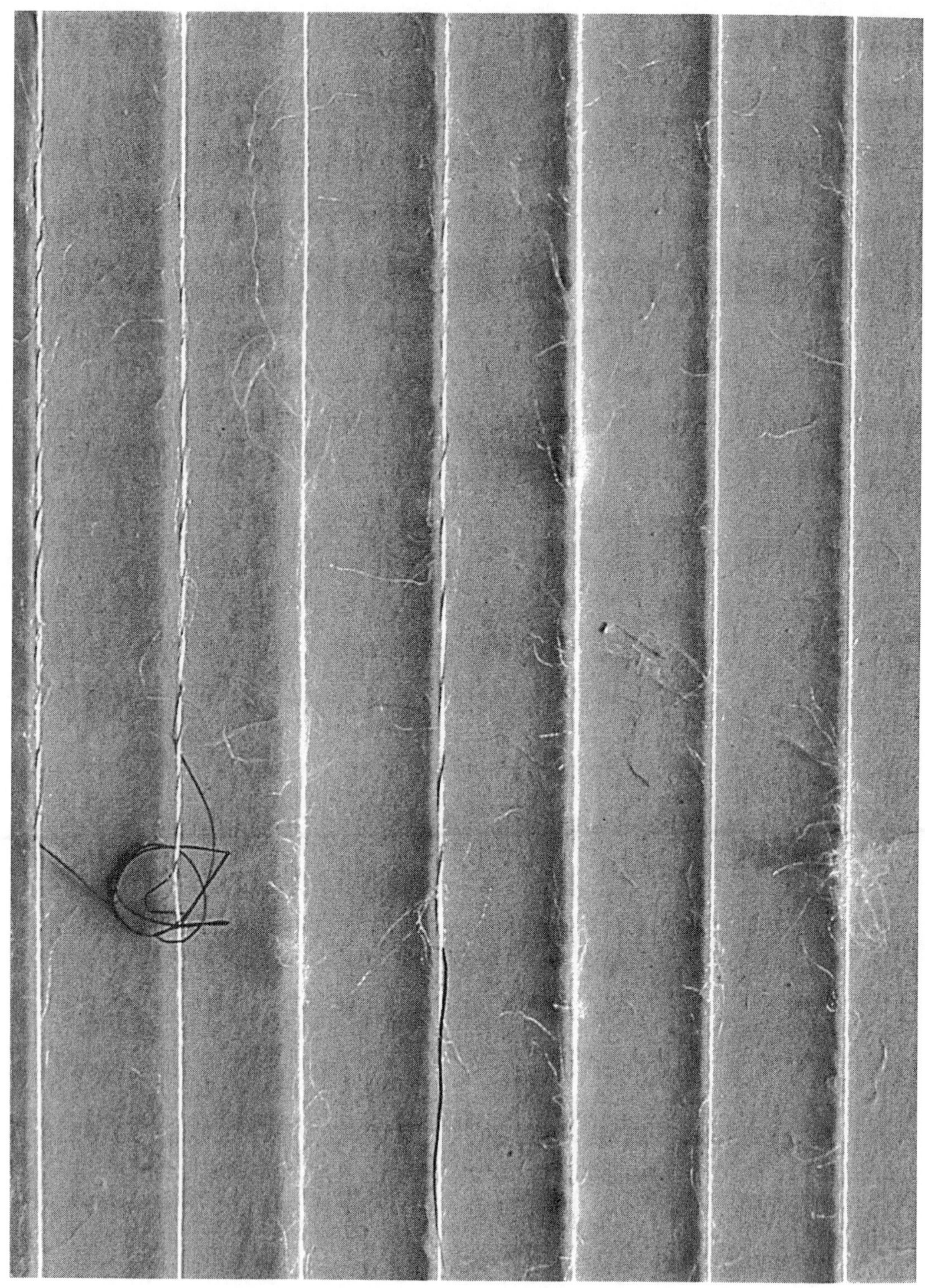

FIGURE 4.4 Yarn faults (courtesy of Zellweger Uster).

side withdrawal:

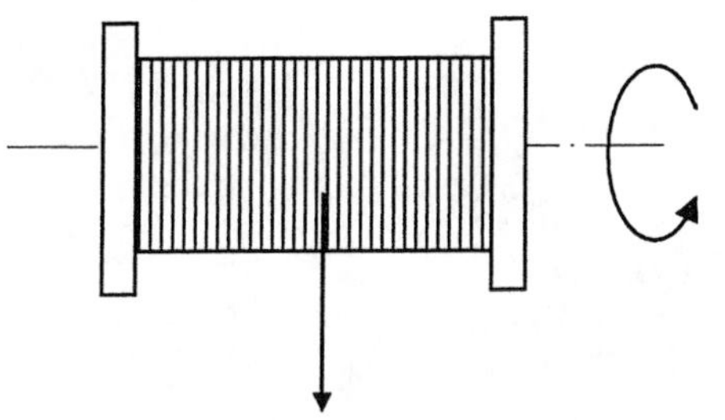

over-end withdrawal:

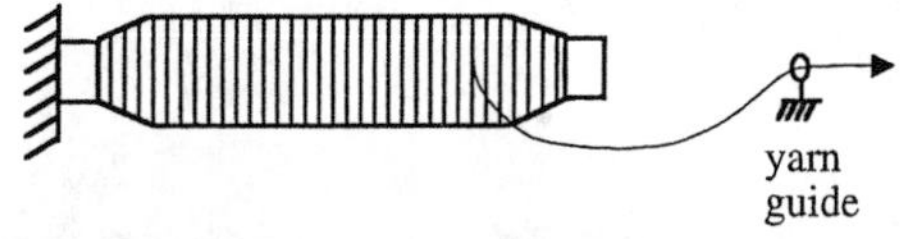

FIGURE 4.5 Yarn withdrawal.

(2) Region 2. The tensioning and clearing region—In this region, proper tension is given to the yarn for a desired package density and body. The typical components of this region are a tension device, a device to detect thick and thin spots in the yarn (clearing device) and a stop motion. Figure 4.7 shows the tension and clearing zone of a modern winding machine. The stop motion causes the winding to stop in case of yarn breakage or the depletion of a supply package. The yarn is directed into this region by a guide.

There are two types of guides (Figure 4.8): closed and open. Closed guides require a yarn end to thread, and open guides do not. Open guides, however, give less positive guiding. Engineering issues here are guide smoothness, abrasion between yarn and guide causing yarn damage. If the guide is too rough, damage of yarn due to abrasion will occur. On the other hand, if the guide is too smooth, friction may develop. Guides are usually made from hard stainless steels or from ceramics.

Wire guides are easier to manufacture to any shape. The chromium layer can be satin finished or mirror polished depending on the need. Ceramic coated metal guides are especially good for synthetic fibers. These guides combine wear resistance of ceramic compounds with ductility of metals while allowing complex shapes to be made. As a result, there is no need for inserts, clamps or gluing. Alumina sintered yarn guides with mat surfaces are recommended for synthetic and mixed yarns (nylon, polyester, etc.) while alumina sintered yarn guides with polished surfaces or ground polished surfaces are generally used for natural fibers (silk, wool, cotton, etc.). Porcelain yarn guides are produced with mat or mirror glazes. They are resistant to wear of natural or synthetic fibers and yarns.

A. Tension device. The tension device maintains a proper tension in the yarn to achieve a uniform package density. It also serves as a detector for excessively weak spots in the yarn that break under the added tension induced by the tension device.

There are three major types of tension devices (Figure 4.9).

a) Capstan (or multiplicative) tensioner [Figure 4. 9(a)]. The output tension depends on the input tension, coefficient of friction between the yarn and the post (μ), and the total angle of wrap (α):

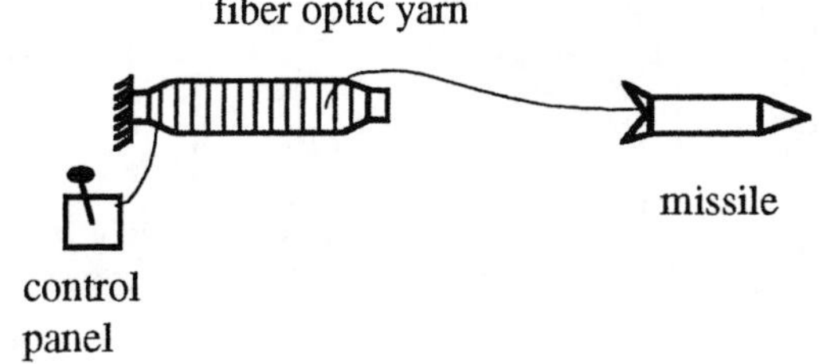

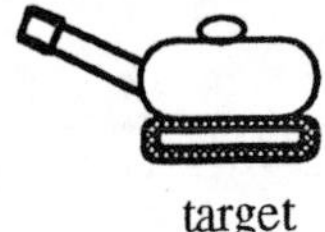

FIGURE 4.6 Fiber optic guided (FOG) missile.

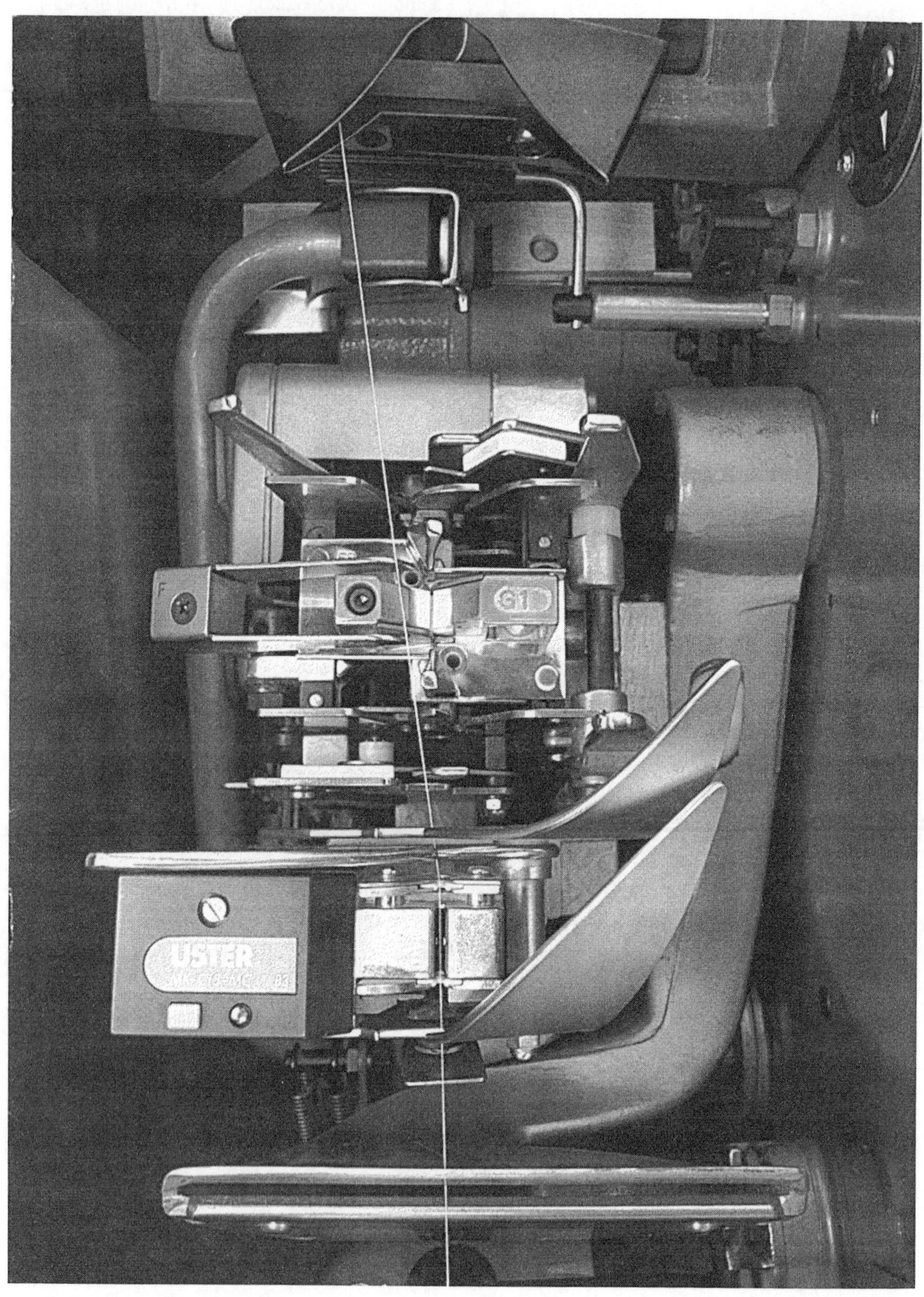

FIGURE 4.7 Tension and clearing zone of a modern automatic winding machine (courtesy of Zellweger Uster).

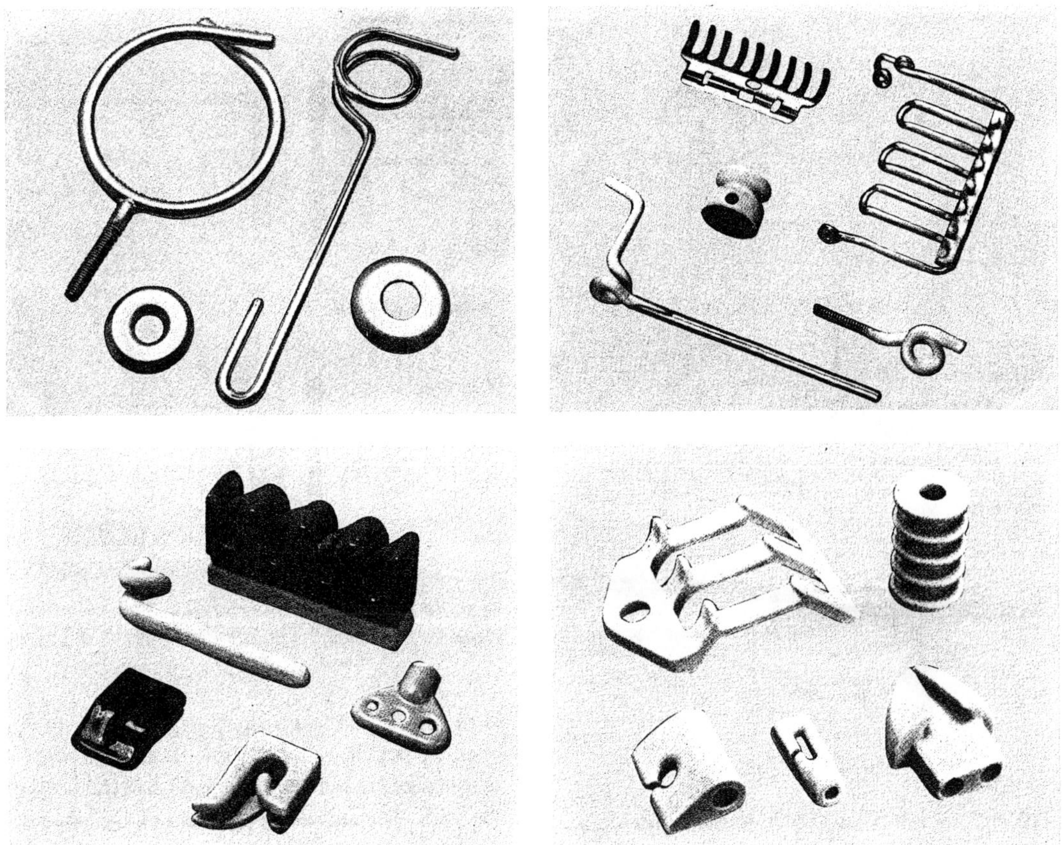

FIGURE 4.8 Various types of yarn guides, top left: wire hard chromed, top right: plasma ceramic coated, bottom right: alumina sintered, bottom left: hard porcelain (courtesy of Italian Textile Machinery Manufacturers' Association).

$$T_{out} = T_{in}\, e^{\mu\alpha} \quad (4.1)$$

Since μ, α and e are constants, T_{out} is a constant multiple of the incoming tension T_{in} (this is the reason why Capstan is called multiplicative). If T_{in} is zero, so is the T_{out}.

Changing μ, α, the number of guides and/or T_{in} changes the output tension. μ can be changed by changing the post material or yarn surface characteristics.

b) Additive tensioner. In this system, a deadweight or spring is used to apply a normal force (N) to change the tension [Figure 4. 9(b)]. The output tension is calculated by:

$$T_{out} = T_{in} + 2\mu N \quad (4.2)$$

Since μ and N are approximately constants for a given system, T_{out} is obtained by simply adding a constant to T_{in}. If T_{in} is zero, there is still an output tension $T_{out} = 2\mu N$. T_{out} may be changed simply by changing the normal force N.

c) Combined tensioner. This is the most common type which consists of at least a disc and Capstan type tensioner. Tension is changed by normal force and/or wrap angle.

$$\begin{aligned} T_{out} &= T_{in} + T_{in}e^{\mu\alpha} + 2\mu N \quad (4.3) \\ &= T_{in}\,(1 + e^{\mu\alpha}) + 2\mu N \end{aligned}$$

B. Yarn clearers. The purpose of a yarn detector is to remove thin and thick places. Yarn detectors are usually two types: mechanical and electronic.

A mechanical clearer may be as simple as two parallel blades (Figure 4.10). The distance between the plates is adjustable to allow only a predetermined yarn diameter to pass through. A thicker spot on the yarn (slub) will cause the tension on the yarn to build up and eventually

a) Capstan tensioner (top view)

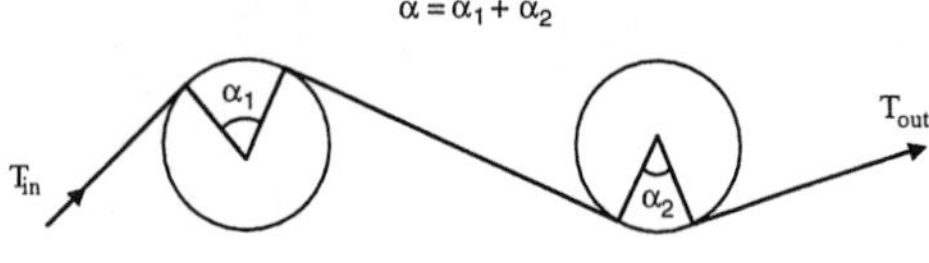

b) Additive tensioner (side view)

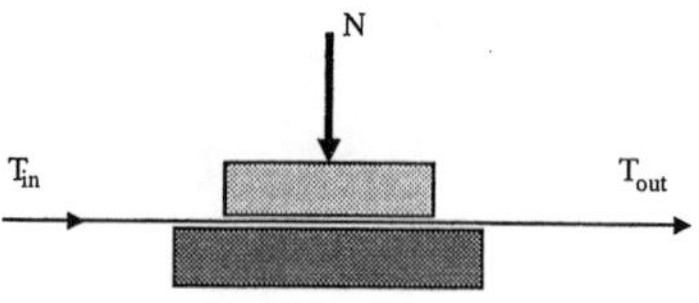

c) Combined tensioner

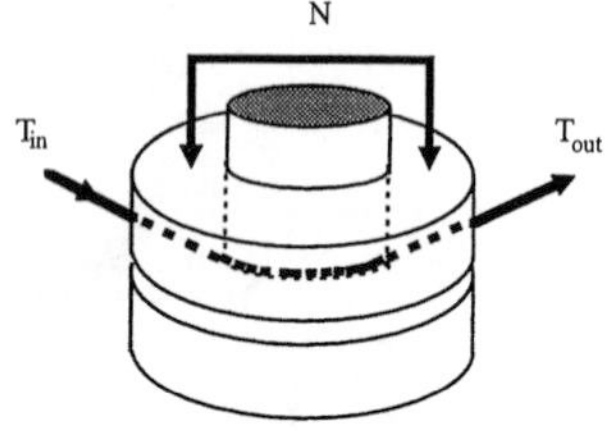

FIGURE 4.9 Principles of basic tensioning devices.

break the yarn. Consequently, this type of device can only detect thick places in the yarn.

The clearers of today's technology are more sophisticated and contain electronics which continuously monitor the yarn to detect thin and thick places. Electronic detectors are mainly two types: capacitive and photo-electric (Figure 4.11). In a capacitive type detector, the variation

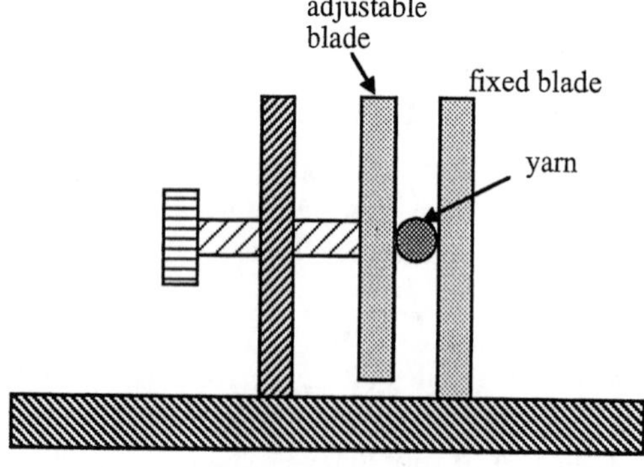

FIGURE 4.10 Schematic principle of mechanical yarn clearer.

a) Capacitive detector

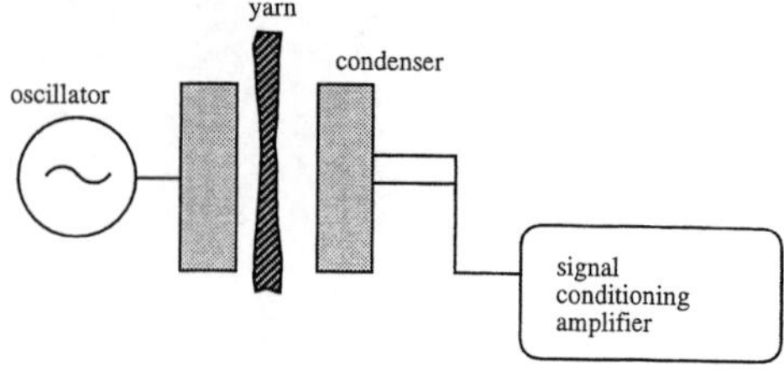

b) Photo-electric detector

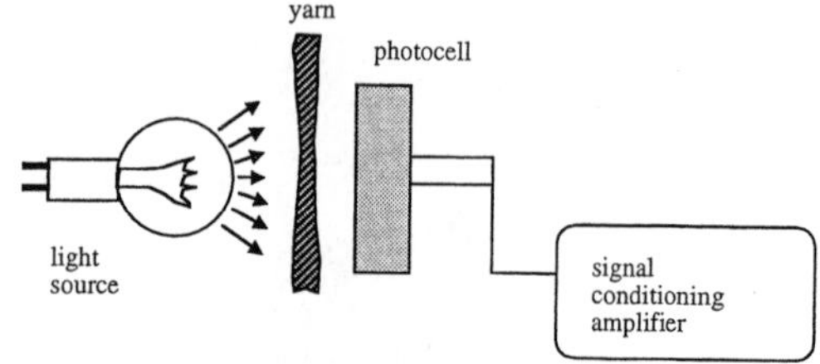

FIGURE 4.11 Schematic principles of electronic yarn clearers.

in the mass of the yarn passing through the plates changes the capacitance of the unit. It should be emphasized that the system measures the mass of the yarn. The signal is not based on the physical dimensions of the yarn. When the generated signal reaches a certain value, the yarn is cut.

In a photo-electric detector, the yarn passes between a light source and a photocell. Any fluctuation in yarn thickness causes the fluctuation of light coming to the photocell, which changes the resistance of the photocell. This resistance change is detected by a signal conditioning amplifier which can be set to send a signal to cut the yarn and stop the winding process.

The latest yarn clearing systems can also detect foreign fibers. These fibers are classified and eliminated during the winding process. As a result, the quality of the yarn can be improved during the winding process.

C. Stop motion. The purpose of a stop motion is to stop winding when the yarn breaks or runs out. Stop motions vary from machine to machine. In general, a mechanical stop motion consists of a counter weighted or spring loaded sensing device which is held in an inactive position if the yarn is present. Breakage or running out causes the absence of this restraining yarn and allows the sensing device to activate. Electronic

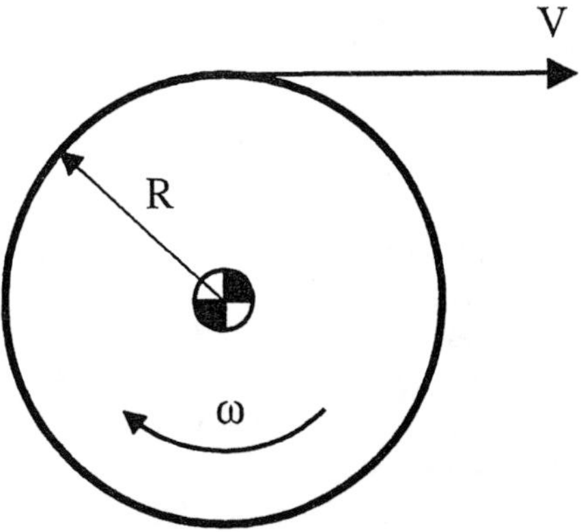

FIGURE 4.12 Rotating package.

stop motions simply sense the existence of the yarn without mechanical contact.

(3) Region 3. The winding region—In this region, the yarn package which is suitable for further processing is wound. Many types of package configurations can be obtained including cone, tube or cheese, dye tube or spool depending on the next stage of processing.

The basic requirement of winding is uniform tension on the yarn. Uniform tension is necessary for consistent winding and yarn uniformity with respect to properties that are functions of tension. If the tension on yarn passing the tension device is constant, the tension in the package should be constant provided that the yarn speed is constant, i.e., the tension on the package is only a function of the yarn speed.

The yarn is wound on the package by only rotating the package. Consider a disc of radius R, rotating at an angular velocity ω (Figure 4.12). Then, the linear velocity (or the tangential speed) of any point on the circumference of the package is:

$$V = \omega R = \text{the yarn linear velocity} \quad (4.4)$$

Therefore $V = f(\omega \text{ and } R)$

The rotation of the package may be accomplished in two ways: spindle drive and friction drive.

1) Spindle drive winder (Figure 4.13). In this system, the spindle, which holds the package, is driven directly. There are two variations of this system: constant speed winders and variable speed winders.

A. Constant speed winders. The spindle is driven at a constant speed, i.e., ω = constant.

Since $\omega = 2\pi n$, then n (rpm) is constant.

Therefore, $V = \omega R = f(R)$

As more yarn is wound on the package, R increases, hence V increases. This is not a desired situation, as explained below.

Since T = tension = $f(V)$, a change in yarn velocity causes a change in tension. Therefore, the tension will vary throughout the package. This problem can be overcome by using the second type of the spindle drive systems in which the spindle speed is varied.

B. Variable speed winder. In the equation $V = \omega R$, this time ω is variable. As R increases (i.e. more yarn on the package), ω will change to keep V = constant. Although R and ω are variables, the product $\omega R = V$ = yarn velocity = constant.

To change ω, a variable speed motor or a variable speed connection is needed which increases the cost. Therefore, this system can be justified only for very delicate yarns. A simple way to achieve this is to use the second type of winder.

2) Friction drive winder (Figure 4.14). In this system, the spindle, that carries the package, is free to rotate and the package is driven through surface friction between the package and a driven drum or roller.

At the point of contact A (assuming no slippage), yarn, friction drum and package have the same velocity, i.e.

$$V_y = V_d = \omega_d R_d \quad (4.5)$$
$$= \text{constant } (\omega_d, R_d \text{ are constants})$$

Thus, a constant surface speed on the package

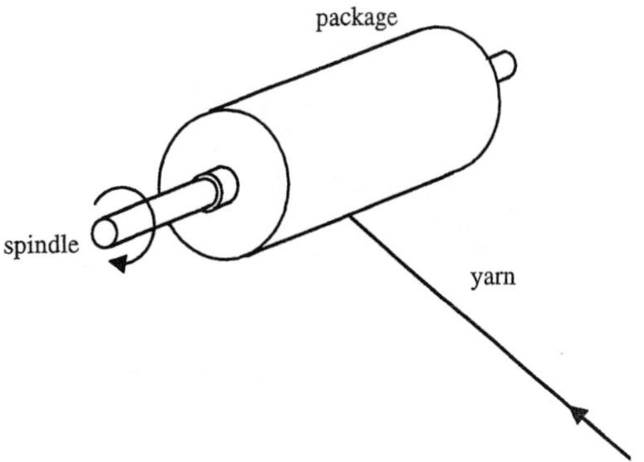

FIGURE 4.13 Spindle drive of a package.

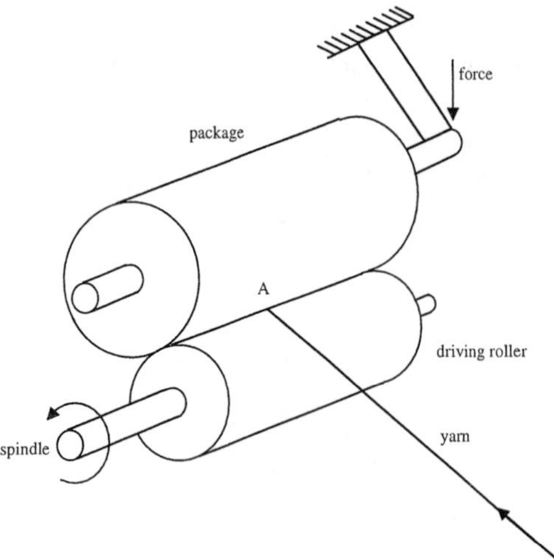

FIGURE 4.14 Friction drive of a package.

and therefore an almost constant yarn winding speed are obtained. This system is widely used for staple yarns.

Traversing Mechanisms

A traversing mechanism is used to distribute the yarn axially along the package. The distribution of the yarn should be done evenly on the package.

In the friction drive winder (only), a traversing groove cut into the friction drum is used (Figure 4.15). The yarn will fit into the groove and travel back and forth along the length of the package as the drum rotates.

In the spindle drive winder (also in some friction drive), a reciprocating traverse is used, i.e. an externally driven guide carries the yarn back and forth across the package (Figure 4.16).

Types of Packages (Figure 4.17)

Based on the winding pattern, yarn packages can be grouped under three categories: parallel, near-parallel and cross-wound packages.

a) Parallel wound packages [Figure 4.17(a)]. These packages are similar to warp beams; there are many yarns which are parallel to each other. For these packages, flanges or shoulders are necessary to prevent yarn instabilities. The application of this type of package is limited.

b) Near-parallel packages [Figure 4.17(b)]. In this type of package, there is usually one yarn end that is wound on the package. A near parallel-wound package is not self-supported. Therefore, for stability, the ends of the package need tapering, flanges or shoulders.

c) Cross-wound packages [Figure 4.17(c)]. A single yarn end is wound on the package at a considerable helix angle which is generally less than 80°. This type of winding provides package stability and, therefore, there is no need to taper or flange the edges. Thus, a cone or tube could be used in the winding process.

The ratio of winding speed (V_w) and traversing speed (V_t) determines the package type for near-parallel and cross-wound packages. If V_t is very large, relatively fast successive layers of yarn will be laid at distinct angles to each other, producing a cross-wound package. If V_t is slow, successive layers will be very close to parallel to each other, producing a near parallel-wound package.

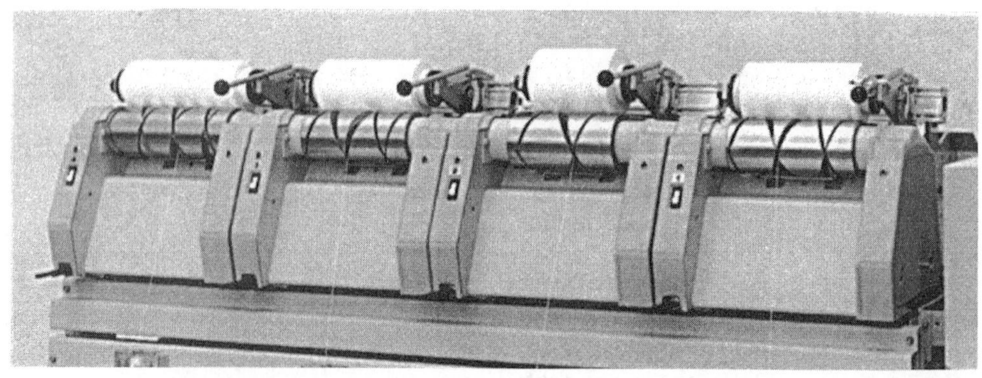

FIGURE 4.15 Grooved roller for yarn traverse (courtesy of SSM).

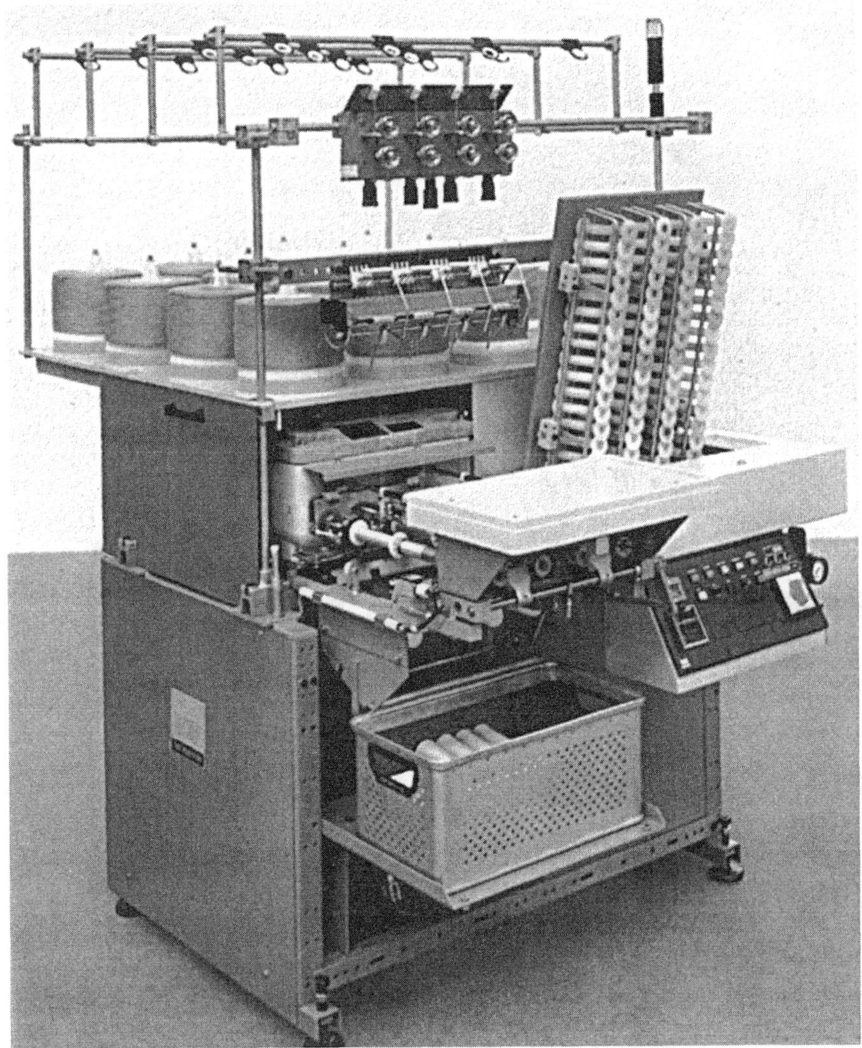

FIGURE 4.16 Near parallel winding machine with reciprocating traverse guide (courtesy of Hacoba).

Sloughing-off is a condition where many coils of yarn unwind from the package at a time. It depends on what is called a critical winding angle.

The package forms can be conical or cylindrical, as required by the subsequent processes. Table 4.1 shows some characteristics of common package types.

4.1.2 Quill Winding

A quill or pirn is a filling bobbin that is placed inside a shuttle in shuttle weaving (Figure 7.2 in Chapter 7, Shuttle Weaving). As the shuttle travels back and forth across the width of the shuttle loom, the filling yarn is unwound from the quill through the eye of the shuttle and laid in the shed. The yarn on the quill is tapered at one end such that the yarn withdrawal takes place continuously without entanglement.

Winding of a quill is different from the regular winding process. In quilling, the yarn is transferred from a larger package to the smaller quill (Figure 4.18). Also, the inspection of yarn is not part of the process, therefore, there is no yarn clearing zone.

(a) Parallel wound package

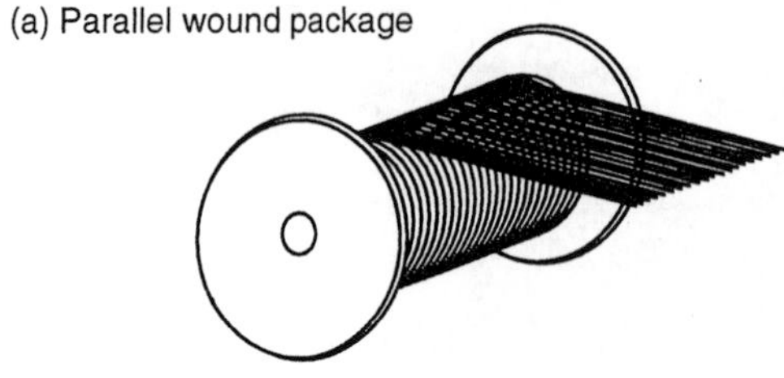

(b) Near parallel wound package

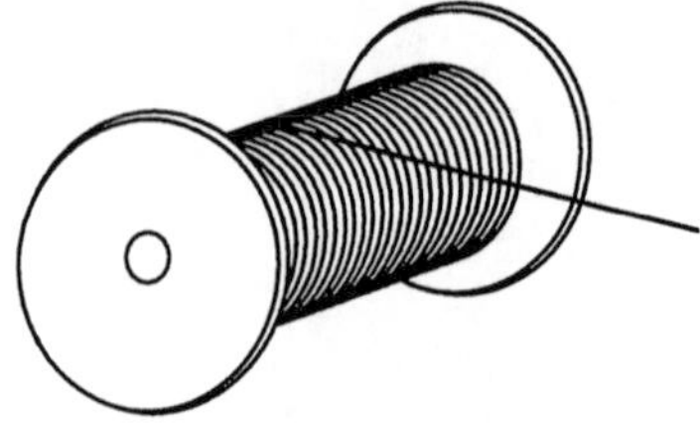

(c) Cross wound package

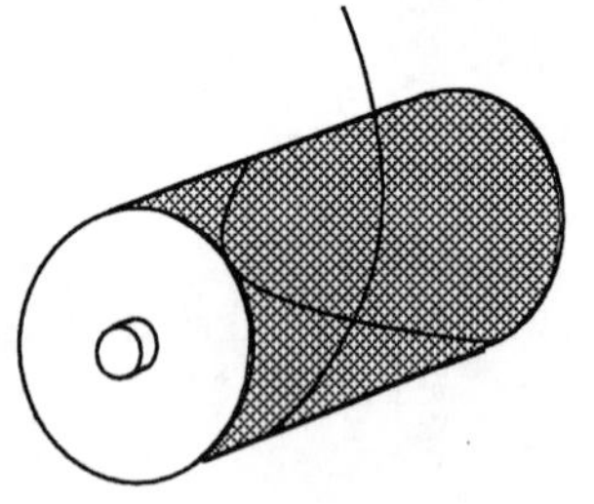

FIGURE 4.17 Types of packages.

The traverse mechanism is also different because of the different geometry of the quill. The traverse here does not go back and forth along the package. It only builds yarn on one part of the package at a time as shown in Figure 4.19. Therefore, quill building is somewhat similar to the building of a bobbin on a ring spinning frame. This type of winding helps reduce ballooning effects, maintain uniform tension, and reduce the possibility of slough-off.

The machines that are used to wind quills are called "quillers" or pirn winding machines. These machines are automatic, which means that when the quill is filled, it is doffed and an empty quill is placed on the spindle automatically. With the elimination of shuttle looms, the quill winding process is also disappearing.

TABLE 4.1 Characteristics of yarn packages and the process.

Cone Winding Process	
Conicity	4°20′ or 5°57′
Package traverse	150 mm (6″)
Diameter	240–300 mm
Winding density	400–450 g/dm^3
Winding speed	up to 1500 m/min
Transfer tails	50 to 70 cm
Integrated length measurement (maximum)	+2%
Cylindrical Package	
Package traverse	150 mm (6″)
Diameter	300 mm
Winding density	400 g/dm^3
Transfer tails	50 to 70 cm
Cylindrical Packages for Filament Yarn	
Package traverse	200–260 mm
Package diameter	320 mm
Tube inside diameter	56/94 mm
Winding angle	12–13°
Winding speed	up to 900 m/min
Transfer tails	50 to 70 cm

4.1.3 Winding Machines

Cross winding machines are used for cross winding of tubes, cones and bobbins with one or two flanges. Yarn laying and package drive are achieved by a grooved drum. In cross-winding, the stability of the package is provided by the acute crossing angle. The package ends can be tapered as well. Figure 4.16 shows a near parallel winding machine with four winding positions and automatic doffing. The yarn traverse is controlled by a cam driven gear. Today's winding machines allow use of different size bobbins with different flange diameters, overall lengths and winding widths on the same machine. For winding of industrial yarns such as aramid, carbon or glass yarns and monofilaments, specially designed yarn guide elements are used. A spindle speed of 5000 rpm is possible.

Figure 4.20 shows a yarn singeing machine with gas burners of stainless steel, traveling blower and gas/air mixing station with variable mixing ratio.

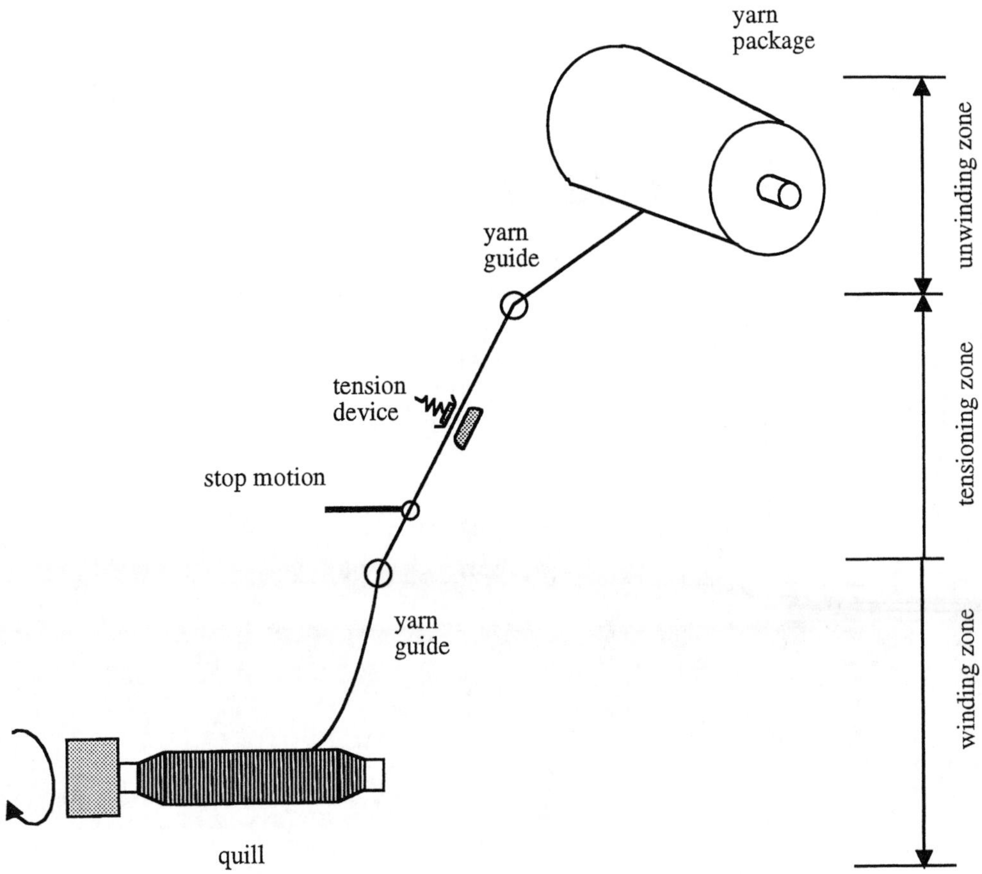

FIGURE 4.18 Schematic of quill winding.

4.1.4 Precision Winding

In precision winding, the position of the yarn as it is laid on the package is controlled very precisely to increase the density of the package. Figure 4.21 shows a precision winding machine. In this particular machine, the yarn positioning system is all-electronic. With the electronic system, freely programmable package building is possible (Figure 4.22).

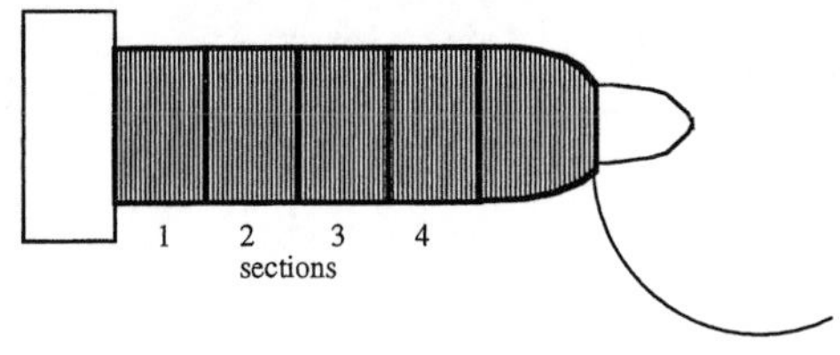

FIGURE 4.19 Yarn deposition on a quill.

Example:

How long will it take for a winder to wind 2.75 lbs of 17 Ne yarn if the winder operates at 750 yd/min with the efficiency $\eta = 93\%$?

$W = 2.75$ lbs
Ne = 17s
$V = 750$ yd/min
t = length/speed = L/V
Ne = $L/(sW)$ where s is the standard hank (standard unit of length)
$s = 840$ yards for cotton

Therefore $L = \text{Ne} \times s \times W = 17 \times 840 \times 2.75 =$ 39,270 yd.

t = 39,270/750
= 52.36 minutes
(assuming no breaks or stops)

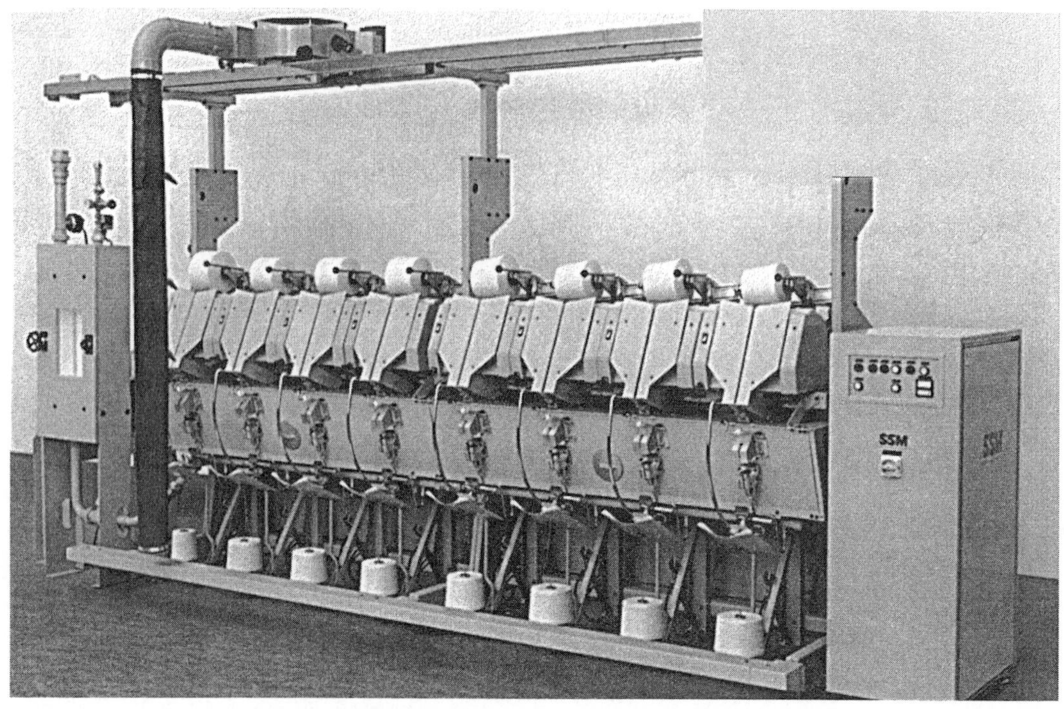

FIGURE 4.20 Yarn singeing machine (courtesy of SSM).

FIGURE 4.21 Precision winding machine with a servomotor (courtesy of SSM).

FIGURE 4.22 Examples of packages made with precision winding (courtesy of SSM).

To consider the effect of η,

$$t = 52.36/0.93 = 56.30 \text{ minutes}$$

4.2 WARP PREPARATION

The preparation of warp yarn is more demanding and complicated than that of the filling yarn. Each spot in a warp yarn must undergo several thousand cycles of various stresses applied by the weaving machine. Weaving stresses include dynamic extension/contraction, rotation (twist/untwist), and clinging of hairs. Additionally, there are metal-to-yarn and yarn-to-yarn flexing and metal-to-yarn and yarn-to-yarn abrasion stresses. Modern weaving machines have placed increased demands on warp preparation due to faster weaving speeds and the use of insertion devices other than the shuttle. Warp yarn must have uniform properties with sufficient strength to withstand stress and frictional abrasion during weaving. The number of knots should be kept to a minimum. The knots should be standard type and size such that they fit through the heddle eyes and reed dents. Size agent must be applied uniformly on the surface of the yarn. The yarns on the warp sheet must be parallel to each other with equal tension.

Warp preparation involves winding, warping, slashing and drawing-in or tying-in. The winding process is explained in Section 4.1 of this chapter. The purpose of warp winding is to form a package of good quality yarn that is large enough to be used in the creel of a warping machine. Winding of yarn for warping is usually done at relatively high tension.

4.2.1 Warping

In general terms, warping is transferring many yarns from a creel of single-end packages forming a parallel sheet of yarns wound onto a beam or a section beam (Figure 4.23). Today's warping machines can process all kinds of materials including coarse and fine filament and staple yarns, monofilaments, textured and smooth yarns, silk and other synthetic yarns such as glass. Usually a static eliminator device is recommended for yarns that can generate static electricity.

The warp beam that is installed on a weaving machine is called a weaver's beam. A weaver's beam can contain several thousand ends and for different reasons it is rarely produced in one operation. There are several types of warping processes depending on the purpose. It should be noted that the warping terminology is quite different in different regions and sometimes the same term may be used to identify different processes in different regions or industries. In this book, an attempt is made to use the warping terminology based on the physical process.

Direct Warping

In direct warping, the yarns are withdrawn from the single-end yarn packages on the creel and directly wound on a beam (Figure 4.24).

Direct warping is used in two ways:

a) Direct warping can be used to directly produce the weaver's beam in a single operation. This is especially suitable for strong yarns that do not require sizing such as continuous filaments or monofilaments and when the number of warp ends on the warp beam is relatively small. This is also called direct beaming.

b) Direct warping is used to make smaller, intermediate beams called warper's beams. These smaller beams are combined later at the slashing stage to produce the weaver's beam. This process is called beaming. Therefore, for example, if the weaver's beam contains 9000 warp ends, then there would be—say—9 warper's beams of 1000 ends each. If this weaver's

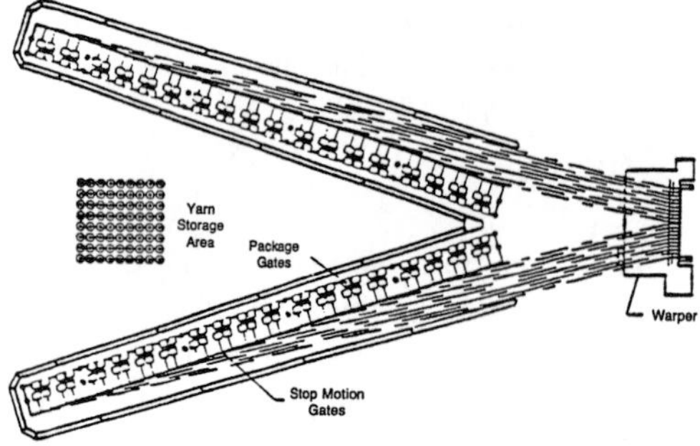

FIGURE 4.23 Schematic top view of warping process (V-creel).

beam were to be made at one stage, the creel would have to have 9000 yarn packages, which is hardly possible to manage and accommodate (Note: The warper's beams may also be called—confusingly enough—section beams. Therefore, this method of producing weaver's beam may also be called section beam warping). Usually 3 to 4 ends per cm (8 to 10 ends per inch) are recommended on section beams for slashing purposes. Beam hardness is recommended to be 50–60 (Shore O-2 durometer); hardness should be achieved with tension, not from packing roll pressure.

Direct warpers are used to warp all conventional staple fibers, regenerated fibers and filaments. In direct warping, a flange beam is used. Since all the yarns are wound at the same time, the flanges provide sufficient yarn stability on the beam. The typical beam flange diameters are 800, 1000, 1250 and 1400 mm with working widths of 1400 to 2800 mm. Machine specific options include tape applicator, static eliminator, windscreen, comb blowing and dust extraction devices, yarn storage and inspection units, oiler, tension roller unit, beam removal unit and control platform. Figure 4.25 shows an expanding zigzag comb which is used to control the width of the beam and keep the yarns parallel and straight.

Indirect (Section) Warping

In indirect warping, a section beam is produced first as shown in Figure 4.26. Other names

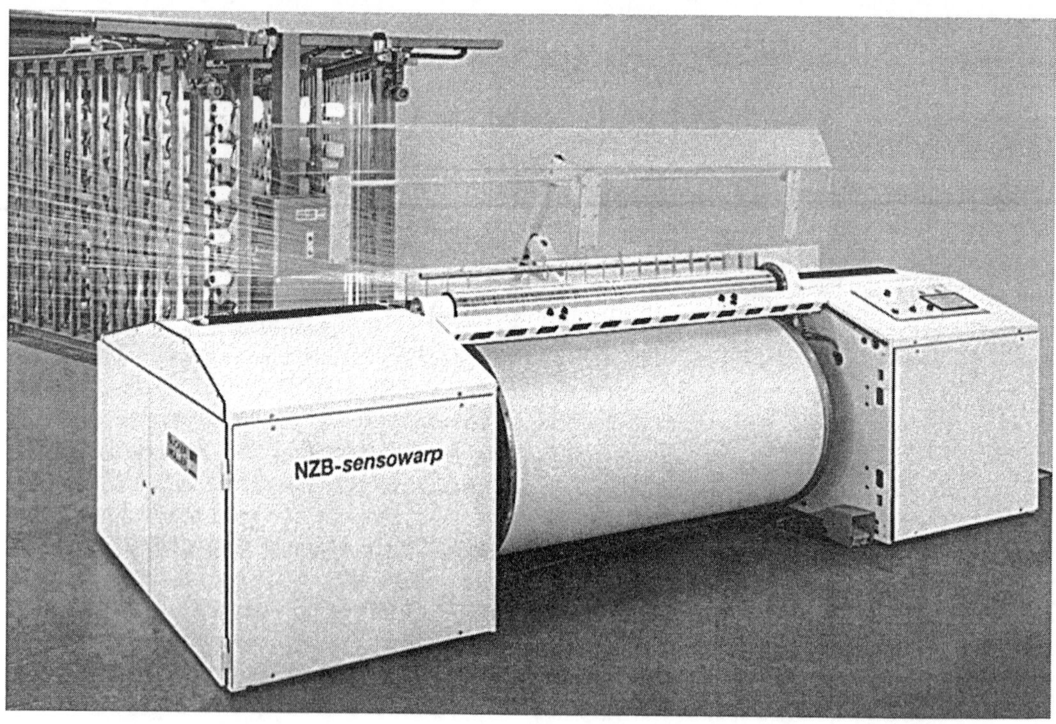

FIGURE 4.24 Direct warping (courtesy of Sucker Muller Hacoba).

FIGURE 4.25 Expanding zigzag comb (courtesy of West Point Foundry and Machine Co., Inc.).

used for section warping are pattern warping, band warping or drum warping. The section beam is tapered at one end. Warp yarn is wound on the beam in sections, starting with the tapered end of the beam (Figure 4.27). Each section has multiple ends that are traversed together slowly during winding along the length of the section to form the angle. Due to the geometry of the yarn sections, the last section on the beam will have a tapered end that will make the whole yarn on the beam stable. It is important that each layer on the beam contain the same number of yarns. The same length of yarn is wound on each section which is measured by a measuring roller. The warping speed can be adjusted in the range of 20 to 800 m/min; however, residual elongation will be reduced at high speeds.

The capacity of the warping drum is dictated by the required beam diameter. The drum is round with a cone which has a slight angle that prevents yarns from slipping off. Figure 4.28 shows typical cone angles and beam diameters used in practice. The circular cross section of the cone eliminates differences in yarn lengths in the first and subsequent sections. Adjustable cone angles are possible; however, a fixed cone prevents pressure marks on the yarn as well

FIGURE 4.26 Indirect warping (section warping) with parallel creel (courtesy of Sucker Muller Hacoba).

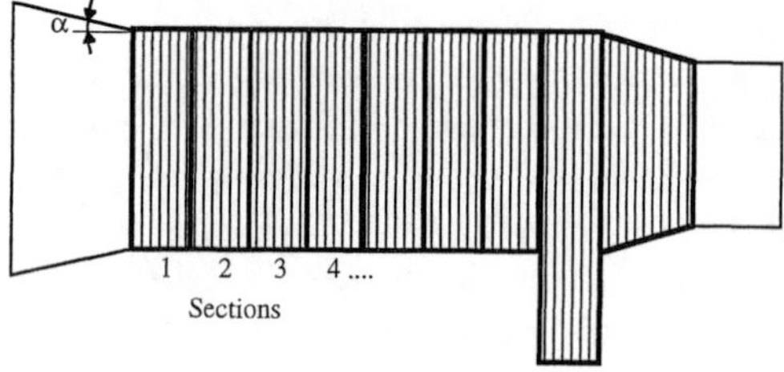

FIGURE 4.27 Schematic of yarn sections on tapered section beam.

as incorrect angle settings. High angles tend to reduce package stability. The amount of yarn wound on the beam is proportional to the length of each section and the conical angle, α (Figure 4.27). The drum is generally made of composites, such as synthetic resin bonded glass fiber, that withstand extreme pressures.

Constant yarn tension in the creel does not necessarily provide an exact uniform height on the warping drum. Exact uniform height can be achieved using a feeler roller. Before the warping starts, the feeler roller is moved firmly against the warping drum (Figure 4.29). As soon as the warping starts, the roller starts measuring the buildup of the yarn with an electronic sensor. The feeler roller is resisted by a force that depends on the density of the yarn. Also, each subsequent section exerts less pressure on the feeler roller at the same height. The feeler roller is pushed back by the yarn as it builds up. The

Ratio: cone height: cone length	cone angle	Beam dia. max. 800 mm	Beam dia. max. 1000 mm	Beam dia. max. 1250 mm
1 : 4	14°	175, ⌀ 800, 700	250, ⌀ 1000, 1000	350, ⌀ 1000, 1400
1 : 6	9° 30'	175, ⌀ 800, 1100	250, ⌀ 1000, 1500	350, ⌀ 1000, 2100
1 : 8	7°	175, ⌀ 800, 1400		

FIGURE 4.28 Typical cone angles (courtesy of Sucker Muller Hacoba).

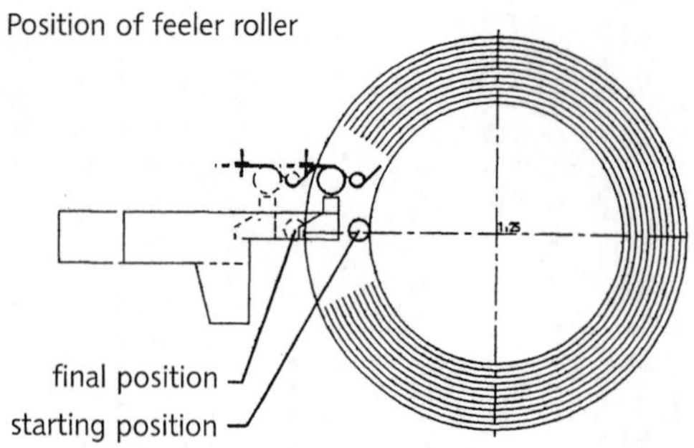

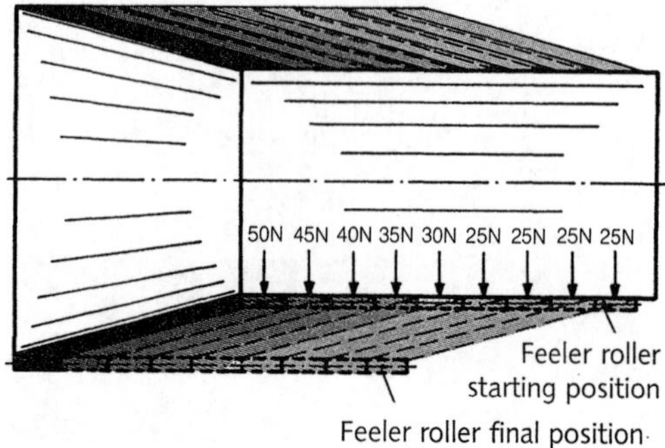

FIGURE 4.29 Position of feeler roller (courtesy of Sucker Muller Hacoba).

measured data is fed to a computer that calculates the exact traverse per revolution.

The maximum speed during winding of the beam depends on the necessary winding tension (Table 4.2). With the standard, single speed range version, one of the stages is selected based on the application. If a double speed range drive is supplied, one of the data pairs can be chosen from the table.

To obtain a uniform winding hardness from inside to outside, the winding tension needs to be held constant throughout the warp length, which can be done by varying the braking force. This method also reduces the tension variations during starting and stopping operations as shown in Figure 4.30.

Lease bands are used in section warping. After a few turns at the beginning, a lease band is inserted in the axial direction of the beam (Figure

TABLE 4.2 Beaming speed and warp tension pairs (courtesy of Benninger).

Max. beaming speed (m/min)	Max. warp tension (N)	Max. beam diameter (mm)
40	15000	1000
80	7500	1000
60	10000	1000
120	5000	1000
80	7500	1000
160	3750	1000
120	5000	1000
220	2700	1000
160	3750	1000
300	2000	1000
50	12000	1250
100	6000	1250
75	8000	1250
150	4000	1250
100	6000	1250
200	3000	1250
150	4000	1250
280	2220	1250

4.31). In unwinding, the lease bands are useful to identify the number of yarn layers. Figure 4.32 shows the operation of an automatic leasing device. With a traditional leasing reed, warping speeds of up to 600 m/min are possible. Figure 4.33 shows the insertion of sizing split components. Figure 4.34 shows how an automatic section change is made.

After all the sections on the beam are wound completely, then the yarn on the beam is wound onto a regular beam with flanges, before slashing (Figure 4.35). This process is called beaming. Sometimes a section beam is also used in the slashing stage. Figure 4.36 shows various possible configurations of a modern sectional warping and beaming machine.

With today's computerized sectional warping systems, once the basic style information is entered, the computer automatically calculates the following [1]:

- number of sections on the beam and width of each section
- carrier lateral movement speed and automatic positioning of each section start point
- automatic stops for leasing
- calculation of the correct feed speed irrespective of the material and warp density

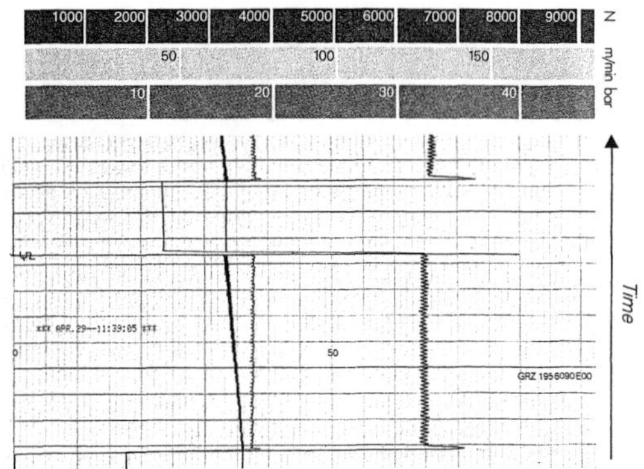

FIGURE 4.30 Warp tension during starting and stopping operations (courtesy of Benninger).

The computer can also monitor the following:

- automatic stops for predetermined length
- operating speed regulation of +/– 0.5% between warping and beaming
- beaming traverse motion
- memory of yarn breakage during warping for beaming

FIGURE 4.31 Lease band application (courtesy of Sucker Muller Hacoba).

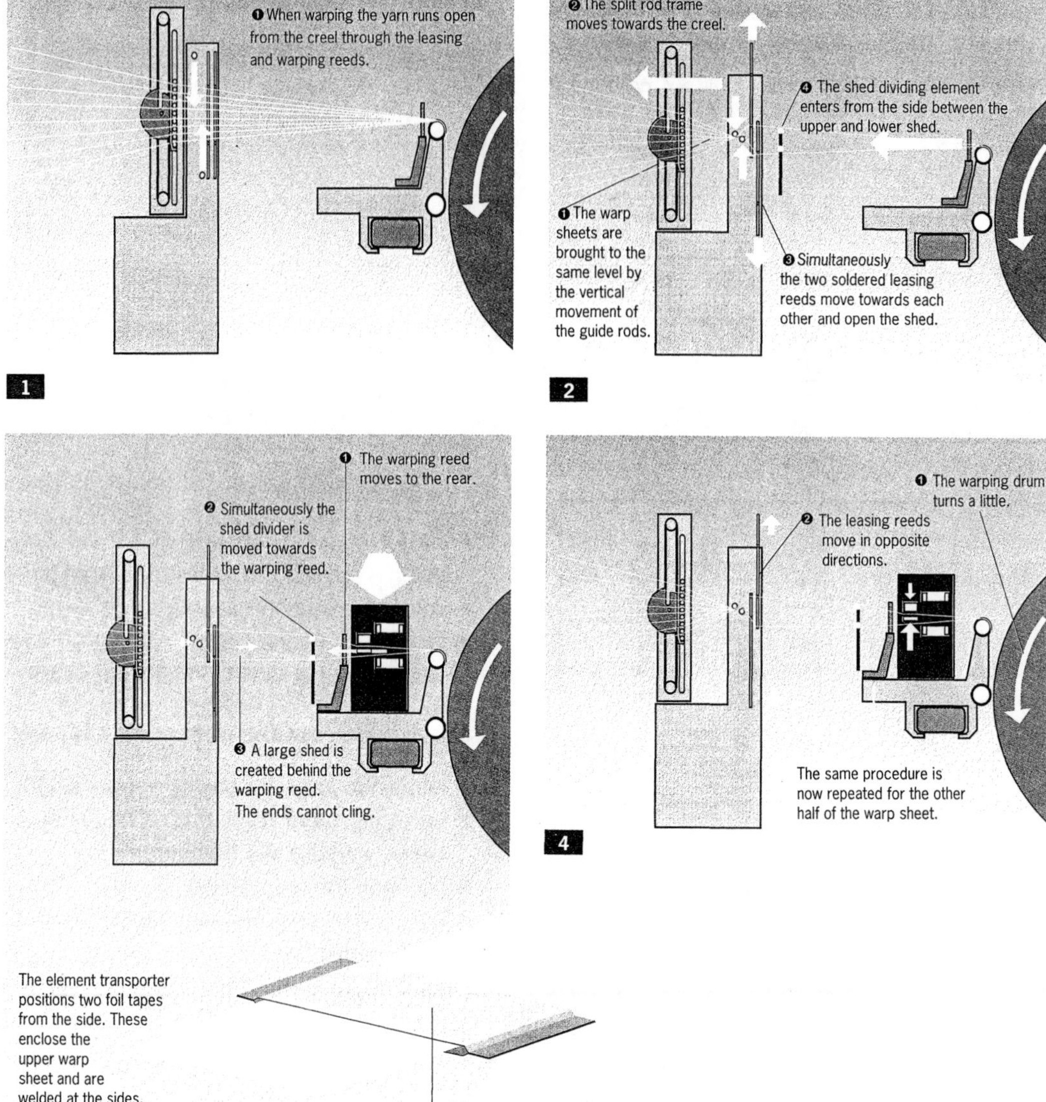

FIGURE 4.32 Operation sequence of an automatic leasing device (courtesy of Benninger).

Other typical features of a modern sectional warper are:

- feeler roller to apply material specific pressure to obtain exact cylindrical warp buildup
- lease and sizing band magazines
- constant warp tension over the full warp width
- automatic section positioning with photo-optical section width measurement
- pneumatic stop brakes
- warp tension regulation for uniform buildup
- automatic warp beam loading, doffing and chucking

Ball Warping

Ball warping is mainly used in manufacturing of denim fabrics. The warp yarns are wound on a ball beam in the form of a tow for indigo

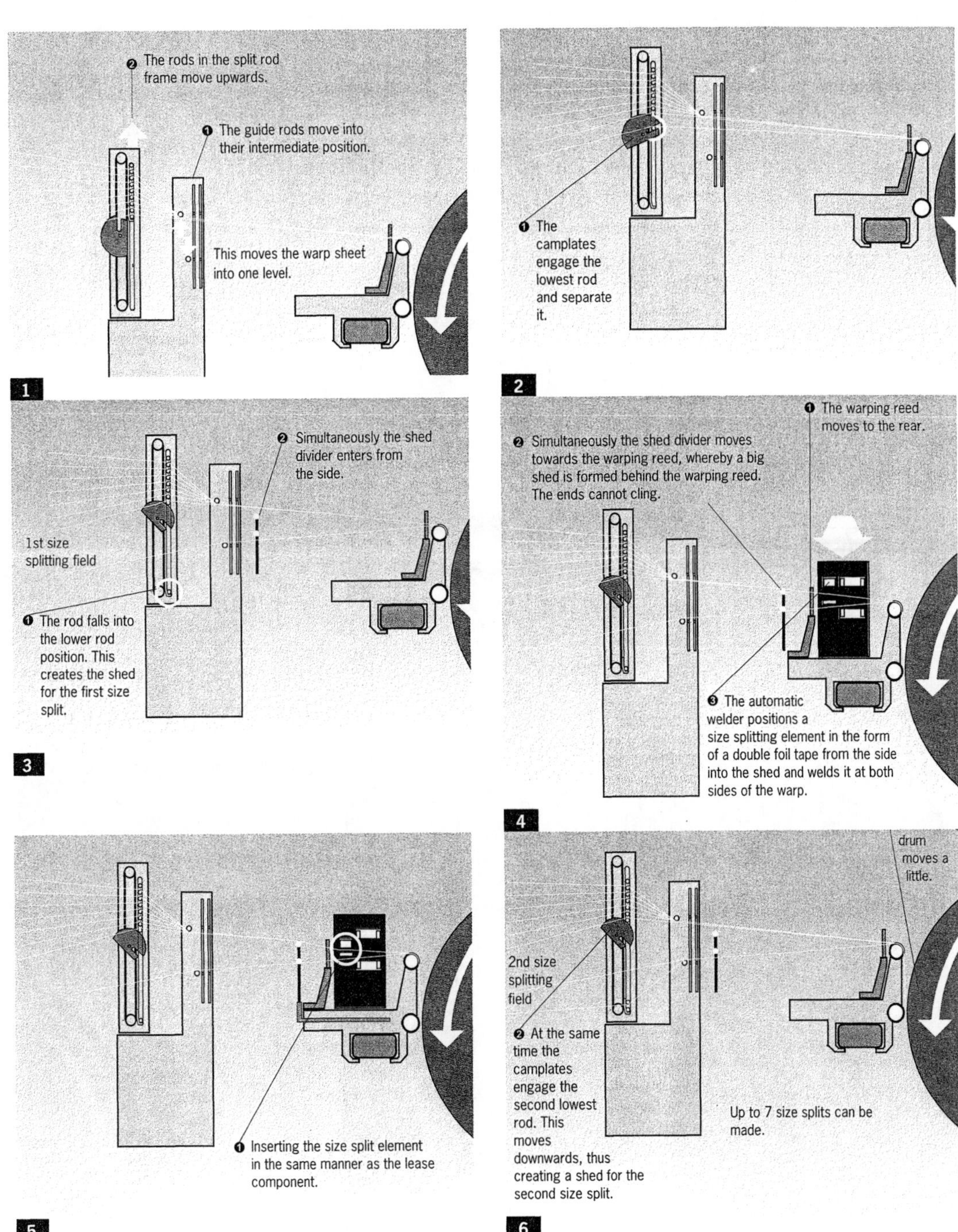

FIGURE 4.33 Insertion of sizing split components (courtesy of Benninger).

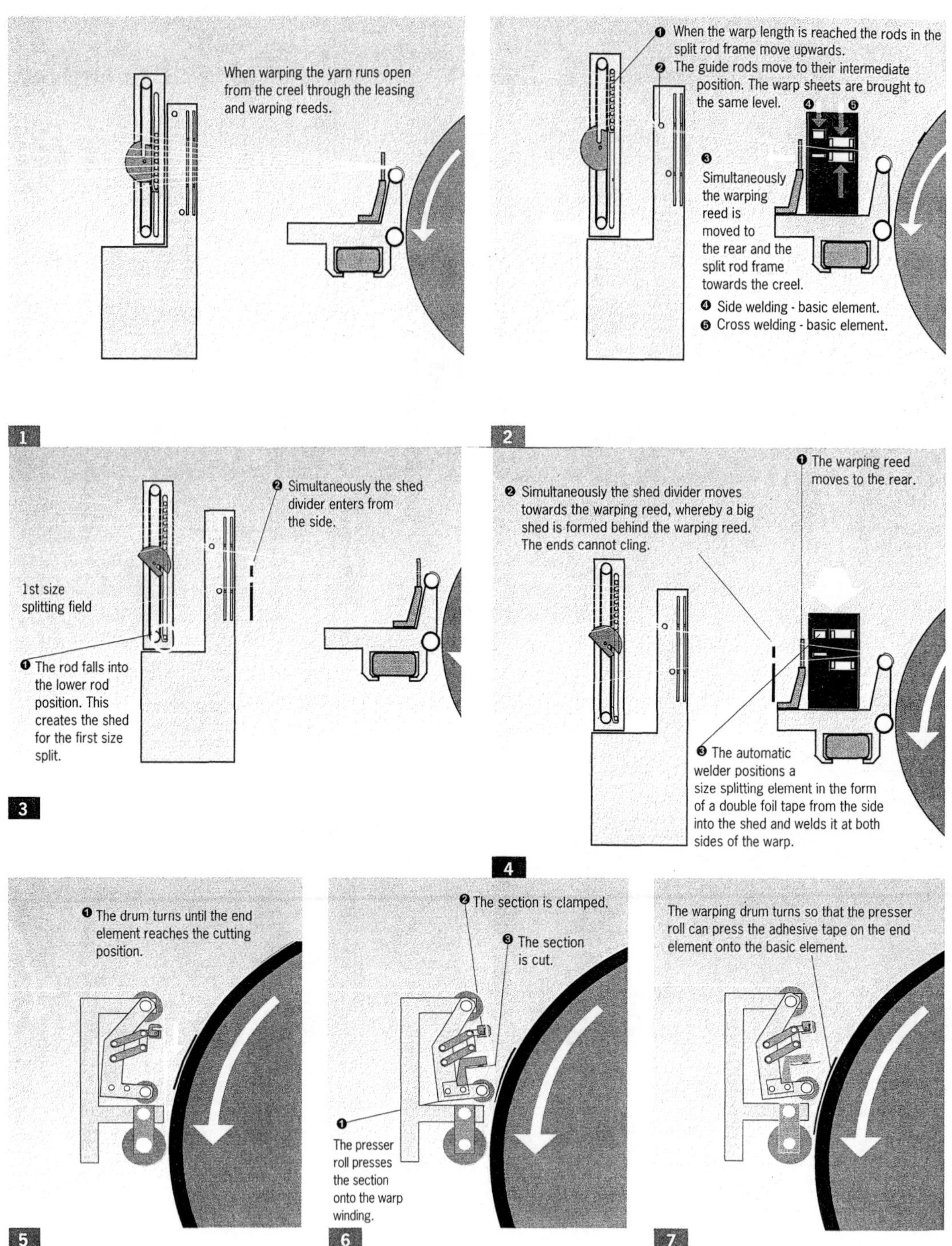

FIGURE 4.34 Schematic of automatic section change (courtesy of Benninger).

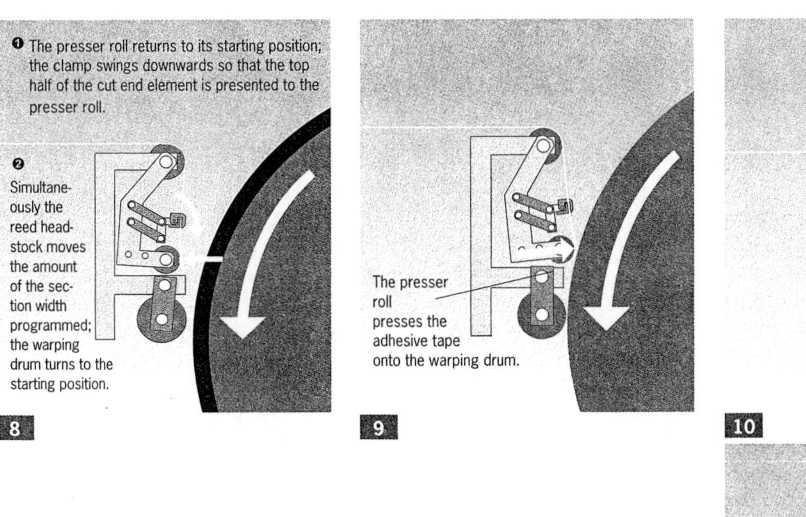

FIGURE 4.34 (continued) Schematic of automatic section change (courtesy of Benninger).

FIGURE 4.35 Transfer of warp from section warper to a flanged beam or to a weaver's beam in case no slashing is required (courtesy of McCoy-Ellison).

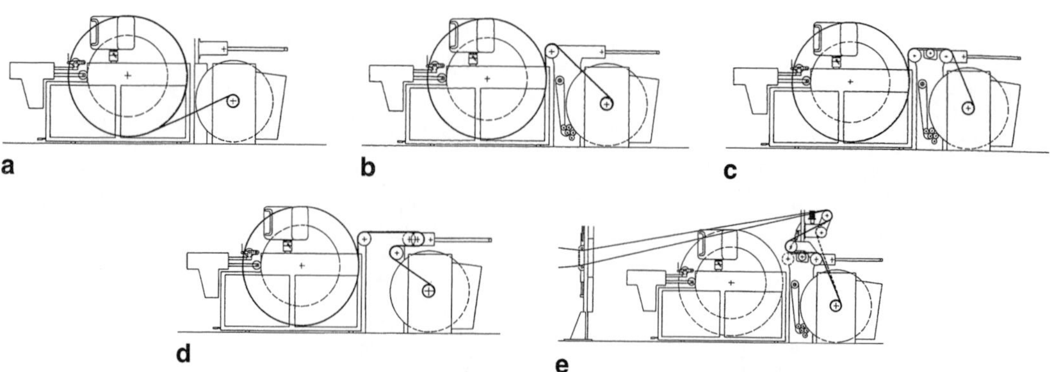

FIGURE 4.36 Possible configurations of a sectional warping and beaming machine (courtesy of Sucker Muller Hacoba). (a) Direct warp path from the warping drum to the warp beam, (b) one guide roller for pressing device, (c) additional guide roller for lubricating or softening device, (d) warp tension regulator, (e) direct beaming unit to produce warps with a small number of ends.

dyeing (Figure 4.37). After the dyeing process, the tow is separated and wound on a beam as shown in Figure 4.38. This stage is also called long chain beaming or rebeaming. Usually a lease comb and a collecting reed are used to achieve tangle free lease insertion.

Draw Warping

Draw warping is combining the drawing of filament yarns with heat setting and warping processes to achieve uniform stretching and heating for improved dye uniformity, end to end. It is used for weaving of thermoplastic yarns. Figure 4.39 shows schematic of draw warping process. Typical speed is up to 1000 m/min. Single or two phase drawing is possible.

4.2.2 Warping Machines

A typical warping machine has three major components: creel, headstock and control devices.

Creel

There are various types of creels. The most common creel types are:

- parallel standard creel with fixed package frame (single end creel)
- parallel creel with package trucks
- parallel creel with swivelling package frame sections (for cotton, viscose, polyester/cotton, wool colored)

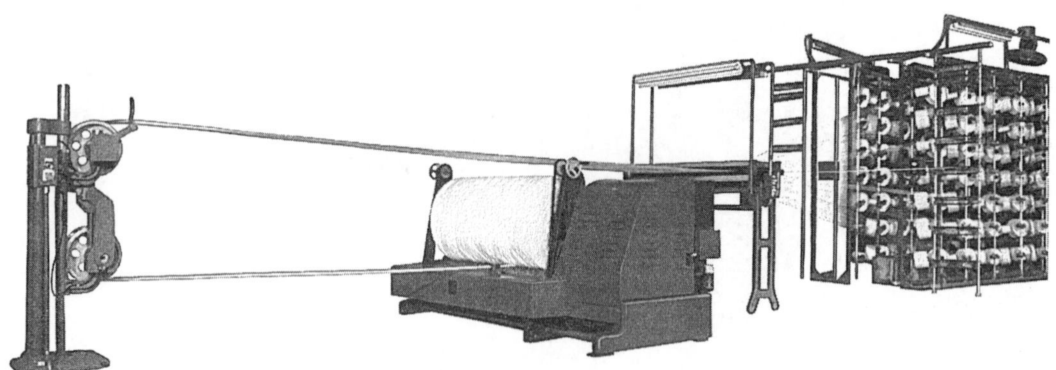

FIGURE 4.37 Winding on ball beam (courtesy of West Point Foundry and Machine Company).

FIGURE 4.38 Denim warp yarn beaming system (courtesy of West Point Foundry and Machine Company).

- parallel creel with reserve packages (magazine creel, for synthetic filaments)
- parallel creel with unrolling draw-off for polypropylene, monofilaments
- V-creel with reversible frames
- V-creel with reversible frames and automatic knotter (for cotton, viscose, polyester/cotton)
- V-creel with traveling packages

Parallel creels are used for sectional warping and direct warping; V-creels are used for direct warping.

In single end creel, there is only one package for each warp end. Since creeling takes a considerable time, the package size should be such that a number of beams can be made from one creel. Also, usually more than one creel is used such that once a creel is depleted, the next one would be readily available to continue warping. Depending on the space requirements, this is done either by moving the headstock or by moving the creels. If the headstock is movable, then usually two creels are used which are called duplicated creels. If the headstock is fixed, again two creels will be enough but a third creel place is needed in which to move the empty creel.

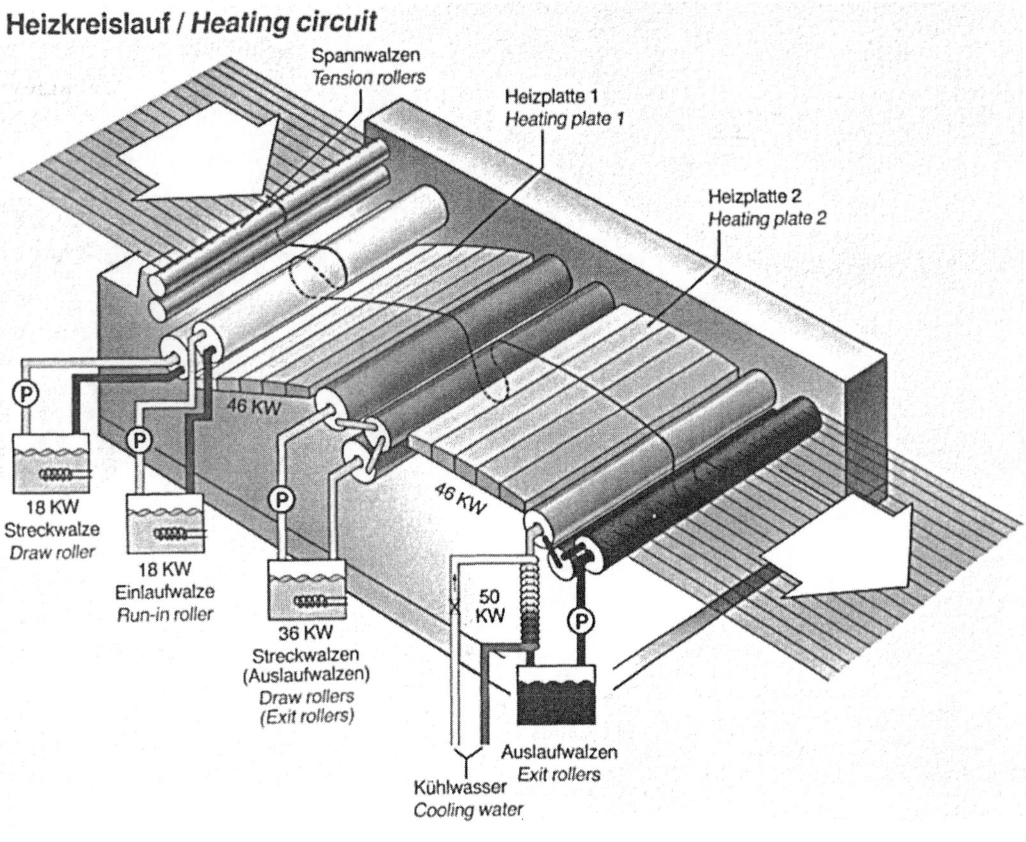

FIGURE 4.39 Schematic of draw warping process (courtesy of Karl Mayer).

This is known as a truck creel or trolley creel (Figure 4.40).

Trolley creels are suitable for both sectional and direct warping. The creel generally has a rectangular tube construction. The trolley creels have wheels for easy maneuvering; however, they are stabilized to prevent tipping over.

In a magazine creel, usually a two package creel is used. The tail end of the running package is attached to the leading end of the reserve package. This allows continuous warping operation. With a yarn splicer, the undesirable effects of knots can be avoided. Figure 4.41 shows a magazine creel with two pivoting spindles: a working spindle and a reserve spindle. When one set of spindles is in operation, the empty packages are removed from the reserve set, which is then filled with new packages. The creel can be loaded from the center aisle or from the outside. They are ideal if long yarn lengths are to be unwound, if the packages do not have measured yarn lengths or if residual packages are used.

In the swivel frame creel, empty packages can be replaced on either side from the center aisle. This creel is suitable for confined spaces. A foot

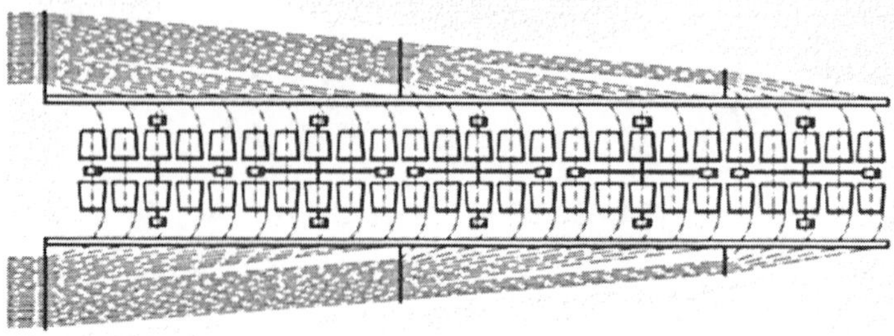

FIGURE 4.40 Trolley creel (courtesy of Sucker Muller Hacoba).

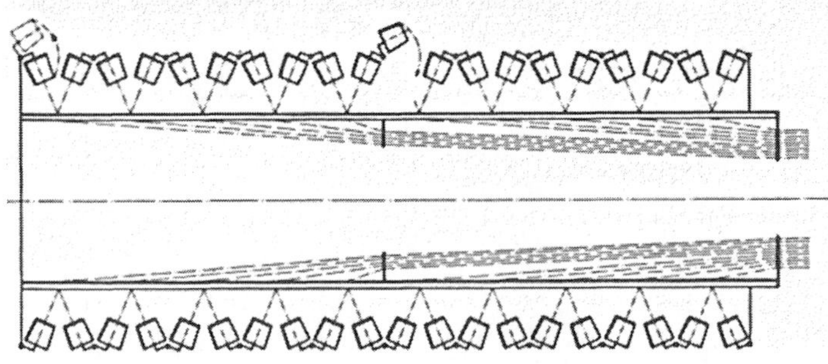

FIGURE 4.41 Magazine creel (courtesy of Sucker Muller Hacoba).

pedal is used to swivel the frame 180° to allow the empty side to be recreeled (Figure 4.42). Swivel creels can have a V shape as well.

In traveling package creels, the creel is like a continuous belt as shown in Figure 4.43. Usually two creels form a V shape. When the full packages are being used for warping on the outside position (active or run position), the empty inner side can be filled with packages. When the full packages are emptied, the side with the full packages is brought to the warping area (outside) by rotation and the warping continues without much interruption except for threading of the warp ends. After rotating the creel, the groups of yarn from the vertical rows are threaded and pulled to the warper where each yarn is positioned in the designated comb dent. The V configuration is especially suitable for warping of staple yarns at high speeds. Other advantages of V-creel are:

- no need for yarn guide
- uniform yarn tension across the whole beam
- free yarn run from the creel to the warping machine
- low yarn tension

In cases where overhead unwinding cannot be tolerated, a roller creel is used. In the roller creel, the package rotates and side withdrawal

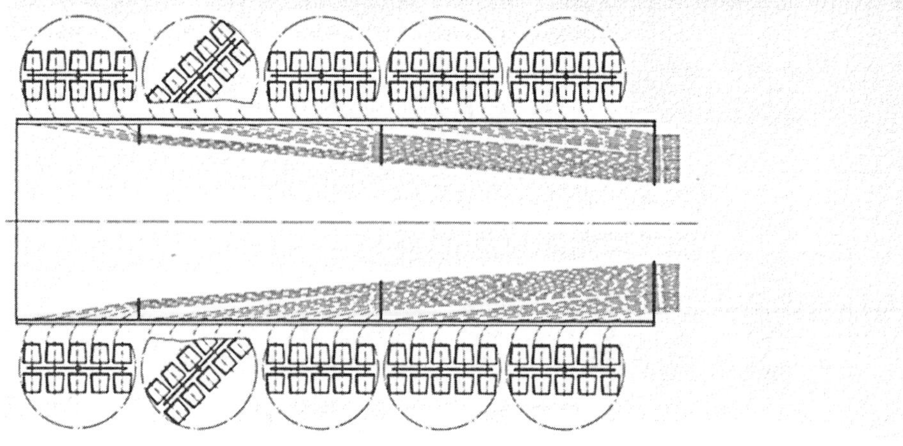

FIGURE 4.42 Swivel frame creel (courtesy of Sucker Muller Hacoba).

of yarn takes place. This type of creel is used especially for carbon filaments, aramid fibers, tape yarns and monofilaments (Figure 4.44).

Warp quality is affected through warp section guidance between the creel exit and the warping drum. To prevent the crossing ends of warp yarns, a "single skip" draft is proposed as opposed to a "straight through" draft. Figure 4.45 compares the two drafts. In single skip, the vertical distribution in the yarn array corresponds to a 1 : 1 lease. It is reported that warp quality is improved with the single skip system.

Headstock

The yarn speed should be kept as constant as possible during warping. In indirect (section) warping, a constant speed drive is generally sufficient in providing approximately uniform yarn speed on the surface of the beam. This is because the thickness of the yarn built on the beam is relatively small compared to the beam diameter such that the surface speed does not change much. In direct warping, the change due to yarn buildup on the beam is significant. Therefore,

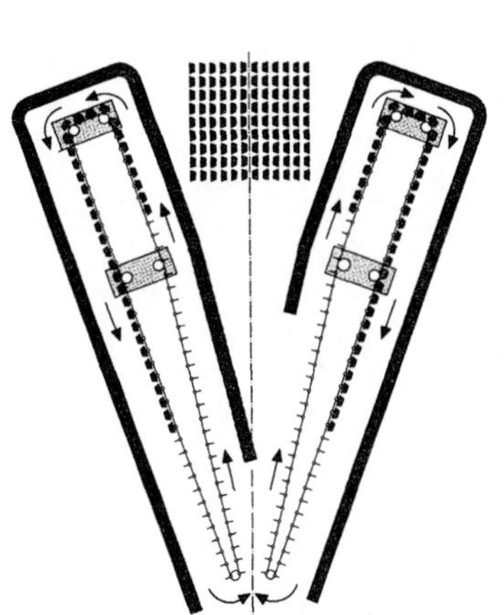

FIGURE 4.43 Traveling package V-creel (courtesy of Benninger).

in direct warping, mechanisms that are similar to the ones used in winding are utilized to attain uniform yarn speed; surface friction drive and variable speed drive are commonly used. For some filament yarns, variable speed drive is chosen since friction drive would cause problems.

Today's headstocks are equipped with advanced design features such as precision direct drive, advanced electronics, smooth doffing and programmable breaking. Automatic hydraulic doffing is accomplished with the operation of one button. Programmable pneumatic braking provides a constant stopping distance regardless of the operating speed or beam diameter. The length of the yarn wound on the beam is controlled with a measuring roller and counter de-

FIGURE 4.44 Roller creel (courtesy of Sucker Muller Hacoba).

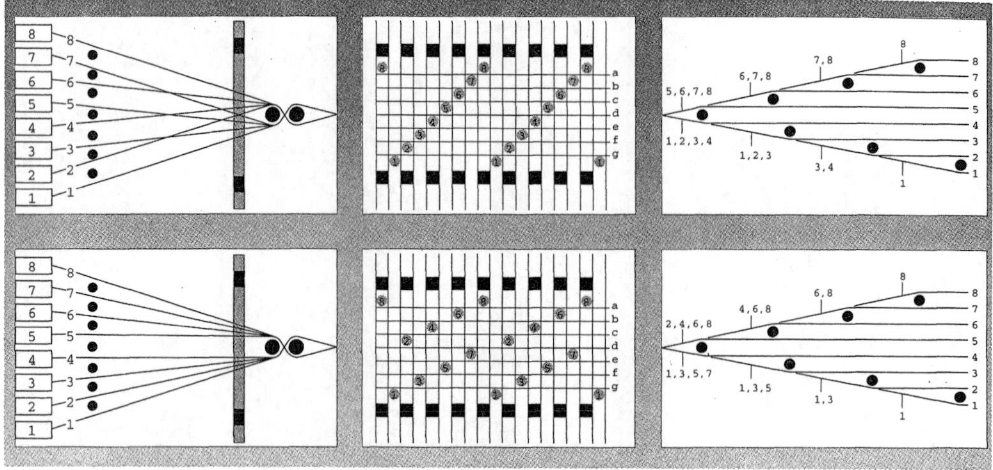

FIGURE 4.45 "Straight through" and "single skip" drafts (courtesy of Benninger).

vice. The density of the yarn can be controlled by tension, pressure or both. Frictional drive usually results in higher yarn density. In spindle drive, yarn tension and a hydraulically activated pressure roller are used to control density. Figure 4.46 shows a typical headstock. Some headstocks are designed to run more than one beam width.

Control Devices

Similar to winding, warp yarns are threaded through tension devices, stop motions, leasing rods and the reed. Uniform tension is necessary so that all the warp ends behave the same way. The tension on the warp yarns is kept relatively low. Every end requires a tension controller which is usually located close to the package.

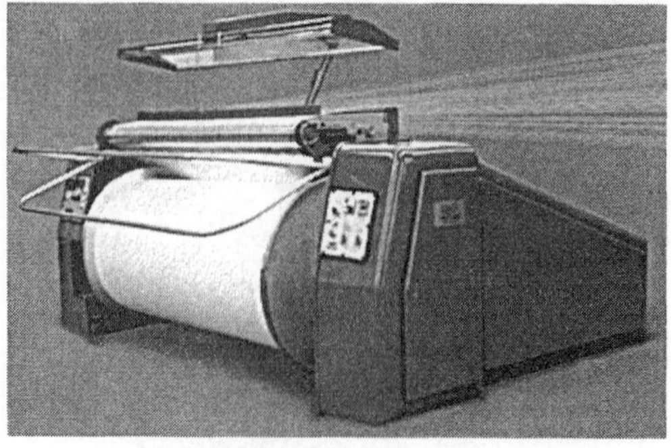

FIGURE 4.46 Headstock (courtesy of West Point Foundry and Machine Company).

A quick response, advanced stop motion is necessary for warping. Due to the high inertia of the beam, it is difficult to stop the beam suddenly once an end is broken. However, the beam must be stopped before the broken end reaches the beam. The stop motion electrically links each warp end to the warper braking system; when a warp end breaks, the warper stops. Powerful brakes are used for this purpose. A light indicates the location of the broken end. The warping process is generally irreversible, unwinding of the beam would cause yarn entanglement. The stop motion device, which can be mechanical or electronic for quick response, is usually located near the creel.

Figure 4.47 shows an electronic, motion sensitive stop motion device. The electronic eye detects movement of individual ends to trigger a warp stop when there is no yarn movement. Figure 4.48 shows mechanical stop motions, drop wire and faller wire.

Figure 4.49 shows the schematic of a vibration sensitive yarn stop motion for creels. The sensor detects the motion of yarn, not the presence of the yarn. The yarn is guided onto a vibration sensitive beam by an entrance and exit eyelet. The pressure of the tensioned yarn moving on the beam generates an electrical signal that is monitored by the processor, this will detect a yarn break or yarn hang-up.

Mechanical clamping devices are used to positively control the yarn during stops and initial

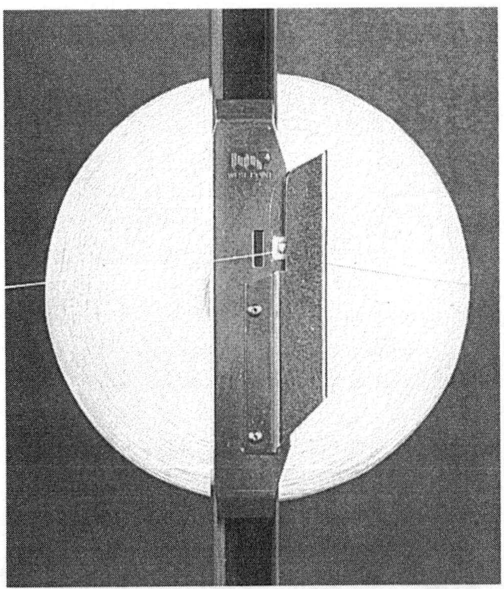

FIGURE 4.47 Electronic yarn stop motion device (courtesy of West Point Foundry and Machine Company).

warper acceleration. Once the warper reaches the regular speed, then the stop motions are activated and the clamps are released for high speed warping. Figure 4.50 shows a mechanical stop motion and yarn clamping system.

Creels have a yarn cutting assembly. Once the active yarn packages are consumed, all yarn clamps are engaged and the yarns are cut. Then the clamp is released and the creel is rotated. Manual cutting and knotting takes an average of 8 seconds; for a 640 package creel, it takes 85 minutes for one person to complete the whole creel. In modern machines, yarn cutting and knotting are done automatically. Figure 4.51 shows an automatic yarn cutting and knotting device. In an 8 tier creel, an automatic knotting and cutting device requires an average 2 seconds per package which totals 21 minutes for the whole creel. The automatic knotting and cutting devices are mounted on rails that are integrated in the creel. There are as many cutting heads as there are tiers. The devices are controlled by a PLC (programmable logical controller). Each package row is approached exactly at traverse/creep speed by means of two proximity initiators. The oscillating suction and gripping tubes offer the yarn ends to the knotting heads, where they are knotted and trimmed. The tails are removed by suction.

Flat spots on the beam are prevented with a warper drive interrupter and a rubber covered pressure roll. The warper drive interrupter automatically controls the speed to minimize machine vibration during warping. The rubber covered pressure roll reduces the effects of machine vibration on the beam. Contractible reeds are used to accommodate varying widths and warp densities (Figure 4.25).

To avoid static buildup, especially with manmade fibers, different methods can be used including chemicals, ionization of air or humidification of air. Fans are used to prevent lint accumulation when warping staple yarns.

Figure 4.52 shows an automatic warp sampling machine which can be used to warp framing samples, design samples and small lot fabrics for testing purposes. The creel has typically 8–10 bobbins and up to 10 colors can be selected. The warp length can be from 7 to 133 m.

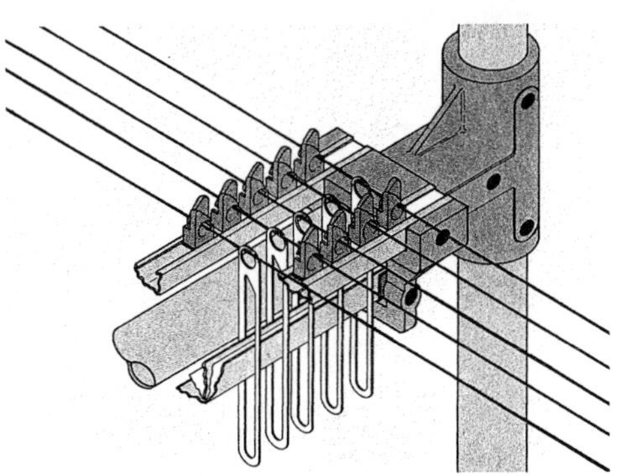

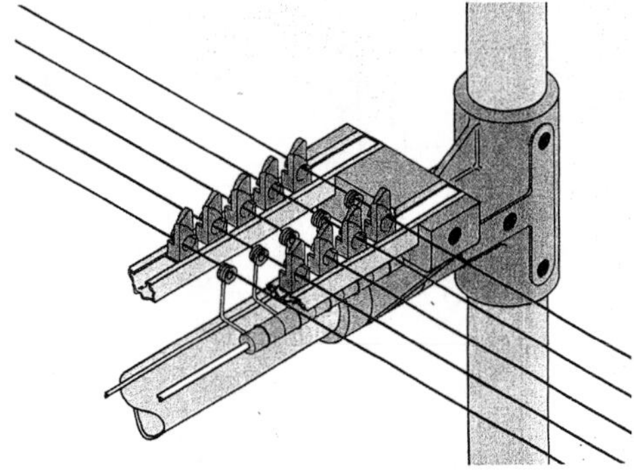

FIGURE 4.48 Schematics of drop wire (left) and faller wire stop motions (courtesy of West Point Foundry and Machine Company).

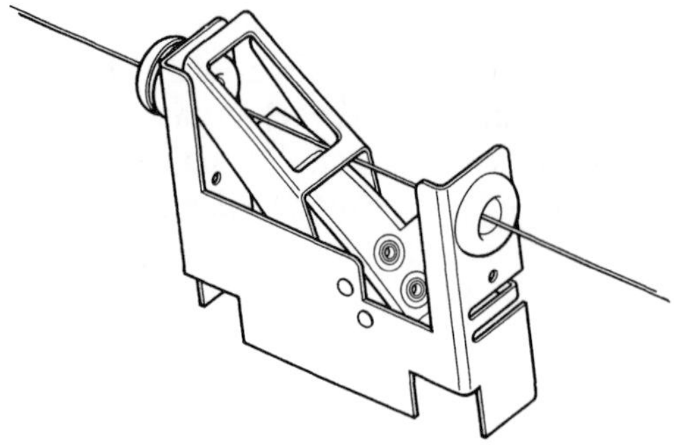

FIGURE 4.49 Schematic of vibration sensitive yarn stop motion (courtesy of West Point Foundry and Machine Company).

4.2.3 Tension Control in Warping

Several tensioning devices are used for unwinding of packages in warping. Figure 4.53 shows a roller tension unit that is used for warping.

This tension unit consists of a fixed rubber roller and a movable rubber roller. The yarn passes through a closed eyelet into the nip of the rollers and then deflected about 90° towards the warper. The yarn tension is generated by a tension spring pressure on the movable roller. The yarn causes both rollers to rotate. The roller tension unit is suitable for yarns of 180 dtex or higher. All staple yarns and continuous filaments with or without twist can be handled. The maximum working speed is around 800 m/min.

Sometimes, during warping of heavy industrial yarns, the tensions on the yarns coming from the creel are not adequate. In this case, a tension roller is used between the creel and warper which increases the tension by a factor of three (Figure 4.54).

Figure 4.55 shows a capstan tensioner for warping of fine yarns. The yarn guidance element is a capstan roller that is rotated by several winds of unwinding yarn wrapped around it.

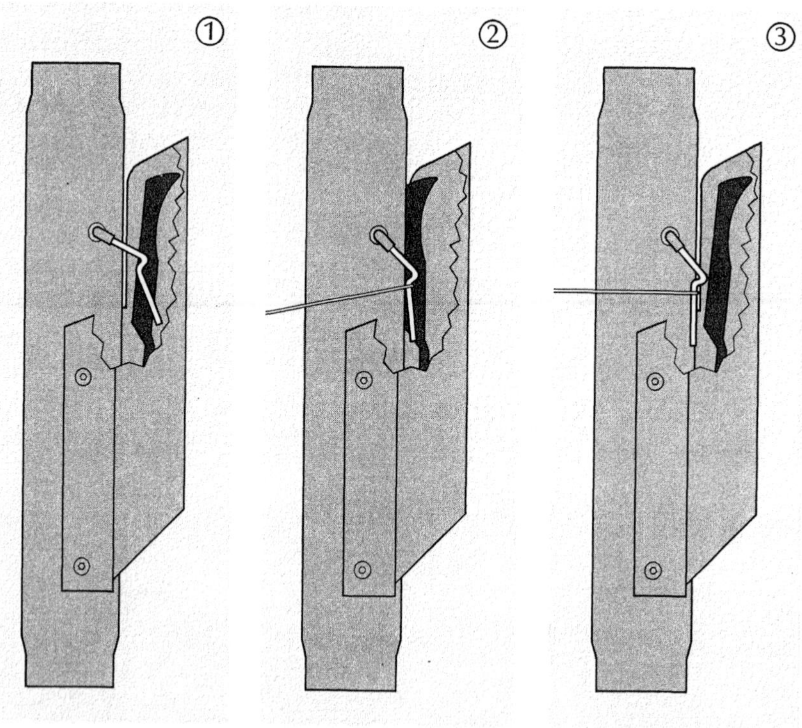

FIGURE 4.50 Faller wire stop motion and yarn clamping system: (1) Clamp in open position for creeling, faller wire in stop position. Tension bars at minimum setting—unit ready to creel. (2) Clamp closed for warper stop to maintain yarn control. Same position used for yarn cutting. Clamp also closed for initial warper acceleration. (3) Clamp open in the production speed run position. Tension bars in control position (courtesy of West Point Foundry and Machine Company).

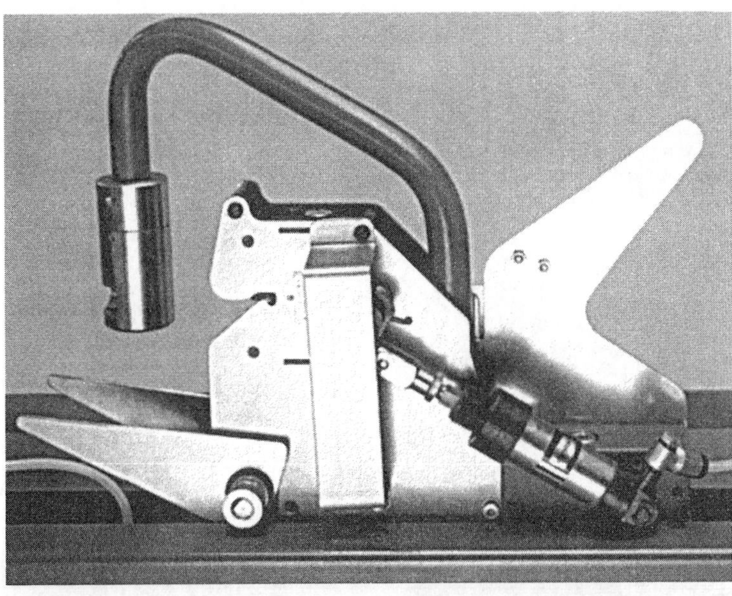

FIGURE 4.51 Automatic yarn cutting and knotting device (courtesy of Sucker Muller Hacoba).

Yarn tension is generated by pressing a mobile rubber coated roller against the axis of the capstan roller. The major advantage of this tensioner is that it does not wear the yarn; it can be used for 10–830 dtex yarns. The maximum working speed is around 800 m/min.

The disc tensioner is good for all types of yarns, 100 dtex and higher, at high speeds, up to 1200 m/min (Figure 4.56). The yarn tension is adjusted by a tension spring with the aid of an angle lever.

With the staple yarns, fibrils tend to accumulate in the disc tensioner. Figure 4.57 shows the schematic of a patented internal intermittent dust blowing system to prevent fly accumulation in the tension unit.

The systems explained above are static systems and do not provide positive tension control. Automated tension control systems provide dynamic tension control in the sense that they can adjust the tension to prevent fluctuation and keep the tension in a tight range. Depending on the

FIGURE 4.52 Warp sampling machine (courtesy of Suzuki's).

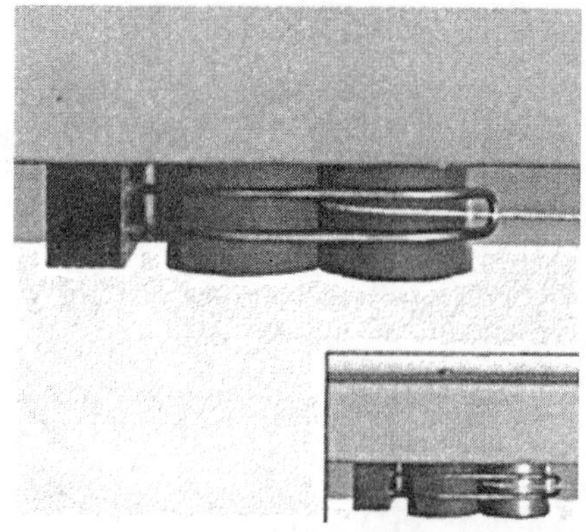

FIGURE 4.53 Roller tensioner (courtesy of Sucker Muller Hacoba).

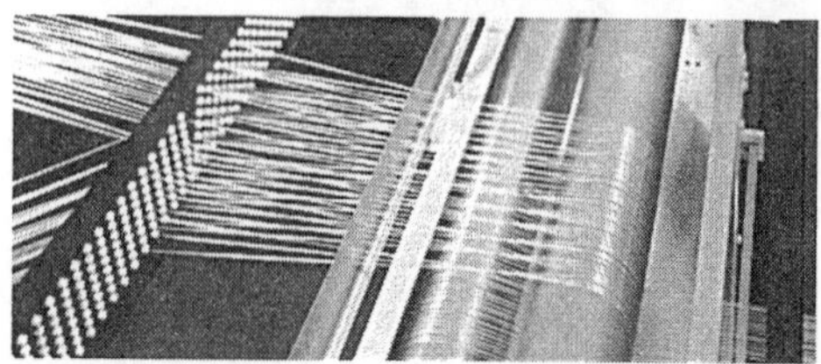

FIGURE 4.54 Tension roller unit (courtesy of Sucker Muller Hacoba).

FIGURE 4.55 Capstan tensioner for warping of fine yarns (courtesy of West Point Foundry and Machine Co.).

FIGURE 4.56 Disc tensioner (courtesy of Sucker Muller Hacoba).

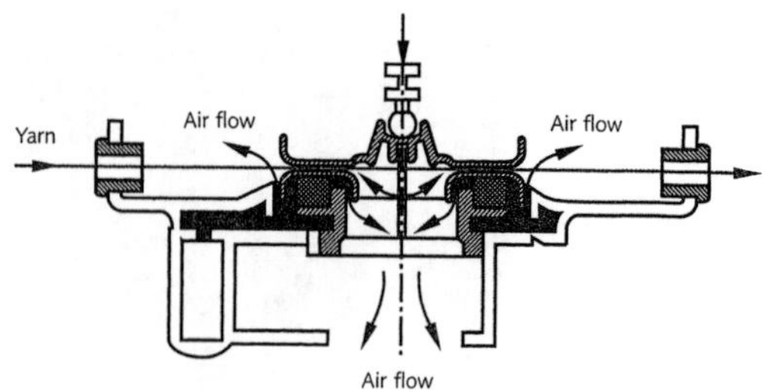

FIGURE 4.57 Schematic of dust blowing system (courtesy of Sucker Muller Hacoba).

low or high incoming tension, the tension control device varies its response. In the ideal case, the outgoing tension should be exact and constant. There are several varieties of this type of tensioner. Figure 4.58 shows a relatively simple system to control the tension. In this device, a plastic cylinder is used with one or more plastic or steel balls inside. Yarn enters the tensioner at the bottom and runs through the surface of the ball(s). As the incoming tension is increased, the yarn has to lift the balls higher against the gravity. As a result, the yarn will use up its extra energy to move the balls inside the cylinder. This will result in compensation of the high tension.

FIGURE 4.58 Tension control device based on gravity (courtesy of Zollinger).

FIGURE 4.59 Compensating yarn tension regulator (courtesy of Karl Mayer).

Figure 4.59 shows another compensating yarn tension regulator, which compensates yarn tension differences between full and empty yarn packages, in a speed range of 200–1200 m/min. Tension peaks are absorbed by means of permanent magnet damping action of the compensating lever.

Figure 4.60 shows a patented closed-loop control mechanism to maintain yarn sheet tension at various speeds and at the stop mode. The output of strain gauges, which are attached to the lower nip roll, is monitored by the PLC continuously. Then, using special formulas within the PLC, the delivery-roll drive system is controlled to provide constant yarn sheet tension.

Whichever tension control system is chosen, it must satisfy several conditions. It should not introduce or magnify tension fluctuations. The yarn twist should not be altered by the tensioner device. Reliability, ease of threading, adjustment and cleaning are other important considerations. Figure 4.61 shows various configurations of tensioning devices.

Figure 4.62 shows examples of warp beams. Some beams are made of sections. The width of the sections can be changed or the beam can be one piece cast beam. Alternately, canisters can be placed on a shaft to produce sections. The current trend in weaving is towards larger warp beam diameters. Figure 4.63 shows a weaving beam of 1600 mm diameter.

4.3 SLASHING (SIZING)

4.3.1 The Need for Slashing

Although the quality and characteristics of the warp yarns coming out of the winding and warping processes are quite good, they are still not good enough for the weaving process for most of the yarns. The weaving process requires

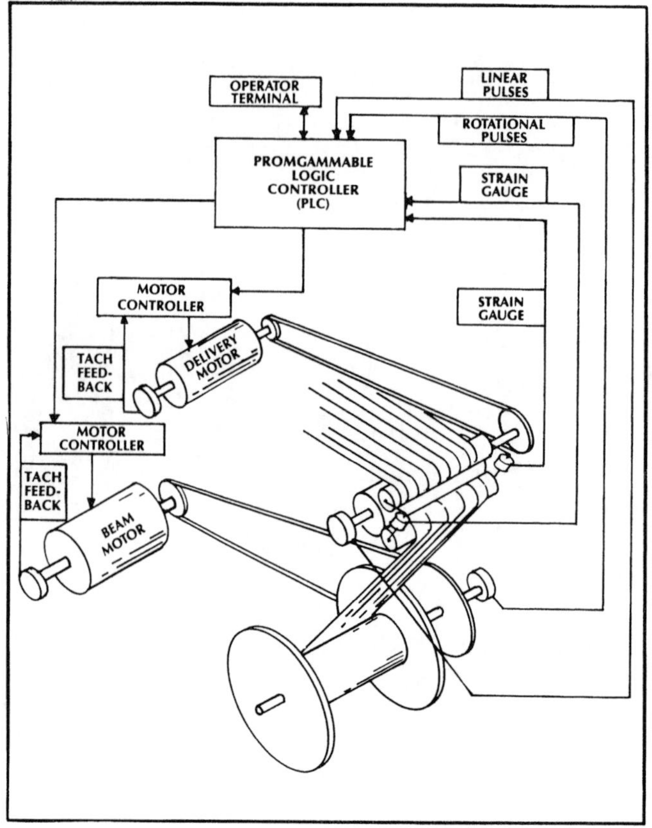

FIGURE 4.60 A patented closed loop mechanism to maintain constant yarn sheet tension in warping (courtesy of West Point Foundry and Machine Company).

the warp yarn to be strong, smooth and elastic or extensible to a certain degree. To achieve these properties on the warp yarns, a protective coating of a polymeric film forming agent (size) is applied to the warp yarns prior to weaving; this process is called slashing or sizing. Sizing is not a value added process in woven fabric manufacturing. This is because, after the fabric is woven, the size materials will be removed from the fabric during the finishing operation, which is called desizing.

The main purposes of slashing are as follows:

- to increase the strength of the yarns
- to reduce the yarn hairiness that would cause problems in weaving process

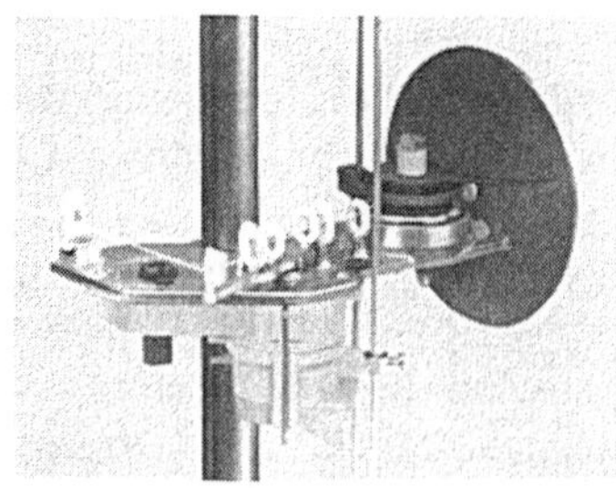

FIGURE 4.61 Various yarn tensioning devices (courtesy of Karl Mayer).

FIGURE 4.62 Examples of warp beams (courtesy of American Scholze).

- to increase the abrasion resistance of the yarns against other yarns and various weaving machine elements
- to reduce fluff and fly during the weaving process for high speed weaving machines.

The ultimate goal of sizing is to eliminate or reduce warp breaks during weaving. Warp breaks are caused either by high tension or by low strength in the yarn. High tensions in the warp are caused by large shed openings, lack of proper tension compensation, high beat-up force and inadequate let-off. Knots, yarn entanglement and high friction also cause tension buildup.

There are three types of tension on a warp yarn during weaving: constant mean tension, cyclic tension variations and random tension variations. The warp is under constant tension on the loom and the magnitude of this tension is generally determined by the take-up/let-off rate,

FIGURE 4.63 Weaving beam of 1600 mm diameter (courtesy of Mallein).

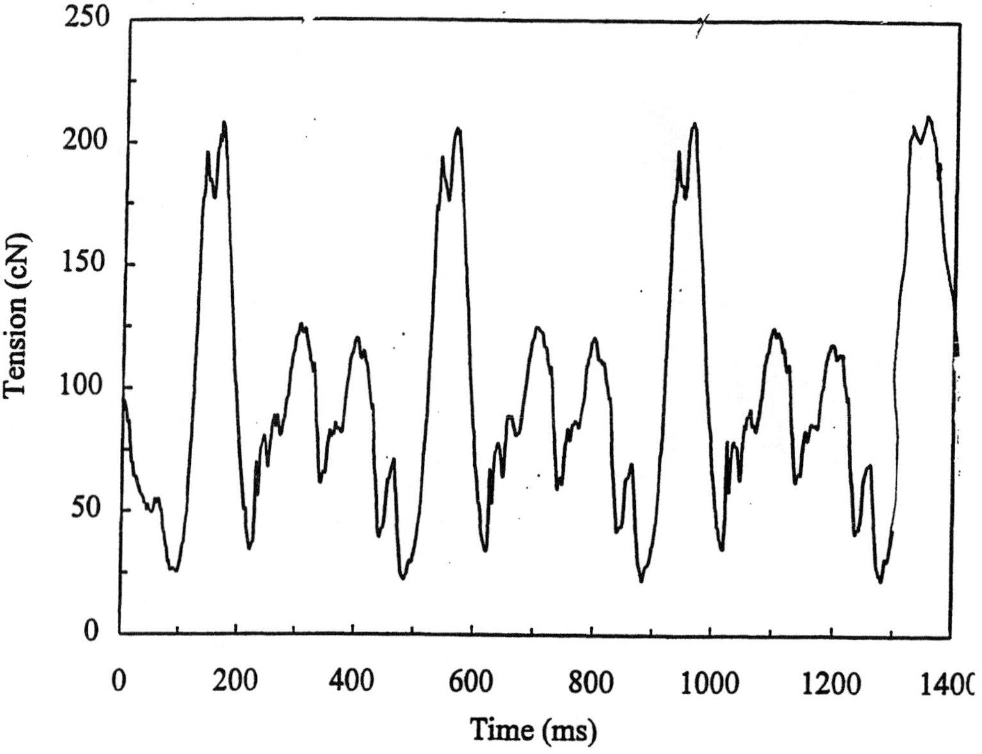

FIGURE 4.64 Warp tension cycles for a 3/1 denim fabric (Adanur and Qi).

crimp developed during weaving and elasticity rate of the warp yarn. Mean tension is usually not the cause for warp breakage.

Cyclic tension variations are caused by shedding and beat-up. Tension patterns due to shedding and beat-up depend on the fabric design and structure. Warp tensions due to shedding and beat-up are so high that they have to be compensated by tension compensation devices; however, it is hardly possible to eliminate them entirely. Figure 4.64 shows a typical warp tension cycle for a 3/1 denim fabric. The highest peak tension is critical which may cause the weak yarns to break.

Random tension variations are caused by different reasons such as an improper knot, entanglement of warp yarns due to protruding fibers, etc. A thick knot may not pass through the heddle eye or reed easily which causes the tension to build up. The warp yarns follow a tortuous path that includes tension and abrasion zones on the loom during fabric formation as shown in Figure 4.65.

It should be noted that only warp yarns need to be sized. This is because, as mentioned earlier, warp yarns are subject to harsher treatments than filling yarns during the weaving process on the weaving machine. Therefore, the filling yarns will be free of size and no special finishing considerations are necessary for these yarns in the fabric. Often, around 80% of yarn failures in weaving are caused by 20% or less of the yarns in a warp which are called repeater ends. The slashing process deals with enhancing individual warp yarn properties not with improving the characteristics of the warp sheet. If done improperly, slashing can worsen yarn sheet characteristics.

Several spun yarn properties are positively affected by slashing. Figure 4.66 shows the effect of sizing on a typical staple yarn sheet. Good sizing should reduce hairiness, improve strength and abrasion resistance while keeping the yarns separated. Elongation is reduced in a controlled manner. Flexibility is reduced but reasonably maintained. If the sizing is not done correctly,

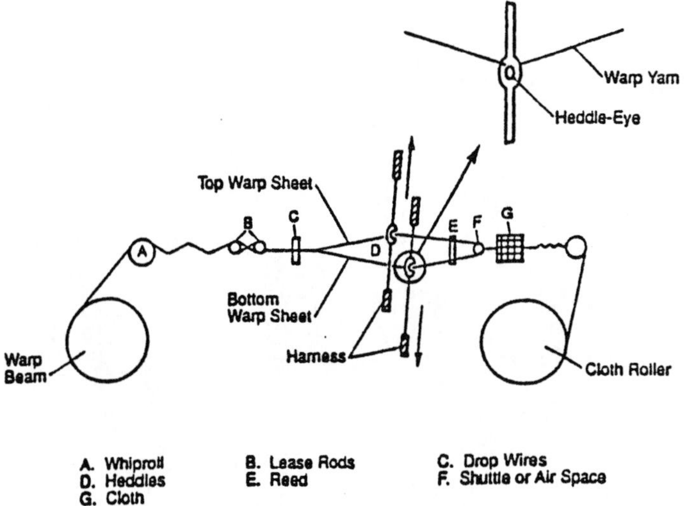

FIGURE 4.65 Tension and abrasion zones in a typical loom [2].

the long hair fibers protruding from one yarn will be glued together with the fibers from the other yarns. This will cause damage of the size film when the yarn sheets are separated back into individual yarns at the separator rods on the slasher which will reduce the strength and cause a yarn break. The fibers should be kept to the body of the yarn such that hairs and fibrils do not interfere with the weaving process.

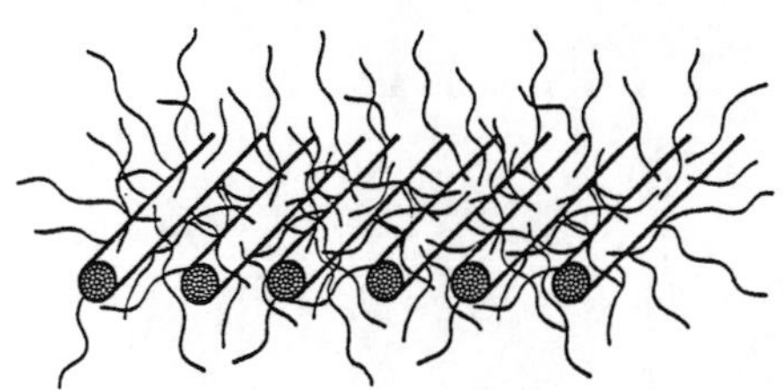

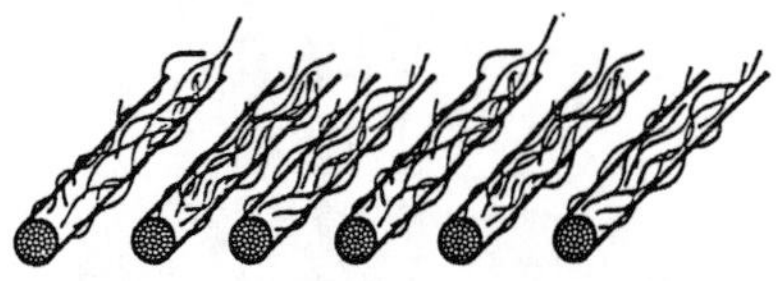

FIGURE 4.66 Control of yarn hairiness with sizing, top: unsized; middle: improperly sized; bottom: properly sized [2].

Factors influencing yarn hairiness include hairiness generated by the winding process, spinning tensions, location of the yarn on the spinning package, yarn balloon shape, yarn twist, spindle speed, yarn size (count), % synthetics in a blend, end spacing at slashing, size add-on, slasher creep speed and bottom squeeze roll cover.

A practical understanding of the importance of size penetration, size encapsulation, yarn hairiness, residual yarn elongation and yarn abrasion resistance is essential to good slashing practice. It is important that the size film must coat the yarn surface without excessive penetration into the body of the yarn, because if the size material is penetrated deep in the yarn, complete desizing would not be possible. Therefore, only enough penetration should occur to achieve bonding of the size film to prevent removal during weaving. Figure 4.67 shows photomicrographs of properly sized yarns. The following terms are used related to sizing:

- *Size concentration:* the mass of oven dry solid matter in size paste
- *Size take-up (size add-on):* the mass of paste taken up in the size box per unit weight of oven dry unsized yarn

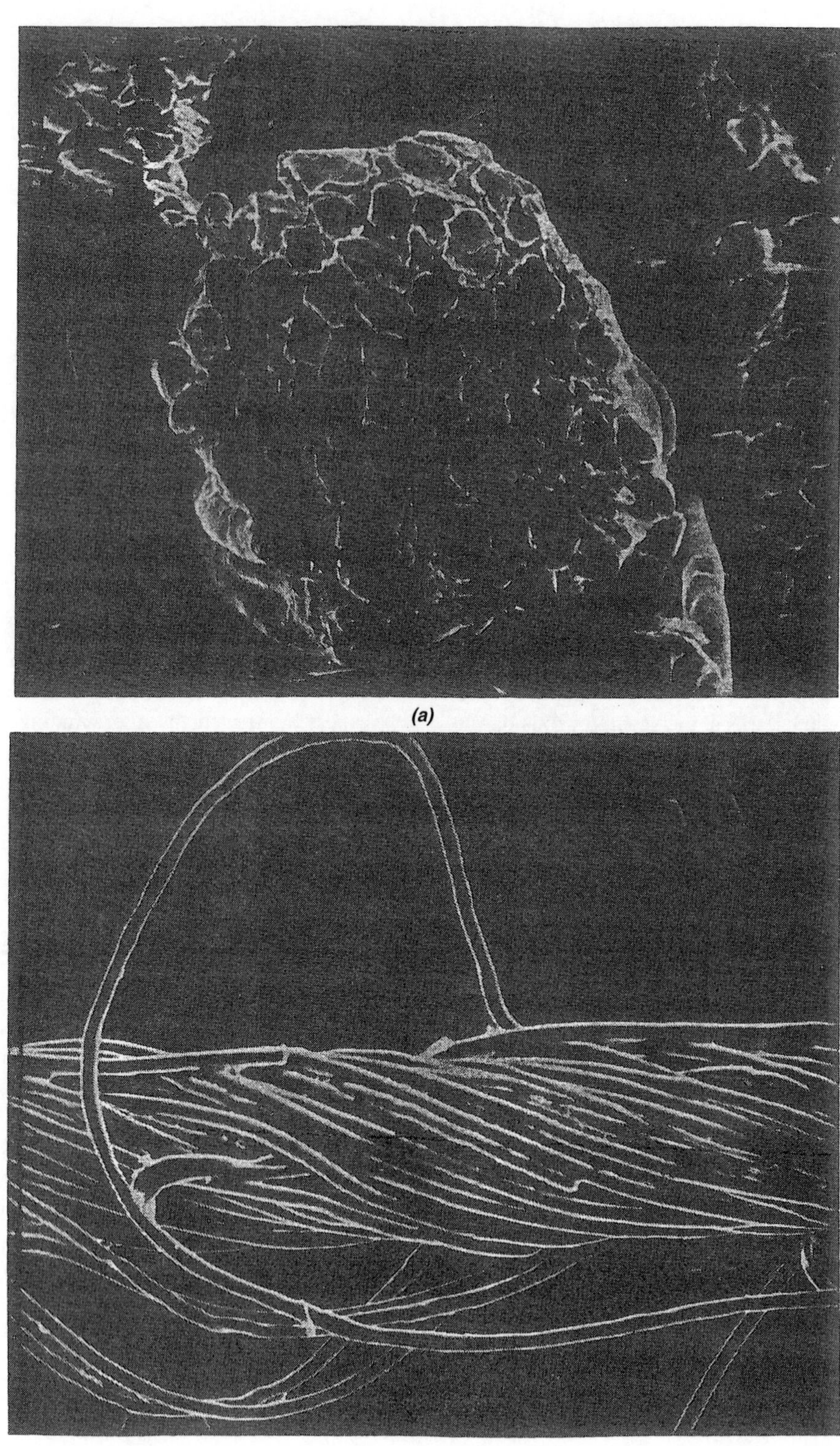
(a)
(b)

FIGURE 4.67 Longitudinal and cross-sectional photomicrographs of properly sized yarns [2].

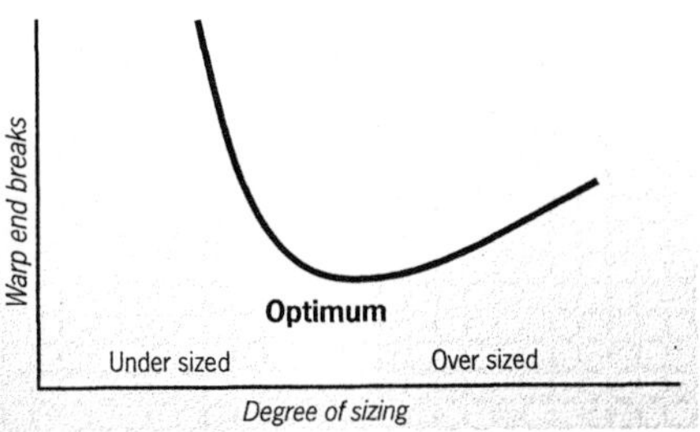

FIGURE 4.68 Typical relation between size add-on and warp breakage rate (courtesy of Benninger).

- *Size percentage:* the mass of oven dry size per unit weight of oven dry unsized yarn

There is an optimum level of size add-on that gives the minimum warp end breakage as shown in Figure 4.68. Excessive size makes the yarn stiffer and less extensible; yarns with too little size will not be strong and smooth enough for weaving. Therefore, too little or too much sizing causes an increase in warp end break. Optimum size add-on gives the best results for weaving.

Although sizing is done mainly to increase the strength of the yarn, some strong yarns such as continuous filaments still need sizing. This is because sizing keeps the slack and broken filaments together in low twist yarns which otherwise would protrude from the body and rub against the machine elements, leading to entanglement, development of fuzz balls and end breaks.

Other points to consider in slashing:

- Slasher creel tension control is critical especially with MJS and open-end yarns. Maximum tension should not exceed 5% of breaking strength (15–20 g for ring spun yarns and 12–15 g for open-end, MJS and MVS yarns). With coarse yarns, sometimes 30 g is allowable.
- The amount of size picked up is affected by the viscosity of the size mix as well as the yarn structure. The viscosity of the mix is controlled by the recipe, amount of solid content in the size liquor and the type of sizing product, mechanical mixing level, temperature and time of boiling. Flat filaments, textured and spun yarns pick up size differently.
- Yarn spacing at the slasher size box and on the drying cylinders is very important.
- The choice of size for staple yarns is usually based on cost. For filament yarns, the size material is chosen based on the compatibility with the fiber.
- Running the slasher at creep speed, which is sometimes necessary, generates a very undesirable condition for proper sizing and should be minimized in every way possible.
- Stretch of warp yarns during slashing should be controlled accurately to maintain residual elongation in the yarn which is needed for good weaving. Back beam—size box stretch should not exceed 0.5%.
- Water soluble sizes can cause problems in water jet weaving.
- Process studies to determine causes for inefficiency should be conducted with strict cause analysis techniques by an experienced practitioner and not as part of a typical stop frequency study for job assignments. Table 4.3 shows some of the causes of waste at the slasher.
- Guide rollers should be kept free from nicks, burrs and sharp edges, especially for MJS yarns. They should be sanded/polished frequently.
- Pre-wetting yarns prior to sizing can reduce the amount of required size add-on for the same performance, especially for cotton yarns.

4.3.2 Slashing Machines

A slasher machine is used to apply the size material to the yarns. The first sizing machine

TABLE 4.3 Some of the points to consider at the slasher.

• uneven length of the section beams
• rough beam heads
• crossed ends
• mixed yarn
• wild yarn
• mechanical reasons such as bad creel bearings, section beams not aligned
• uneven beam tension
• improper control of size level
• improper pressure on squeeze rolls
• improper steam pressure in dry cans
• ends out of lease
• break outs which result in short doffs

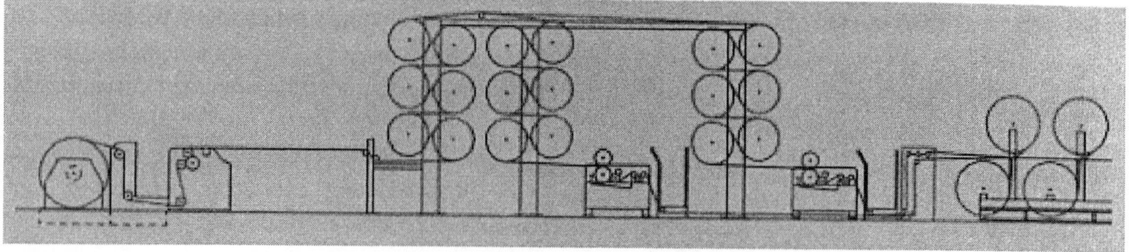

FIGURE 4.69 Schematic of a typical sizing machine (courtesy of West Point Foundry and Machine Company).

was built in 1803 in England. The major parts of the slasher are the creel, size box, drying units, beaming and various control devices (Figure 4.69). Figure 4.70 shows a slasher.

The size box is probably the most important section of the sizing machine. During the slashing process, the sheet of yarns is passed through the size box which contains the hot water solution or mixtures of sizing agents. The yarns pick up the required quantity of size solution in the size box, any excess size is squeezed off as the yarns pass through squeeze rolls. Depending on the size material, warp quality and density, single and double immersion rolls and single squeeze and double squeeze configurations are used. Multiple size boxes can also be employed. In general, single box sizing machines have two squeezing rollers and two box machines have a single roller in each box. It is important that the rollers provide uniform squeezing pressure. The squeezing system determines the degree of size pick up to a large extent. While providing size consistency, the roll pressure should be adjusted to get around 125–130% wet pick up for cotton yarns, 110–115% for poly/cotton and 95–105% for polyester. On average, MJS and open-end yarns pick up around 10–15% more wet size than a comparable ring spun yarn. Therefore, about 10% more water should be added to get the same add-on [3]. The bottom rollers are usually made of steel and the top rollers are rubber coated. For sizing of fine yarns, especially ring spun yarns for air-jet weaving, textured or orange peel roll can reduce the amount of hairiness generated at creep speed. Figure 4.71 shows a double squeeze size box. Loading can be classified as conventional pressure (up to 9 kN or 2000 lbs), medium pressure (up to 27 kN or 6000 lbs) and high pressure (up to 90 kN or 22,000 lbs). Figure 4.72 shows the optimum

FIGURE 4.70 Slasher (courtesy of West Point Foundry and Machine Company).

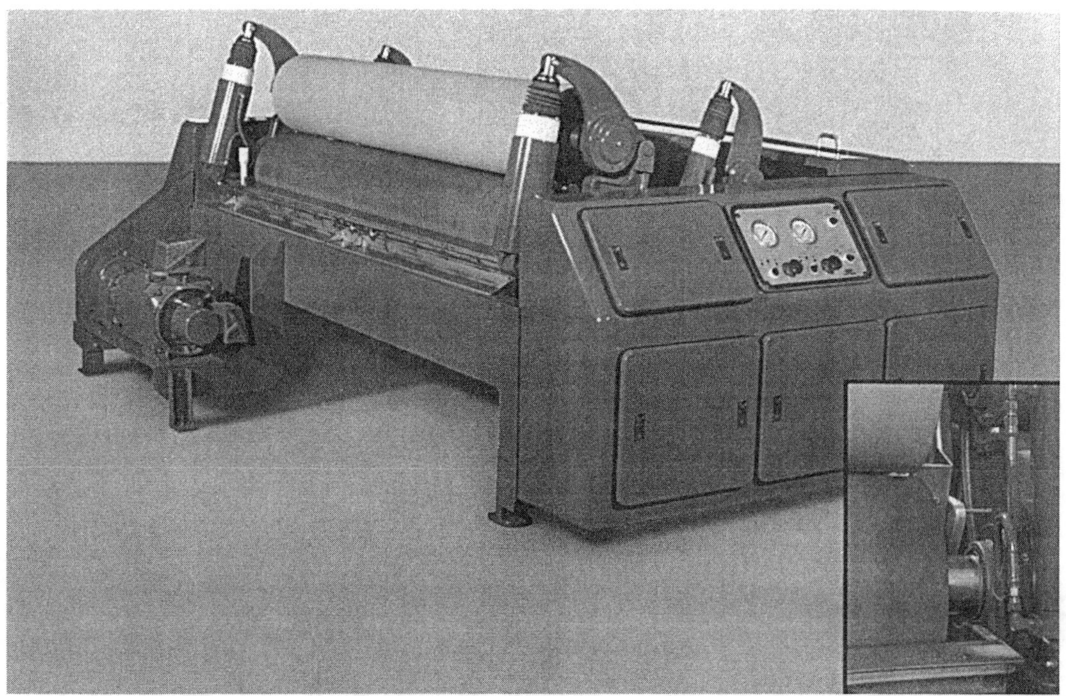

FIGURE 4.71 Double squeeze size box (courtesy of West Point Foundry and Machine Company).

warp exit from squeeze roll nip using adjustable position squeeze roll.

The critical parameters to watch in the sizing process are size homogeneity, constant speed of the sizing machine, constant size concentrations and viscosity. Flooding or dry zones should be prevented in the size box. Temperature of the size box is important for proper size pick up. For 100% polyvinyl alcohol (PVA) sizing, a temperature of 160–170°F is recommended. Constant size temperature can be obtained in two ways:

1. Direct heating in which steam is injected into the size.
2. Indirect heating in which steam flows in pipes around the double walled size box.

A cooker is used to prepare the size and the shearing action in the cooker is important for uniform mixing. Powdered size from silos, big-bags or sacks is metered into weighing stations and then transferred to the cooker.

Today's modern sizing machines dynamically adjust the degree of sizing. Expert software packages calculate sizing values on-line as a function of the warp weight. Size application measuring and control systems are used to measure and calculate sizing parameters automatically instead of time-consuming laboratory test procedures. Based on the calculated parameters, the squeezing pressure at creep and normal speed is controlled via computer. A byrometer measures the density of the mix and controls the supply rate of the ingredients, to keep the warp

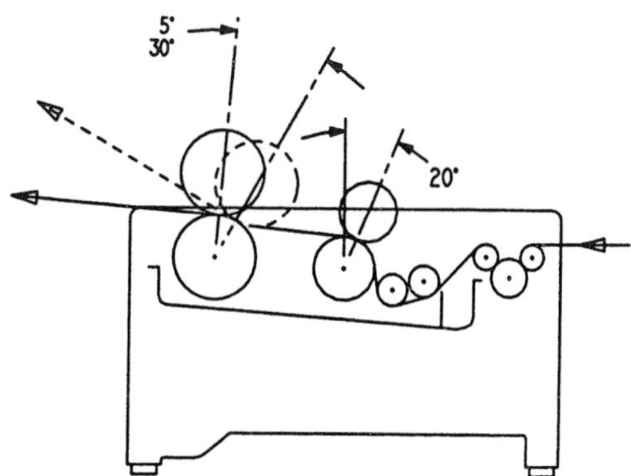

FIGURE 4.72 Optimum warp exit from squeeze roll nip (courtesy of West Point Foundry and Machine Company).

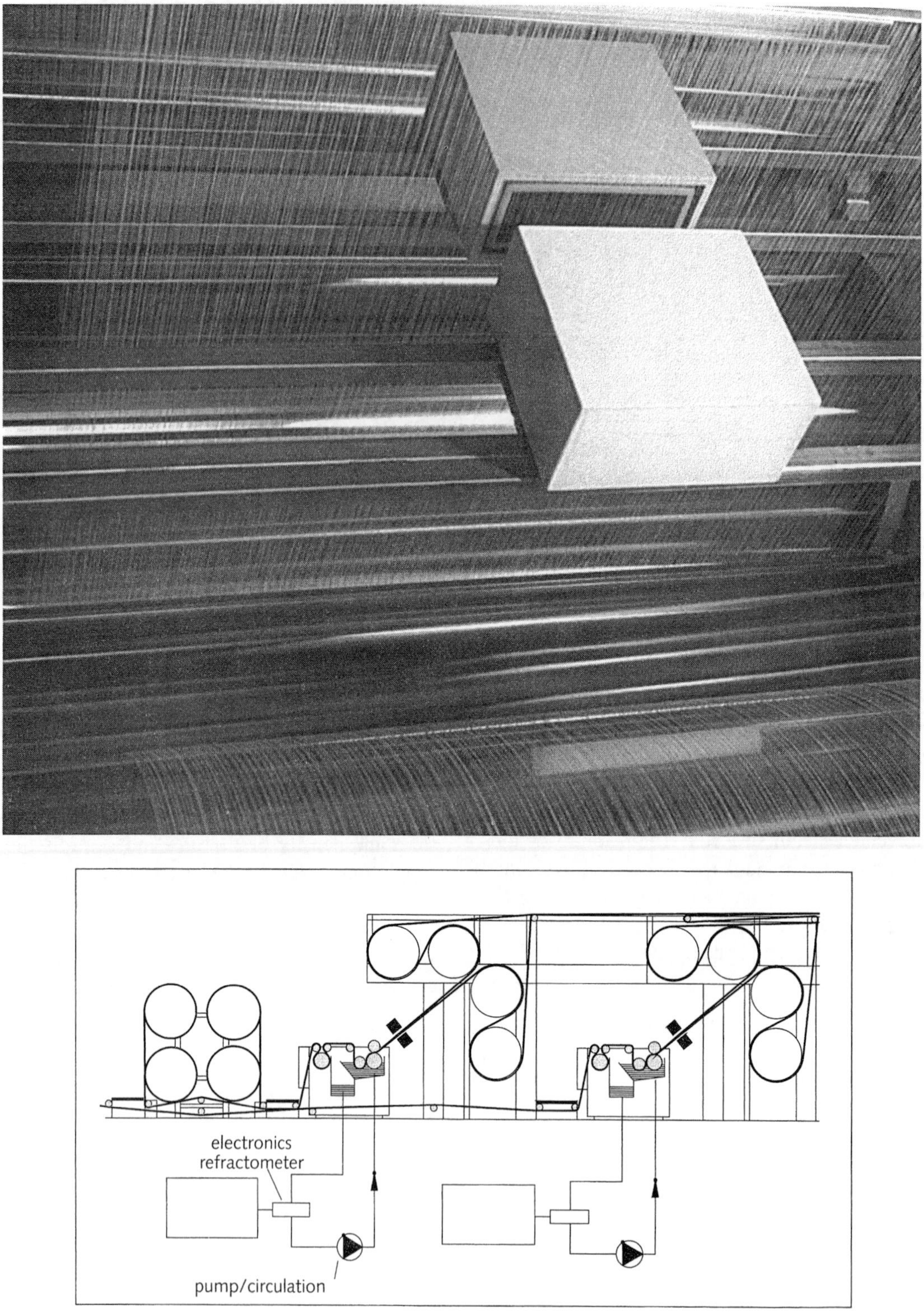

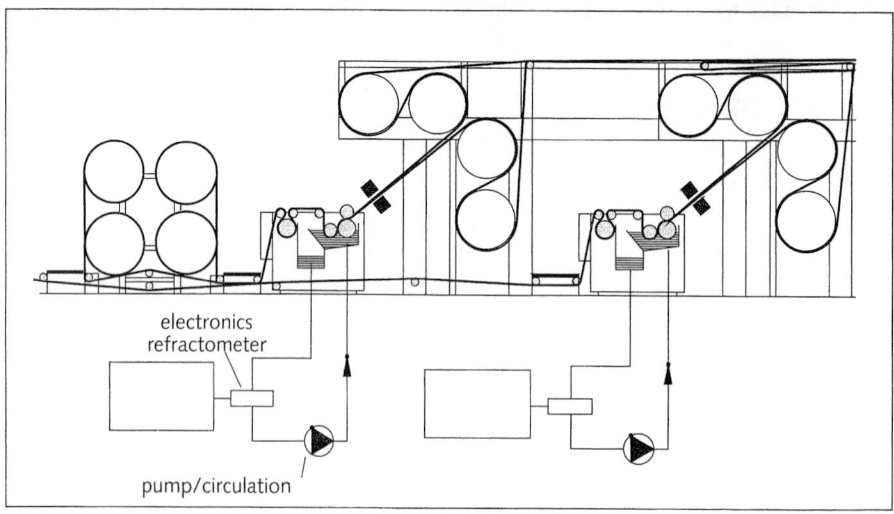

FIGURE 4.73 Size application control unit and its position in the slasher (courtesy of Sucker Muller Hacoba).

sizing degree constant. Figure 4.73 shows a typical size application control unit.

Size mix is also called size liquor. Liquor concentration and the amount of water absorbed by the warp after the squeezing rollers are measured using a microwave measuring head, which enables the degree of sizing to be calculated. The degree of sizing can be calculated as the absolute degree of sizing or as a relative value. If the measured degree of sizing is different than the set data, then the squeeze pressure in the box is changed with a controller. The squeeze pressure is also adjusted according to speed of the system.

Figure 4.74 shows a rotor cooker and size preparation units. In a rotor cooker, the size is prepared without pressure. The ingredients are added cold and mixed with a rotor and raking agitator. Additional breaking elements can be used to obtain high shearing effect. The size ingredients are automatically fed-in.

High speed weaving machines require minimum hairiness in warp yarns. During slashing, yarn hairiness is affected mainly by the spacing between adjacent yarn ends in the size box and the slasher/dryer configuration. In practice, the size box occupation may be used to determine yarn spacing. The size box percentage of occupation is given by [4]:

$$\%\ \text{Occupation} = \frac{a}{b \times c} \times 100 \qquad (4.6)$$

where a = Number of ends being run
b = Yarns/unit length at 100%
c = (section beam distance between flanges

"Yarns/unit length at 100%" is the number of yarns that will occupy the specified unit length when placed side-by-side. The "section beam distance between flanges" is the total working space available within the size box since this is the width of the yarn sheet.

After the size box, the yarns go through the dryer section. The wet yarns are dried by using hot air, infrared radiation or cylinder drying.

FIGURE 4.74 Rotor-cooker and size preparation units (courtesy of West Point Foundry and Machine Company).

Cylinder drying is done using steam heated hot rolls which are called drying cylinders. Sometimes, a combination of drying methods is used on the same machine. Quite often, the wet yarns (usually filament) are predried using hot air or infrared and drying is completed with drying cylinders. There is a wide variety of dry can designs that are used for efficient drying of the yarns. Figure 4.75 shows a sizing line in which the air dryer is used before the cylinder dryer. This helps to keep the yarn cross section round. The drying temperature is critical; excess temperature increases the penetration of size into the yarn, which can lead to excessive hairiness and even yarn breakage. The typical temperature range is 80–105°C (176–221°F). If there is polyester in the system, the temperature should not exceed 280°F to prevent crystallization and dye variability. In cylinder machines, the evaporation rate can be calculated as the mass of water evaporated per unit contact area between warp and drying cylinder per unit time. A typical evaporation rate in a modern slasher is around 13 kg/hr/m^2. The maximum recommended machine speed is 120 m/min (400 ft/min).

The Teflon® coating on all the dryer cans should be in good shape to prevent dry can sticking (also called shedding), which may be a problem. Since open-end and MJS yarns have high wet pickups, the slasher may have to be slowed down to eliminate dry can sticking [3].

Splitting the warps after sizing and separate drying reduces the risk of adjacent yarns sticking together which reduces the number of yarn breaks. For air-jet weaving, a minimum of 75% open space on the dry cans is recommended for any type of yarn. Figure 4.76 shows different drying configurations. In the top configuration, the warp is divided into half and separately predried before drying in a single layer in the final dryer. In the middle configuration, each warp yarn is predried after the size box before they are final-dried together. In the bottom configuration, the warp is divided into two sections (wet split) after each size box. After predrying, the four sections are combined and final-dried together.

Due to the nature of sizing, the yarns in the sheet may be stuck together at the exit of the dryer section. Therefore, they are separated into individual ends by using bust rods. First, the individual sheets of yarns from each section beam are separated followed by pins in the expansion comb to separate the yarns within each sheet. Then the yarns are wound onto a loom beam for weaving (weaver's beam).

Beam arrangements in the creel are usually two types (Figure 4.77):

a) Groups of 2, 4, 6 or 8, one to four tiers
b) staggered, two-tier arrangement

Some sizing machines can have up to 24 beam positions. The beams can be controlled in groups or individually. The let-off can be individual let-off, single group let-off or wrap-round let-off.

Slashers are classified based on the method of drying (cylinder, hot air or infrared) or according to the method of yarn supply (single end, direct and indirect). In single-end slashers, yarns are fed to the size box directly from the supply packages (sometimes, "single-end" sizing is also used to indicate the sizing of a single warp end). This type of creel is generally used for a small number of yarns and textured yarns. In direct sizing, yarns are fed to the size box from a single creel beam or warper's beam. In indirect sizing, several warper's beams ("section beams") are combined sheet to sheet, forming a final beam for weaving (weaver's beam). It is also possible to draw the yarns from a combina-

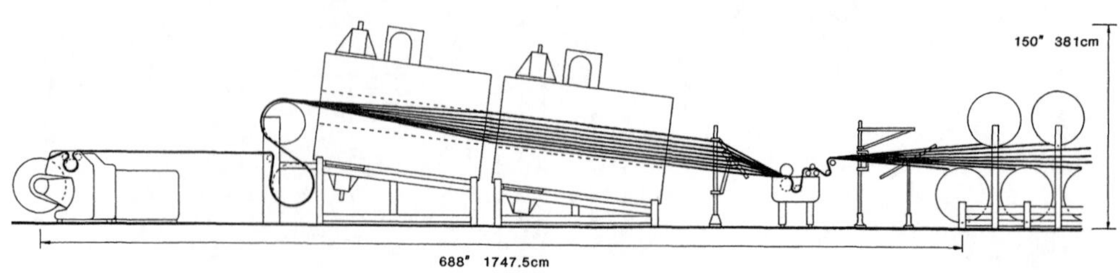

FIGURE 4.75 Sizing line with air and cylinder drying (courtesy of West Point Foundry and Machine Company).

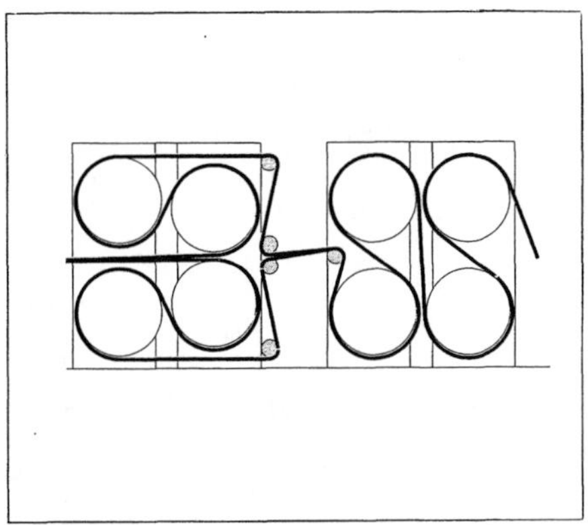

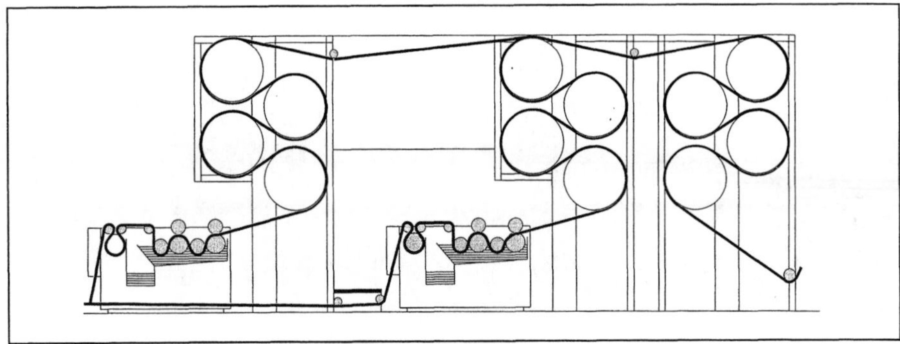

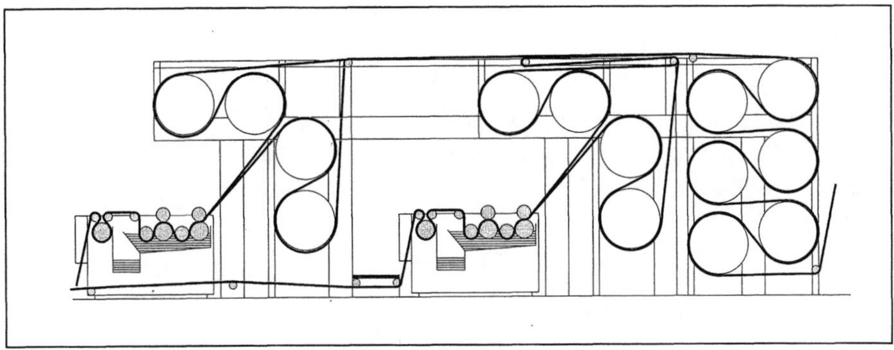

FIGURE 4.76 Examples of drying configurations (courtesy of Sucker Muller Hacoba).

tion of package creels and warp beams simultaneously, as shown in Figure 4.78.

Selection of a sizing machine depends on several factors, including warp specifications, weaving requirements and production volume. The output of the sizing machine is determined by the size of the dryer. Figure 4.79 shows typical sizing machine configurations.

Figure 4.80 shows a sizing winder machine. Each spindle is driven by an individual motor. Typical yarn speed is 150 m/min to 350 m/min. Size permeation into yarn is provided by a two-step roller system and sponge roller; squeeze type roller can also be used. Sized yarn is dried while yarn proceeds from left to right on a double reel that is placed vertically, with a special finish.

In the so called "walk-through head-end" beam winder, the beam support/drive unit is independent of the delivery/comb unit. There is a "walk-through" platform in between that allows better access to the comb, delivery roll and beam. Automatic hydraulic beam loading

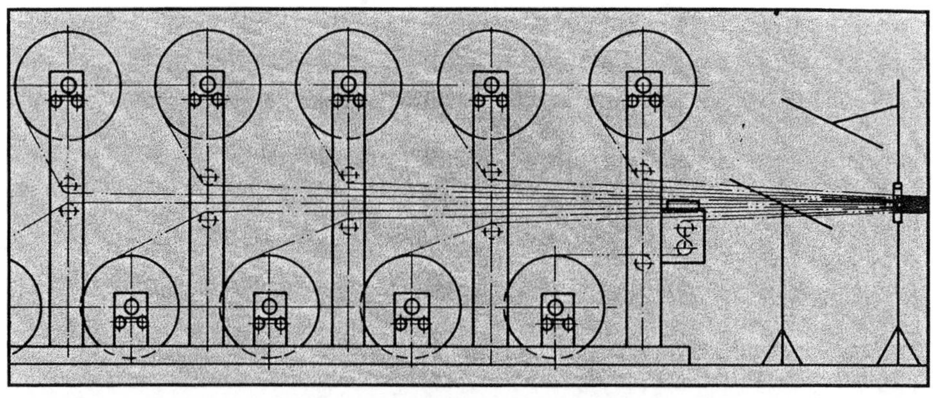

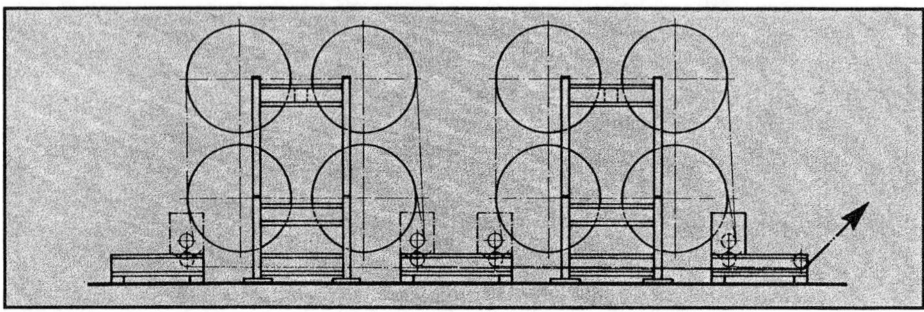

FIGURE 4.77 Beam creel arrangements. Top: staggered beam creels for filament yarns, Bottom: beam creel for staple yarns with modular groups of 4 (courtesy of Sucker Muller Hacoba).

and unloading, independent hydraulically lifted delivery nip rolls, pneumatically operated expansion, contracting and shifting of the comb are some of the other features of this new system. This concept was developed for large warp beams.

The quality of woven fabrics depends to a great extent on the quality of warp preparation. Therefore, sizing machines are usually incorporated in weaving room control and monitoring systems as shown in Figure 4.81. For trouble-free weaving, a well-slashed warp is a must. Poor slashing may increase loom stops which in return increase the cost of weaving. Appendix 2 lists some factors to consider during the sizing operation for trobleshooting.

4.3.3 Film Formers

Table 4.4 shows the film forming polymers that are commonly used in sizing; these polymers can be natural materials or synthetic. Natural products include starch and derivatives of cellulose. Synthetic polymers are derived from petroleum such as polyvinyl alcohol (PVA),

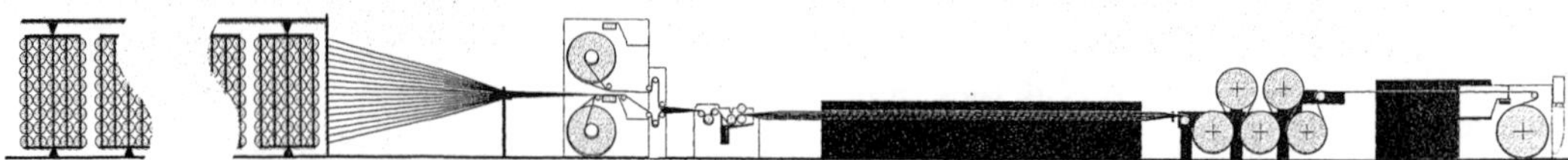

FIGURE 4.78 Sizing yarns from package creels and warp beams (courtesy of Benninger).

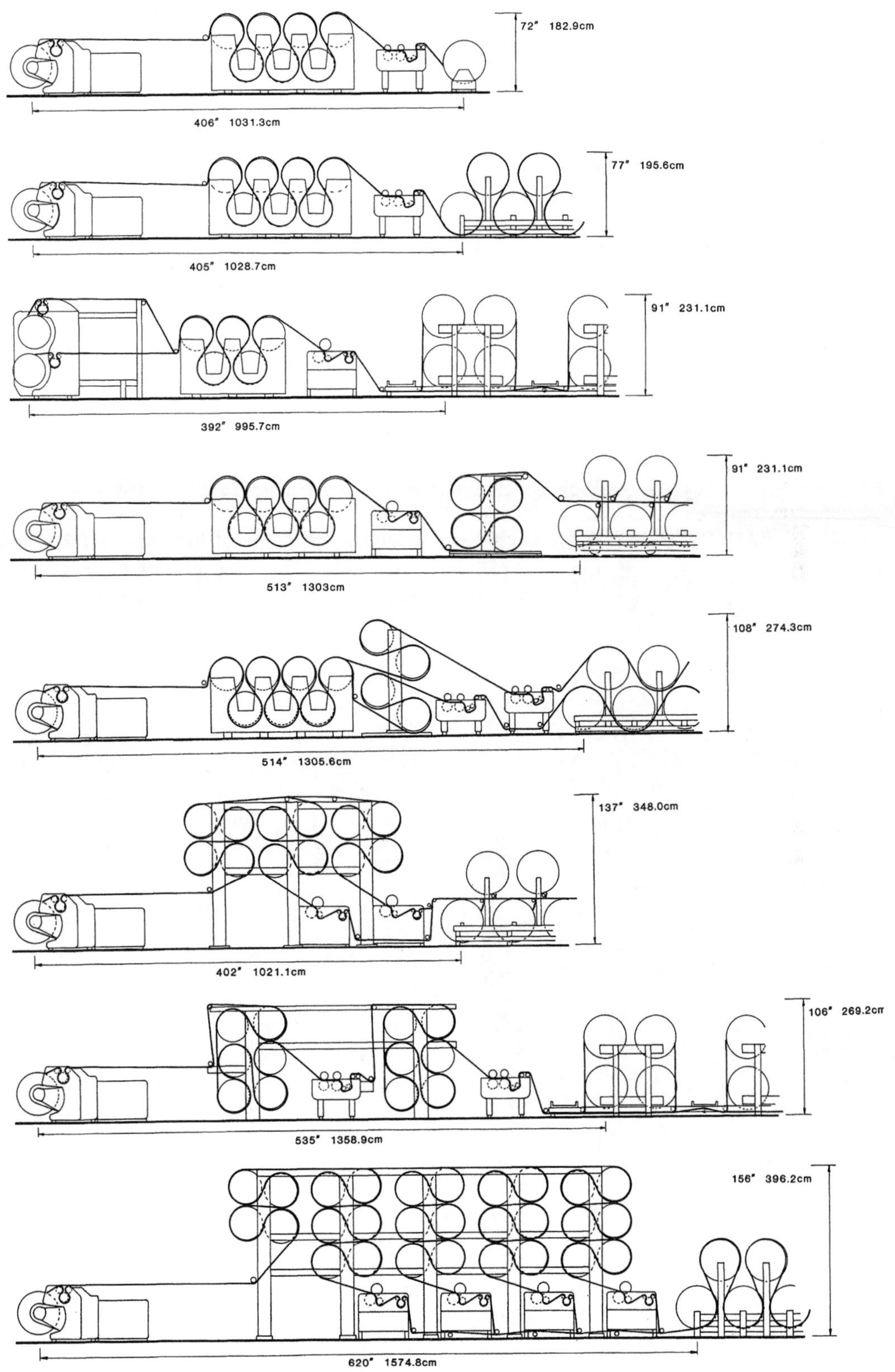

FIGURE 4.79 Examples of sizing machine configurations (courtesy of West Point Foundry and Machine Company).

FIGURE 4.80 Sizing winder (courtesy of Yamada Ironworks Company Limited).

other vinyl polymers such as acrylates and acrylamides or addition polymers such as polyester resins. One of the major requirements in choosing a sizing material is good adhesion to fiber. Fibers and size materials that are similar in their constitution will have good bonding. For example, starch and cotton have good adhesion because they both have cellulosic structures.

Starch

Starch is the oldest film forming material used in sizing of cellulosic fibers. It is also the most widely used in the world due to its low cost and ease of availability. In the United States, mostly cornstarch derived from yellow dent corn is used. In Europe, potato starch is more popular (Figure 4.82). Potato starch that is used in the U.S. is mostly imported. Other starches are also used in different parts of the world. For example, yucca starch is still used in Central America and to a certain extent in South America, West Africa and Asia; long root starches, e.g., tapioca and manioc, are still used in Brazil and Thailand; and rice starch is used in parts of the Far East. Some sizing conditions may require using certain starches. It is expected that the use of corn starch will increase at the expense of other starches.

Each starch type has unique characteristics. Figure 4.83 shows the viscosity curves of different starches; each starch type usually has different variations due to varying molecular weight of the starch for better viscosity control on the slasher. Chemical modifications also cause differences between the variations, they are done to improve flexibility, film forming, adhesion to polyester and water solubility. The most widely used grades of cornstarch are [5]:

- pearl (unmodified)
- acid modified (lower viscosity)
- oxidized (lower viscosity)
- hydroxy ethylated (chemical change)
- carboxy methylated (chemical change)
- penflex (chemical reaction)
- gums (cold water soluble)

Table 4.5 shows the properties of cornstarch types.

Pearl or unmodified starch is the basic and lowest cost starch since it is the pure starch straight from the corn kernel. It has high viscosity in solution and tends to become a gel upon cooling. The film flexibility, strength and elongation are low; it cracks and splinters under

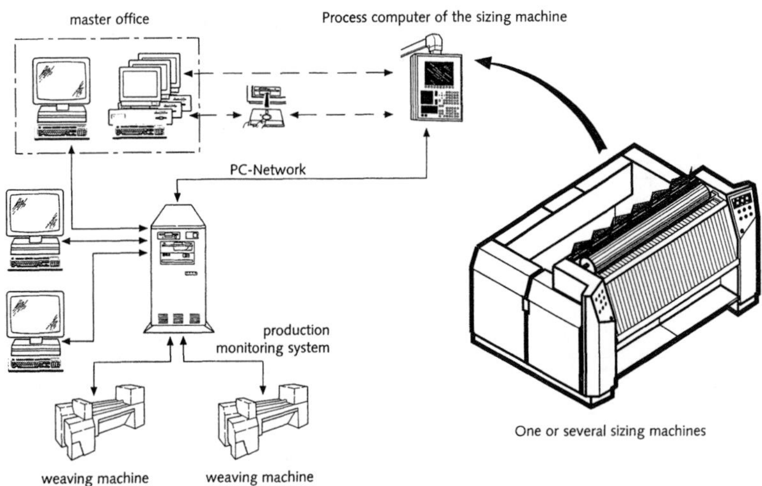

FIGURE 4. 81 Sizing machines as part of the control and monitoring systems (Sucker Muller Hacoba).

bending and there is no adhesion to polyester. To obtain acid modified and oxidized starches, the pearl starch is broken down by an acid or oxidizing technique, respectively, which reduces the viscosity of the starch. In practice, the term "fluidity" is used to indicate the viscosity of a starch, which is opposite to the "viscosity". Unmodified pearl starch has zero fluidity. As the fluidity number increases, the starch gets thinner; a 90 fluidity starch is as thin as water. Although acid modified and oxidized starches have better viscosity control than pearl starch, their film properties are only slightly better than

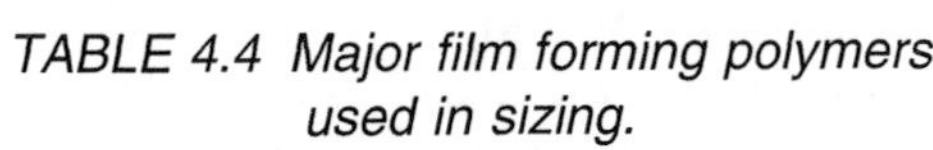
TABLE 4.4 Major film forming polymers used in sizing.

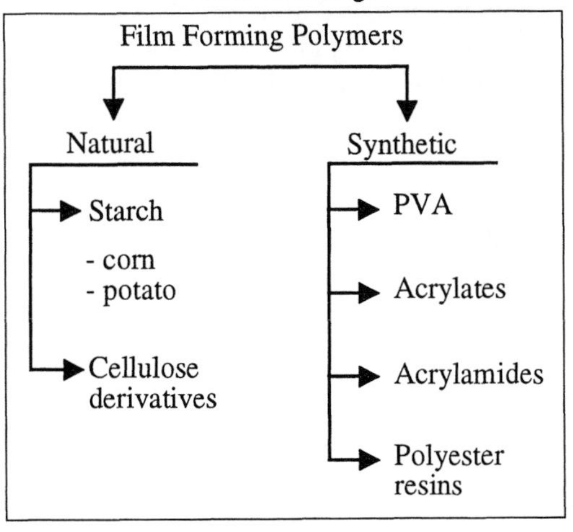

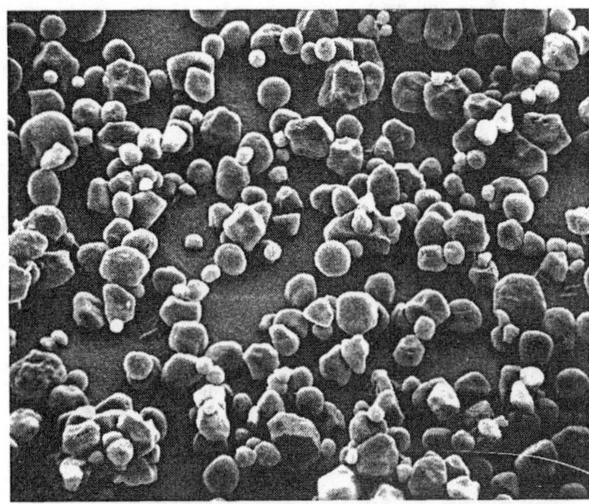

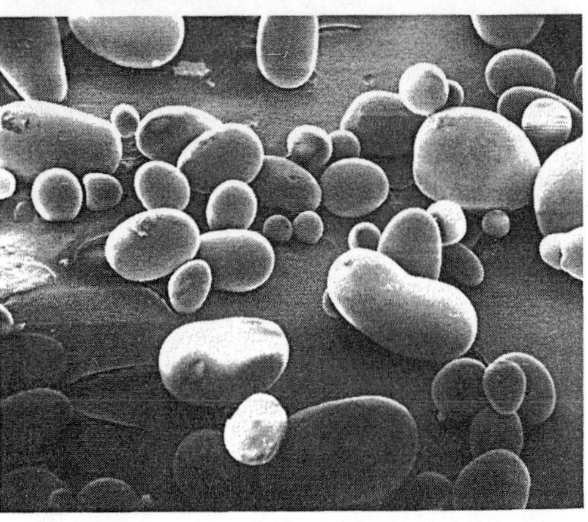

FIGURE 4.82 Cornstarch granules (top) and potato starch granules [2].

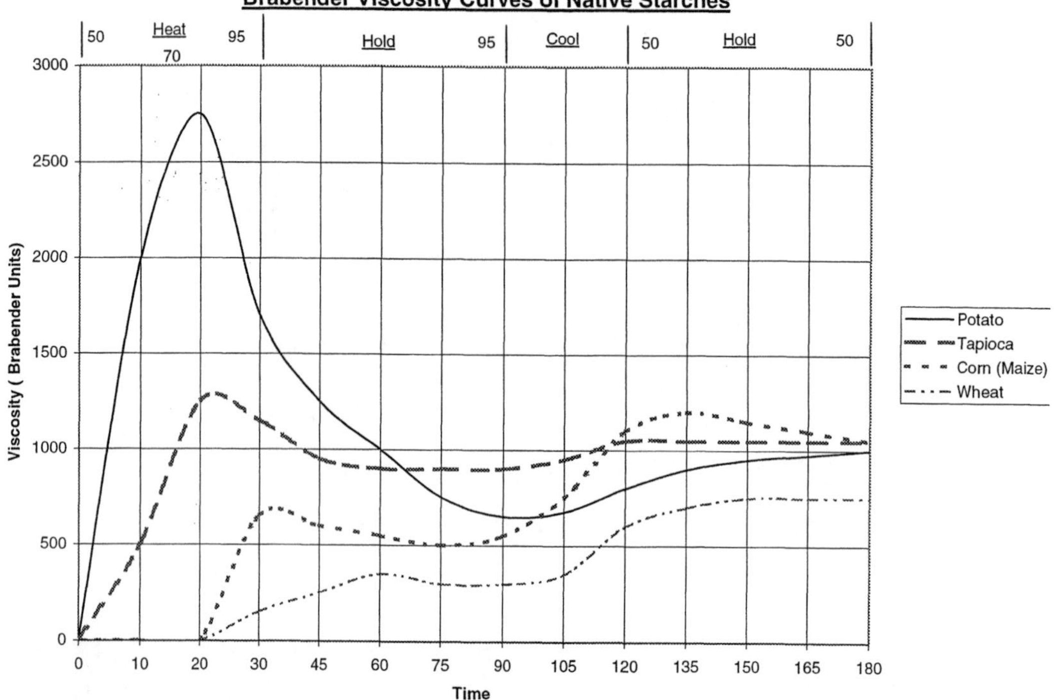

FIGURE 4.83 Starch viscosity curves [5].

pearl starch. Pearl, acid modified and oxidized starches require an enzyme to desize. To obtain the necessary adhesion and flexibility, hydroxy ethyl starches have been developed by chemical modification. Specific chemicals impart improved film strength, flexibility and solubility, and better adhesion to polyester. Hydroxy ethylated starches can be desized with hot water; no enzymes are required. They can be blended with PVA to reduce cost. Patented penflex starch is obtained by reacting a synthetic polymeric material onto the starch backbone. The resulting starch reacts as if it has already been blended with PVA. The penflex starch film is transparent, has good adhesion to polyester, film strength and flexibility. It is also easy to remove in hot water without any requirement for enzymes.

There are two types of chain structures in starch granules: linear and branched. For most starches, approximately 25% of the chains are linear assemblies of glucose molecules (called amylose) and the remaining chains are highly branched chains (called amylopectin) (Figure 4.84 and Table 4.6). Since the starch granule is not water soluble, the granules must be cooked at boiling temperatures to release the compressed chains within the granule into the size solution.

The monomer of starch, which is the building block of the chain structure, is identical to that of cellulose. Therefore, the adhesion between

TABLE 4.5 Properties of cornstarches [5].

Starch type	Viscosity Stability	Retrogradation	Film		
			Surface	Strength	Flexibility
Pearl	Very poor	Very high	Very poor	Very poor	None
Acid modified	Poor	High	Very poor	Poor–medium	None–medium
Oxidized	Medium	Medium	Good	Medium	Medium
Hydroxyl Ethyl	Excellent	Very low	Excellent	Good	Good
Penflex	Excellent	Very low	Excellent	Excellent	Excellent

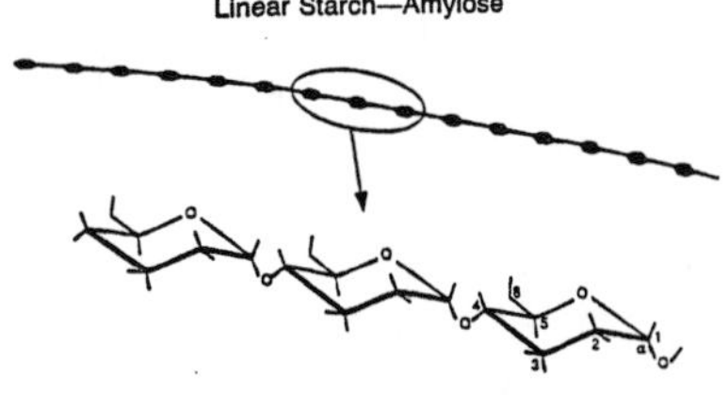

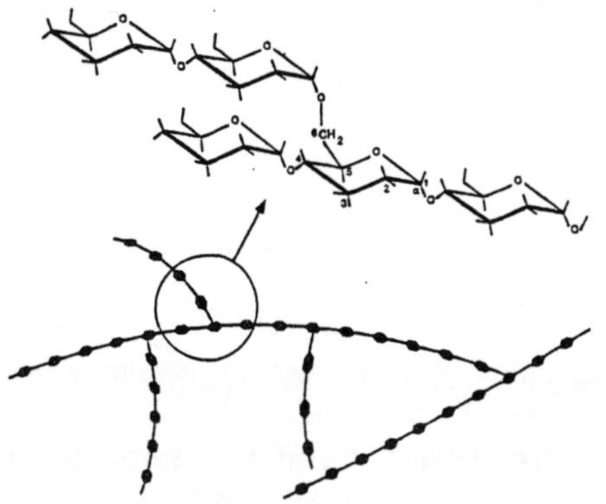

FIGURE 4.84 The starch granule contains about 25% linear chains (amylose) and about 75% branched chain (amylopectin) [2].

TABLE 4.6 Percentages of amylose and amylopection in starches [5].

Starch	Amylose (%)	Amylopectin (%)
Potato	21	79
Maize (corn)	28	72
Wheat	26	74
Mandioca	17	83
Waxy maize	0	100

starch films and cotton, viscose, linen or other cellulosic fibers or their blends is good. The polymer chain structure of cellulose and starch is compared in Figure 4.85. The way the glucose units of the polymers are attached to each other determines the properties of these materials.

Retrogradation of starches is determined by measuring the viscosity increase when starch paste is cooled from 95°C to 50°C. Amylose is the main cause of the retrogradation process; amylopectin is less prone to retrogradation. Cereal starches (e.g., corn, wheat and rice), retrograde quicker than tuber (potato) and root (menioc) starches. Retrogradation of waxy starches is very slow.

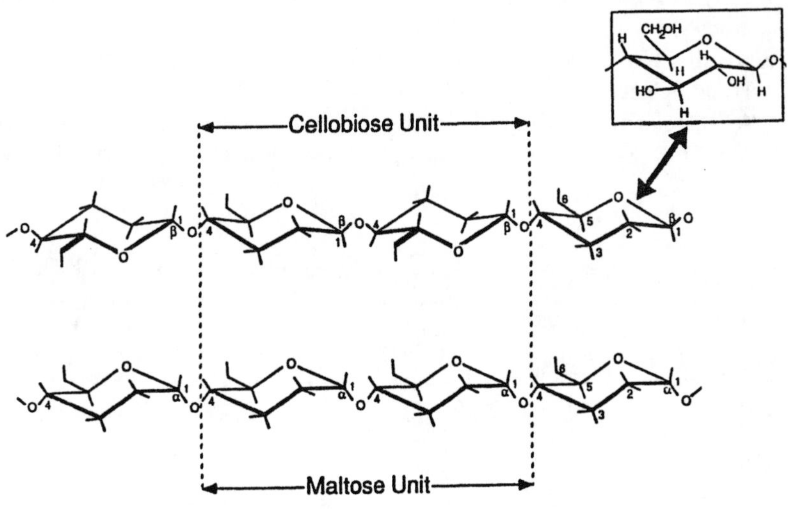

Note: See insert for locations of hydrogens (H) and Hydroxyls (OH)

FIGURE 4.85 Comparison of the glucosidic chain structures of cellulose (β) and starch (α) [2].

Polyvinyl Alcohol (PVA)

PVA is the second largest film former used in sizing. It is mostly used on synthetic yarns such as polyester and poly/cotton blends. It is also suitable for conventional wool sizing where low desizing temperatures are required.

PVA coating is strong, abrasion resistant and can easily be desized in hot water. Its strength is greater than starch and also more flexible than most standard starches. PVA is less prone to set-up in the size box compared to pearl or acid modified starches. PVA can form foam in the size box which is controlled with a defoamer. When using PVA, the slasher should be left running at creep speed even during doffing to prevent "skin" in the size box. If the dry cans are not anti-stick coated, the PVA may stick to them. PVA may be too strong for some sizing applications. In this case, some weaker film forming polymers such as starch are added to modify the mixture, which also reduces the cost, since PVA is more expensive than starch. Figure 4.86 shows the compatibility of starch with PVA.

Since PVA is synthesized from petroleum, specific properties can be engineered into the molecule such as hydrolysis and viscosity. PVA is synthesized from polyvinyl acetate and can be manufactured in different grades. The amount of hydroxyl (OH) groups determines the hydrolysis of the PVA. There are four hydrolysis levels: super (+99%), full (98–99%), intermediate (94–98%) and partial (86–90%). If the acetate groups are removed from the vinyl chains after the polymerization, a PVA grade called "Fully Hydrolyzed" (FH) is obtained; "Partially Hydrolyzed" (PH) grades can be obtained if some acetate groups are left on the molecule. Different levels of hydrolysis impart different properties on the final PVA. In general, PVAs with intermediate or partial hydrolysis are used in textile sizing. The PVA can also be co-polymerized with methyl methacrylate among other co-monomer groups, this gives PVA unique sizing properties and recovery stability. Figure 4.87 summarizes the various reactions.

As the molecular weight increases, the viscosity of the PVA increases. There are three viscosity levels of PVA: high (25 cps), medium (15 cps), and low (5 cps). Mostly medium viscosity PVAs are used in textile sizing.

In sizing with PVA, most of the formulations are similar for all spinning methods, i.e., ring, open-end, MJS and MVS. For weaving machine speeds of over 800 ppm, it is recommended that the strongest, most abrasion resistant formulation that the slasher can handle be used. The solid levels need to be adjusted to compensate for the higher wet pickup. For a 125% wet pick up of ring yarn, MVS would pick up 125–130%, open-end 130–140% and MJS 135–145%.

PVA, as a size material, became popular in the 1960s with increasing use of polyester-cotton blends. Use of PVA in sizing is growing worldwide. However, in Western Europe, some coun-

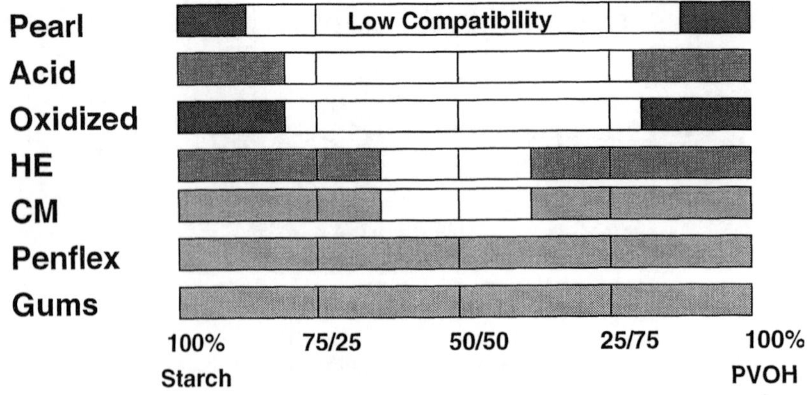

FIGURE 4.86 Starch compatibility with PVA [5].

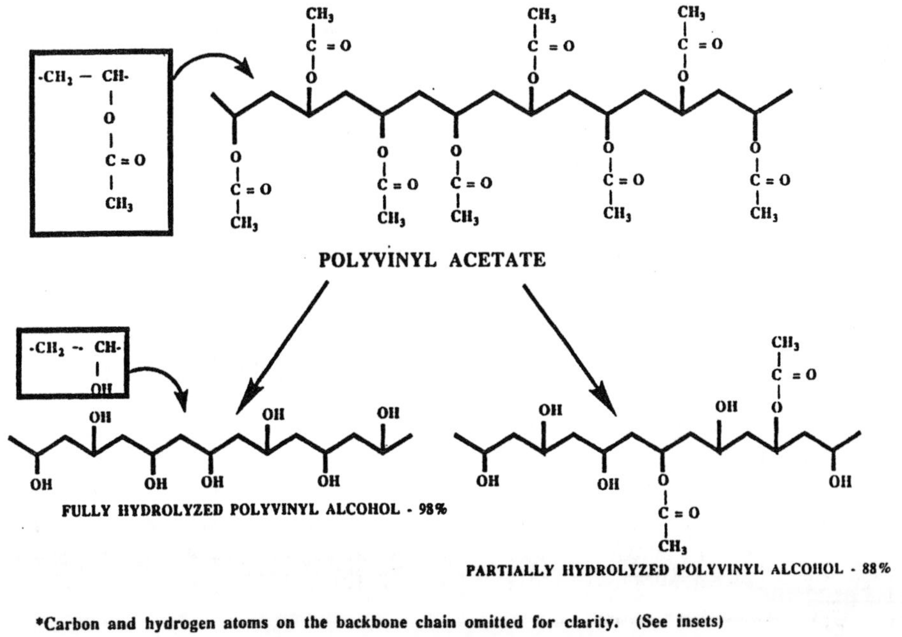

POLYVINYL ACETATE

FULLY HYDROLYZED POLYVINYL ALCOHOL - 98%

PARTIALLY HYDROLYZED POLYVINYL ALCOHOL - 88%

*Carbon and hydrogen atoms on the backbone chain omitted for clarity. (See insets)

FIGURE 4.87 The chemistry of polyvinyl alcohol [2].

tries have restricted the use of PVA as size material for environmental and effluent reasons.

Other Film Formers

The most widely used size materials are starch and PVA. However, other size materials have been developed and used for specific purposes.

Carboxymethyl cellulose (CMC) was used a lot in the past (Figure 4.88). CMC is produced from wood pulp and cotton lint/scrap and it is harder to prepare size solutions with CMC. CMC has good adhesion to cotton. It has moderate film strength and abrasion resistance. This size was easily desized since it had high water solubility and resolubility. It also had low Biological Oxygen Demand (BOD) type pollution. Nevertheless, CMC has been largely replaced by PVA. Table 4.7 shows the ranking of sizing materials based on adhesion to base fibers.

Polyacrylic acid based sizes (polyacrylates and polyacrylamides) are used to size hydropho-

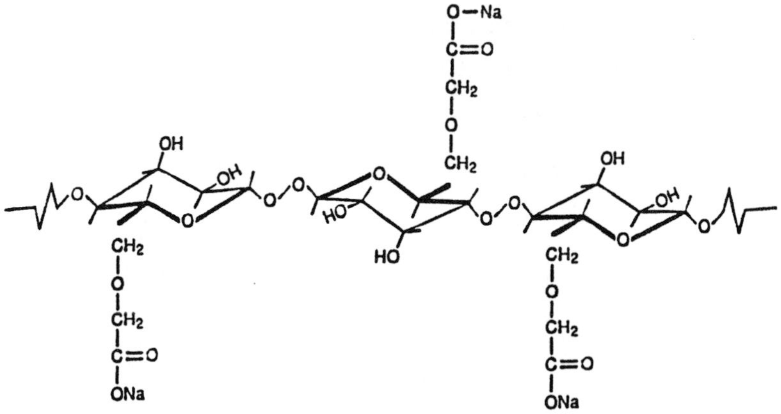

FIGURE 4.88 The structure of carboxymethyl cellulose (CMC) [2].

TABLE 4.7 Adhesion of sizing materials to base fibers [6].

Adhesion to polyester		Adhesion to Cotton	
Material	Force (kP)	Material	Force (kP)
Eastman WD	15	PVA	4.9
PVA	10	CMC	3.9
Polyacrylate	5	Starch	3.6
CMC	4	Polyacrylate	3.5
Starch	3	Eastman WD	0.3

bic fibers and their blends such as nylon, acrylics, acetate, polyester, etc., because of their good bonding. These sizes can be made water insoluble such that they can be used in water-jet weaving. Figure 4.89 shows the structures of various acrylic type sizes.

Figure 4.90 shows the other vinyl polymers and copolymers. Figure 4.91 shows the synthesis and structure for polyester type binders. Table 4.8 compares the properties of various size materials and Table 4.9 compares their BOD contents.

Water dispersable (WD) polyester sizes have been used for some time as filament sizes and as binders for spun yarn sizing. Polyester resins are produced from recycled polyethylene terephthalate (PET). A recent development is a polyester that behaves as a lubricant which increases

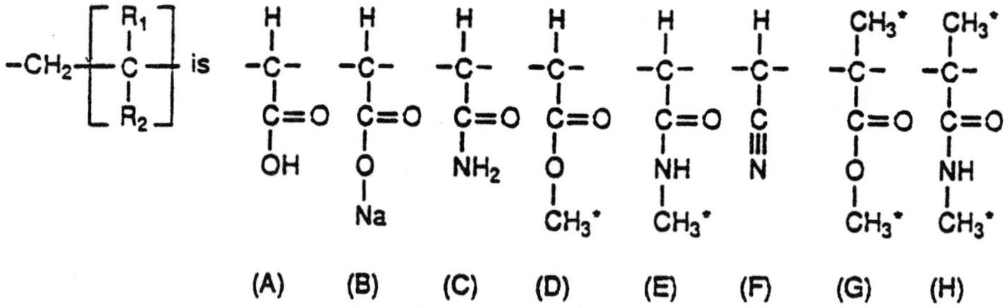

(A) Poly Acrylic Acid (PAA)

(B) PAA Sodium Salt (or Ammonium Salt $-\overset{O}{\overset{\|}{C}}-O-NH_4$)

(C) Acrylamide

(D) Methyl Ester (Methyl Acrylate)

(E) Methyl Amide (Methyl Acrylamide)

(F) Nitrile (Acrylonitrile)

(G) C2 Substituted Ester (N-Methyl Methacrylate)

(H) C2 Substituted Amide (N-Methyl Methacrylamide)

Water Soluble (A), (B), & (C)

Water Insoluble (D), (E), (F), (G), (H)

*Derivatives other than methyl include ethyl, propyl, n-butyl, isobutyl, and 2-ethylhexyl among others

FIGURE 4.89 Some of the acrylic family of copolymers useful in sizing [2].

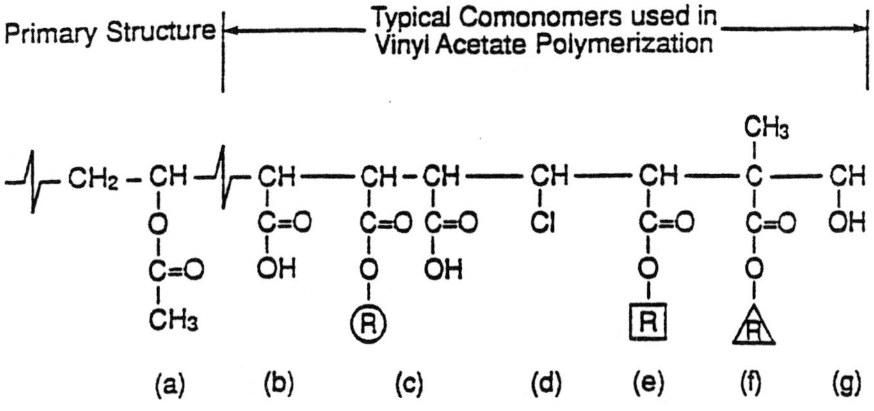

Primary Structure | Typical Comonomers used in Vinyl Acetate Polymerization

(a) (b) (c) (d) (e) (f) (g)

R = Methyl, Ethyl, etc.
a) Basic Vinyl Acetate Monomer (Homopolymer)
b) Acrylic Acid (Ammonia or Sodium Salt)
c) Monoester of a Dicarboxylic Acid
d) Vinyl Chloride
e) Acrylate Ester
f) Methyl Acrylate Ester
g) Vinyl Alcohol

FIGURE 4.90 Combined acetate (PVAc) polymers and copolymers [2].

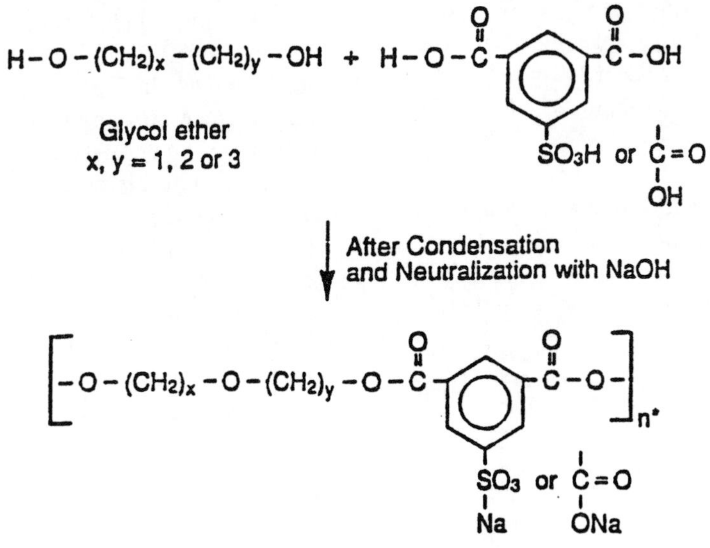

$H-O-(CH_2)_x-(CH_2)_y-OH$ + $H-O-C(=O)-C_6H_3(SO_3H \text{ or } COOH)-C(=O)-OH$

Glycol ether
x, y = 1, 2 or 3

After Condensation and Neutralization with NaOH

$[-O-(CH_2)_x-O-(CH_2)_y-O-C(=O)-C_6H_3(SO_3Na \text{ or } COONa)-C(=O)-O-]_{n^*}$

*n = 3–5.5 (oligomer types)
n = 18–26 (medium MW type)

FIGURE 4.91 Synthesis of a typical polyester resin [2].

TABLE 4.8 Comparison of the properties of some film forming sizes [2].

Size	Tensile Strength (PSI)	Elongation %	Moisture Content at		
			50% RH	65% RH	80% RH
PVA*	7000 to 15,000	100 to 150		8–9**	16–17**
Starch	600 to 900	8 to 12		15–20	19
CMC	2000 to 4000	10 to 15	14	15–20	30.5
Acrylic	1000 to 2000	100 to 600			17–21

*At 70°F and 65% RH.
**Depending upon the degree of hydrolysis.

the endurance of the sizing film. It is claimed that the new technology polyester has a synergy with the size film and tends to withstand abrasion better than its tallow wax counterparts [8].

4.3.4 Lubricants and Other Additives

In addition to a film former polymer, lubricants and other various additives are added to a size mix.

Lubricant is almost always added to increase abrasion resistance of the yarn which is especially useful for rapiers and projectile machines. Lubricants with anti-sticking agents (lecithin) also prevent sticking of PVA to dry cans. Emulsifiers are added to the wax to improve desizability. Paraffin or marine glycerides are added to harden the wax and better lubricate the yarns; however, if not removed properly during finishing, some lubricants can cause problems in later operations such as heatsetting or dyeing.

Like film formers, different lubricant types are used in different parts of the world. In the U.S., tallow based wax is the most widely used. Its basis is hydrogenated tallow glycerides (HTG) and bleached tallow. The wax level is 4–12% and the majority of the plants run around 7–8%. There are different wax grades as well. Some people believe that waxes actually degrade the size film and reduce its tensile strength and flexibility. In international markets, synthetic lubricants are used the most. This is due to finishing concerns, particularly in Europe where many plants experienced quality issues related to the removability of tallow waxes in finishing. The U.S. mills use caustic to saponify the wax for easy removal. In Europe, the mills expect the wax to come off during the desizing step with water or enzymes. In the U.S., a small amount of synthetic lubricants, such as polyethylene glycol (PEG) and polypropylene glycol (PPG), is used. The price is a limiting factor for wider use of synthetic waxes.

Various other additives may be included in the size mix depending on the particular weaving machine requirements or if a particular type finishing is required once the fabric is woven. These additives include humectants, anti-static and anti-foam agents, removable tints (for warp or style identification), binders, preservatives (if the warps or the fabric is to be stored for long periods of time), penetrating agents (to allow the size to penetrate into tightly constructed styles), viscosity modifiers, weighting agents (to make the cloth heavier, usually used on weaving machine finished goods), anti-mildew, anti-skin, etc. Softeners such as lubricants, soaps and

TABLE 4.9 Waste treatment requirements for various sizes [7].

Size Material	BOD_5 (ppm)*
Starch	
Pearl corn	500,000
B2 gum (starch dextrins)	610,000
Keofilm No. 40	550,000
Penford Gum 300 (starch ether)	360,000
Wheat starch	550,000
Polyvinyl Alcohol (PVA)	10,000 to 16,000
Carboxymethyl Cellulose (CMC)	30,000
Polyvinyl Acetate (PVAc)	10,000
Hydroxyethyl Cellulose (HEC)	30,000
Sodium Alginate	360,000
Acrylic	205,800

*5 days, parts per million.

waxes are used to make the yarn sufficiently extensible; they also prevent cracking of size during weaving.

Humectants such as urea, sugar and glycerin, are used to retain some moisture in the size product. Moisture makes the size film more flexible and less brittle. High starch level may result in film dryness which can cause shedding. The humectant level depends on the starch used, its grade and the mill conditions. The typical humectant level is 2–10%.

4.3.5 Choosing the Proper Chemistry for Sizing

With all the ingredients available, a size mix can get quite complex. The key is to make the size mix as simple as possible. Several factors should be considered when choosing the size mixture:

- yarn material (cotton, poly/cotton, polyester, rayon, wool, etc.)
- yarn hairiness
- yarn structure (ring spun, open-end, jet spun)
- water to be used for cooking (recycled or fresh)
- type and speed of weaving machines to be used (projectile, rapier, air-jet, water-jet)
- % add-on (and % solids) required
- yarn occupation in the size box and on the dry cans
- desizing procedures
- reclamation of size and use of enzymes in the finishing plant
- slasher design and number of size boxes
- environmental restrictions

As mentioned in Chapter 2, there are several spinning methods and each method produces a different yarn structure. Figure 4.92 shows yarns spun by air-jet, open-end and ring spinning methods. Even for the same yarn count, the sizing requirements are different for each yarn. As an example, the diameter of the open-end yarn is 10–15% larger than the equivalent ring spun yarn. The diameter determines the number of yarns that can fit in a single size box. Open-end spun yarn has more open and porous surface structure which requires higher viscosity size mixture to avoid excessive size penetration into the yarn.

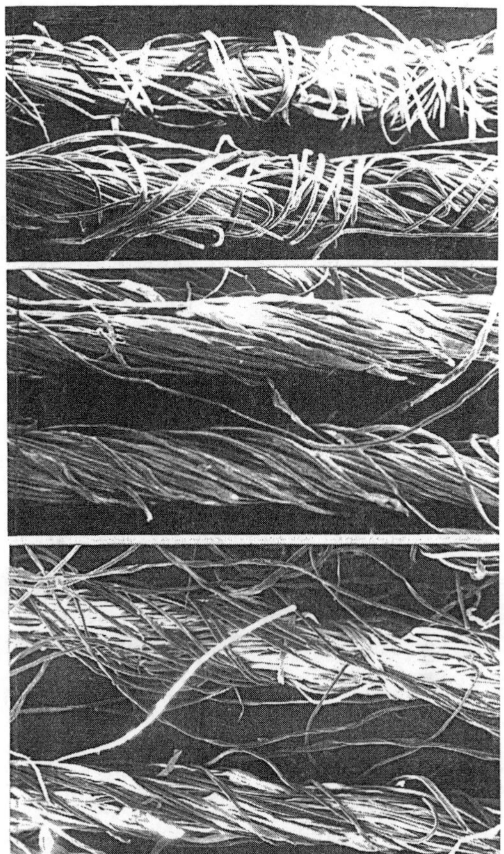

FIGURE 4.92 Comparisons of open-end (top), ring spun (middle) and air-jet spun yarns [2].

The penetration of size into the yarn depends on the amount of twist (twist per inch, Chapter 2) for ring spun yarns. High twist yarns generally require less viscous size solutions or the use of a penetrating agent for proper penetration into the yarn. Blend levels of fibers should be considered carefully in sizing. Appendix 3 gives typical sizing examples of common fabric styles. The range of formulation for film forming polymers are given for each case. Percent of wax and additives are based on total of PVA, starch and CMC.

Example:

50% PVA	100 kg PVA
50% Starch	100 kg starch
8% wax	16 kg wax
5% acrylate (dry)	10 kg acrylate dry (40 kg wet)

TABLE 4.10 Sizing considerations for various weaving machines [6].

Weaving Machine	Preferred sizing
Shuttle	Starch or PVA/Starch
Rapier	PVA/Starch or PVA
Projectile	PVA/Starch or PVA
Air-jets	PVA/Starch (high speeds, more PVA)

In each case in the Appendix, the PVA component could be replaced by recovered PVA. All of the formulations can be run with 100% PVA for use in PVA recovery system.

Other fiber characteristics in the yarn such as shape, denier and delusterant level also affect the size pickup of the yarn.

The type of weaving machine should also be considered for choosing the sizing material. Table 4.10 shows the size choices for different machines.

Since sizing is not a value added process, minimizing the cost of sizing is extremely important. However this should not be done by using cheap size materials at the expense of productivity and performance in the weave room. The ultimate goal is to optimize weaving performance with the cost of sizing.

4.3.6 Filament Sizing

Twisted and zero-twist filament yarns can be sized. The continuous multifilament yarns are generally smooth and have a surface finish that protects the yarns against abrasion and static during processing. Because of their smooth and lubricated surfaces, high twist filaments may not require sizing. However, low twist multifilaments should be sized because if a single filament breaks, it can develop a fuzz ball, float or skip that will ultimately cause a loom stop.

Size considerations for filament yarns are shown in Table 4.11. Size requirements for filament yarns are as follows [2]:

- The size solution must sufficiently penetrate the filament bundle. This may be achieved by using binders and additives such as emulsifiers, wetting agents, etc. The size solution should have a low enough viscosity to allow adequate penetration into the filaments.
- The adhesion between the filament and size must be good.
- The sizing agent must dry quickly enough without producing a tacky surface.
- The elasticity and flexibility of the size film should match to those of the yarns under weaving stresses.
- The size should not cause static build up.
- The size should not shed excessively causing a build up on the heddles, reeds or other weaving machine parts.
- Size film properties should not be affected drastically by extreme humidity changes.
- The size should be easily removable during desizing.

TABLE 4.11 Size considerations for filament yarns [2].

Size	Fiber	Type of Loom		
		Shuttle	Air-jet	Water-jet
Dispersible Polyester	Polyester	S	S	S
Polyacrylates (esters)				
Sodium	Polyester	S	S	N
Ammonium	Polyester, Nylon	S	S	S
Polyacrylic acid	Nylon	S	S	S
Polyvinyl Alcohol				
FH	Viscose Rayon	S	S	N
PH	Nylon*, Acetate, Polyester	S	S	N
Styrene/maleic anhydride				
Sodium	Acetate	S	S	N
Ammonium	Acetate	S	S	S
Polyvinyl acetate	Acetate, Polyester	S	S	S

S: suitable — N: not suitable — *Fabrics requiring a neutral pH

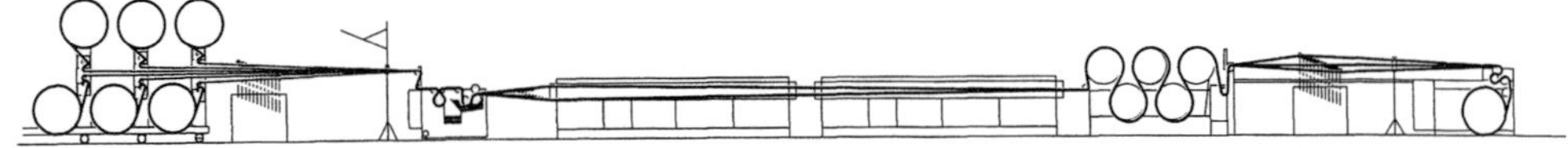

FIGURE 4.93 Configuration for sizing of filament yarns (courtesy of Sucker Muller Hacoba).

- The size should not cause adverse effects on the yarn, processing equipment or human health.
- The size should be easy to process and use.
- The producer spin finish oils should not affect the size properties.
- Sizes may foam at high machine speeds. To prevent this, anti-foaming agents can be used.

The concentration of the size, which controls the number of spot welds, depends on [10]:

- yarn denier—lower deniers have more fiber surface area and require more size.
- number of ends/inch
- type of weave—plain weaves are more difficult to weave than those with long floats, e.g., satins.
- the weaving machine type and condition—some shuttleless machines are more abrasive to the warp.
- type of slasher and drying arrangement employed—each type (conventional, pre-dryer type, single end beamer) has different constraints.

Figure 4.93 shows the configuration of a sizing machine for sizing of filament warps for industrial fabrics made of glass, polyamide, polyester, aramide, carbon, etc. The unit is equipped with air predrying and cylinder drying. Figure 4.94 shows schematic of a warping-sizing machine for industrial yarns in which warping and sizing are carried out in a single work cycle. Warp yarns are sized straight from the creel and the warp ends are guided individually through the machine.

4.3.7 Desizing

After weaving, the size must be removed from the fabric in the finishing process unless it is a loom finished material such as denim. If the size is not recovered then the effluent from the finishing plant will contain the size and should be treated before it can be discharged. Ease of size removal and the cost of desizing are different for each size material. The type of ingredients in the size mixture is also critical. They affect the finishing process since these materials should be completely removed before other finishing and dyeing processes.

Fabrics with starch sizes are treated with chemicals that break down both the linear and branched chains into shorter fragments. These fragments become soluble in hot water and are removed from the fabric. The chemicals used during desizing should not affect the fibers of the fabric.

Starch is a carbohydrate. Therefore, it can be broken down similar to a starch eaten by animals and bacteria. Weak acids and enzymes are used to break down the starch chain structure without damaging the cotton cellulose (Figure 4.95). As a result, the chains are broken down into smaller water soluble fragments that are washed away. Amylose and amylopectin are broken down in a similar fashion but the amylopectin residues may still have some branching. Reagents are used to allow the desizing process to take place at room temperature; however, the desizing process can proceed at a faster rate at higher temperatures.

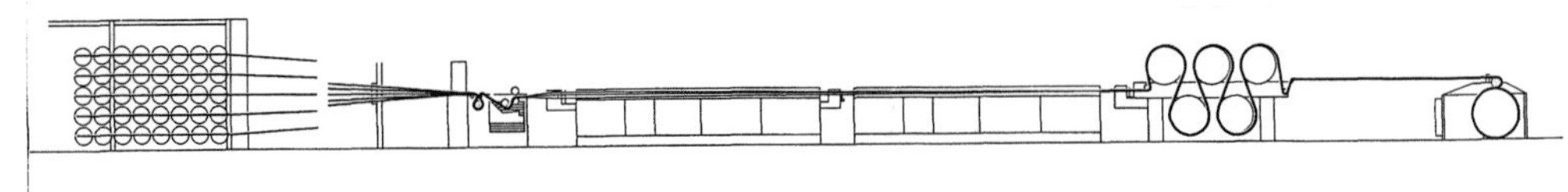

FIGURE 4.94 Schematic of warping-sizing process (courtesy of Sucker Muller Hacoba).

FIGURE 4.95 Desizing of starch molecules is accomplished by treatment of the cloth with acids, enzymes or oxidizing agents [2].

Since starch consists of polymerized sugar (glucose), the waste water contains nutrients having a high Biological Oxygen Demand (BOD) that can affect the ecological balance of rivers. Therefore, the water must be treated to destroy these materials before being released into the waterways. This increases the cost of slashing process.

Since PVA forms true solutions with water, it is only required that the polymer redissolves in the hot water during desizing. That is, it is not necessary to degrade the PVA chains in order to remove the size film, allowing the recovery of the size by one of several recovery processes for reuse. This is an important advantage of PVA because some textile companies recover and reuse PVA sizes.

During the desize operation of polyacrylic acid based sizes, the size is resolubilized using an alkaline desize. A solvent desize may be necessary for acetate yarns.

Water pollution is a concern with the disposal of size materials. Therefore, materials removed during desizing are desired to be recycled. However, most of the textile industry uses water soluble sizes to protect the environment.

Some industrial fabrics need not to be desized. For example, for some coated fabrics, the adhesion of the PVA size is needed as a primer coating for adhesion to the coating. However, this requires a more uniform size application than normal.

The fabric must be free of size and auxiliaries prior to dyeing and printing because faulty desizing can cause dyeing problems such as dye spots, streaks, etc. Residual size contamination can also affect permanent press, soil release and other finishes. Figures 4.96–4.98 show steps to identify the common types of sizes in desizing unknown fabrics. Figure 4.99 shows the desizing efficiency of various PVAs at different temperatures.

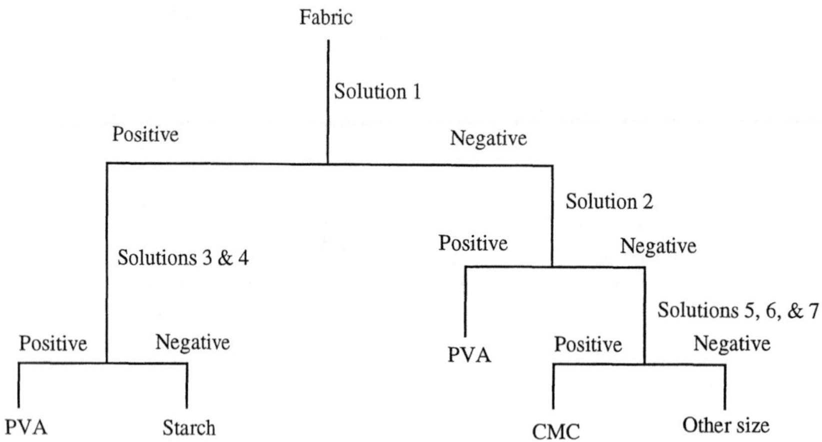

Solution 1: 2.4 g of potassium iodide, 1.3 g of iodine dissolved in water and diluted to one liter
Solution 2: 0.13 g of iodine, 2.6 g of potassium iodide, 4.0 g of boric acid added in that order and diluted to 100 ml
Solution 3: 11.88 g of potassium dichromate 25 ml of concentrated sulfuric acid, diluted with 50 ml of water
Solution 4: 30 g of sodium hydroxide in 70 ml of water
Solution 5: 0.012 M ferric chloride
Solution 6: 0.06 M potassium thiocyanate
Solution 7: 0.005 M potassium ferrocyanide

FIGURE 4.96 Flow chart to identify starch, CMC and PVA sizes [9].

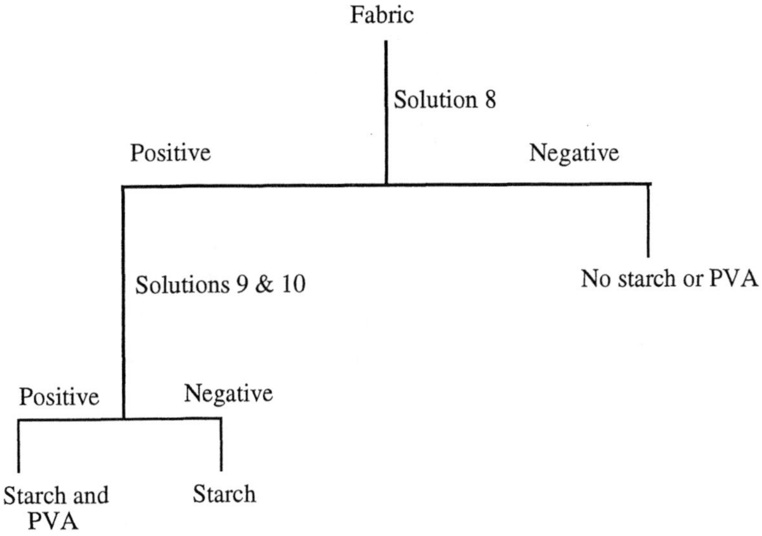

FIGURE 4.97 Flow chart to identify starch/PVA blends [9].

4.3.8 Cold and Pre-Wet Sizing

Recently, a new cold size application process for sample warps and short warps was developed [11]. In this process, the warp surface is treated gently without squeezing. It is claimed that in comparison to waxing on the sectional warper, cold sizing produces better yarn compaction, smoother surface, and less hairiness which improves the weaving efficiency. Cold size products from the suppliers can be used for cold sizing of single yarn, siro yarns and two ply yarns of wool and blends, terry warps and synthetic yarns. The chemical products used are water soluble, recyclable and biologically degradable. The main advantages of this system have been reported to be as follows:

- less liquor pickup and high liquor concentration
- 30–50% less application compared to normal sizing

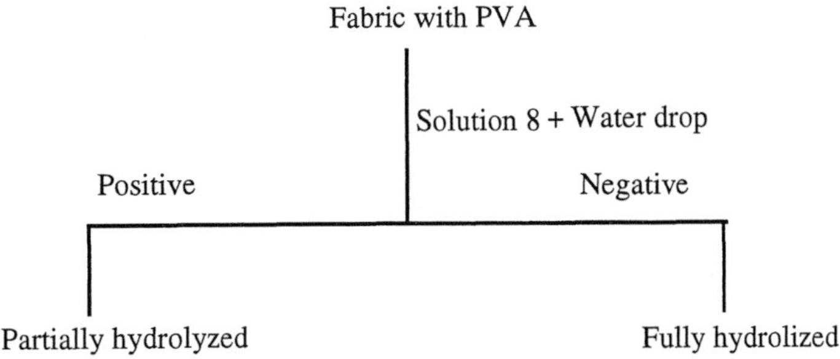

FIGURE 4.98 Flow chart to identify partially or fully hydrolyzed types of PVA [9].

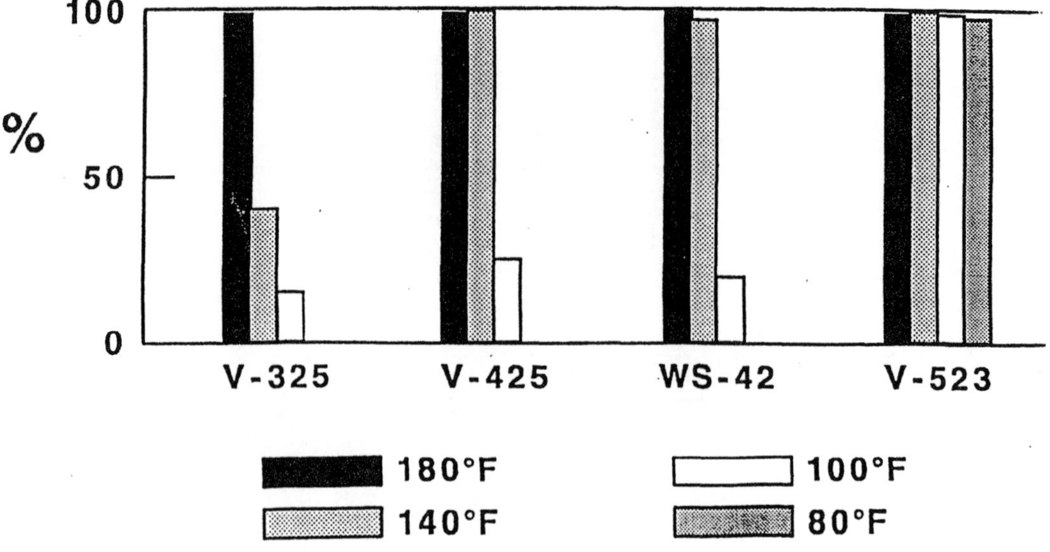

FIGURE 4.99 Desizing efficiency of various PVAs tested at different temperatures [9].

- savings in energy in drying
- less machine space
- higher modularity and productivity

Pre-wet Sizing Technology

In another recent development, called pre-wet sizing technology, the yarns are wetted and washed with hot water prior to entering the size box. It is claimed that by doing this, the size add-on can be reduced by 20–40%, size adhesion is improved, abrasion resistance is increased and hairiness is reduced.

Figure 4.100 shows the schematic of pre-wetting process. Figure 4.101 shows the effect of pre-wetting on weaving behavior of yarns. The improvement in weaving performance is attributed to better encapsulation of the yarn by the sizing agent and better adhesion of the sizing agent to the yarn. The preliminary results showed the following advantages of this system [12]:

- 15–19% increase in tensile strength
- 50% reduction in hairiness
- increase in abrasion resistance by 70–200%
- clinging tendency reduction

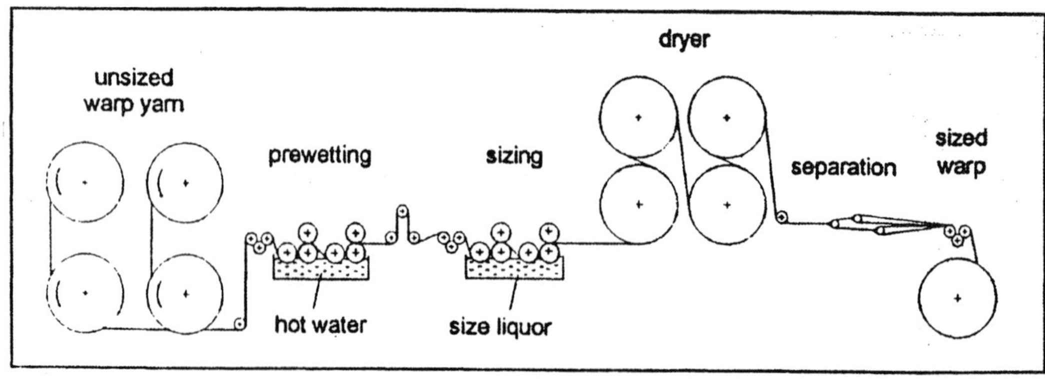

FIGURE 4.100 Schematic of pre-wetting sizing process [12].

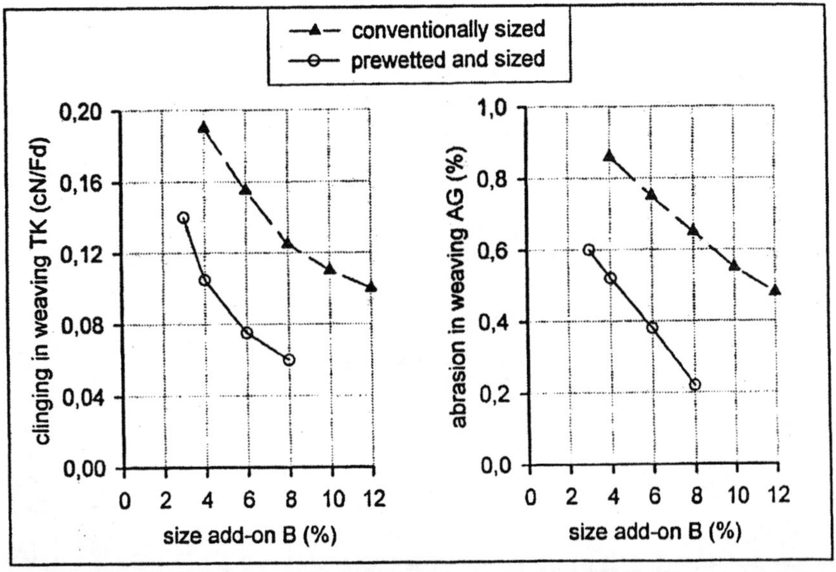

FIGURE 4.101 Comparison of the weaving behavior of conventionally sized and pre-wet sized yarns [12].

- 30–50% reduction in lint formation during weaving

4.4 DRAWING-IN AND TYING-IN

4.4.1 Drawing-In

After slashing, the sized warp beam is prepared to be placed on the weaving machine. Figure 4.102 shows a typical warp flow system before weaving starts.

High fashion fabrics generally have high density which increases the demand on the quality of shed opening. As a result, warp leasing is becoming more popular. Figure 4.103 shows a warp leasing machine. Different lease combinations can be selected with the automated leasing machines.

Drawing-in is the entering of yarns from a new warp into the weaving elements of a weaving machine, namely drop wires, heddles and reed, when starting up a new fabric style (Figure 4.104). Tying-in the new warp ends to the depleted warp is done when a new pattern is not required.

A drop wire is a narrow metal sheet that is hung in the air by the tensioned warp yarn. If the warp yarn is broken, then the drop wire drops and touches a metal bar that extends along the width of the machine. This contact between the drop wire and metal bar closes an electrical circuit and shuts down the machine immediately (Figure 4.105). There is a drop wire for each warp yarn.

Pinning machines are used to pin open drop wires on warps. Since the pinning speed is high (up to 200 wires per minute), these machines are economical for more than 3000 warp ends.

After drop wire, the warp yarn goes through the heddle eye (there is only one warp yarn per heddle eye). This is done according to a plan called drawing-in-draft, DID (Chapter 3). Then, the yarn is threaded through the reed spaces. A reed space is the opening between two dents (metal) in a reed. In general, one, two or three warp yarns are passed through one reed space. The reed plan specifies the number of yarns per reed space. The number of yarns depends on the diameter of the yarns and the dent opening; each yarn should be able to move freely up and down in the reed space independent of the other yarn(s).

In the manual mode of drawing-in, one person sorts the warp yarn and the other draws it through from the other side. The sorting step can be

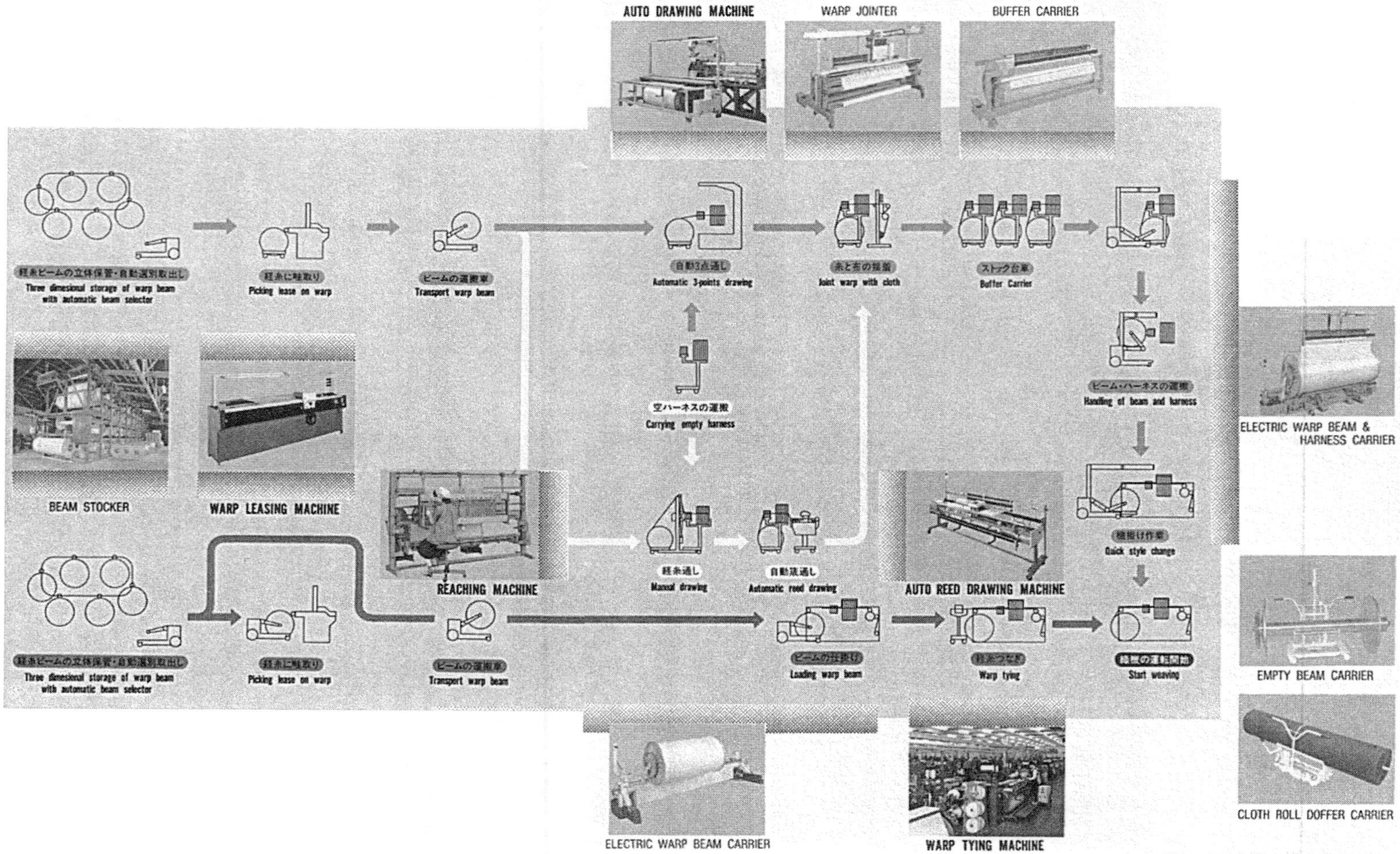

FIGURE 4.102 Warp flow system after slashing (courtesy of Todo Seisakusho Ltd.).

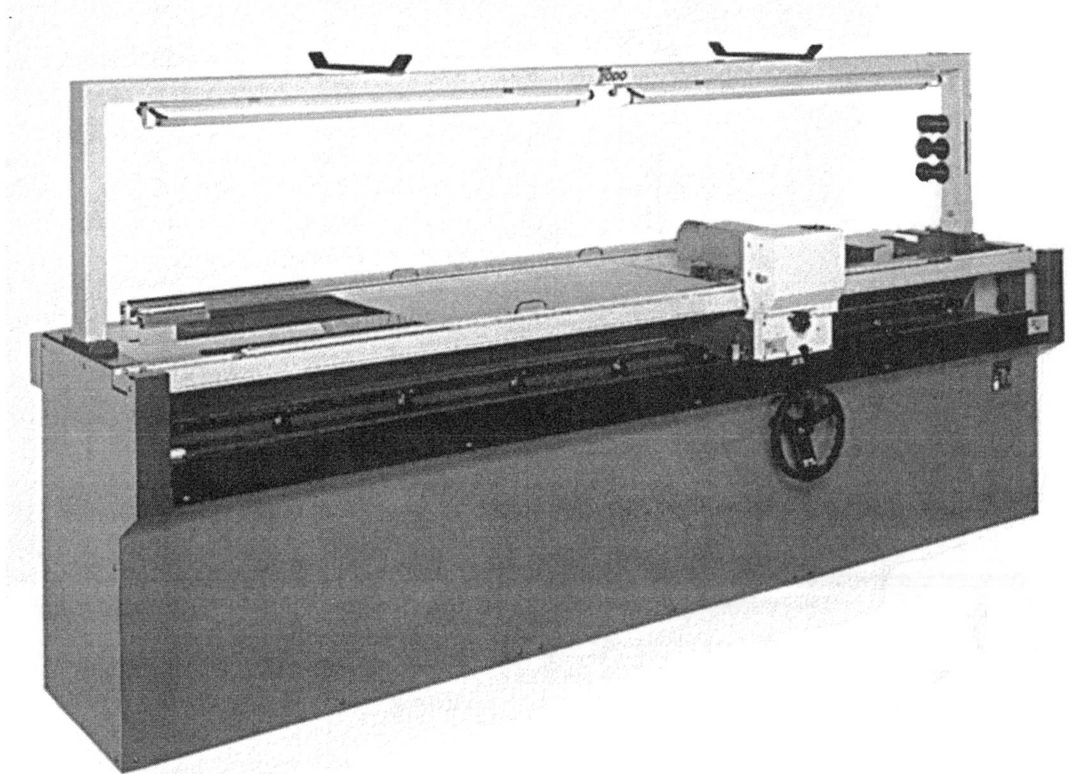

FIGURE 4.103 Warp leasing machine (courtesy of Todo Seisakusho Ltd.).

automated by a reaching machine (Figure 4.106).

Today, the drawing-in and tying-in processes are fully automated. Drawing-in is done using robot-like machines. A special type of heddle is needed for automated drawing-in. Figure 4.107 shows an automated drawing-in machine. The warp ends, taken from the warp sheet, are fed individually to the drawing-in element; heddles are separated from the stack and brought to the drawing-in position; a plastic knife opens a gap in the reed and a hook draws-in the warp end through the heddle and reed in one step (Figure 4.108).

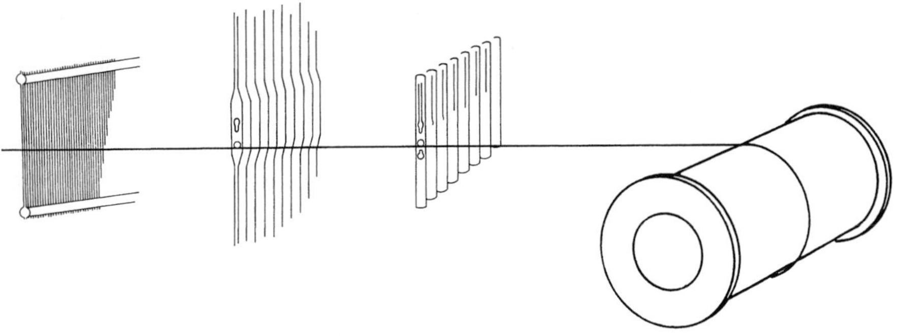

FIGURE 4.104 Schematic of drawing-in (courtesy of West Point Foundry and Machine Company).

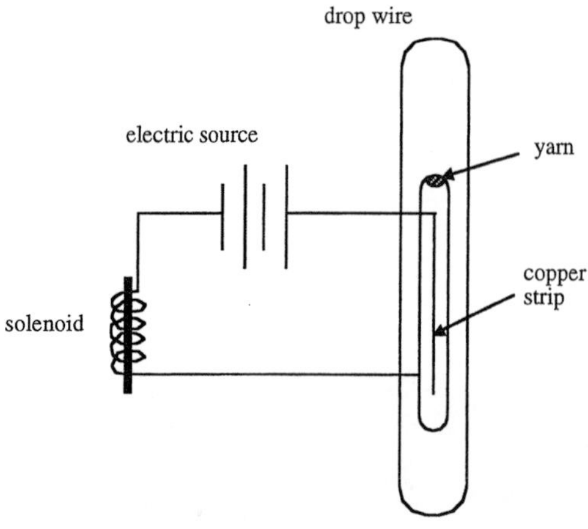

FIGURE 4.105 Electrical diagram of warp stop motion.

Automatic drawing-in increases speed, flexibility and quality in weaving preparation compared to manual drawing-in. A drawing rate of 50,000 warp ends per 8 hours (200 ends per minute) is possible.

4.4.2 Tying-In

After the depletion of a warp beam on the weaving machine, if there will be no change in design, then the drawing-in process needs not be repeated. The ends of the old warp beam (now a fabric beam) are cut and the ends of the new warp beam are tied to the corresponding ends of the old beam which is called tying-in process. Then, the warp ends are pulled through the heddle eyes and reed until the knots are cleared.

A small portable robot is used on or off the weaving machine for tying-in. Figure 4.109 shows an automatic warp tying machine. A typical warp tying machine can knot single or ply yarns from 1.7 to 80 Ne (340-7 tex). They can knot cotton, wool, synthetic and blend warp yarns as well as yarns of different thicknesses. Typical knotting speed of a knotter is from 60 to 600 knots per minute.

With continuous filaments and bulky yarns, a non-slip double knot is recommended which

FIGURE 4.106 Reaching machine (courtesy of Todo Seisakusho Ltd.).

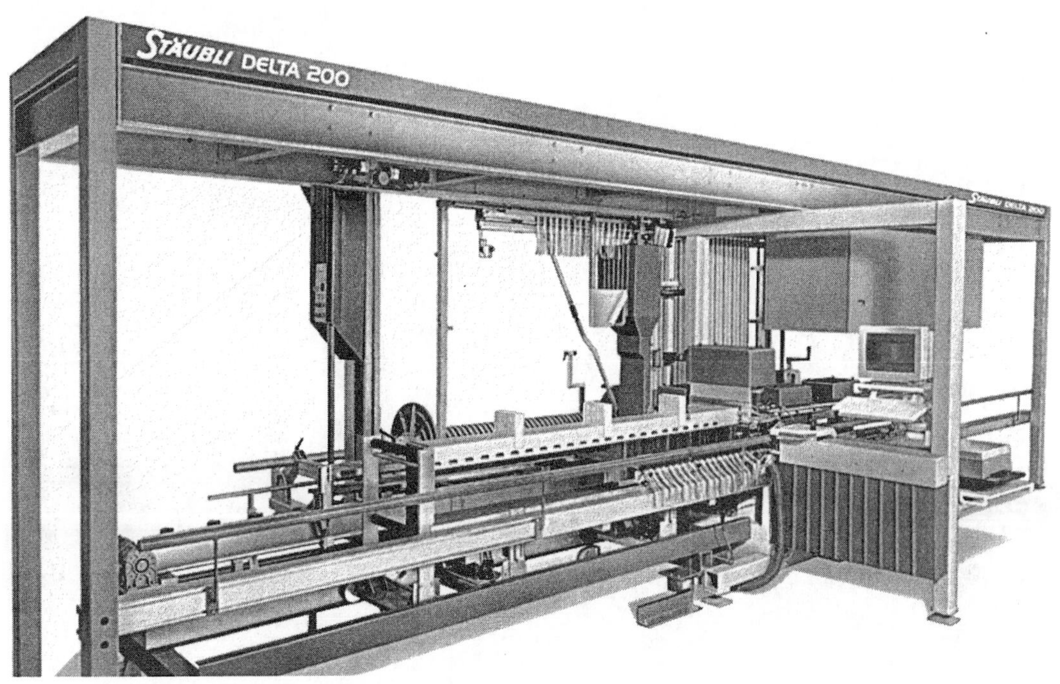

FIGURE 4.107 Fully automated drawing-in machine (courtesy of Staubli).

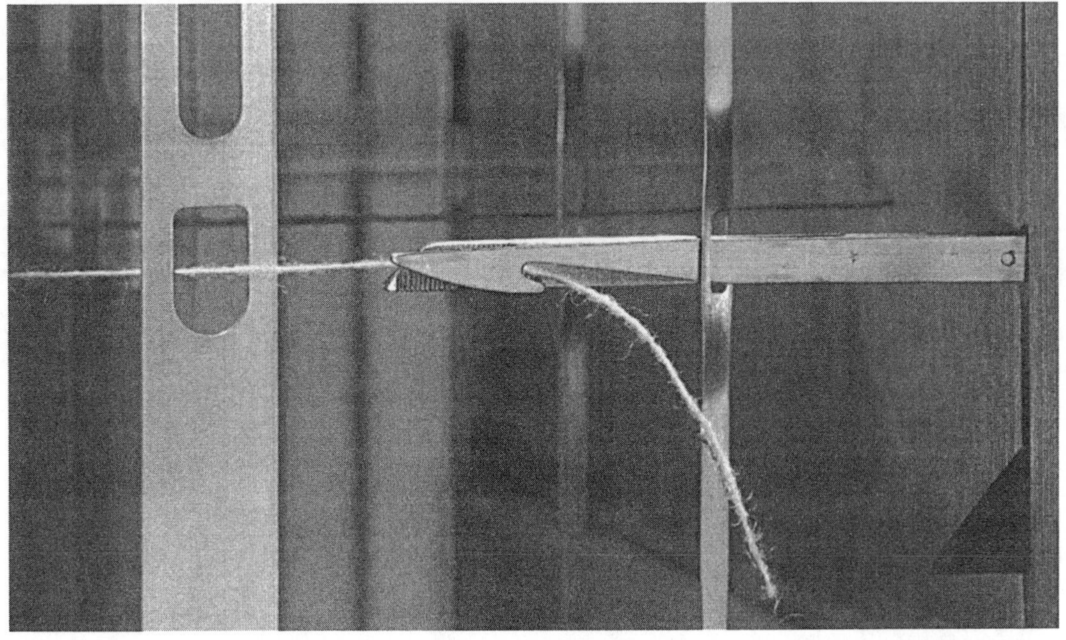

FIGURE 4.108 Automatic drawing of warp end through drop wire, heddle eye and reed (courtesy of Staubli).

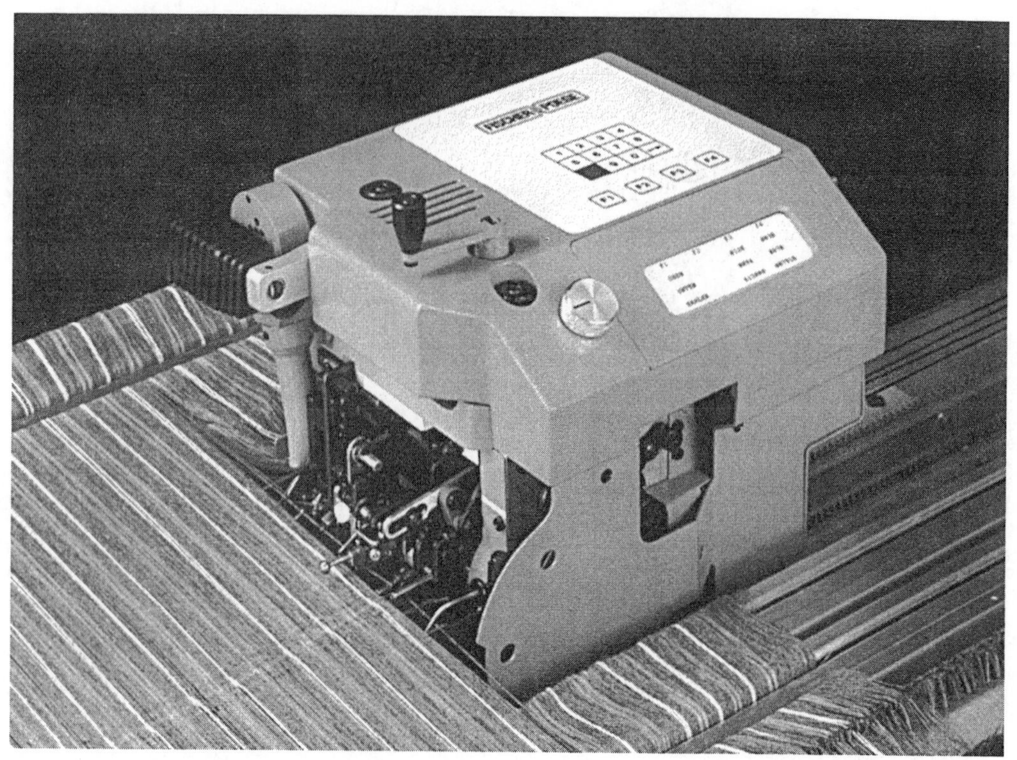

FIGURE 4.109 Warp tying machine (courtesy of Fischer Poege).

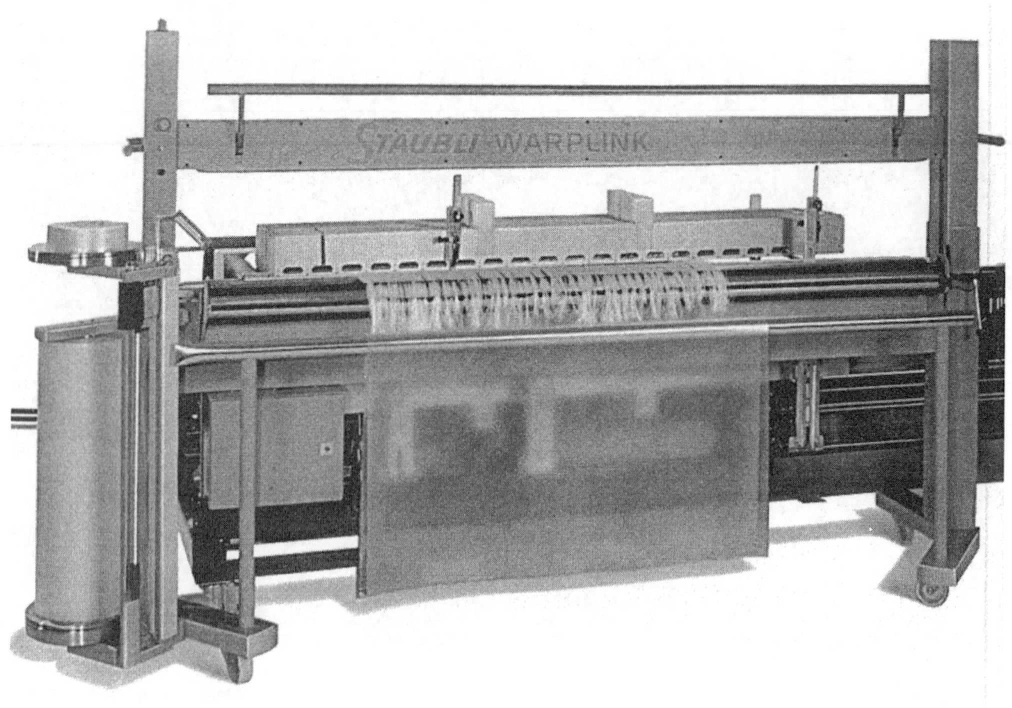

FIGURE 4.110 Warp welding machine (courtesy of Staubli).

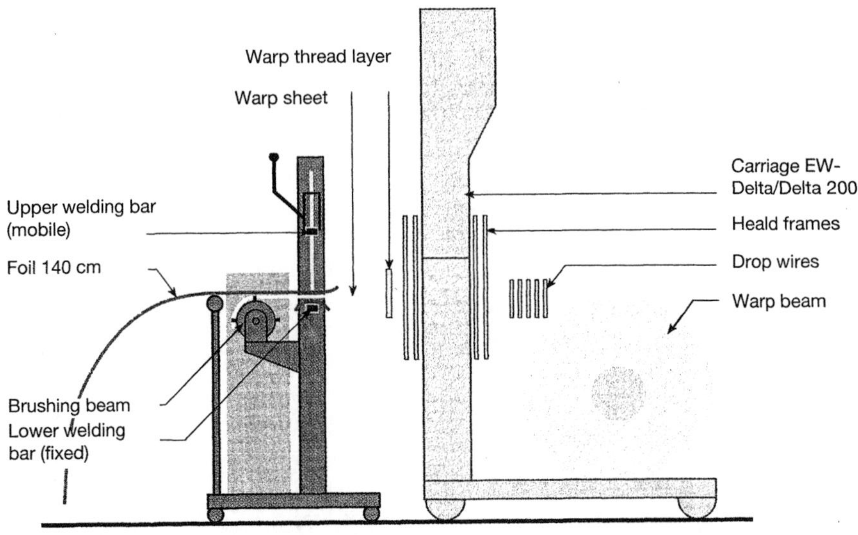

FIGURE 4.111 Schematic of warp welding (courtesy of Staubli).

can be handled by knotting machines. Some automatic tying machines can knot extremely short tails of yarns (5 mm). Tape yarns and monofilaments require a slightly different tying machine. Tape yarns of up to 8 mm width can be tied. The knotting speed is typically 60 to 450 knots per minute. The number of warp ends to be tied together can be preprogrammed; once this number is reached, the knotter stops automatically. A dual knotting system is used on a double beam weaving machine; the knotters work from left to right and from right to left simultaneously.

Figure 4.110 shows a warp welding machine. This machine is used to weld the warp end layer with a plastic foil after drawing-in which provides simple insertion through the weaving machine. This results in time saving at the machine startup. After drawing-in with a brush beam the ends protruding from the reed are aligned parallel and stretched evenly. An approximately 5 cm wide plastic foil is placed on top of the lower welding bar and a longer piece of plastic foil is placed on the warp yarns above the lower piece of plastic foil. By moving the upper welding bar down, the plastic foils are welded together with the warp yarns in between (Figure 4.111).

Several points should be considered during drawing-in and tying-in. Improper splicing and/or knotting can become critical to good weaving performance. The straightness of individual warp yarns and their freedom to act independently as they pass through a weaving machine are important for quality weaving. Yarns that are crossed and tangled cannot proceed without excessive stress and yarns that are restricted or influenced by drop-wire activity, heddle spacing, harness interference or reed spacing will not weave at top performance.

REFERENCES

1. ME-1000 Sectional Warper, McCoy-Ellison, Inc., 1997.
2. Hall, D., "Slashing" in Wellington Sears Handbook of Industrial Textiles, Technomic Publishing Co., Inc., 1995.
3. Robinson, G. D., "The Sizing of Spun Yarns", ATMA Slashing Short Course, Auburn University, Auburn, AL, Sept. 1999.
4. Gibbs, W. A., "Latest Developments in Warp Preparation Equipment", ATMA Slashing Short Course, Auburn University, Auburn, AL, Sept. 1993.
5. Vassallo, J. C., "The Chemistry of Spun Sizing", ATMA Slashing Short Course, Auburn University, AL, Sept. 1998.
6. Robinson, G. D., "An Overview of Spun Yarn Size Formulation Selection", ATMA Slashing Short Course, Auburn University, Auburn, AL, Sept. 1993.

7. McAllister, I., "How to Make Those Sizing Calculations", Textile World, 130, No. 8.
8. Adams, S., "Innovations", ATMA Slashing Short Course, Auburn University, Auburn, AL, Sept. 1998.
9. Davis, G., ''The Effects of Slashing on Fabric Finishing,'' ATMA Slashing Short Course, Auburn University, Auburn, AL, Sept. 1998.
10. Mullins, S. M., "Filament Yarn Sizing", Textile Slashing Short Course Proceedings, Auburn University, 1984.
11. "New Cold Sizing Line Ben-Ecosize for Short Warps", Melliand International (4), 1997.
12. Stegmaier, T., Trauter, J., and Wunderlich, W., "Reducing Effluent Loading in Sizing and Desizing", Melliand International (1), 1998.

SUGGESTED READING

- Proceedings of the Slashing Short Course, Auburn University, Auburn, AL, Sept. 1998.
- Proceedings of the 32nd Annual Textile Slashing Short Course, Auburn University, AL, Sept. 1992.
- Proceedings of the 33rd Annual Textile Slashing Short Course, Auburn University, AL, Sept. 1993.
- Textile Slashing Short Course Proceedings, Auburn Univesity, AL, 1981.
- Proceedings of Meeting on Winding and Weaving, Alabama Textile Operating Executives, Auburn University, Auburn, AL, January 1982.
- Proceedings of Meeting on Winding and Weaving, Alabama Textile Operating Executives, Auburn University, Auburn, AL, January 1984.
- Proceedings of Meeting on Winding Process, Alabama Textile Operating Executives, Auburn University, Auburn, AL, January 1978.
- Proceedings of Meeting on Slashing and Weaving, Alabama Textile Operating Executives, Auburn University, Auburn, AL, May 1972.
- Proceedings of Meeting on Winding and Weaving, Alabama Textile Operating Executives, Auburn University, Auburn, AL, February 1981.
- Trauter, J., "Warp Preparation at ITMA-95", Melliand International (1), 1996.
- Ellis, T., "Effect of Pre-wetting Warp Yarns Prior to Size Application", ATMA Slashing Short Course, Auburn University, Auburn, AL, Sept. 1998.
- Lord, P. and Mohamed, M. H., Weaving: Conversion of Yarn To Fabric, 2nd. Ed., 1982.
- Mohamed, M. H., Production Mechanics and Properties of Woven Fabrics, Class Notes, NCSU, 1983.
- Adanur, S., Fabric Design and Manufacturing, Lecture Notes, Auburn University, 1998.
- Walker, R. P., Fabric Forming Systems, Lecture Notes, Auburn University, 1992.
- Peghini, A., "Minimum Application Process for Sizing", Melliand International (1), 1998.
- Lange, A. and Weinsdorfer, H., "Analysis of Yarn Stresses in Warping and Beaming", ITB International Textile Bulletin, 6/98.
- "New Ben-Tronic Warping Technology", Melliand International (2), 1998.
- Rozelle, W., "Slashing Received New Demands in Yarns, Processes", Textile World, December 1998.
- Maletschek, F., "Mayer: New Concept in Sectional Warping", Textile World, April 1997.
- "Weft Insertion Control by Co-Axial Tensioners", ATI, October 1991.
- "Modern Systems can Control Slashing Yarn Tension", Textile World, January 1991.
- Robinson, G. D., "Size Recovery and Wastewater Treatment", Textile Asia, February 1993.
- Milner, A. J., "Sizing and Cold Bleaching", Textile Asia, March 1993.
- Rozelle, W. N., "Pre-Wet: New Money Maker in Warp Sizing Operations", Textile World, May 1999.
- Maletschek, F., "Sizing Process with Consideration to Yarn Sheet Elongation", Melliand International (4), 1998.
- Wildhaber, J., and Nef, U., "Processing of Elastane Yarns on High-Speed Weaving Machines", Melliand International (4), 1998.
- "New Drawing-in Machine", Melliand International, (4) 1997.

REVIEW QUESTIONS

1. Explain why winding is necessary for some yarns although it is not a value-added process.

2. Compare the advantages and disadvantages of side withdrawal and over-end withdrawal.
3. What are the factors to change yarn tension during winding?
4. Explain the differences between quill winding and regular winding.
5. What are the major differences between the winding and warping processes? Explain.
6. What are the reasons for draw warping? Explain. What are the possible end uses of yarns subjected to draw warping?
7. What are the characteristics given to warp yarns as a result of slashing? How do they affect the weaving process?
8. What are the factors to consider when choosing a size formula for a particular warp yarn? Explain.
9. What are the major components of a slashing machine?
10. What is leasing?
11. Explain the major differences between spun yarn sizing and filament sizing.
12. What are the latest developments in sizing? Explain.
13. In which type of garment is the size material not removed from the fabric?

5

Weaving Fundamentals

Since yarn is the raw material for weaving, one might wonder why the yarn manufacturing process is not followed immediately by the weaving process, i.e., why they are not connected. There are several prohibitive reasons for this. Yarn manufacturing and weaving processes are mechanically not compatible. They cannot be physically connected because yarn manufacturing speed is a lot higher than the consumption rate of warp yarn and a lot lower than the consumption rate of filling yarn. Another important reason is that the yarns—as manufactured—are generally not suitable for weaving immediately; they have to be prepared as explained in Chapter 4. As a result, completely new technology is required to convert the yarns into fabrics.

Figure 5.1 shows a schematic of weaving. The warp yarns are stored on a beam called a weaver's beam or warp beam (also called a loom beam) and they flow to the front of the machine where the fabric beam is located. The filling yarn is withdrawn from a single package and inserted between the sheets of warp yarns which are perpendicular to the filling yarn.

5.1 BASIC WEAVING MOTIONS

Although there are many mechanisms on a modern weaving machine for various purposes, there are five basic mechanisms that are essential for continuous weaving:

- warp let-off
- shedding
- filling insertion
- beat-up
- fabric take-up

5.1.1 Warp Let-off

Warp let-off mechanism releases the warp yarn from the warp beam as the warp yarn is woven into the fabric. The let-off mechanism applies tension to the warp yarns by controlling the rate of flow of warp yarns. The mechanism should keep the proper tension on the warp yarns which controls the crimp rates of warp and filling yarns. Uniform tension is essential in weaving. Increasing the warp tension decreases the warp crimp and increases the filling crimp in the fabric. The crimp ratio of warp and weft affects the fabric thickness (Chapter 13 Fabric Structure, Properties and Testing). Yarn diameters being the same, equal warp and filling crimps result in the lowest thickness of the fabric.

Let-off mechanisms can be classified as negative and positive. In negative let-off mechanism, the tension on the warp yarns provides the driving force against friction forces in the let-off motion. The tension of the warp is regulated by the friction between the chain or rope and the beam ruffle. The negative friction type of let-off mechanisms were mainly used for non-auto-

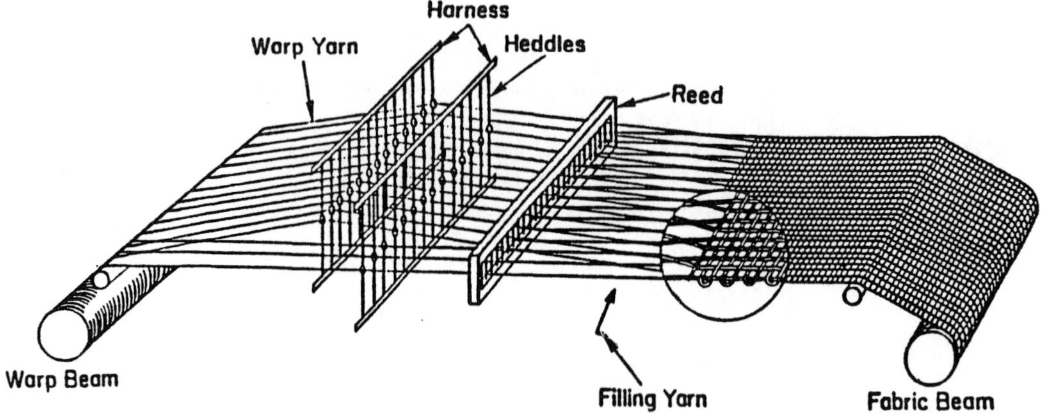

FIGURE 5.1 Schematic of weaving (courtesy of Johnston Industries).

matic weaving. In positive let-off mechanisms, the warp beam is turned at a rate which depends on the yarn length between the warp beam and cloth fell. A separate mechanism is used to apply constant tension on the warp yarns as the warp is depleted.

Let-off mechanisms can also be classified as mechanical or electronic. Most modern weaving machines have electronic let-off. Electronic warp let-off provides a positive and controlled release of warp yarn from full to empty beam which results in a consistent warp tension. It is good for preventing fabric defects such as pick density variation and stop marks. Weaving tensions should be maintained at minimum levels for best weaving performance.

The electronic let-off system can be equipped with a pulley mechanism or a reduction gear mechanism. The linear and positive letting off of the warp beam can be provided by a magnetic reading of the whip roll position. Electronic warp let-off systems have programmable movements with a tenth of pick accuracy to eliminate stop marks. They have the capability to release the yarn tension at the stop of the weaving machine and recover it at the starting of the machine by a number of picks ranging 1/10 of a pick to 50 picks. This way, the overstretching of the yarn, which is the major cause of defects during the standstill time, is prevented. The system can follow any movement of the machine, such as the forward slow motion (jogging) and pick finding motion. With the electronic let-off mechanism, since brake and coupling linings are no longer needed, spare parts cost is reduced. Let-off mechanisms of particular weaving machine types are explained in the subsequent chapters.

5.1.2 Warp Shedding

Shedding is the movement of some warp yarns up and some down to make an angled opening for the filling yarn to be inserted through. This opening is called "shed". Before the insertion of the next filling yarn, the warp sheet has to be rearranged according to the fabric design pattern so that the required fabric structure is produced.

Shed Geometry

Figure 5.2 shows the geometry of a symmetric warp shed. The requirements for the shed opening are determined by the filling insertion and beating motions. Considering the Figure, the maximum shed opening is given by

$$H = B \tan \alpha \tag{5.1}$$

where H is the shed opening, B is the distance between the heddle eye and cloth fell, α is the angle between the warp yarns and fabric plane.

It is desirable to have a small H in order to reduce the lift of the harnesses and therefore to reduce the stress on the warp; however, the magnitude of H is mostly determined by the size of the filling insertion device. The timing of the shed opening is critical. The tension in the upper

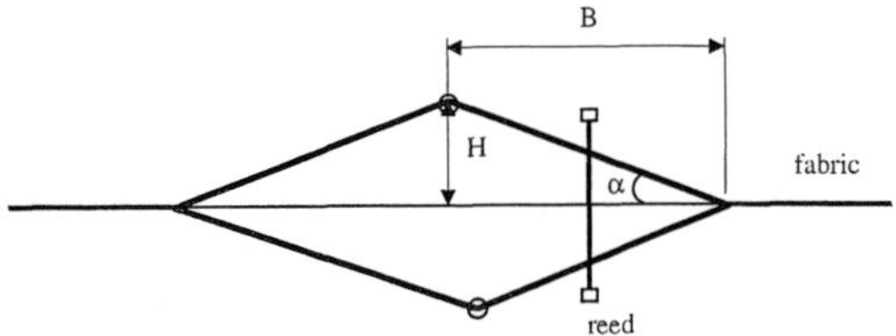

FIGURE 5.2 Warp shed geometry.

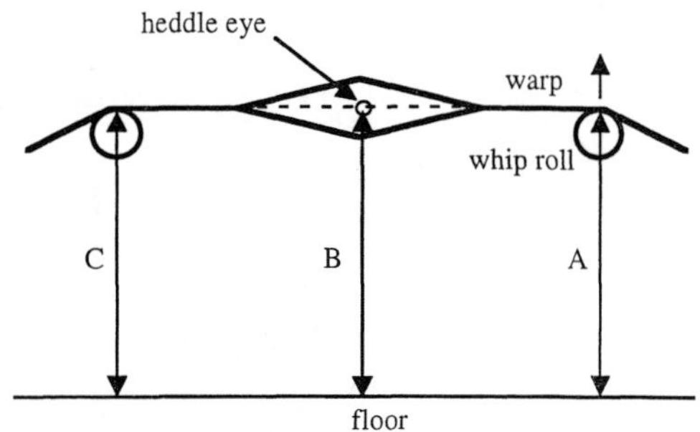

FIGURE 5.4 Zero line or "perfect" shed line.

and lower warp yarns at the shed opening must be kept uniform to ensure high quality fabrics; variations affect the fabric structure.

Warp yarns controlled by lifting devices closest to the back of the weaving machine receive more stress since these yarns have to be raised and lowered a greater distance to form a clear shed for filling yarn insertion (Figure 5.3). An unclear shed would have the danger of the filling yarns being caught by the warp yarns or breaking of the warp yarns by the insertion device.

In normal conditions, the warp yarns and the fabric should be parallel to the ground as shown in Figure 5.4. This is called zero line or "perfect" shed line (A = B = C). Spun warp yarns are sometimes stressed more when they are lowered than when they are raised during weaving because this helps to remove reed marks, giving the fabric a better appearance. Raising whip roll makes bottom shed tighter. The shed geometry changes and becomes unsymmetric.

Figure 5.5 shows the elements of warp shedding mechanism using harnesses for crank, cam or dobby shedding. There is another type of shedding mechanism called jacquard shedding in which there is no harnesses and each warp end is controlled individually. These shedding mechanisms are explained in detail in the following chapters.

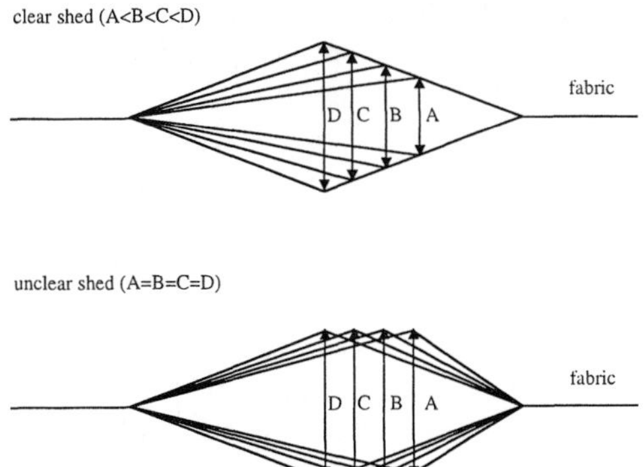

FIGURE 5.3 Clear and unclear shed openings.

As shown in Figure 5.5, each warp yarn is drawn through the eye of a heddle (Chapters 3 and 4) which is a metallic bar. Heddles that guide the warp yarns with identical patterns are attached to the same harness. When a harness moves up or down, all the warp yarns attached to that harness move up or down together. Figure 5.5 shows the arrangement for a plain weave which is the most basic weave. A plain weave requires only two different sets of warp yarns such that during insertion of a filling yarn, one set is up and the other set is down. Since each set of warp yarn is attached to a different harness, plain weave requires only two harnesses. For the next filling, the set of warp yarns swaps their position to allow interlacing of warp and filling yarns.

Order of Entering and Order of Lifting

In Figure 5.5, the first warp end is drawn through harness 1, the second warp end is drawn through harness 2, the third warp end is drawn through harness 1, the fourth warp end is drawn through harness 2. The pattern of drawing warp yarns through the harnesses is called order of entering (Chapter 3). Therefore, for plain weave (Figure 3.13 in Chapter 3), the order of entering the warp yarns is harness 1, harness 2, harness 1, harness 2, and so on.

To make the plain weave with this order of entering, the harnesses are lifted in alternating order, i.e., harness 1 is lifted first while harness 2 is down, and then harness 2 is lifted while

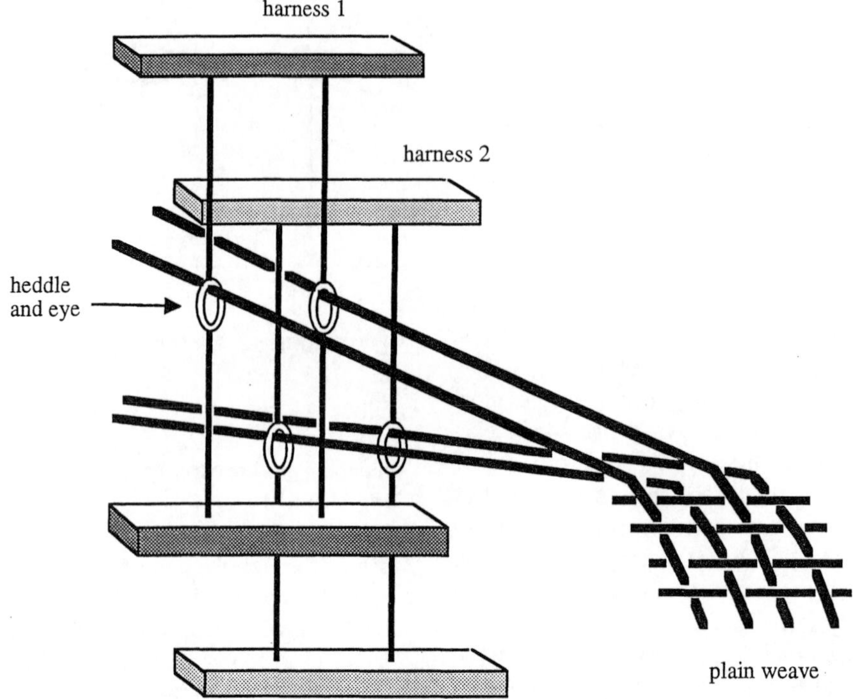

FIGURE 5.5 Elements of warp shedding motion.

harness 1 is lowered. The pattern of lifting the harnesses is called order of lifting. For plain weave, the order of lifting is harness 1, harness 2, harness 1, harness 2, and so on.

Changing the order of entering and/or order of lifting results in different weave patterns. For example, if the order of lifting of Figure 5.5 is kept the same but the order of entering is changed to harness 1, harness 1, harness 2, harness 2, then a different fabric weave is obtained as shown in Figure 5.6. This new design is called filling rib (oxford weave) as shown in Figure 3.16. Note that a different design is obtained using the same number of warp yarns and harnesses.

If the entering order of Figure 5.5 is kept the same but the order of lifting is changed to harness 1, harness 1, harness 2, harness 2, and so on, then a new design is obtained as shown in Figure 5.7. This design is called warp rib which is different than plain and filling rib designs (Figure 3.14).

If we have the entering order of Figure 5.6 and the lifting order of Figure 5.7, then we obtain yet a different fabric design: 2/2 basket weave as shown in Figure 3.18.

The constantly increasing speeds and filling insertion rates increase the mechanical stresses acting on the weaving machine in the form of increasing forces and vibrations. Figure 5.8 shows the four most important characteristics of a typical harness motion, i.e., stroke, velocity, acceleration and momentum. The flow of the stroke is shown twice (mirrored), representing shed opening and shed closing. All four movements are shown to cover the same path from 0 to 1, during the same time from 0 to 1. The different maximum velocities manifest themselves during the stroke flow in the form of different steepnesses in the range of shed closing. The curves in the diagram show clearly that quicker shed opening entails peak values in terms of velocity, acceleration and momentum.

The frame slat profiles to which harnesses are attached can be made of carbon fiber based composite materials which provide dimensional stability and fatigue resistance (Figure 5.9). They also reduce noise and energy consumption.

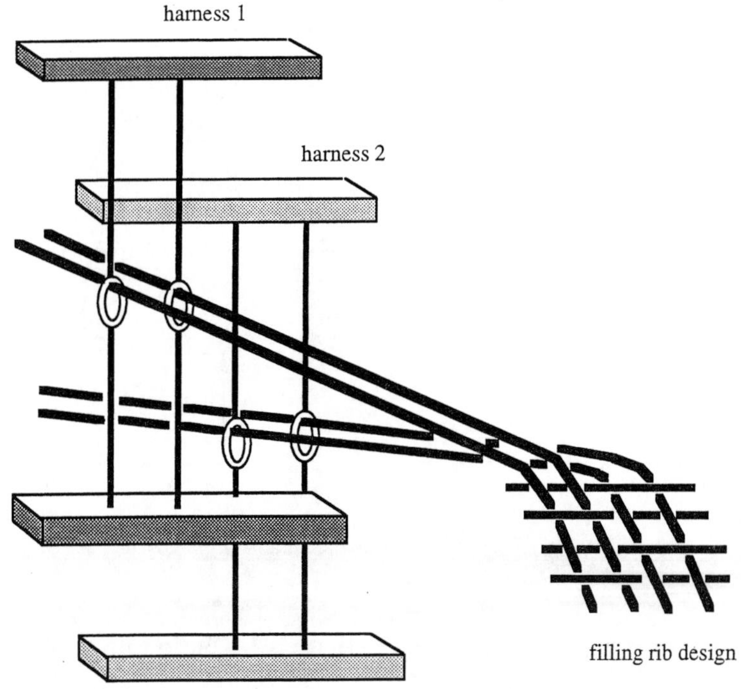

FIGURE 5.6 Filling rib obtained by changing the entering order of Figure 5.5.

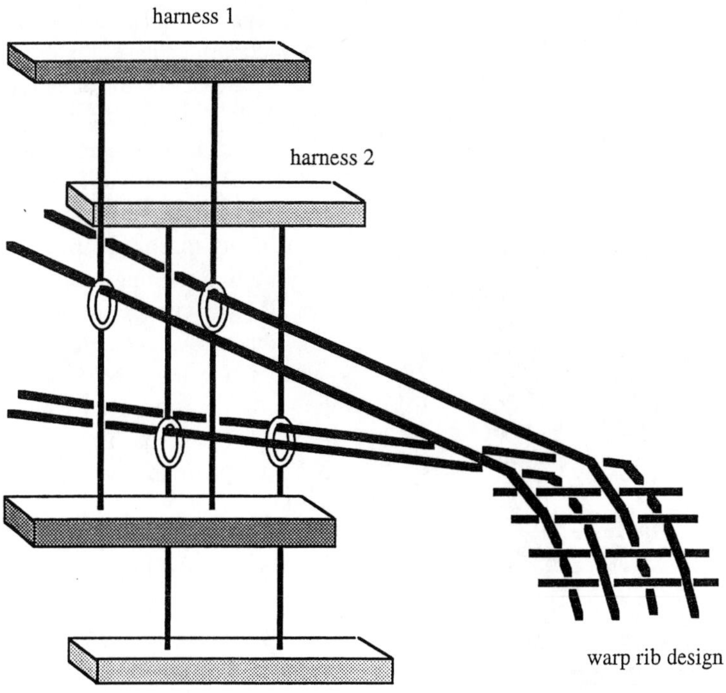

FIGURE 5.7 Warp rib obtained by changing the lifting order of Figure 5.5.

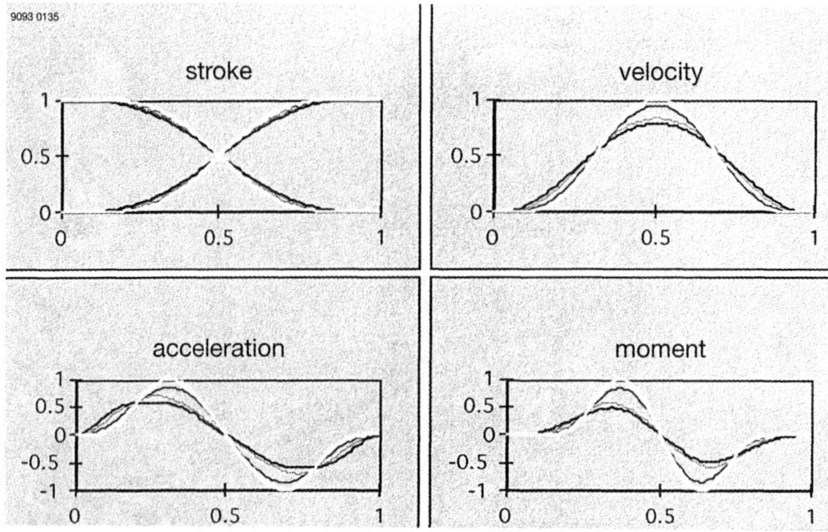

FIGURE 5.8 Motion diagrams of harnesses.

Figure 5.10 shows various types of open-end loop heddles for riderless frames and Figure 5.11 shows heddles with closed end loops for normal frames.

Automatic harness leveling devices are used on weaving machines with cam motions to equalize tension on warp yarns during a machine stop. This eliminates start up marks in the fabric. It also makes warp yarn repairs easier.

Shedding mechanisms of major weaving machine types are included in the following chapters.

FIGURE 5.9 Carbon composite frame slat profiles for harnesses (courtesy of Burckle).

5.1.3 Filling Insertion

After each shed change, the filling yarn is inserted through the shed as shown in Figure 5.12. It is possible to select and insert different filling yarns one after another. These filling yarns can be of different color, weight, etc., and a selection mechanism is used for this purpose. Depending on the machine type, several different filling yarns can be used in the same fabric. The selection mechanism presents the proper filling yarn to the yarn carrier for insertion of each yarn.

Weaving machines are usually classified according to the filling insertion mechanism. Figure 5.13 shows various classifications of weaving machines. The major filling insertion systems that are used today are air-jet, rapier, projectile and water-jet which are called shuttleless weaving machines (Figure 5.14). M8300 is a multiphase air-jet weaving machine (Chapter 11).

A gripper projectile transports a single filling yarn into the shed (Figure 9.2). Energy required for picking is built up by twisting a torsion rod. On release, the rod immediately returns to its initial position, smoothly accelerating the projectile through a picking lever. The projectile glides through the shed in a rake-shaped guide, braked in the receiving unit, the projectile is

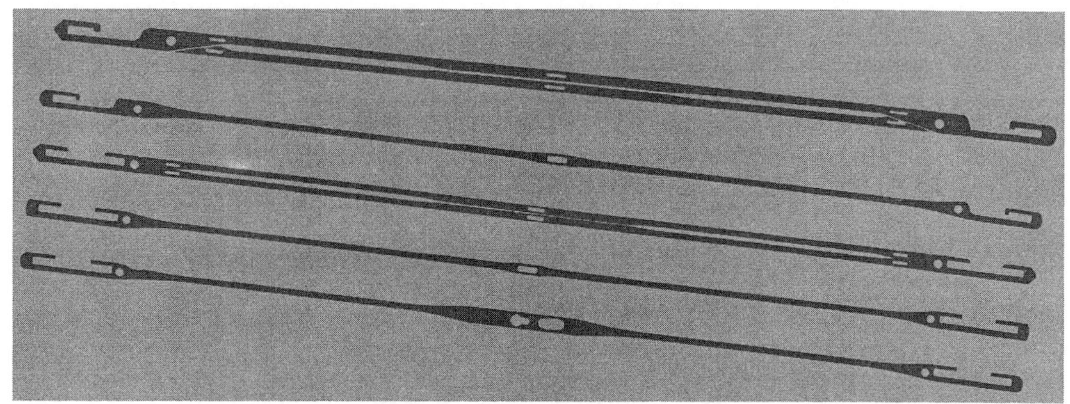

FIGURE 5.10 Open end loop heddles (courtesy of Burckle).

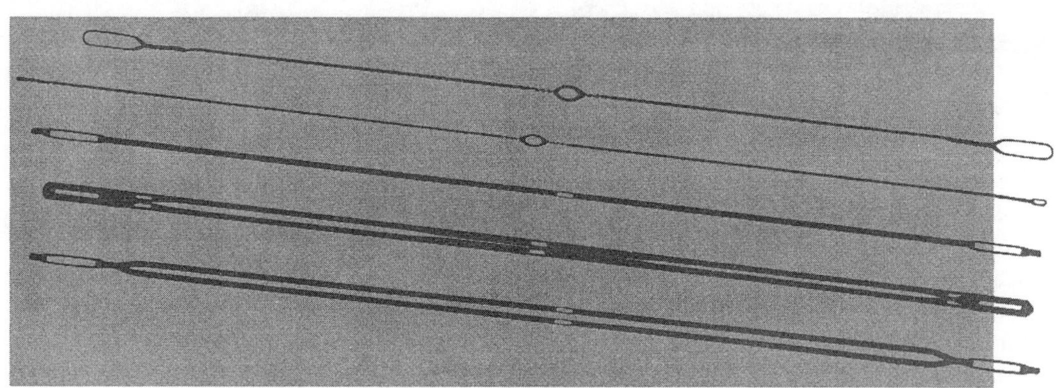

FIGURE 5.11 Closed end loops (courtesy of Burckle).

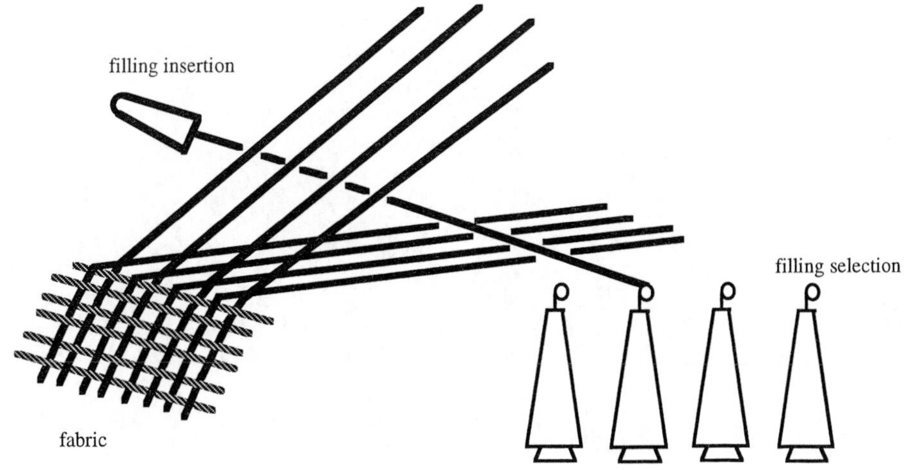

FIGURE 5.12 Selection and insertion of filling yarns.

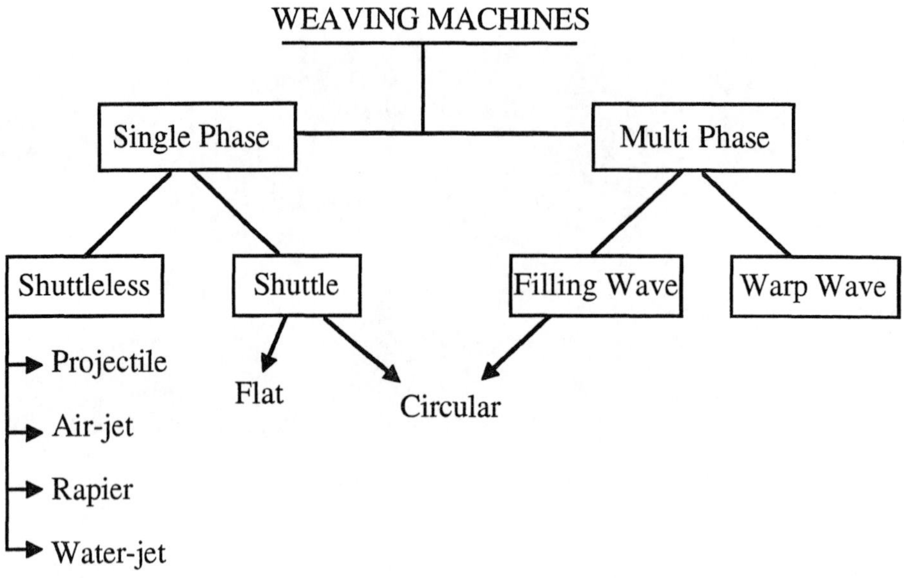

FIGURE 5.13 Classification of weaving machines.

then conveyed to its original position by a transport device installed under the shed. The projectile's small size makes shedding motions shorter which increases operating speeds over wide widths of fabric, often weaving more than one panel of fabric with one insertion mechanism.

Figure 10.4 illustrates filling insertion by two flexible rapiers with filling carriers, a giver and a taker. The filling is inserted half way into the shed by one carrier and taken over in the center by the other carrier and drawn out to the opposite side of the fabric. A spatial crank gear drives the oscillating tape wheels to which the rapier tapes are attached. In the shed, the tapes move without guides. The grippers assume the correct clamping position automatically. Different versions of rapier insertion systems are explained in Chapter 10.

The most popular method of filling insertion is illustrated in Figure 8.1 where a jet of air is used to "blow" the filling yarn into the shed. This small mass of insertion fluid enables the mechanism to operate at extremely high insertion rates. The picks are continuously measured and drawn from a supply package, given their initial acceleration by the main air nozzle and

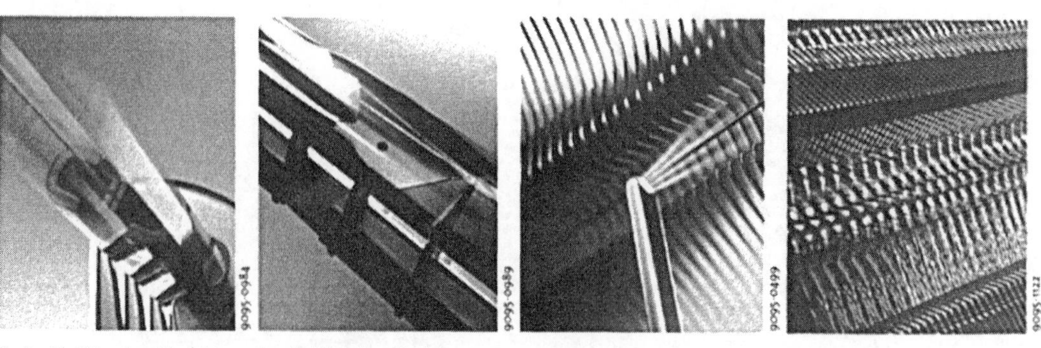

FIGURE 5.14 Major shuttleless weaving systems.

boosted or assisted across the fabric width by timed groups of relay air nozzles. The other fluid system uses water as the insertion medium, but the use of a water-jet is generally limited to hydrophobic yarns such as nylon or polyester filament.

A shuttle loom uses a shuttle to store and carry the yarn back and forth across the loom (Chapter 7). Shuttle looms have become obsolete in manufacturing of traditional woven fabrics due to several reasons, including low production rate, high noise, safety concerns, limited capabilities, etc. Nevertheless, the shuttle loom is still used as a reference point for the modern shuttleless weaving machines. Besides, some industrial woven fabrics are still being made on specially designed shuttle looms.

Filling insertion mechanisms of major weaving machine types are included in the following chapters.

Yarn Feeders

Yarn feeders or accumulators are used to wind a predetermined yarn length to make it ready for insertion. Their main purpose is to supply filling yarn to the weaving machine smoothly and at a constant and proper tension. There are various types of feeders used (Figure 5.15). Typical characteristics of these machines are:

- one-step pneumatic threading up
- integrated yarn break sensor
- floating element yarn store sensors
- stepless yarn separation adjustment
- microprocessor controlled speed and motor effect
- sealed spool body
- S/Z rotation shifting
- insertion speeds up to 2000 m/min
- serial communication interface

The selection of a feeder depends on several factors:

- maximum speed delivered
- yarn count
- winding direction (S or Z)
- yarn reserve control

Maximum speed depends on the yarn count range. Reserve control can be done mechanically or electronically by means of photocells. The threading through the feeder can be done manually or pneumatically. The tensioning of the yarn is controlled by a breaking device which can be of different types including bristle, metal lamella, flex brake and coaxial output tensioner (CAT). Figure 5.16 shows the membrane and the endless beryllium copper tensioning strip. The flex is used to replace the brush ring and output tensioner in conventional brake systems. In CAT system (Figure 5.17), the yarn travels through two tensioning discs mounted in the feeder nose. An adjustable tensioning spring regulates the base force exerted by the discs which

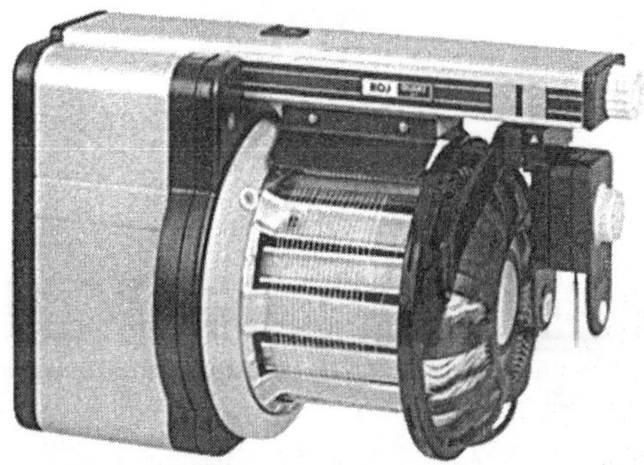

FIGURE 5.15 Yarn feeders for rapier and projectile weaving machines (courtesy of Nuova Roj Electrotex).

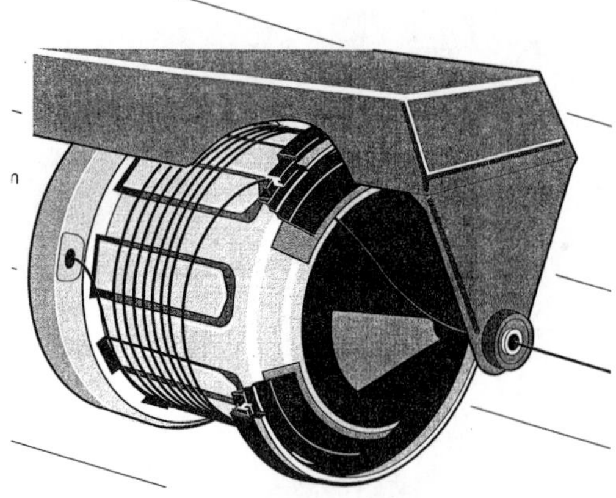

FIGURE 5.16 Flex brake (courtesy of IRO).

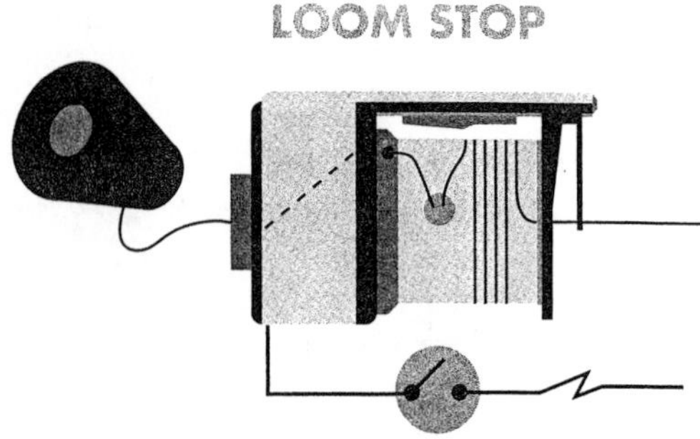

FIGURE 5.18 Schematic of feed detector to stop the weaving machine (courtesy of Nuova Roj Electrotex).

allows the setting and maintaining of tension levels. During the filling insertion process, the CAT compensates yarn tension fluctuations. Filling breakage at the feeder entry is detected electronically to stop the weaving machine (Figure 5.18).

For heavy yarns, a balloon breaker can be fitted in front of the feeder instead of the normal eyelet (Figure 5.19). Figure 5.20 shows the path of the yarn inside a typical feeder.

During weaving of fine woolens and linen yarns, usually a lubricant is used which is supplied by a liquid dispenser. The purpose of the lubricant is to reduce filling breakages and increase weaving machine speed and weaving efficiency. A liquid dispenser is placed between filling package and feeder as shown in Figure 5.21 which allows an even distribution of liquids, wax, oil, moisturizers and anti-static lubricants on filling yarns during weaving. The filling yarn is coated when it passes over a motor driven rotating cylinder that is immersed in a liquid reservoir.

To improve the fabric appearance, i.e., to compensate yarn count fluctuations and color irregularities, a 1–1 filling insertion from two bobbins instead of filling insertion from only one bobbin is recommended.

5.1.4 Beat-Up

When the filling yarn is inserted through the shed, it lies relatively far from its final position. This is because the insertion device (air-jet, projectile, rapier, etc.) cannot physically fit at the acute angle of the shed opening. This final position is called fell which is the imaginary line where the fabric starts. Therefore, the newly inserted filling yarn needs to be brought to its

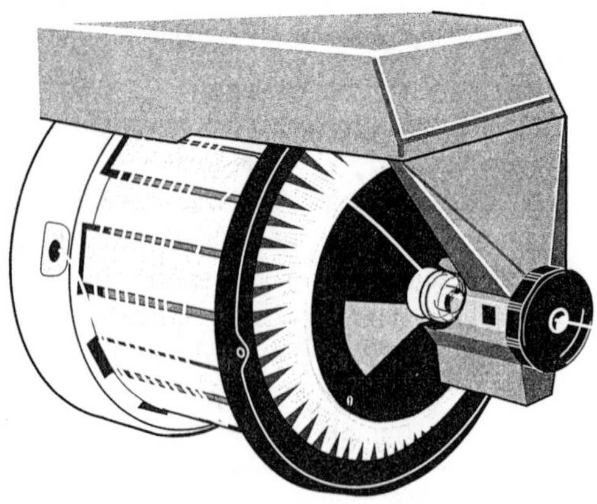

FIGURE 5.17 Coaxial output tensioner (courtesy of IRO).

FIGURE 5.19 Balloon breaker for heavy yarns (courtesy of Nuova Roj Electrotex).

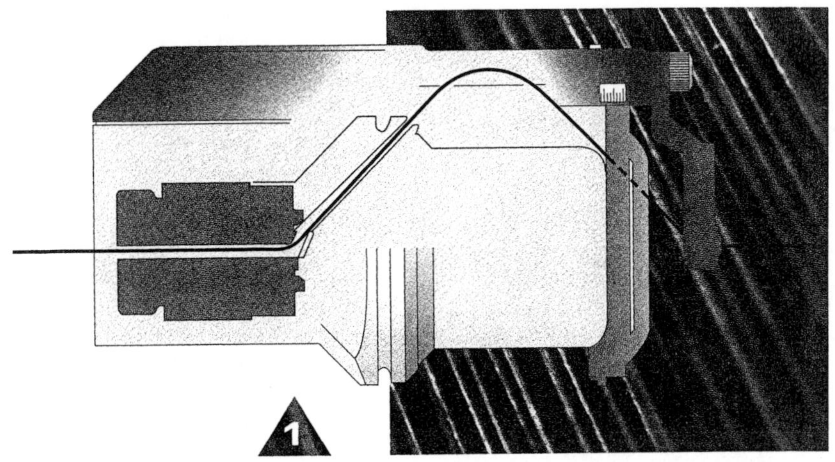

FIGURE 5.20 Yarn path inside a feeder (courtesy of Nuova Roj Electrotex).

FIGURE 5.21 Liquid dispenser (courtesy of Savitec).

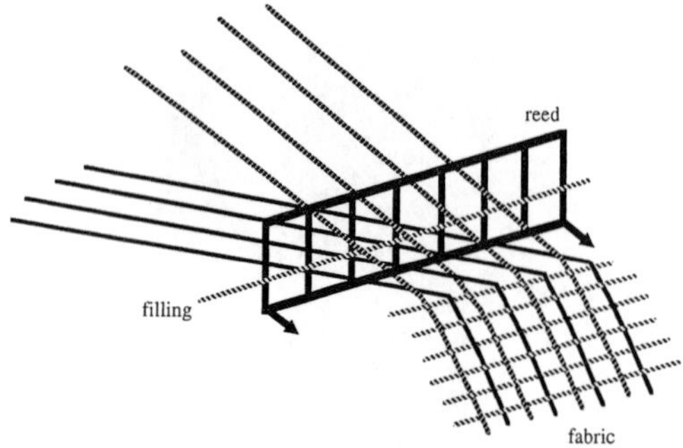

FIGURE 5.22 Schematic of beat-up process.

final position by pushing through the warp sheet. Beat-up is the process of pushing the last inserted filling yarn to the cloth fell by using a device called reed as shown in Figure 5.22. For all practical purposes, the fabric is not formed until beat-up occurs.

Reed is a closed comb of flat metal strips (wires). These metal strips are uniformly spaced at intervals that correspond to the spacing of warp ends in the fabric; therefore, the reed is also used to control warp yarn density (closeness) in the fabric. Warp density is expressed as either ends per inch (epi) or ends per centimeter (epc), which affects the weight of the fabric. The spaces between the metal strips are called "dents". The reed holds one or more warp yarn(s) in each dent and pushes them to the cloth fell. After beating up the filling, the reed is withdrawn to its original rest position before the insertion of the next pick. Figure 5.23 shows a regular reed and a profiled reed. Profiled reed is used in air-jet weaving machines. In shuttle looms, the reed also guides the shuttle.

The shape and thickness of the metal wires used in the reed are important. Reed selection depends on several considerations including fabric appearance, fabric weight (ends per unit width), beat-up force, air space requirement and weave design.

Reeds are identified by a "reed number" which is the number of dents per unit width. Specifying the number of ends per dent with a certain reed number dictates the construction (density) of ends per inch in the fabric on the loom. It should be noted that interlacing causes a natural contraction of yarns in the fabric such that density of warp ends off the loom will be higher than in the reed; generally about 5% higher depending on the weave, tensions and yarn sizes involved.

There is a close interaction between shedding and beating which may be dictated by the yarn type and weave. The shedding and beating actions need to be properly timed and synchronized for the most effective filling insertion with minimum warp tension. In general, the beat-up is done on an open shed for filament yarns as shown in Figure 5.24 which also shows that the beat-up is usually done on a crossed shed with staple fiber warp yarns. It is rare to beat-up the filling yarn at the time the warp sheets cross.

Figure 5.25 shows the beating mechanism of a projectile weaving machine. For high speed operation, light-weight parts are used for beating motion. The rigidity of the parts should be high and the beating stroke should be short. In some

FIGURE 5.23 Regular reed (bottom) and profiled reed (courtesy of Burckle).

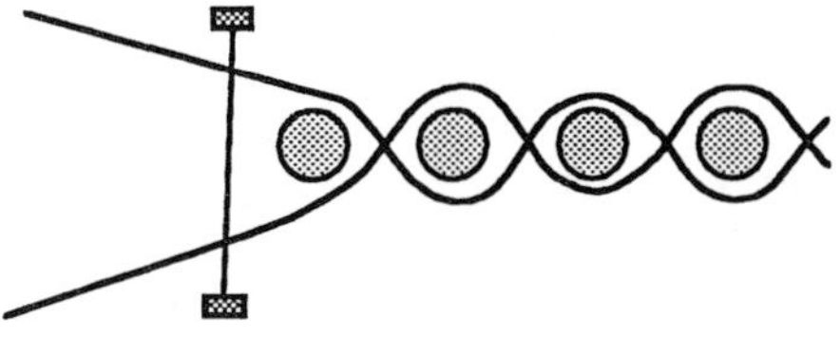

a) beating up on an open shed

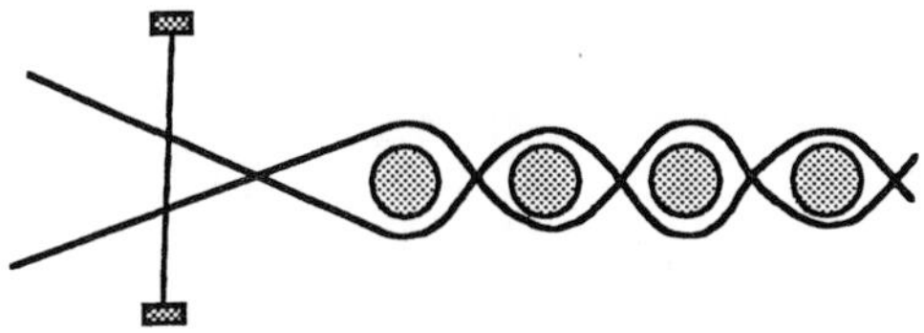

b) beating up on a closed shed

FIGURE 5.24 Types of beating-up.

terry machines, the technique of two beat-up positions of the reed is used as shown in Figure 5.26.

During beat-up, the filling yarn is pushed against frictional forces of the warp. Beating-up of filling yarn requires considerable amount of force. The frictional force depends on the coefficient of friction between the warp and filling. Another reaction arises from the bending of warp and filling yarns due to crimp interchange. In order to overcome these reactions, the pushing of the filling yarn into the cloth fell is done in a rather harsh manner which gives the action its name, i.e. beating-up. The beat-up force increases as the beating proceeds towards the cloth fell.

The beat-up process is very complex. The warp tension increases and the fabric tension decreases as the filling yarn is being pushed into the cloth fell. Figure 5.27 shows the movement of yarns during beat-up. The fabric area near the cloth fell does not represent the actual fabric structure because the distances between the yarns are not uniform. When the reed returns after beat-up, warp tension exerts a force on the last few picks and these picks tend to go back toward the warp beam. This is balanced out by the frictional restraint of the filling yarns. When the next beat-up cycle occurs, the new filling yarn is inserted and pushed into the cloth fell by the reed. The yarns which have previously slipped back are pushed into the fell again. The

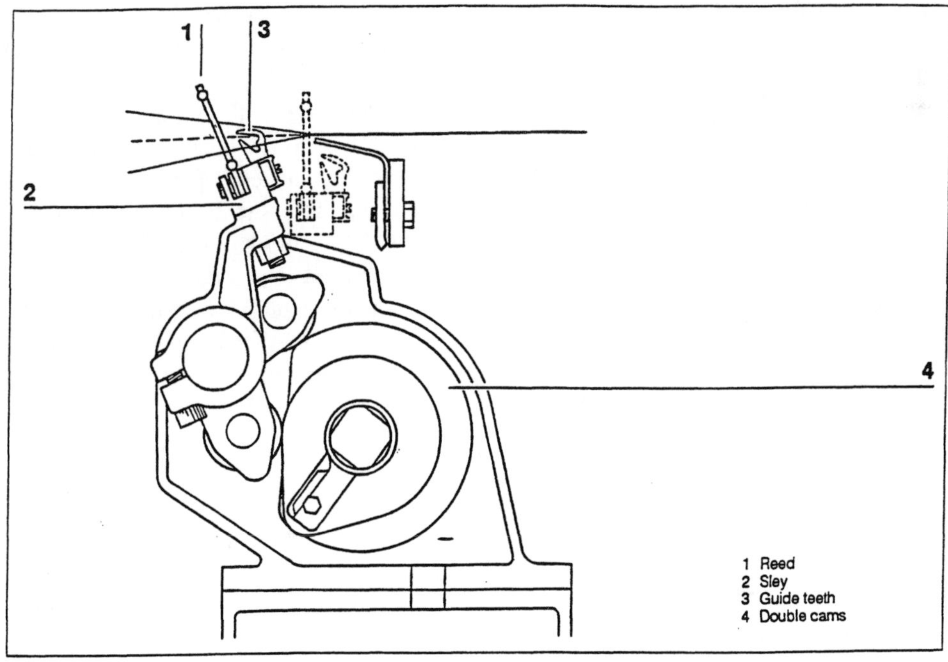

FIGURE 5.25 Beating mechanism of a projectile machine.

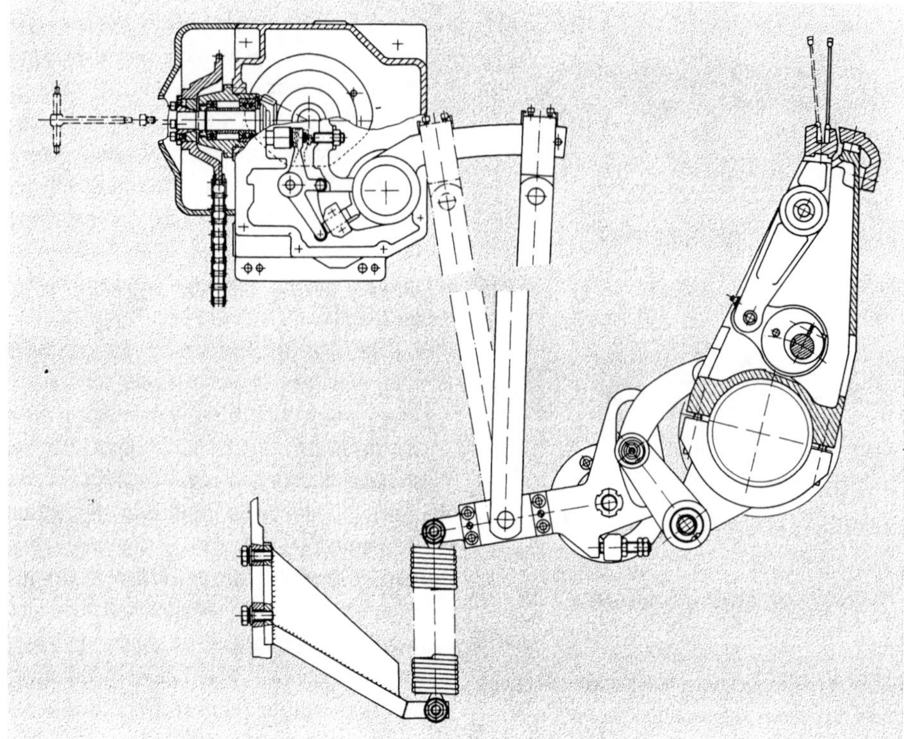

FIGURE 5.26 Double beat-up position for terry weaving.

shed is unbalanced in most situations to obtain closer pick spacing and a better fabric appearance. This unbalanced shed develops different tensions in the upper and lower sheds. The yarns in the upper shed are shorter and have less tension and the yarns in the lower shed are longer which creates more tension. This could ultimately affect the quality of the fabric.

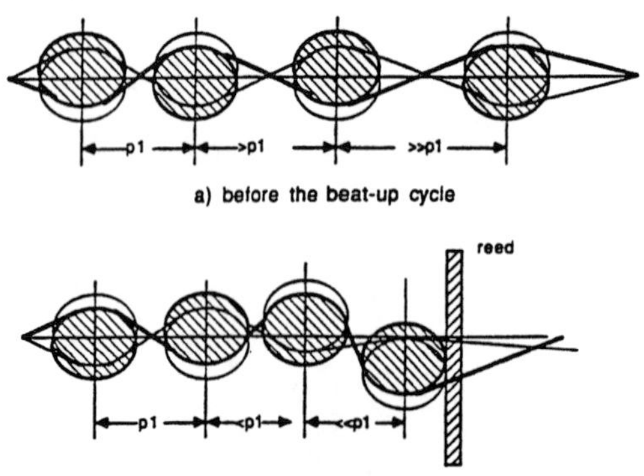

FIGURE 5.27 Filling yarn spacing during beat-up [1].

Normally, the fabric fell is in equilibrium and the position of the fell does not change from pick to pick. However, any interruption in weaving will cause the fell position to change and this in turn distorts the motion of beat-up. Variations in the cloth fell cause irregularities in the pick spacing in woven fabrics, causing a defect.

Several factors tend to affect the position of the cloth fell during beat-up. These include warp and fabric tensions, weaving machine speed and shed motion during beat-up. Also, there are forces that evolve during the beat-up process. Some of these are the beat-up force, warp and fabric tensions and weaving resistance.

Figure 5.28 shows the forces in beating [2]. T_1 and T_2 are the tension forces on the warp yarn. R is the reaction force. R' is the reactions from the opposite interlacings. As the filling moves towards the fell, the angle β steepens, the magnitudes of R and R' increase, and α gets smaller. When the filling is moved far enough into the fell, the angle α becomes so acute that the filling would be squeezed out if it were not

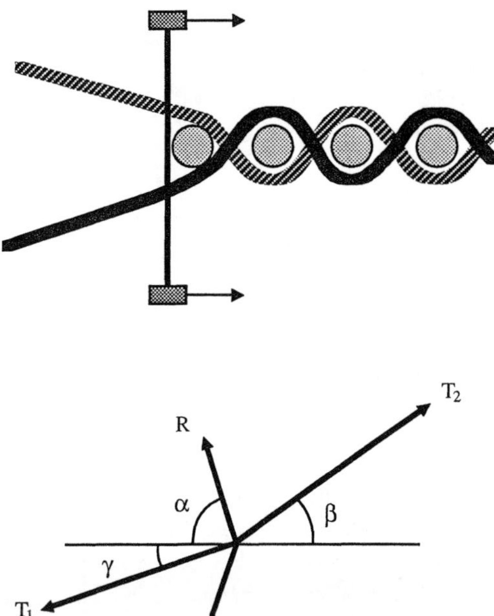

FIGURE 5.28 Forces in beating on an open shed [2].

restrained. Thus, there is a critical value for α beyond which the filling yarn cannot be pushed any more forward in case of open shed. This critical value depends on the coefficient of friction between the filling and warp and it determines the minimum pick spacing that can be obtained by beating on an open shed. When beating is done on a crossed shed, there is a smaller reaction to squeeze the filling yarn out. As a result, closer pick spacings can be obtained in a cross shed.

The beat-up forces are affected by crimp levels, the yarn dimensions and the filling spacing. As the filling yarn is forced into the cloth fell, the beat-up becomes more and more difficult. If the beat-up force is increased beyond a certain magnitude to obtain a close pick spacing, the beat-up force is taken by the warp and the fabric becomes slack. This situation is known as "bumping" which indicates a "jammed" fabric. If jamming condition is reached, the pick spacing cannot be reduced any further.

In beat-up, energy is required to beat the last pick into position. The energy required is dependent upon many factors. The main factors are fabric construction and width, but tension is also an important factor. An increase in the warp or filling tension will lead to an increase in the amount of energy required for beat-up. Also, the shed timing is important in determining the required energy. It determines the tension distribution between the top and the bottom sheds. The closer the distribution is to being equal the more energy required [3].

Spun warps are timed to interlace the filling yarn earlier, increasing the force of beat-up and the stress on warp yarns to enhance fabric appearance (Figure 5.29). These extra stresses generally are not applied to filament warp yarns since the uniformity of yarn dyeability would be detrimentally affected.

Pick spacing is an important factor in beat-up. As stated earlier, irregular pick spacing caused by variations in the fabric fell causes defects. These defects result from loom stoppages. If a loom stops for some reason such as a broken end, and, if it is started without the fell being corrected, a start-up line may be developed. This is where the previous picks were not fully beaten into the fabric fell [1]. It was also shown that beat-up force increases when pick spacing decreases [4].

5.1.5 Take-Up

As the fabric is woven, it should be removed from the weaving area. This is achieved by the take-up motion. The fabric take-up removes cloth at a rate that controls filling density [picks per inch (ppi) or picks per centimeter (ppc)]. Two factors determine filling density: weaving machine speed and rate of fabric take-up. Gener-

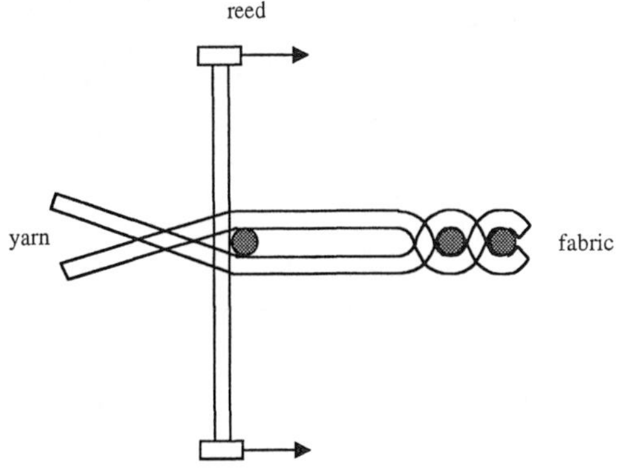

FIGURE 5.29 Early shed timing.

ally, the pick insertion rate of a weaving machine is fixed at the time of purchase based on the range of fabrics it is intended to produce, the type of insertion mechanism and the weaving machine width. Weaving machine speed is expressed as picks per minute (ppm) and rate of take-up as inches per minute (ipm) or centimeter per minute (cm/min). Warp density and filling density together are referred to as the "construction" of the fabric. Figure 5.30 shows the main parameters of fabric construction and manufacturing. The following relations exist:

Reed Number = dents/inch or dents/cm (5.2)

Weft Density (picks/in or picks/cm) =
Machine Speed (picks/min)/
Take-up Speed (in/min or cm/min) (5.3)

Warp Density (end/in or end/cm) =
Reed Number × (ends/dent) (5.4)

Construction =
Warp Density × Weft Density (5.5)

It should be emphasized that both the ends and picks contract because of interlacing causing construction on the loom and off the loom to be different. Subsequent fabric finishing steps also introduce changes in the fabric construction which must be considered in setting up loom specifications.

Take-up mechanisms of major weaving machine types are included in the following chapters.

5.2 AUXILIARY FUNCTIONS

In addition to the five basic motions of a loom, there are many other mechanisms on typical weaving machines to accomplish other functions. These include:

- a drop wire assembly, one wire for each warp yarn, to stop the machine when a warp end is slack or broken (Chapter 4)
- a tension sensing and compensating whip roll assembly to maintain tension in the warp sheet
- a mechanism to stop the machine when a filling yarn breaks
- automatic pick finding device reduces machine downtimes in case of filling yarn breakages
- filling feeders to control tension on each pick
- pick mixers to blend alternate picks from two or more packages

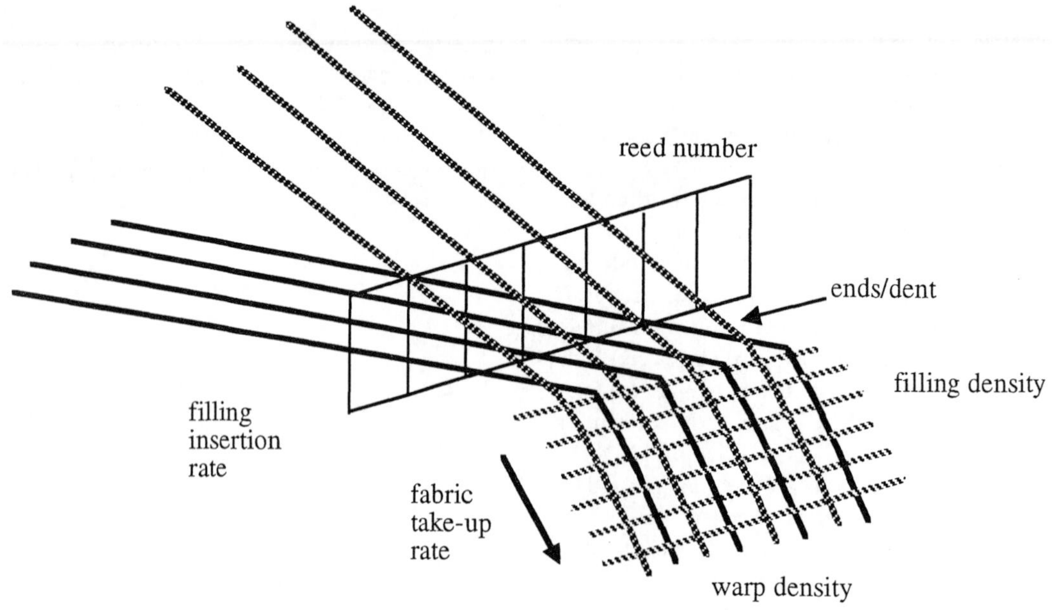

FIGURE 5.30 Fabric construction and manufacturing parameters (courtesy of Johnston Industries).

- filling selection mechanism for feeding multi-type filling patterns
- filling selvage devices such as trimmers, tuckers, holders and special weave harnesses for selvage warp ends
- filling replenishment system to provide uninterrupted filling insertion by switching from a depleted to a full package
- a temple assembly on each selvage to keep fabric width at the beat-up as near the width of the warp in the reed as possible
- sensors to stop the machine in the event of mechanical failure
- a centralized lubrication control and dispensing system
- a reversing mechanism to avoid bad start ups after a machine stop
- a color coded light signal device to indicate the type of machine stop from a distance
- a production recording system

These auxiliary functions are described for each type of weaving machine in the following chapters.

5.3 FABRIC CONTROL

5.3.1 Fabric Width

At the moment it is woven, the fabric width is equal to the reed width as shown in Figure 5.31. However, as the weaving continues and fabric gets away from the reed, the fabric starts narrowing due to several factors (it should be noted that there are certain fabrics which do not get narrower, e.g. glass fabrics). These are weave design, fabric construction and weaving tensions. The interlacing pattern of the weave design affects the crimp level in the fabric and crimp on the filling yarn causes the fabric to contract in width direction. Fabric construction, i.e., the number of weft and warp yarns per unit length, also affects fabric crimp and therefore fabric width. High weaving tensions, especially in the warp yarns, cause fabric to shrink. Warp yarns closest to the selvages of the fabric undergo more stress due to widthwise contraction of the fabric toward the center, causing linear angular displacement of these outermost yarns.

The narrowing of fabric width should be prevented, by using a temple on each side of the machine. Control of fabric contraction by the temples of the machine is another critical aspect of good weaving performance. A temple is a metallic device that keeps the fabric stretched by applying a force along the filling direction. There are various temple types as shown in Figure 5.32. It is also possible to have a temple across the full width of the fabric as shown in Figure 5.33. Full width temples ensure uniform fabric quality over the entire weaving width with delicate fabrics and easier operation. The full temple has the following advantages:

- uniform warp and weft tension over the entire width
- uniform fabric characteristics over the entire width
- no fabric drawing defect
- no damage to fabric by needle rings
- rapid changeover from full width to cylindrical temples

Appendix 4 shows more temple types used in the industry.

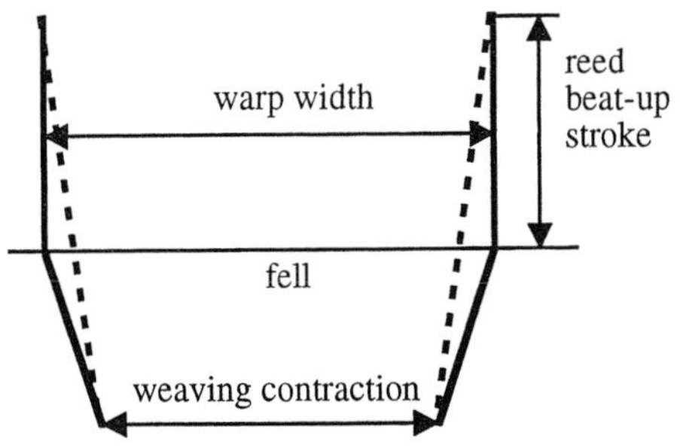

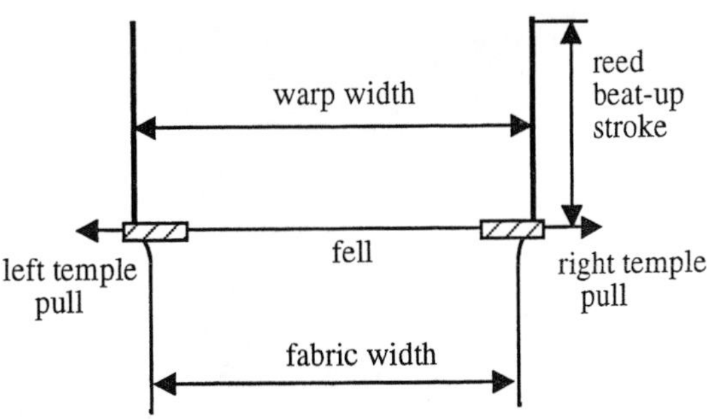

FIGURE 5.31 Temple function in weaving.

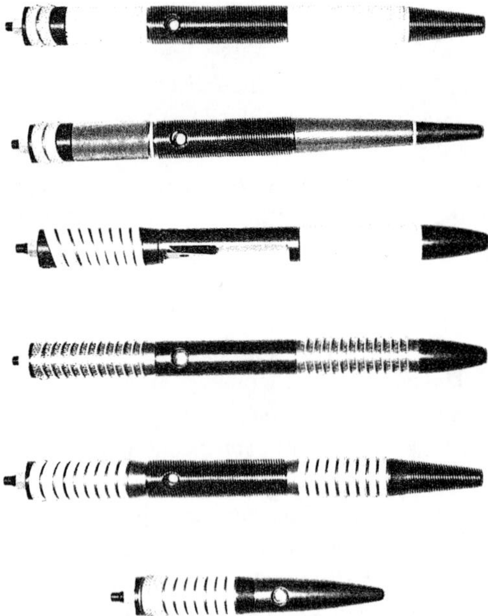

FIGURE 5.32 Examples of temples used in weaving machines (courtesy of Broll).

5.3.2 Selvages

Selvages (also called selvedges) provide strength to fabric for safe handling of the fabric. Selvage should not curl. In shuttle looms, there is no need for special selvage; since the yarn is not cut after each filling insertion, the edges of the fabric are smooth and strong (Figure 5.34).

In shuttleless weaving, since the pick yarn is cut after every insertion, there is fringe selvage on both sides of the fabric (Figure 5.34). In this case, special selvages are needed to prevent slipping of outside warp yarns out of the fabric. There are several types of selvage designs that are used for this purpose with shuttleless looms.

FIGURE 5.33 Full width temple.

In tucked-in selvage, the fringed edges of the filling yarns are woven back into the body of the fabric using a special tuck-in mechanism. As a result, the filling density is doubled in the selvage area (Figure 5.34). Tucked-in selvage was being only used for projectile weaving machines in the past, however, it is now also applied to other shuttleless weaving machines. Pneumatic tucking units are also available.

When setting up for the selvages on a projectile weaving machine, the following points must be noted.

- The selvage must be drawn into the reed 15 mm wide.
- The selvage must not be thinned too much.
- The reed must be filled with yarns up to the last dent.

If possible, the selvages are always drawn-in on separate harnesses. The selvage harnesses are always behind the ground harnesses, so that the front shed is shorter. This arrangement enables the shed to be adjusted smaller.

In leno selvage, a leno design at the edges of the fabric locks the warp yarns in (Figure 5.34). Half cross leno weave fabrics have excellent shear resistance. They are made with special leno weaving harnesses.

Electronically controlled thermal cutters are used to cut and fuse selvages of synthetic fabrics on weaving machines. The temperature of the cutters is reduced when the machine is stopped. Figure 5.35 shows schematic of a thermal cutter application.

Fabric Inspection Lines

After weaving, some fabrics are inspected on the weaving machine for quality purposes (Figure 5.36). Inspection speed can be varied between 0 to 100 linear meters per minute. Inspection machines have a lighted diffusion screen. Fabric alignment is controlled by a mobile trolley operated by photocells to sense the cloth.

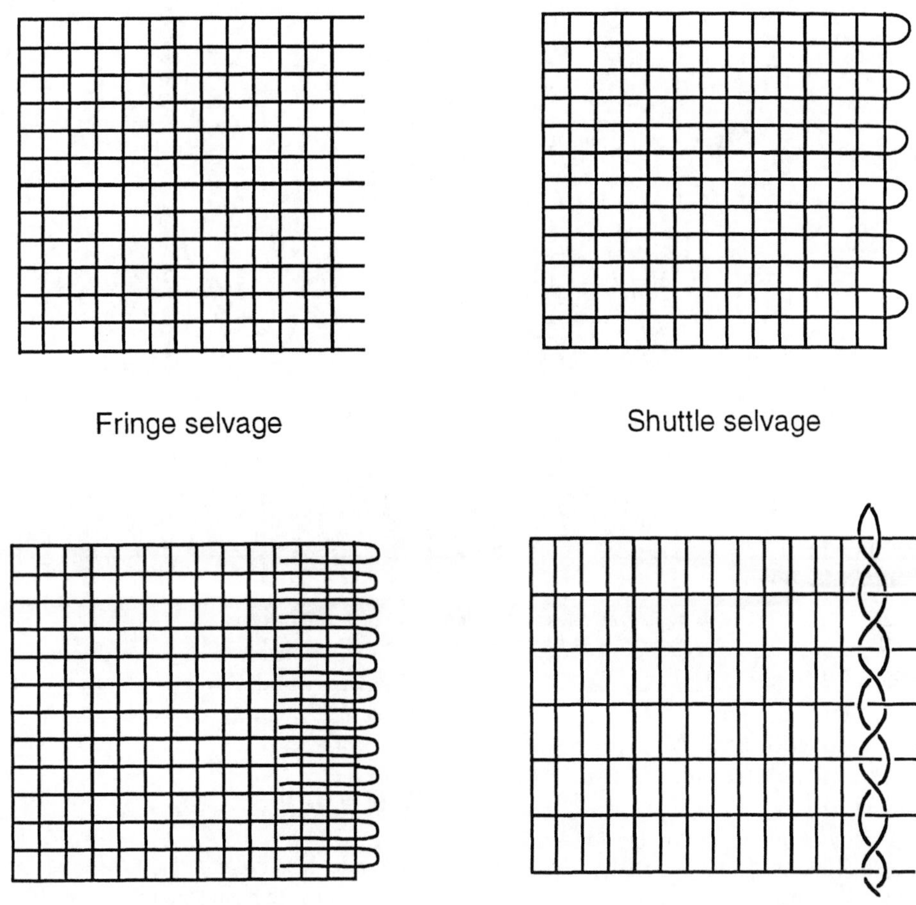

FIGURE 5.34 Fabric selvages.

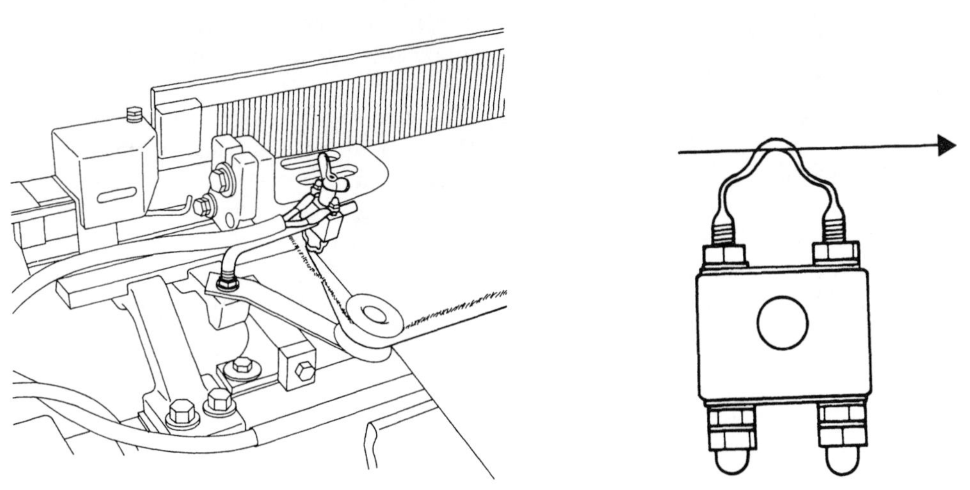

FIGURE 5.35 Thermal cutter (courtesy of Loepfe).

FIGURE 5.36 Fabric inspection (courtesy of Formia Nuova s.r.i.).

REFERENCES

1. Zhang, Z., and Mohamed, M. H., "Theoretical Investigations of Beat-up", Textile Research Journal, July 1989.
2. Lord, P., and Mohamed, M. H., Weaving: Conversion of Yarn To Fabric, Merrow Technical Library, 1982.
3. Greenwood, K., and McLoughlin, W. T., "The Design and Operation of a Loom with Negative Beat-up", Shirley Institute Memoirs, Vol. XXXVIII, Shirley Institute, Didesbury, Manchester, 1965.
4. Shih, Y. et al, "Analysis of Beat-up Force During Weaving", Textile Research Journal, December 1995.

SUGGESTED READING

- Marks, R., and Robinson, A., Principles of Weaving, The Textile Institute, 1976.
- Weinsdorfer, H., and Salama, M., "Measuring the Movement of the Fell of the Cloth During Weaving with the Aid of a High Speed Video System", Textil Praxis International, September 1992.

REVIEW QUESTIONS

1. Explain the advantages and disadvantages between negative and positive let-off mechanisms.
2. Derive the formula for harness lifts in an unsymmetric shed.
3. What are the effects of the position of the zero line on fabric properties? Explain.
4. Explain how order of entering and order of lifting affect the fabric structure on a weaving machine.
5. Why are the yarn feeders necessary? Explain.
6. Derive a formula to calculate the beat-up force. Make the necessary assumptions.
7. Name five auxiliary motions in a typical weaving machine.
8. Why does a fabric continuously narrow on a weaving machine if no temple is used? Explain.

6

Shedding Systems

Every weaving machine provides a control device for each warp yarn. Heddles controlling warp yarns that always follow the same interlacing pattern are grouped together into a common frame called a harness. There must be a different harness provided for each group of warp yarns that follow a different weaving pattern. In the case where every end weaves a different pattern, a harness cord is provided for each heddle.

There are four systems used to provide manipulation to the warp yarns:

- crank shedding
- cam (tappet) shedding
- dobby shedding
- jacquard shedding

Crank, cam and dobby mechanisms control the harnesses; jacquard system provides control of individual warp yarns. Each shedding mechanism can be mounted on any weaving machine. Dobby and jacquard systems can be mechanical or electronic. There are many variations of cam, dobby and jacquard shedding mechanisms. For the purpose of this book, only the major groups of these mechanisms will be discussed which will be limited to the elementary principles of these shedding mechanisms. Table 6.1 shows the major characteristics of the most common shedding systems.

6.1 CRANK (TREADLE) SHEDDING

This is the simplest and least expensive shedding system. In this system, the harness motion is provided by the crank shaft of the weaving machine. A wheel is rotated a half turn for each crank shaft revolution. The harness is linked to the wheel through a lever arm and a drive pin. In each weaving cycle, the harness changes its position, therefore, this system is used only for plain weave and its derivatives. These systems are used in air-jet and water-jet machines where speed is generally high.

6.2 CAM SHEDDING

Cams with weave pattern profiles rotate to deliver lifting and/or lowering instructions to harnesses. A typical cam system can handle weave patterns with up to 14 different harnesses. Cam shedding mechanisms are relatively simple and inexpensive to design and maintain, they are more reliable for producing fault free fabric and they do not restrict the weaving machine speed. A pair of cams is sufficient to weave a plain fabric. The main disadvantage of the cam shedding mechanisms is their restricted patterning possibilities. Another disadvantage is that, when the weave has to be changed, it is

TABLE 6.1 Characteristics of shedding mechanisms (Staubli).

	Repeat Length	No. of Harnesses	Pitch (mm)
Cam:			
1. Positive cam motion	up to 8 picks	12	18
2. Negative cam motion			
Dobby:			
1. Positive mechanical dobby	6000 picks	28	12
2. Negative mechanical dobby	150 picks	16	12
3. Mechanical rotary dobby	4700 picks	28	18
4. Electronic negative dobby	6400 picks	16	12
5. Electronic rotary dobby	6400 picks	28	12
	Program Memory Capacity		No. of Hooks
Jacquard:			
1. Mechanical double lift jacquard			2688
2. Electronic double lift jacquard	1,800,000 picks		2688
3. Electronic CX jacquard			6144

usually necessary to change or rearrange the cams which is time-consuming and not practical for frequent pattern changes.

6.2.1 Cam Design

A cam is a disk that transforms a rotational motion of its own to a reciprocating motion of a follower. The transfer is done by means of the cam's edge or a groove cut in its surface as shown in Figure 6.1. The design and functions of cams are extensively explained in mechanical engineering literature [1], therefore, the discussion of cam design will be limited to the weaving machines only for the purpose of this book.

The size of the weave repeat in cam shedding is limited by the maximum practicable number of picks to the repeat. This can be explained using Figure 6.2. Assuming that a fabric with eight yarns in a unit cell will be woven with a negative cam system, a total of eight cams will be required which will be mounted on a shaft as shown in the figure. In the diagram, one pick occupies one-eighth of the revolution or 45°. If

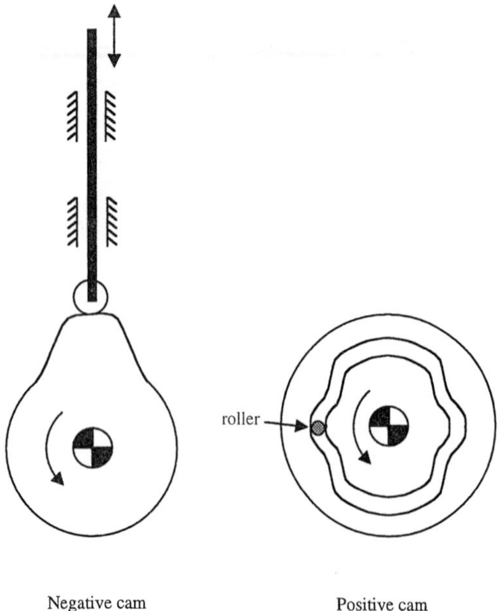

FIGURE 6.1 Cam design.

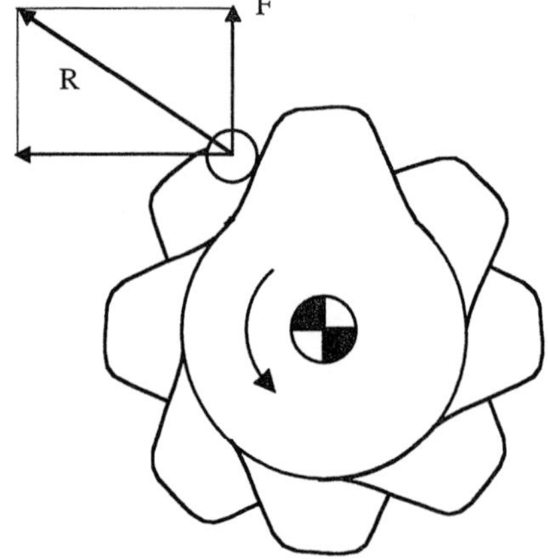

FIGURE 6.2 Arrangements of cams on a shaft.

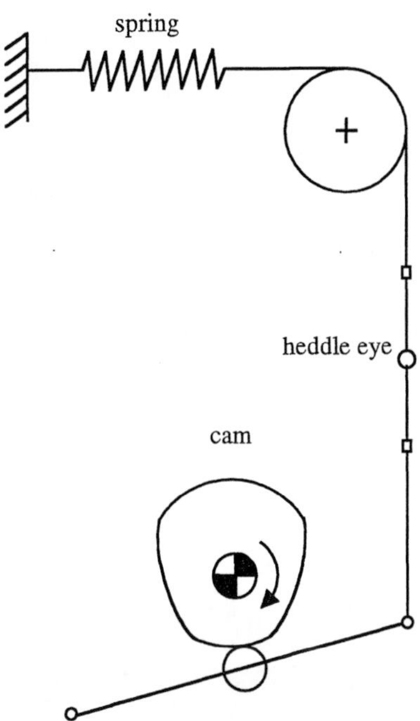

FIGURE 6.3 Schematic principle of negative cam shedding.

the number of cams is increased, then there would be less space for each cam for dwell. This also means that the slope of the cam contour will increase which will increase the maximum force acting in the system. In the figure, to produce a vertical force F to lift the harness, the cam must apply a force R on the cam follower. As the slope of the cam contour increases, the force R also increases. The cams are designed to give a simple harmonic motion to the cam follower for a smooth operation. The cam follower moves in a vertical imaginary line that passes through the axis of the cam shaft. To avoid excessive force in the system, for a given cam size, the maximum slope of the cam contour should be minimized which requires low number of yarns per unit cell in the fabric.

There are two types of cam mechanism in weaving machines: negative cam and positive cam systems.

6.2.2 Negative Cam Shedding

The harnesses are either raised or lowered by the cam mechanism but they are returned by the action of some external device. Figure 6.3 shows the principle of a negative cam mechanism. In negative cam shedding mechanisms, some form of spring reversing motion is used with separate springs for each harness. Negative cam shedding is being used less and less in modern weaving machines.

6.2.3 Positive Cam Shedding

In positive cam shedding, the harnesses are both raised and lowered by the cams. There are two main types of positive cams. In the first type, a frictionless roller follows a groove machined in the face of the cam (Figure 6.1). The cam follower, which is attached to one end of a lever, moves up and down and the lower end of the lever moves back and forth in the horizontal direction. Then the motion is carried to the harness frame with various levers. This type of mechanism is not used much any more.

In the second type of positive-cam shedding, a pair of matched cams are used for each harness (Figure 6.4). The frictionless rollers, which are in contact with the cam faces, oscillate the lever about its fulcrum. As a result, a reciprocating movement is obtained in the lever. This type of mechanism is common in modern weaving machines.

Figure 6.5 shows positive cam mechanisms that can be adapted for all weaving machines.

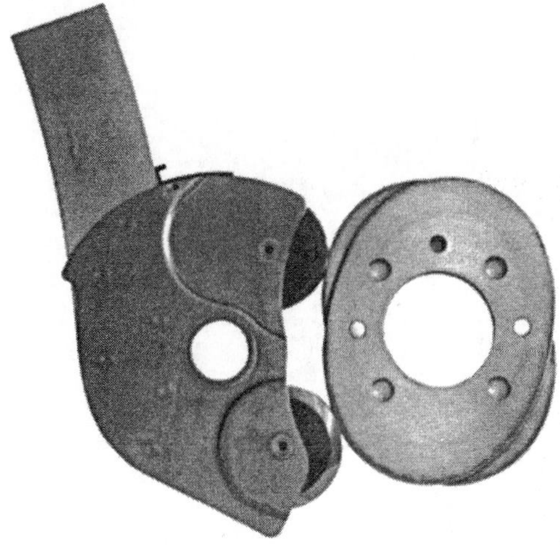

FIGURE 6.4 Positive cam system with two negative cams (courtesy of Fimtextile).

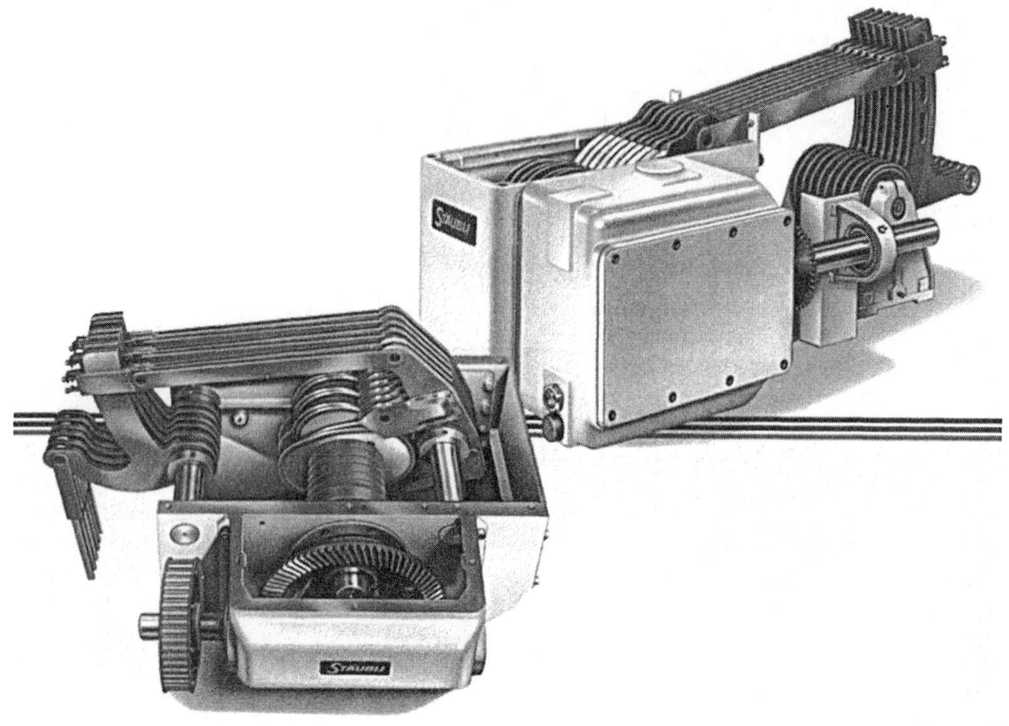

FIGURE 6.5 Positive cam mechanisms (courtesy of Staubli).

Figure 6.6 shows the inside schematic of a positive cam mechanism. Figure 6.7 shows weaving possibilities with cams. Figure 6.8 shows mounting possibilities of positive cam mechanisms on weaving machines. The current trend is to mount the cam mechanism on the floor next to the machine.

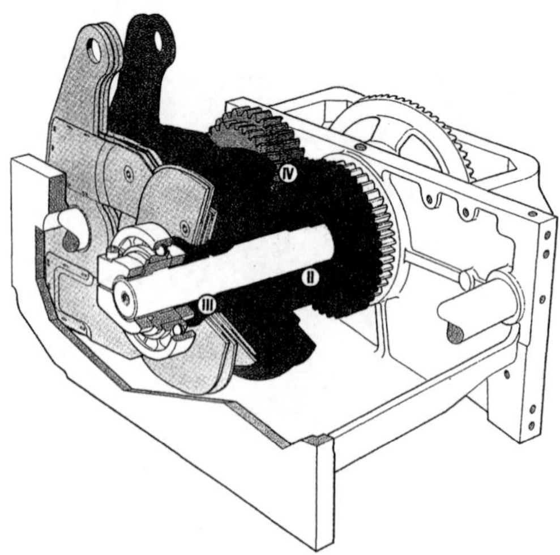

FIGURE 6.6 Inside schematic of positive cam mechanism (courtesy of Staubli).

Automatic harness leveling devices are used in cam shedding to provide even tension on warp yarn during machine stops. The device also helps with eliminating the start-up marks. In modern machines, all moving parts of the cam mechanisms are immersed in an oil bath. The cams are usually made of hardened steel.

6.3 DOBBY SHEDDING

Dobby mechanisms are more complicated than cam systems. They usually have higher initial and maintenance costs. They are normally built to control up to 30 harnesses. Picks per repeat are virtually unlimited in dobby shedding. Due to their complexity, dobby mechanisms are more liable to produce fabric faults than cam systems.

Basically there are two separate functions in a dobby mechanism: 1) power transmission, 2) connection and disconnection of the harnesses

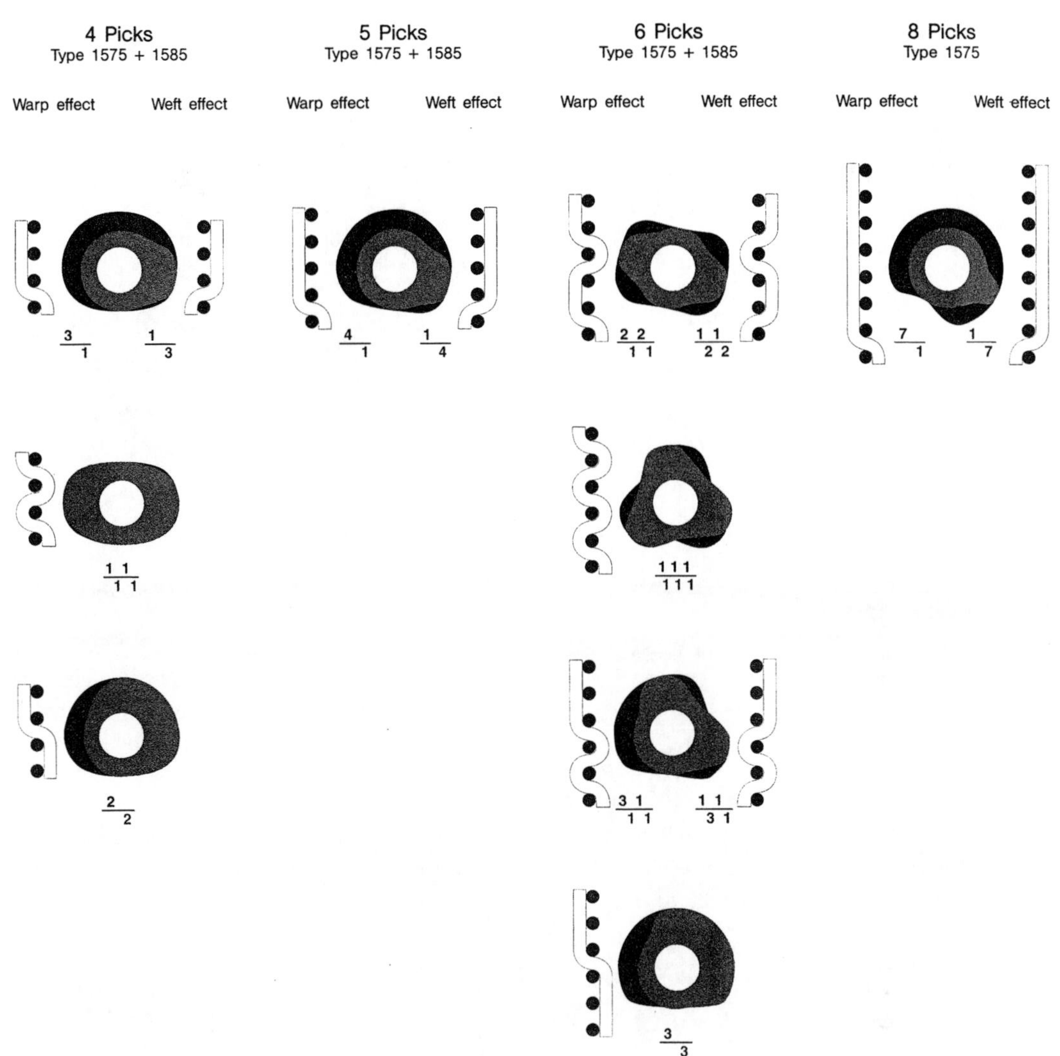

FIGURE 6.7 Weaving possibilities with cams (courtesy of Staubli).

to and from the power source at the proper time. Dobby mechanisms are classified as negative, positive and rotary dobbies, they can be mechanical or electronic. The first wooden lag dobby was made commercial in 1898 (Figure 6.9).

6.3.1 Negative Dobby Shedding

In negative dobby shedding, the harnesses are lifted by the dobby and lowered by a spring motion. Figure 6.10 shows the schematic of a basic double lift, negative dobby mechanism in which a baulk and pairs of feelers, pegs, hooks and knives are used for each harness (single lift dobbies, which have become obsolete, had only one knife per harness). The double lift dobby's cycle occupies two picks and therefore most of its motions occur at half the loom speed which allows higher running speeds. All modern negative dobbies are double lift dobbies. Negative dobbies tend to be simpler than the positive dobbies.

Referring to the figure, the knives (K_1 and K_2) reciprocate in slots along a fixed path (the mechanism to move the knives is not shown). They complete one reciprocation every two picks. When a peg in the lag forming part of the pattern chain raises the right end of feeler F_1,

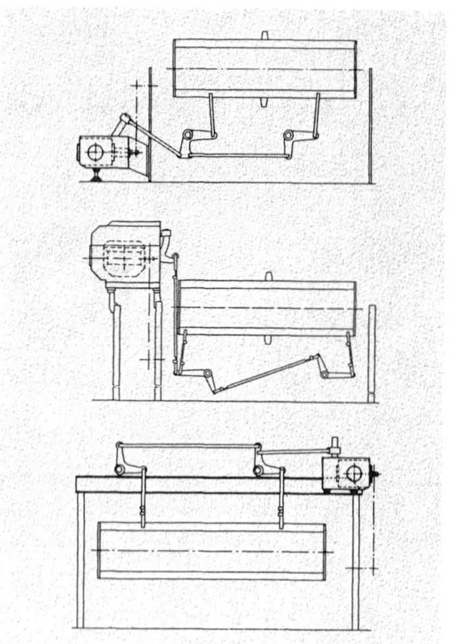

FIGURE 6.8 Mounting possibilities of positive cam mechanisms (courtesy of Staubli).

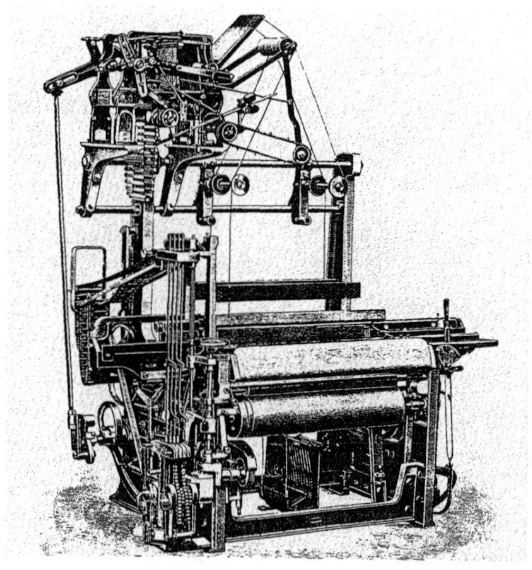

FIGURE 6.9 First wooden lag dobby mechanisms circa 1898 (courtesy of Staubli).

the rod C is lowered which in turn causes the hook (H_1) to be lowered and engaged with the knife K_1. Then, the knife K_1 is moved to the right carrying the hook H_1 with it. This movement is transmitted to D on the baulk (AB). The stop (S_2) acts as a fulcrum. As a result, the jack is moved through the central link. The motion of D is magnified by the lever action of the jack and is transmitted to the harness via straps.

Figure 6.10 shows the method of lag-and-peg for a 3/1/2/1 twill design. Each lag corresponds

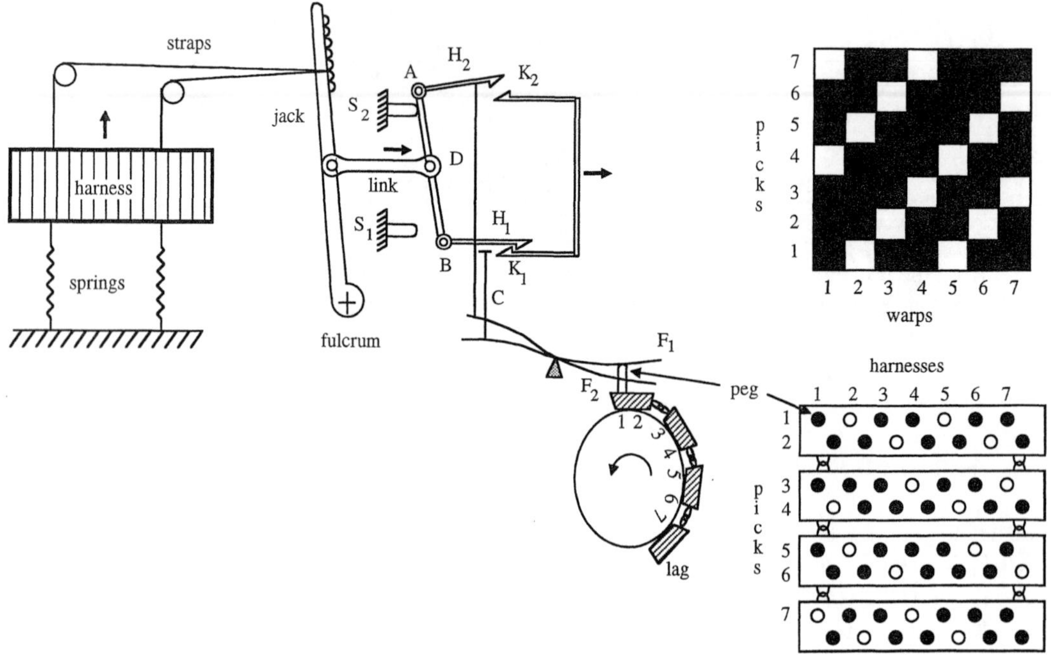

FIGURE 6.10 Schematic of a negative dobby mechanism.

FIGURE 6.11 Card cylinder for a modern dobby mechanism (courtesy of Staubli).

to two picks. The holes in the lags are positioned such that they correspond with the location of the feelers. The pattern chain is turned intermittently by a wheel so that a new lag is presented every second pick. A filled circle represents a peg in the lag. Figure 6.11 shows a card cylinder for a mechanically programmable dobby shedding machine.

As an alternative to the lag-and-peg chains, punched paper or plastic pattern cards are used. A punched hole in the paper corresponds to a peg in the lag, thus a hole causes the corresponding shaft to be lifted. Light feelers such as needles are used to detect the presence or absence of a hole. The force needed to move the hooks is not supplied by feelers. Auxiliary knives and hooks are used to engage and disengage main knives and hooks. Holes on paper or plastic cards are punched on special punching machines. These cards are especially suitable for very long patterns.

In the early mechanical dobby mechanisms, the knives were actuated from a crank mounted on the end of the bottom shaft on the loom. In modern dobbies, the knives are actuated from cams mounted on a shaft in the dobby.

Recently, more modern dobbies utilize electronic systems for input of the harness lifting and lowering patterns. These dobbies can weave patterns requiring up to about 30 harnesses and repeating on as many as 6400 picks.

Figure 6.12 shows a negative dobby with electronic control. It operates in correct pick sequence synchronously with the weaving machine during normal operation as well as during the pick finding process. The electronic control is comprised of an electromagnet for each lifting unit and a control unit or "controller" that activates the electromagnets of the dobby and starts the control functions of the weaving machine according to the weave data. The electronically controlled weave data program controls the retention hooks via intermediate elements and initiates a harness frame motion as soon as the selected baulk hook enters the range of the controlled retention hook. The internal parts consist of baulks to which short hooks are hinged (Figure 6.13). The push bars are directly driven by complementary cams via balance levers pushing the baulks. In the mechanical version of this machine (Figure 6.14), programming takes place by lag-and-peg cards which are made of lightweight, reusable plastics.

Figure 6.15 shows a computer programming system for dobby weaving. Weaving programs can be prepared and archived with computer aided systems for both electronically and card controlled dobbies. With the programming systems, data can be prepared, stored or transferred according to requirements utilizing program carrying cards and/or floppy disks.

FIGURE 6.12 Negative dobby with electronic control (courtesy of Staubli).

Card cutting and copying machines are used for cutting, copying and pasting pattern cards for dobbies, color and function control devices and name weaving machines. Figure 6.16 shows a motor driven card cutting and copying machine, which can cut and copy paper and plastic cards. The machine consists of keyboard, quick reading plate for point paper, drive, copying device, pasting device, setting dial for card transport wheels and cutting device (Figure 6.17). Figure 6.18 shows a cylinder with reading plate which acts as holder for the point paper and facilitates the reading and transferring of the point paper onto the keyboard.

As stated earlier, the negative dobby mechanisms raise the harnesses but can not lower them; spring undermotion mechanisms are used to lower the harnesses. Figure 6.19 shows the application of negative dobby.

FIGURE 6.13 Functional principle of negative electronic dobby (courtesy of Staubli).

FIGURE 6.14 Mechanical negative dobby (Staubli).

6.3.2 Positive Dobby Shedding

In positive dobby shedding, the harnesses are both raised and lowered by the dobby mechanism which eliminates the need for a spring undermotion. Therefore, in any positive dobby some kind of mechanism is necessary to return the ends of the baulks to their stop bars and to hold them there. A locking bar is used for this purpose. A sample mechanism to achieve this is shown in Figure 6.20.

Push bars B_1 and B_2 and the knives K_1 and K_2 reciprocate together. When the knife returns after displacing the hook, it pushes the end of the baulk against its stop bar. Then, the locking bar, L_1 engages a notch in the hook which is pushed up by the selection mechanism. Therefore, the baulk is prevented from moving until the next selection. The locking bar L_1 will hold the baulk against its stop while the knife K_1 and the push bar B_1 make one complete cycle.

Figure 6.21 shows a high performance positive dobby machine. This double lift open shed dobby operates according to the Hattersley principle. There are two separate units in the system: a drive unit for harness frame motion and a reading unit for hook selection. The harness frames move throughout the entire cycle without play, regardless of load or speed. Although this machine is developed specifically for rapier machines, under specific conditions, it can also be

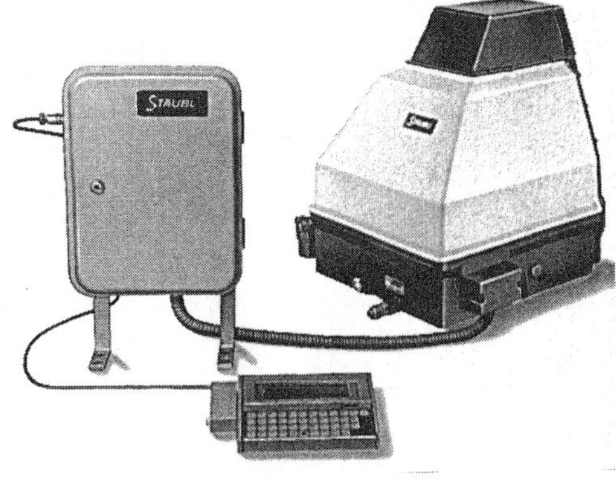

FIGURE 6.15 Computer programming system for dobby weaving (courtesy of Staubli).

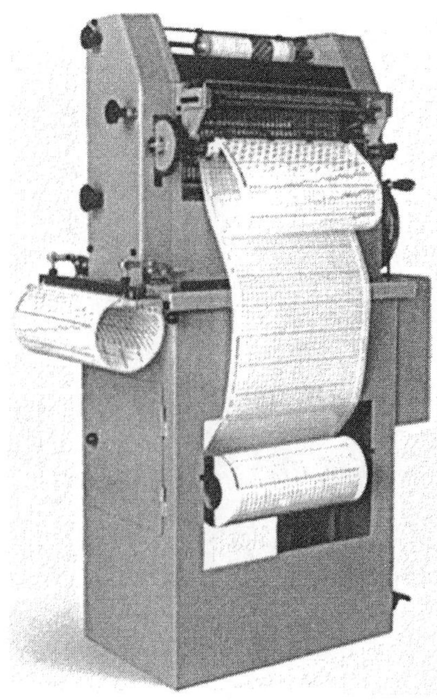

FIGURE 6.16 Motor driven card cutting and copying machine: left, front of the machine; right, back of the machine (courtesy of Staubli).

used for projectile machines. The dobby is driven either directly by the main shaft of the weaving machine or by a back-gear shaft. The reading unit reads the pattern card and transmits the selection to the traction hooks. The cylinder movement is continuous for precise pattern card feed and the cylinder swings out for pattern card changes.

Figure 6.22 shows the functional principle of double lift positive dobby machine. The drive and control unit for the hooks have two functions directly related to the harness frame movement: positive hook control without springs and movement and positioning of harness frames in their end positions in upper or lower shed. Four pairs of complementary cams provide these functions. Figure 6.23 shows the placement of positive dobby systems on weaving machine.

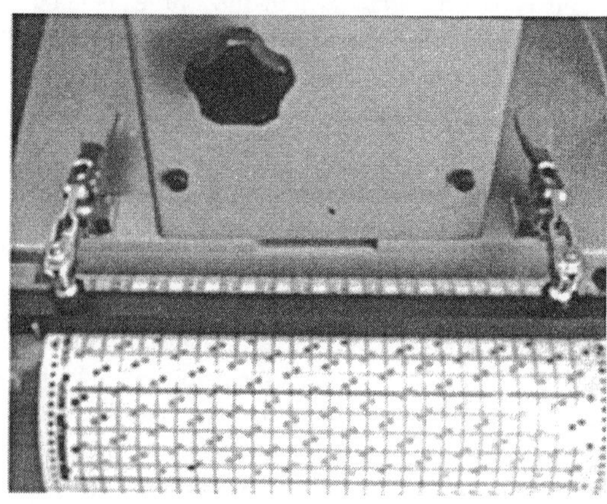

FIGURE 6.17 Copying (left) and pasting devices (courtesy of Staubli).

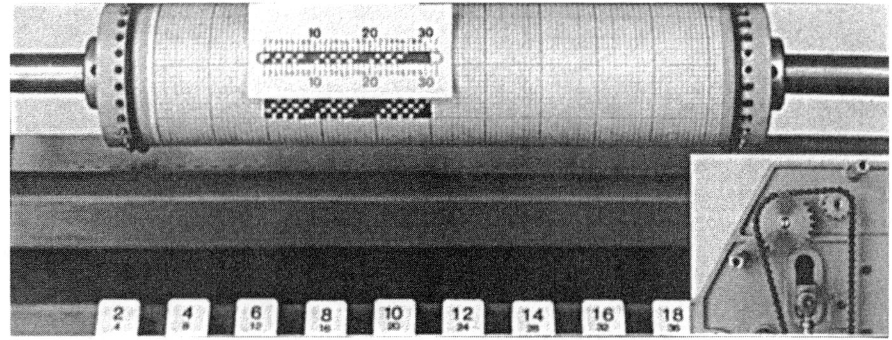

FIGURE 6.18 Cylinder with point paper and reading plate (courtesy of Staubli).

6.3.3 Rotary Dobby

Figure 6.24 shows a rotary dobby with electronic control. It is a positive machine operating according to the rotary principle. The dobby is composed of the following units:

- main drive with modulator and complementary cams
- drive block with cam units for harness frame motion
- control unit with magnet block for transformation of electronic signals
- electronic control box

The dobby is driven either by the weaving machine's main drive shaft or back-gear shaft.

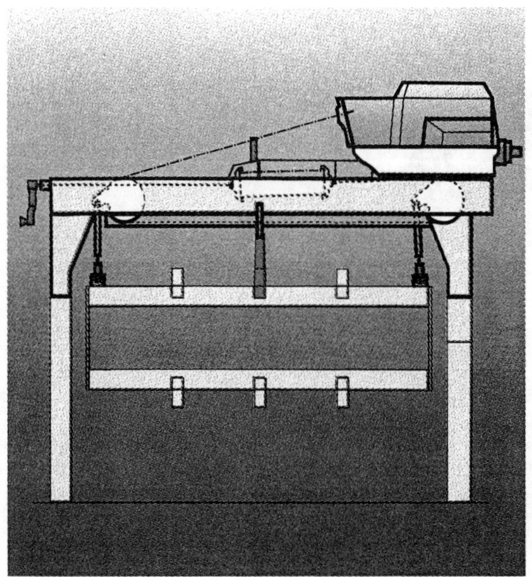

FIGURE 6.19 Placement of negative dobby (courtesy of Staubli).

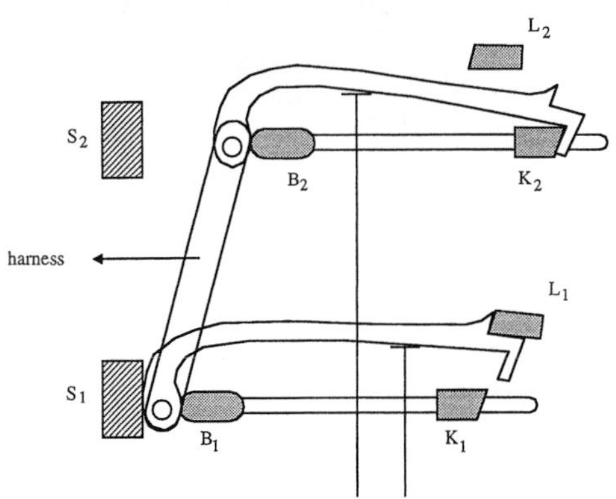

FIGURE 6.20 Schematic principle of positive dobby [2].

FIGURE 6.21 Double lift open shed positive dobby (courtesy of Staubli).

FIGURE 6.22 Functional principle of double lift open shed positive dobby (courtesy of Staubli).

The control is composed of a magnet bar with 20 or 28 electromagnets and a controller which activates the electromagnets of the dobby and starts the control functions of the weaving machine according to the weave data. The controller carries out the following operational functions:

- leveling of harness frames in lower, middle and upper shed
- individual lifting of each harness frame
- separation of warp threads

The rotary dobby with electronic control operates according to the rotary principle and is

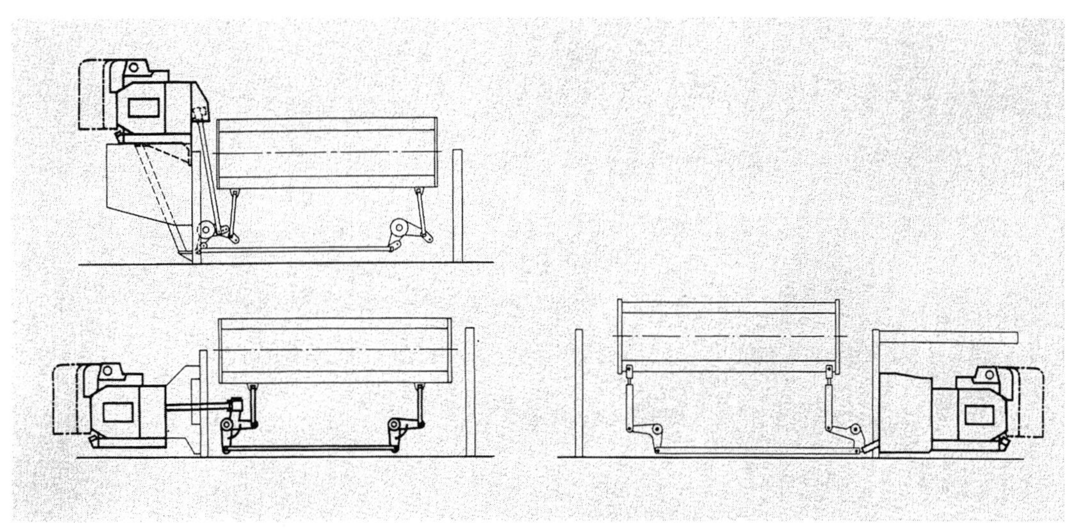

FIGURE 6.23 Mounting positions of positive dobbies (courtesy of Staubli).

FIGURE 6.24 Rotary dobby with electronic control (courtesy of Staubli).

FIGURE 6.25 Functional principle of rotary dobby (courtesy of Staubli).

founded on two elements: cam unit and modulator (Figure 6.25). Each harness frame is controlled by a cam unit only 12 mm wide. This cam unit converts the irregular rotary motion of the main drive shaft directly into the linear motion required for the harness frame drive. The essential element is a crank mechanism enclosing a cam with ball bearings (Figure 6.26). A ratchet placed on the outside of the cam connects it with the driver, and by a 180° rotation of the cam causes a lifting motion. The ratchet is controlled according to the pattern by the control unit. The modulator transforms the regular rotary motion from the weaving machine into an irregular rotary motion. By the use of complementary cams precise laws of motion result. Figure 6.27 shows the possible mounting locations of rotary dobby on weaving machines.

High Performance Rotary Dobby

Figure 6.28 shows a positive dobby machine that operates on the rotary principle with pattern card control. This system can be used for high

FIGURE 6.26 Rotary dobby arm (courtesy of Fimtextile).

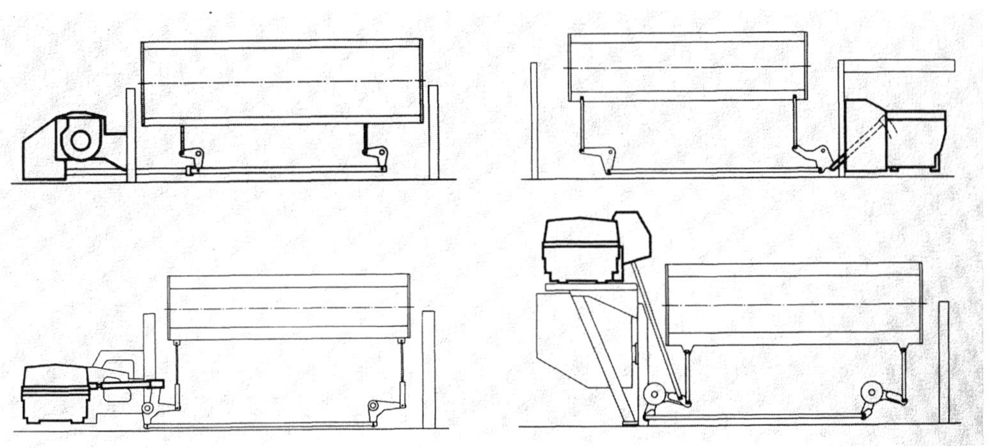

FIGURE 6.27 Possible mounting locations of rotary dobbies (courtesy of Staubli).

FIGURE 6.28 Positive rotary dobby with pattern card control (courtesy of Staubli).

speed rapier, projectile and air-jet weaving machines. In this system, rotary principle is combined with mechanical control and it can have up to 4500 picks repeat and 28 harnesses. The dobby is driven either directly by the main shaft or by the back gear shaft with a drive belt. This system has the following features:

- lower and middle shed leveling device
- automatic pick finding device integrated in the dobby drive
- stitching yarn device, 1/1 weave with separate continuous drive and individually adjustable shed closure
- color latch

Figure 6.29 shows the functional principle of this dobby. Each harness is activated by a cam unit which consists of a crank rod enclosing a cam that is supported by bearings and a movable key which is the only activated part. The link between the cam and the coupling fixed on the main shaft is provided by the key. This causes the cam turning in 180° steps on the end of the crank rod to perform a lifting option. An indexing arm for each cam unit actuates the key according to the pattern. The pattern information, read by feeler needles, is transmitted by pulleys moving back and forth to the key via traction elements and indexing arm.

Rotary Dobby for Industrial Fabrics

Figure 6.30 shows a rotary dobby that operates by the reverse motion method in the open shed, closed shed or mixed mode. This dobby is particularly suitable for manufacturing of industrial fabrics such as screen, felt, wire, multilayer fabrics and carpets. The dobby consists of main drive with servo motor, drive block with cam unit for harness frame motion, control unit with solenoid valves for transformation of electronic signals and electronic control unit.

The lifting units work on the patented rotary principle. The harness motion is based on the open shed and closed shed principle with closed shed continuously adjustable from 50 to 100% (Figure 6.31). The shed dwells can be electronically programmable.

FIGURE 6.29 Functional principle of positive rotary dobby with pattern card control (courtesy of Staubli).

Figure 6.32 shows the functional principle of this rotary machine. A cam unit (20 mm wide) controls each harness frame. The cam unit converts the rotary motion of the main shaft into the linear motion required for the harness frame drive through a crank mechanism. A ratchet placed on the outside of a cam connects it with the driver and activates a lifting motion by means of an intermittent rotation of the cam through 180°. The ratchet is controlled by the electronic control unit via a pneumatic system. A unit modulated by a servo motor produces the required nonuniform rotary motion of the main shaft. Figure 6.33 shows the locations of this shedding mechanism on the machine.

Figure 6.34 shows a closed shed dobby machine for industrial fabrics such as press felts and dryer fabrics used in papermaking. Weaving machines for paper machine clothing are very wide (up to 30 m) which require high loading capacity, precise harness frame motion and minimum maintenance. Figure 6.35 shows the functional principle of this machine which operates according to the closed-shed principle with a two-knife system. The two-knife system is directly driven by tempered and ground cylinder

FIGURE 6.30 Rotary dobby for industrial fabrics (courtesy of Staubli).

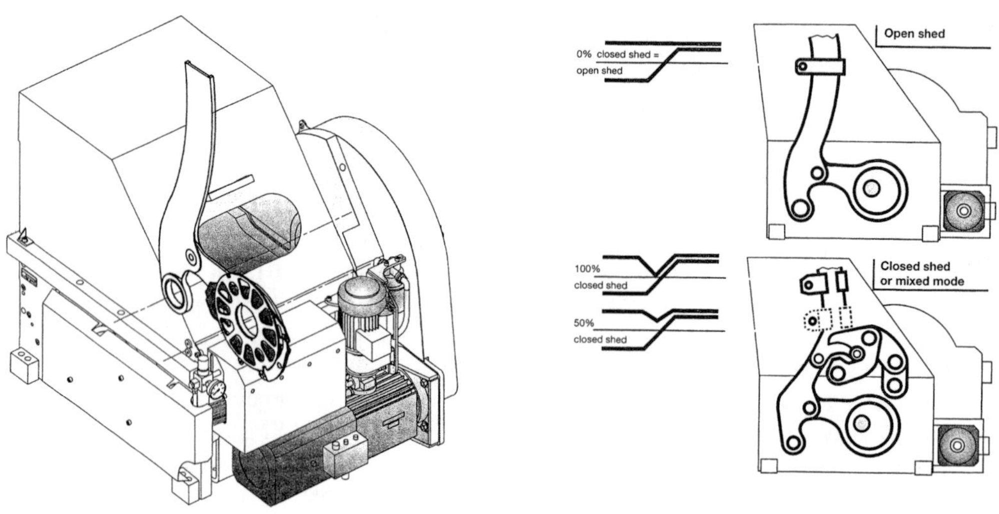

FIGURE 6.31 Schematic of shed motions (courtesy of Staubli).

FIGURE 6.32 Functional principle of rotary dobby for industrial fabrics (courtesy of Staubli).

cams. The lifting unit with the integrated pneumatic control cylinders, the clamping pieces and the twin-hook run on the lever axle and swing around the same axle in the complete knife motion. The pattern control transfers the commands "harness frame to upper shed" respectively "harness frame to lower shed" independently from speed through the solenoid valve and the pneumatic cylinders to the clamped twin hook of the lifting unit. Only in the correct operating position—in closed shed—do the knives release the clamping of the preset twin hook and within only a few milliseconds the twin hook of the lifting unit is read in to the knife for top or bottom shed. The continuous lifting movement of the knives swivels the lifting unit and moves

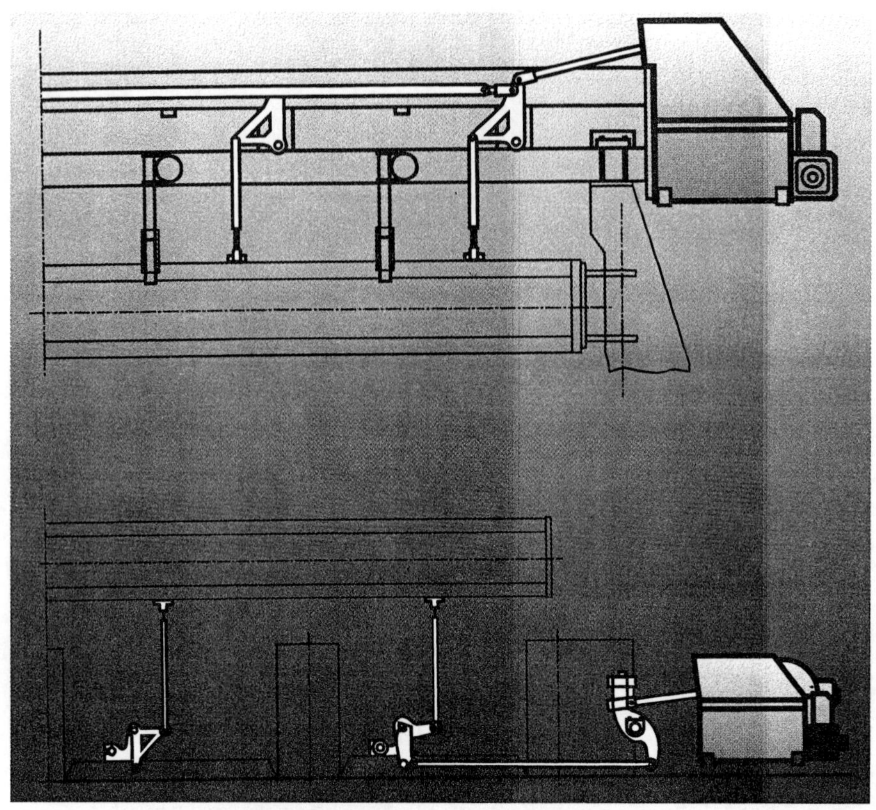

FIGURE 6.33 Locations of the rotary dobby on the machine (courtesy of Staubli).

FIGURE 6.34 Closed shed dobby machine for industrial fabrics (courtesy of Staubli).

FIGURE 6.35 Functional principle of the dobby in Figure 6.34 (courtesy of Staubli).

all harness frames to the required top or bottom shed. By means of corresponding cylinder cams the harness frame motion is adapted to the kinematics of the weaving machines.

Rotary Dobby for Double Pile Weaving Machines

Figure 6.36 shows a positive dobby machine with electronic control for double pile weaving machines. This machine combines the rotary principle with a three-position device. It is possible to select different shed dwells for the pile and base harness frames. The electronic read-in system is synchronized to allow reversal of the dobby motion in harmony with the weave. Figure 6.37 shows the functional principle of this machine.

Auxiliary Devices for Dobby Mechanisms

Figure 6.38 shows a tabletop ultrasonic welding device for welding of plastic pattern cards

FIGURE 6.36 Positive rotary dobby machine with electronic control for double pile weaving machines (courtesy of Staubli).

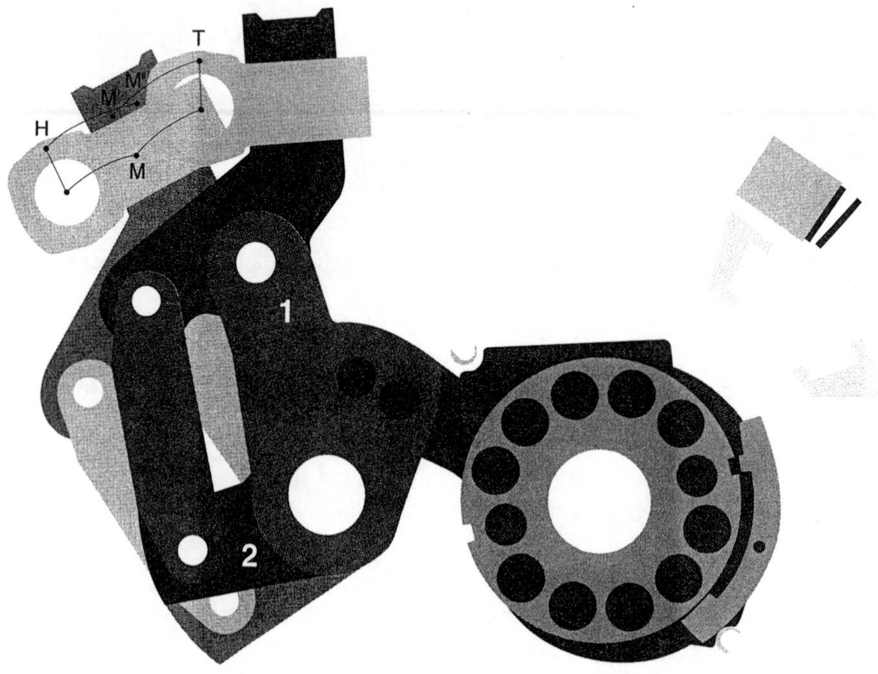

FIGURE 6.37 Functional principle of positive rotary dobby for double pile weaving machine (courtesy of Staubli).

FIGURE 6.38 Ultrasonic welding device for dobbies (courtesy of Staubli).

for dobbies. It makes point and ring welded seams that resist chemical and mechanical influences.

Figure 6.39 shows various harness frame motions and shaft locking systems for dobby mechanisms. In model e12, the harness motions are fitted with roller bearings. Shed setting is carried out at the harness motion jacks underneath the shafts. In model e22, shed setting is done at the actual dobby jacks.

Figure 6.40 shows a multifunctional clamping loop for adjustment of shed height and lift. Adjustment of shed lift (A) is done by changing the lever ratios by moving the clamping loop on

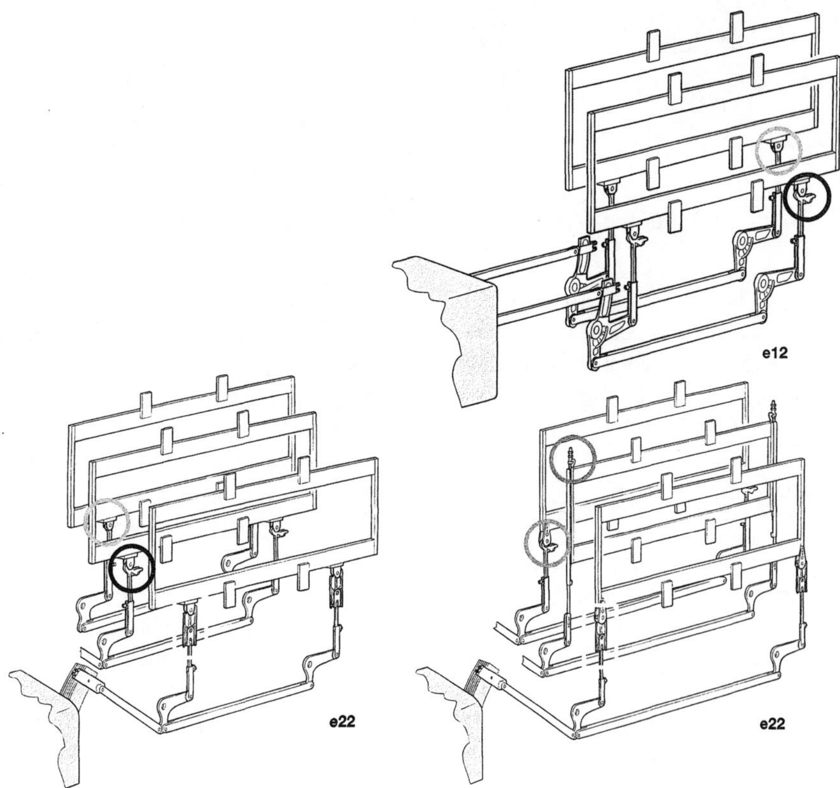

FIGURE 6.39 Positive harness motions (courtesy of Staubli).

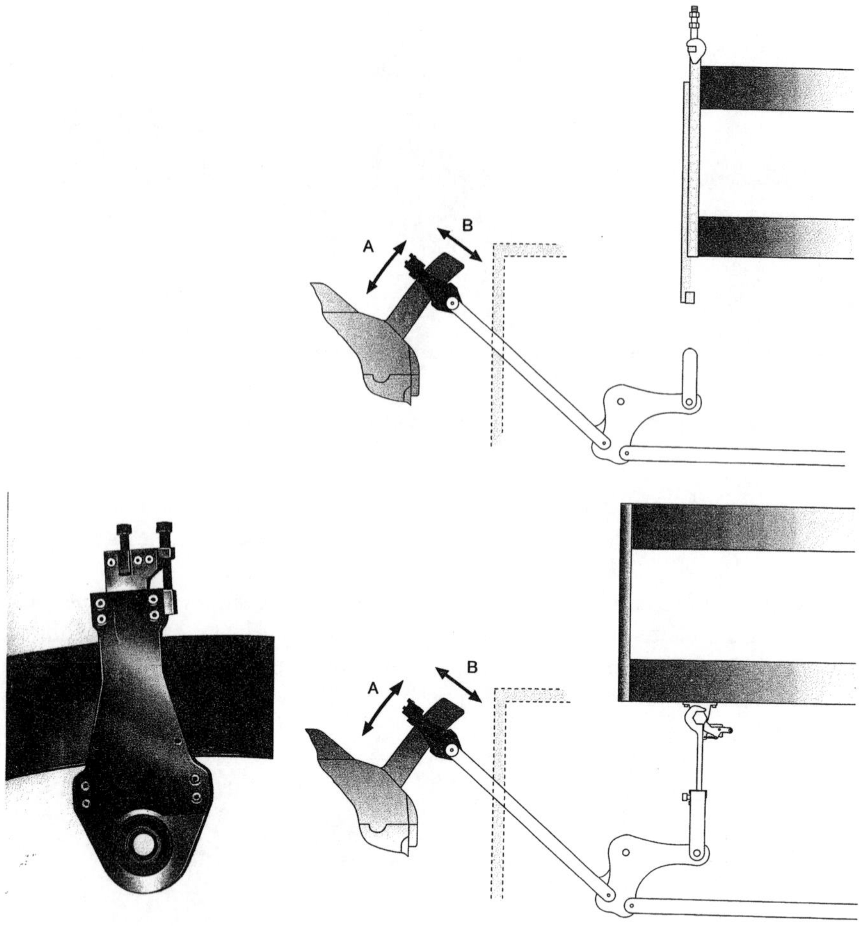

FIGURE 6.40 Multifunctional clamping loop (courtesy of Staubli).

the jack. Adjustment of shaft height (B) is done by varying the length between the dobby jack and the harness motion lever. Figure 6.41 shows examples of shed settings.

6.4 JACQUARD SHEDDING

The original jacquard machine, invented by Joseph Marie Jacquard (1752–1834), is shown in Figure 6.42. Jacquard shedding offers unlimited patterning possibilities. Although they are simple in principle, jacquard machines contain many parts which make them relatively costly to install and maintain. Jacquard mechanisms are more liable than dobby or cam shedding to produce faults in the fabric.

The jacquard machines offer the highest level of warp yarn control. This versatility is due to the separate control of each warp end or groups of similarly interlacing warp ends within the pattern repeat across the fabric width. They enable the most sophisticated patterns, such as pictures, to be produced in the woven fabric (Figure 6.43).

Jacquard machines can be mechanical or electronic with single or double lift mechanisms; the new machines are all double lift. Recently, more modern jacquards utilize electronic systems for input of the harness lifting and lowering patterns. Modern jacquard heads generally are equipped to handle over 1200 harness cords with patterns repeating on about 9000 picks and multiple

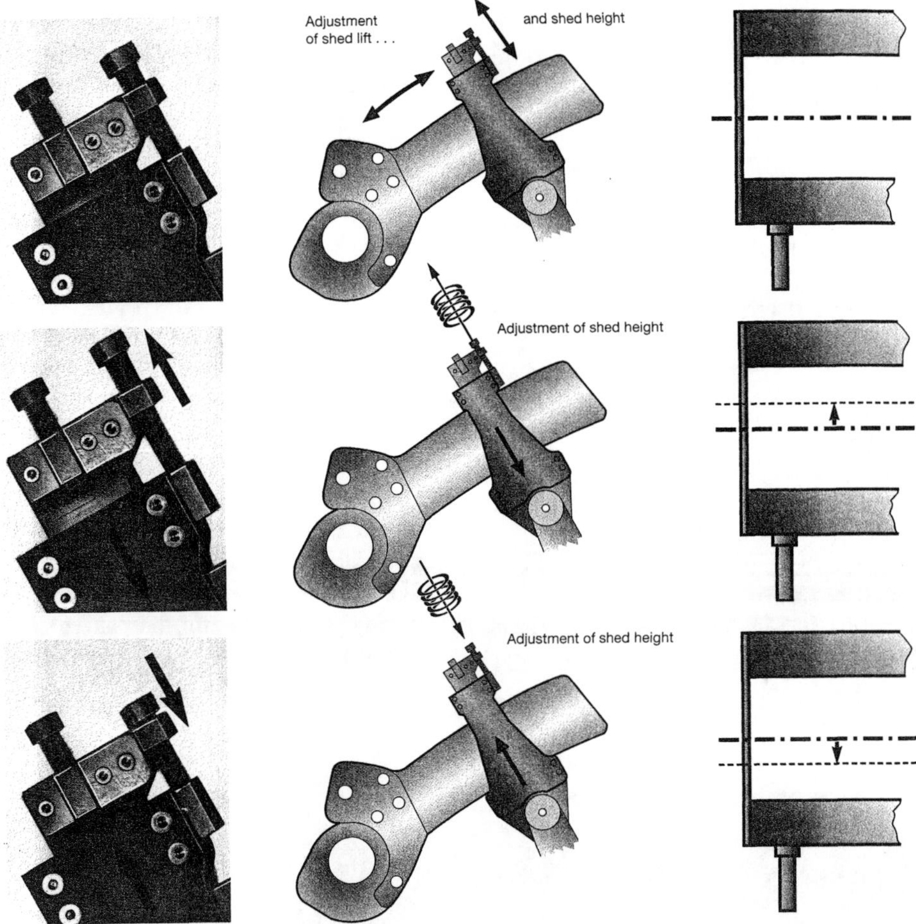

FIGURE 6.41 Shed settings (courtesy of Staubli).

heads can be employed over a single weaving machine to increase the weave pattern capability. Jacquard systems are placed on top of the weaving machine as shown in Figure 6.44.

6.4.1 Single Lift, Single Cylinder Jacquard

This is the simplest jacquard mechanism. Although it is not used anymore, the schematic of a single lift, single cylinder jacquard is shown in Figure 6.45 for clarity. The fabric design is punched in pattern cards which are joined together to form a continuous chain. Each card on the chain corresponds to one pick in the weave. There is one needle and one hook for every end in the repeat. Every cord has holes to mate with the corresponding needles. Pattern cards are presented to a four-sided "cylinder" (it can be pentagonal or hexagonal) in such a way that every card fits one side. Coil springs press the needles towards the cylinder. If there is a hole in the card opposite to a particular needle, the needle will enter the hole and the hook engages with the knife. As a result, the hook and the warp connected to it will be lifted by the knife during its upward movement; if there is no hole for the needle, the hook remains disengaged and is not lifted. After each pick, the cylinder is moved away from the needles, rotated to present a new card and then moved again towards the needles.

The capacity of a jacquard head is indicated by the number of hooks or needles. For example, a 600 needle jacquard would have 12 horizontal

FIGURE 6.42 Jacquard machine of Joseph Marie Jacquard (Historical Museum of Lyon Fabrics, France. Photo courtesy of Staubli).

rows of needles with 50 needles in each row. The typical minimum load per hook is 150 g, and the speed-dependent maximum load is 1.2 kg.

6.4.2 Double Lift, Single Cylinder Jacquard

A double lift, single cylinder jacquard has two sets of knives which move up and down opposite to each other during a two pick cycle. A 600 needle machine has 1200 hooks; each needle controlling two hooks. When no hook is lifted the harness cord remains down. This system can run at higher speeds than the single lift machine.

Figure 6.46 shows a mechanically controlled double lift jacquard machine for rapier, projectile and air-jet weaving machines. The reading-in needles are equipped with springs which are all made of high grade hardened steel. The card cylinder swings out for the pattern cards to be changed. Pattern card repeats may have up to 9000 picks.

FIGURE 6.43 Jacquard shedding offers endless design pattern possibilities.

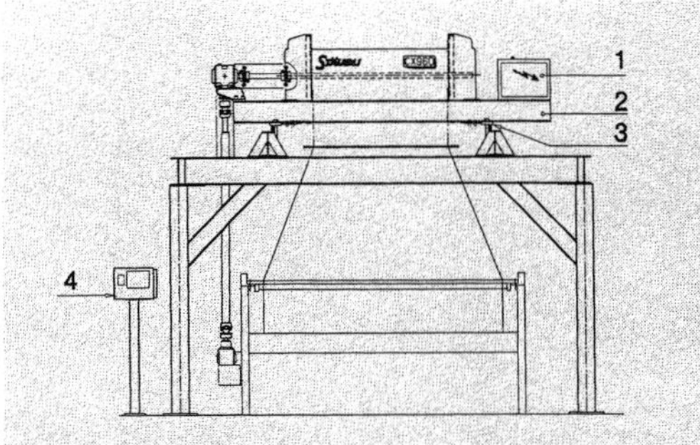

1 Power supply
2 Understructure
3 Adjustable legs
4 Control box

FIGURE 6.44 Schematic of Jacquard placement (courtesy of Staubli).

Figure 6.47 shows functional principle of a double lift open shed jacquard machine. Each rotary hook has 4 nebs (a,b,c,d). Regarding the position of neb (a), neb (c) is placed 45° to the left and neb (d) 45° to the right. The functional principle of this system is illustrated in five operational phases.

1. Lower shed position: Rotary hook is positioned by neb (a) on bottom board (A). Lifting knives (C) and (D) are moving down and up, respectively.
2. Upper shed position: Rotary hook with neb (d) is caught up lifting knife (D), turned 45° to the left and lifted into upper shed. Neb (c) is rotated and thus beyond the catching range of lifting knife (C).
3. Upper shed position: Rotary hook is positioned on open shed knife (B) by neb (a), and turned back 45° to original position (1). Lifting knives (C) and (D) move up and down, respectively.
4. Upper shed position: Rotary hook with neb (c) has been caught by lifting knife (C), turned 45° to the right and lifted into upper shed.
5. Lower shed position: Rotary hook is returned by neb (a) to bottom board (A) and rotated 45° to original position (1). Lifting knives (C) and (D) initiate the following and lowering motion, respectively.

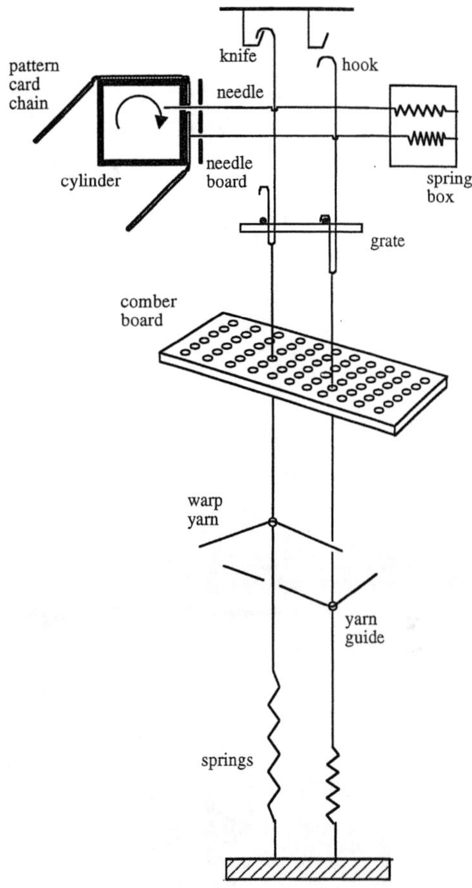

FIGURE 6.45 Schematic principle of single lift, single cylinder jacquard machine.

6.4.3 Double Lift, Double Cylinder Jacquard

The functional principle of the double lift, double cylinder machine is similar to the double lift, single cylinder machine. There are two cylinders, one for odd numbered and one for even numbered cards. There are two needles and two hooks for each harness cord and each end in the repeat. For example, a 600 machine has 1200 heddles and 1200 hooks that control 600 ends in the repeat pattern. This system forms a semi-open shed and runs at higher speeds.

6.4.4 Electronic Jacquard

Figure 6.48 shows an electronically controlled, double lift jacquard machine for rapier,

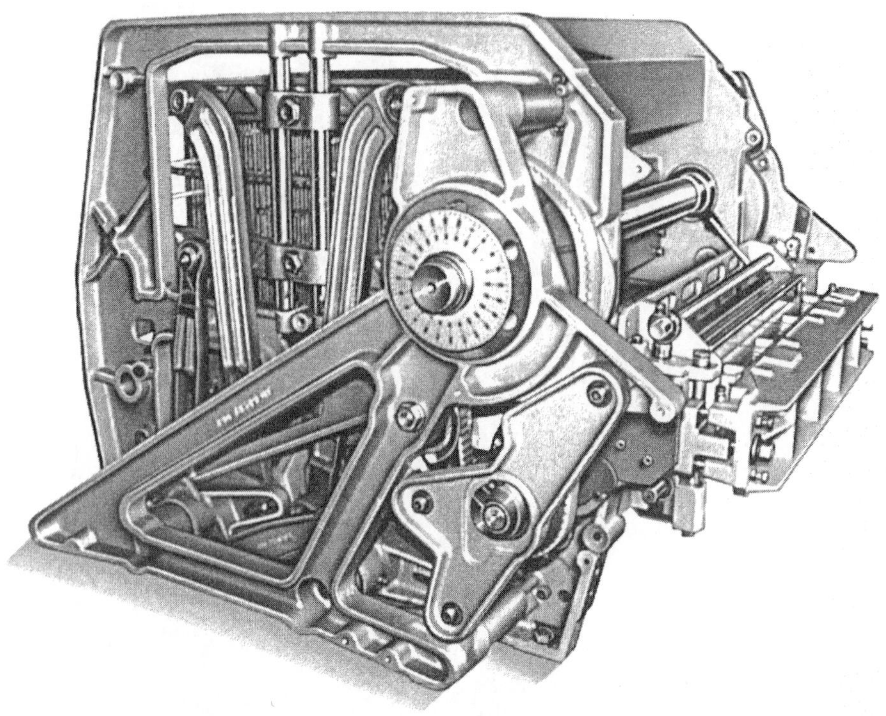

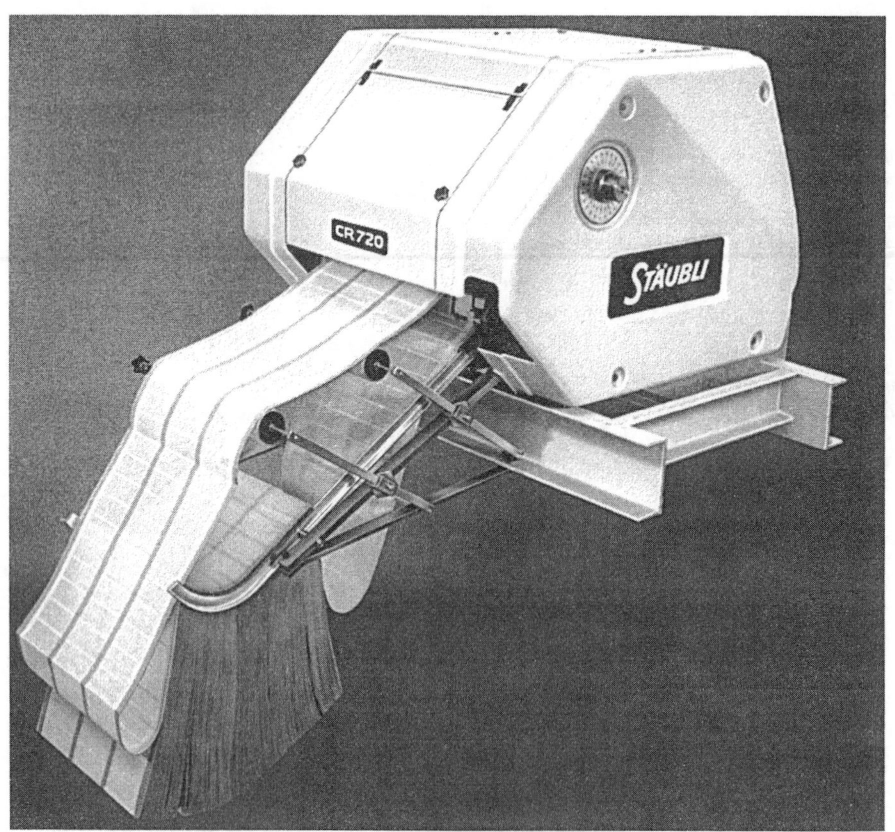

FIGURE 6.46 Mechanically controlled double lift jacquard (courtesy of Staubli).

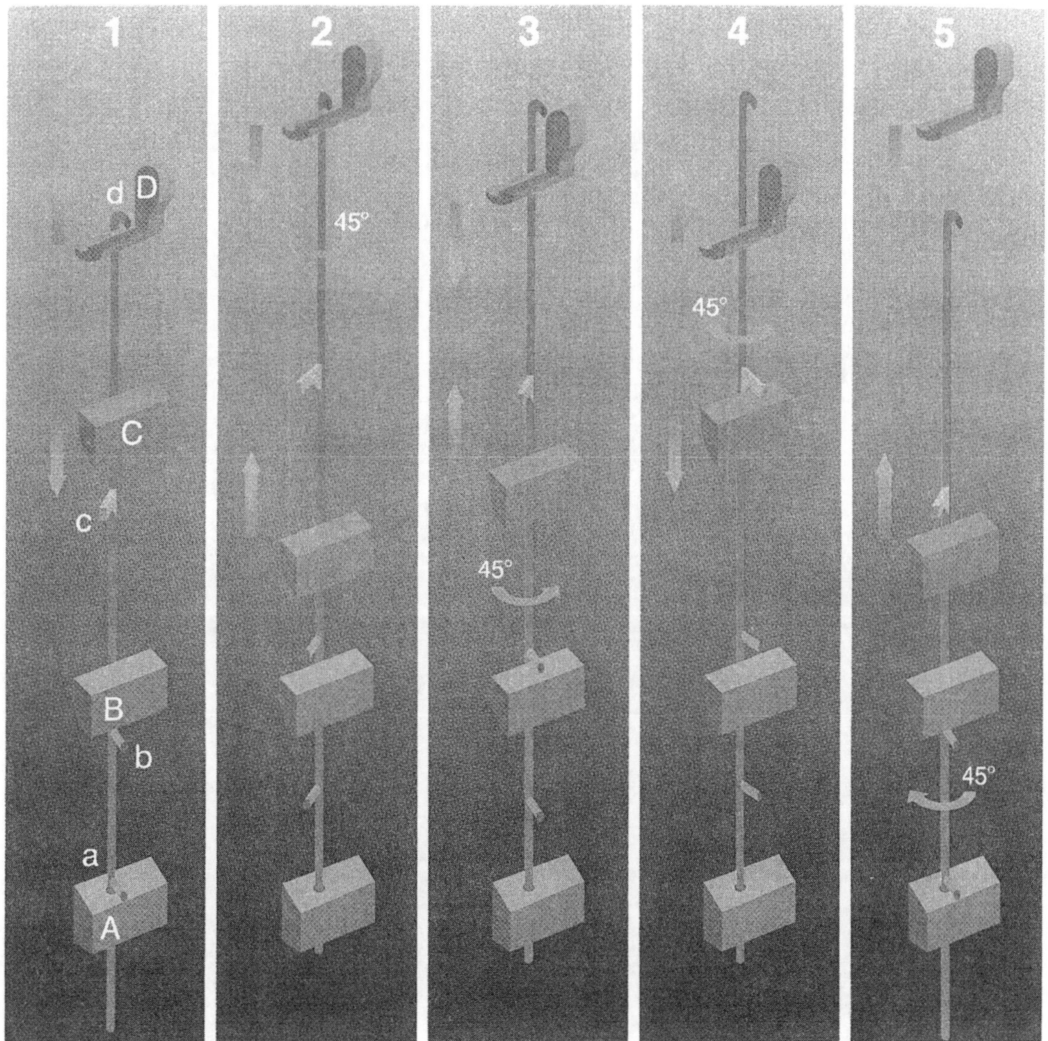

FIGURE 6.47 Functional principle of a double lift open shed jacquard machine (courtesy of Staubli).

projectile and air-jet weaving machines. With the electronically controlled jacquard machines, floppy disks are used as the data carrier or data can be entered directly through a network system.

Figure 6.49 shows the functional principle of this machine. Double rollers (a) move the harness cord to upper and lower shed position.

1. Lower shed position: Hook (b) in its uppermost position has placed retaining hook (d) against electromagnet (h). This magnet is activated according to the pattern, briefly retains retaining hook (d) and prevents hook (b) from hooking onto retaining hook.
2. Lower shed position: Hooks (b) and (c) follow the knives (g) and (f) moving up or down. Double roller (a) offsets the motion of hooks (b) and (c).
3. Lower shed position: By the rising motion of knife (g), hook (c) has placed retaining hook (e) against electromagnet (h). According to the pattern, the magnet is not activated, causing hook (c) to catch on to retaining hook.
4. Shed motion: Hook (c) is caught on retaining hook (e). Hook (b) follows the rising knife (f) and thereby lifts the harness cord.

FIGURE 6.48 Electronically controlled, double lift jacquard machine (courtesy of Staubli).

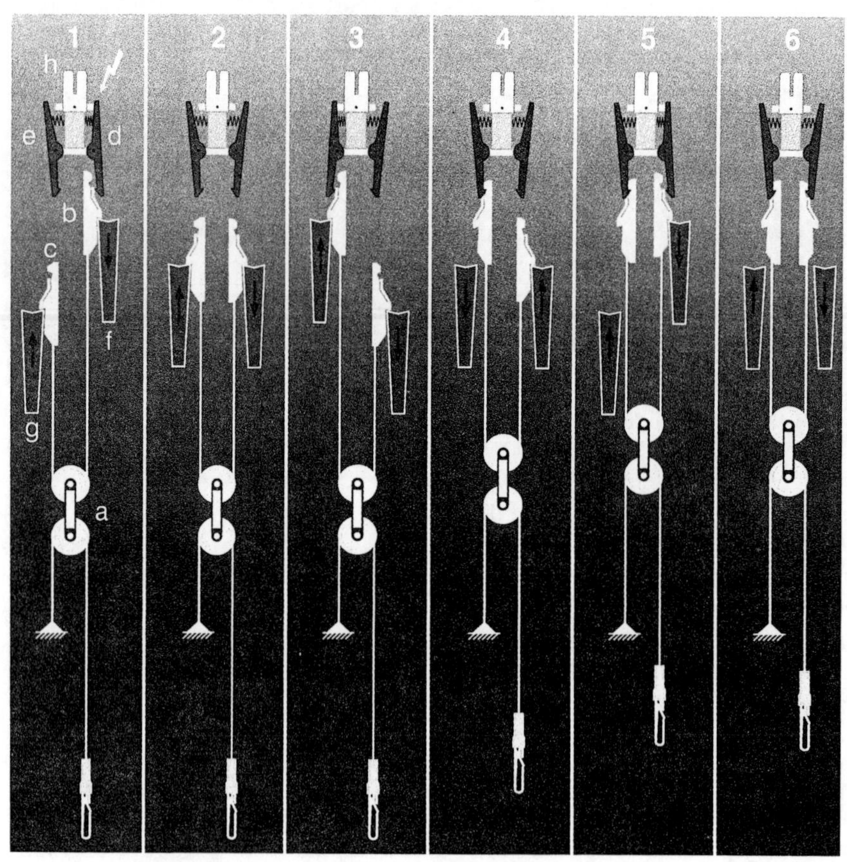

FIGURE 6.49 Functional principle of electronically controlled, double lift jacquard machine (courtesy of Staubli).

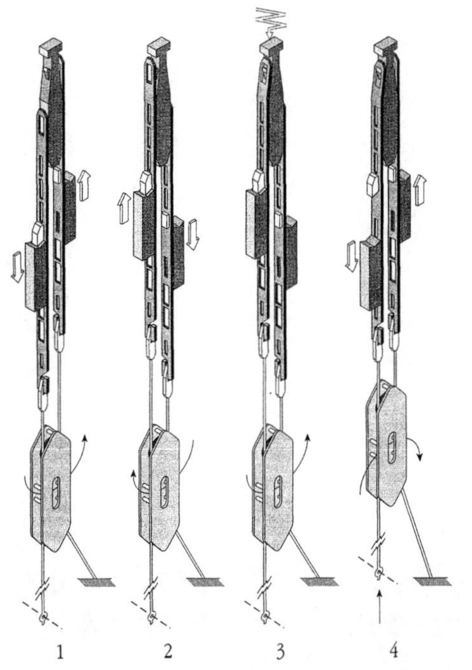

FIGURE 6.50 Electronic jacquard selection mechanism (courtesy of Bonas).

5. Upper shed position: Hook (c) remains hooked onto retaining hook (e). Hook (b) has placed retaining hook (d) against electromagnet (h) by the rising motion of knife (f). According to the pattern, the magnet is not activated, causing hook (b) to be held by retaining hook.
6. Upper shed position: Hooks (b) and (c) remain held by retaining hooks (d) and (e). Knives (g) and (f) are in rising and descending motion, respectively.

Figure 6.50 shows an electronic jacquard selection mechanism by Bonas. Each solenoid (electromagnet) acts on two flexible steel hooks that control the warp yarn through a pulley and cord. The hooks rest on knives which move up and down as shown in stages (1) and (2). The warp stays in the bottom shed until the solenoid is activated by an electronic pulse, attracting the first hook towards the latch. The hook rests gently on the latch as the knives descend (stage 3). The shed is developed when the corresponding hook rises and the warp end is lifted as shown in stage (4).

The first electronic jacquard was introduced at ITMA, Milan in 1983 by Bonas. Modern electronic jacquard machines incorporate the latest computer technology. The fabric patterns can be displayed on screens (Figure 6.51) and the patterns can be stored on hard or floppy disks. Patterns, weaves and programs can be exchanged easily. The machine can also be connected to an external network for data communication. Combining and developing of patterns, weave and program data for a new weaving program is also possible. Other features include modification and correction of pattern and program data, reversing of patterns, mirror-imaging and inserting, changing and deleting of crossing points.

At ITMA-99 in Paris, probably the ultimate control of warp yarns by jacquard machine was exhibited: the lift of each warp yarn can be varied at will individually by a computer.

6.4.5 Special Jacquard Mechanisms

There are other variations of jacquard systems for less conventional fabrics such as industrial fabrics and pile fabrics. For example, specially designed mechanically or electronically controlled double lift jacquard shedding machines are used for double pile weaving. Industrial and heavy fabrics (i.e. screen, wire, felt and multilayer fabrics) which are required to fulfill the rigorous demands in the field, are manufactured using specially designed shedding mechanisms. Mechanical or electronic-pneumatic controlled shedding systems have been developed for this purpose.

Figure 6.52 shows an electronically controlled double lift jacquard machine with patented CX module for double pile weaves. This jacquard is particularly suitable for double rapier weaving machines.

Figure 6.53 shows the CX module with three position unit patented by Staubli for narrow fabric weaving on double needles. This module provides the collective link between the lifting mechanism of the jacquard machine and the harness. The module is made of dimensionally stable composite materials. In each module, there are eight separately guided lifting units. A guide roller in roller bearings integrated in the module provides the three shed positions.

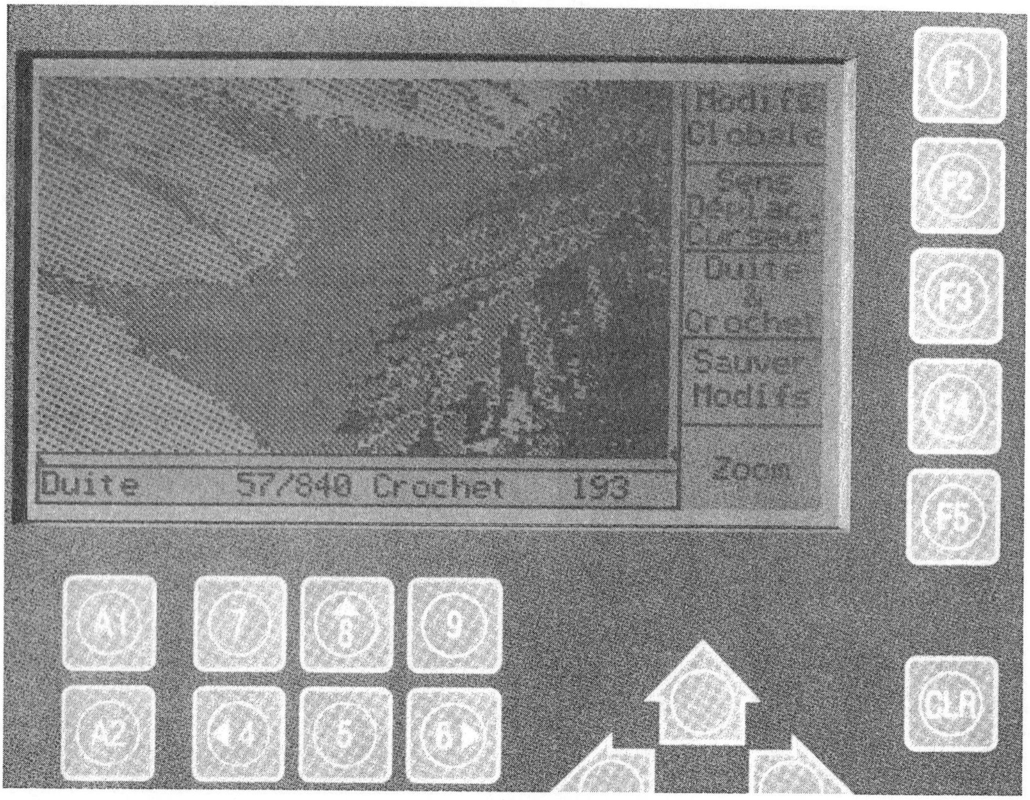

FIGURE 6.51 Jacquard computer screen with the fabric pattern (courtesy of Staubli).

FIGURE 6.52 Electronically controlled double lift jacquard for double pile weaves (courtesy of Staubli).

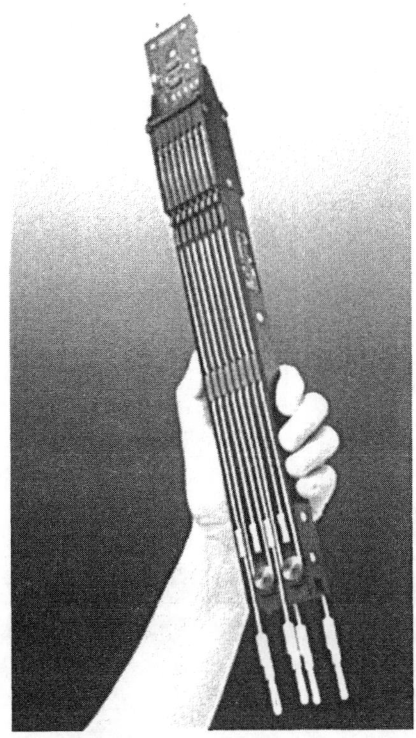

FIGURE 6.53 CX module with three shed position unit (courtesy of Staubli).

While conventional jacquard weaving works only in two positions (high/middle, middle/low, high/low), in the CX module, the warp yarns work as needed in the three positions (high/middle/low). Figure 6.54 shows the patterning possibilities with conventional jacquard weaving and with the CX module.

Figure 6.55 shows the functional principle of the CX module. The double rollers (a) and (k) and the deflection roller (s) determine the upper, lower and middle position of the hook (t) for the harness. Knives (f) and (m), as well as (g) and (l) work in the opposite direction. Referring to the figure, knives (f) and (m) are located at bottom dead center and knives (g) and (l) are located at top dead center. Hooks (c), (b), (n) and (o), which are in their uppermost positions, place the retaining hooks (e), (d), (q) and (p) positively against electromagnet (h) and (r). If electromagnets (h) and (r) are not activated, as determined by the pattern, the hooks (c), (b), (n) and (o) catch onto the retaining hooks (e), (d), (q) and (p). Three working phases are possible as shown in the figure.

Phase 1: Depending on the pattern, electromagnets (h) and (r) are not activated. Retaining

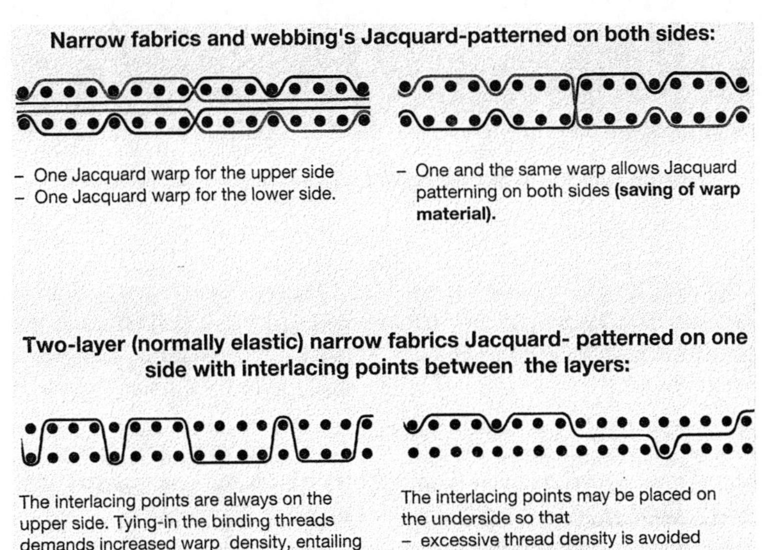

FIGURE 6.54 Patterning possibilities of narrow fabrics and webbings with conventional jacquard (left) and with CX module (right) (courtesy of Staubli).

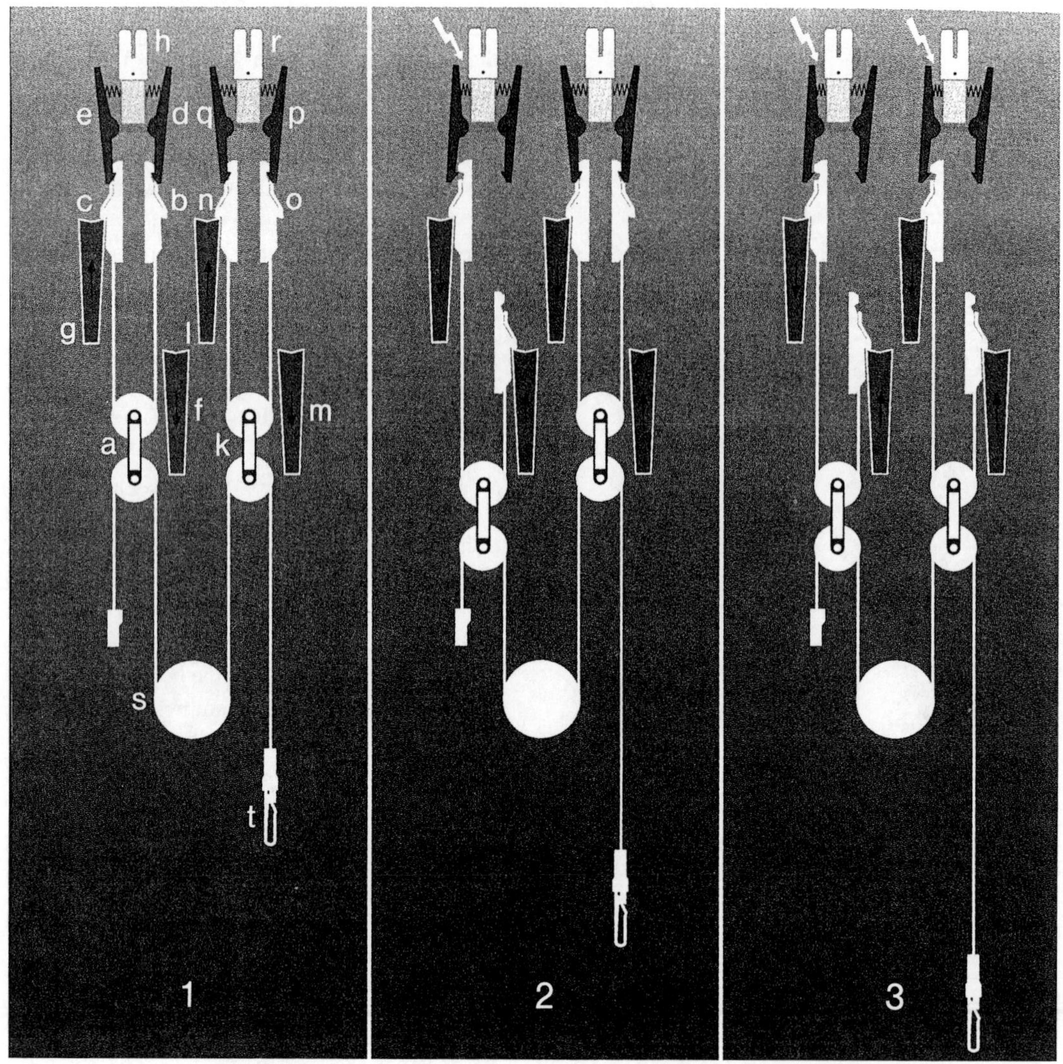

FIGURE 6.55 Functional principle of CX module (courtesy of Staubli).

hooks (e) and (q) lock hooks (c) and (n) in their upper position. The hook (t) for the harness remains still or moves into the upper position.

Phase 2: Electromagnet (h) is activated as required by the pattern; electromagnet (r) is not activated. Hooks (n) and (o) are held in their upper position. The retaining hook (e), held temporarily by electromagnet (h), releases hook (c), which follows the downward motion of knife (g). The hook (t) for the harness remains still or moves into the middle position.

Phase 3: Electromagnets (h) and (r) are simultaneously activated which is determined by the pattern. The retaining hooks (e) and (q), temporarily held by activated electromagnets (h) and (r), release hooks (c) and (n), which follow the downward motion of knives (g) and (l). Hook (t) for the harness remains still or moves into its lower position.

Figure 6.56 shows an electronically controlled double lift name-weaving jacquard machine. It is suitable for air-jet, rapier and projectile weaving machines.

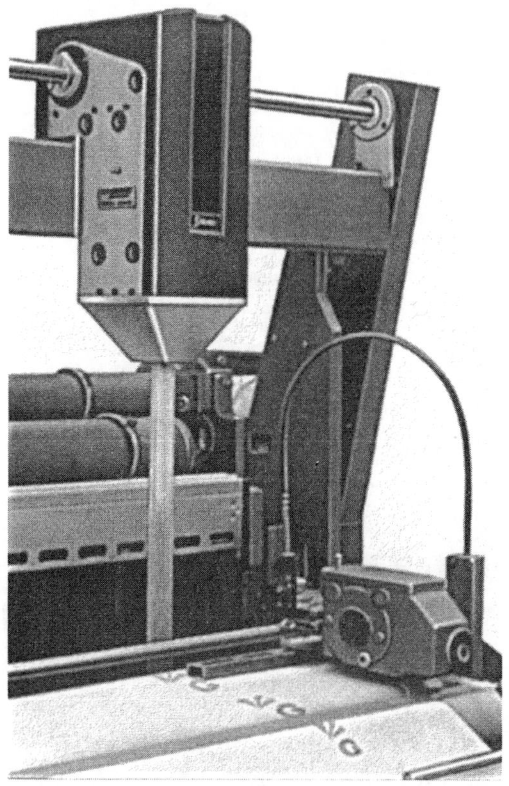

FIGURE 6.56 Electronic name-weaving jacquard machine (courtesy of Staubli).

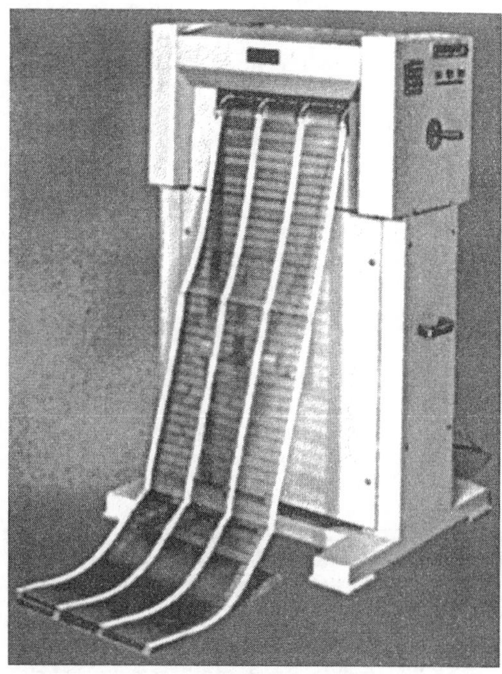

FIGURE 6.57 Electronically controlled card cutting machine (courtesy of Staubli).

6.4.6 Auxiliary Equipment for Jacquard Shedding

Figure 6.57 shows an electronically controlled card cutting machine. Figure 6.58 shows a computerized jacquard card reader. Designs can be read from a card with this device, the information being subsequently transferred onto a floppy disk which can be used for the control of electronic jacquard machine. A typical card reader machine consists of a mechanical assembly unit for driving the pattern cards. Infrared reading units can be used to read the cards fast and with high precision.

The jacquard harness is the system of cords, heddles and downpull elements that transmit the movement of the hooks to the individual warp threads through a harness board (comber board) as shown in Figure 6.59. The lower

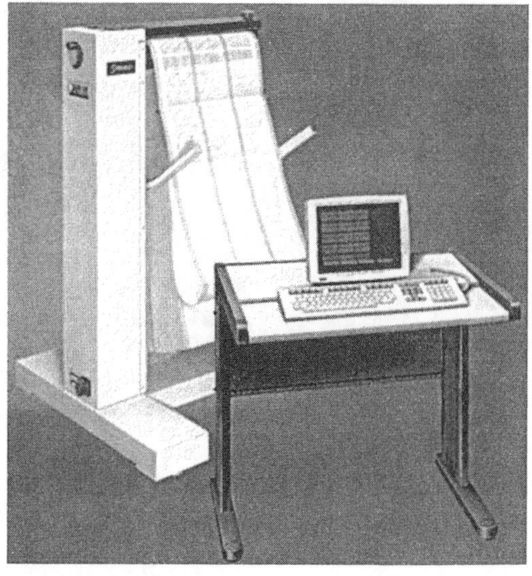

FIGURE 6.58 Computerized jacquard card reader (courtesy of Staubli).

FIGURE 6.59 Components of jacquard harness (courtesy of Staubli).

FIGURE 6.60 Harness (comber) board (courtesy of Staubli).

end of each cord is attached to the heddle which has an eye in the middle for the warp yarn.

Figure 6.60 shows a harness board. Figure 6.61 shows a harness board for terry fabrics. The harness board is divided in depth. Both boards are held in their own frames, this allows the setting of different shed heights for ground and pile warps.

Harness cords are usually braided tubular

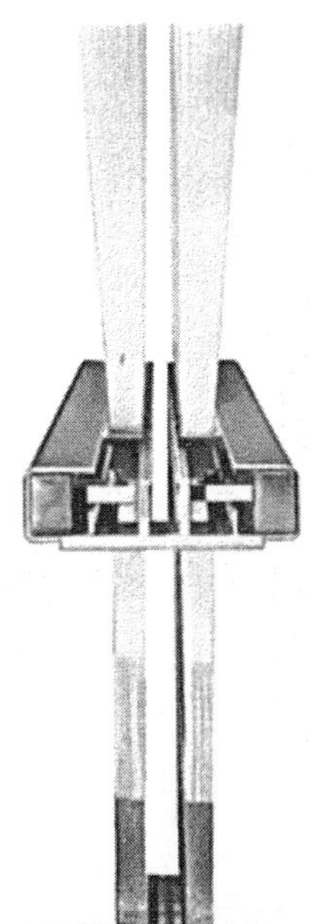

FIGURE 6.61 Harness board for terry fabrics (courtesy of Staubli).

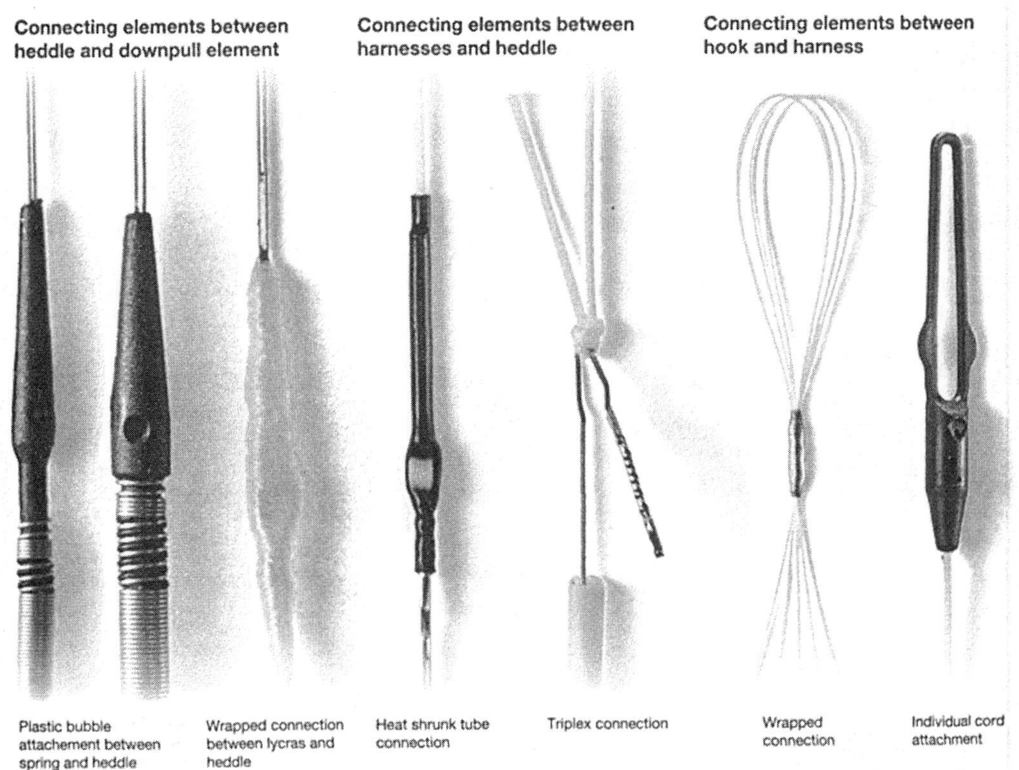

FIGURE 6.62 Various connecting elements for jacquard systems.

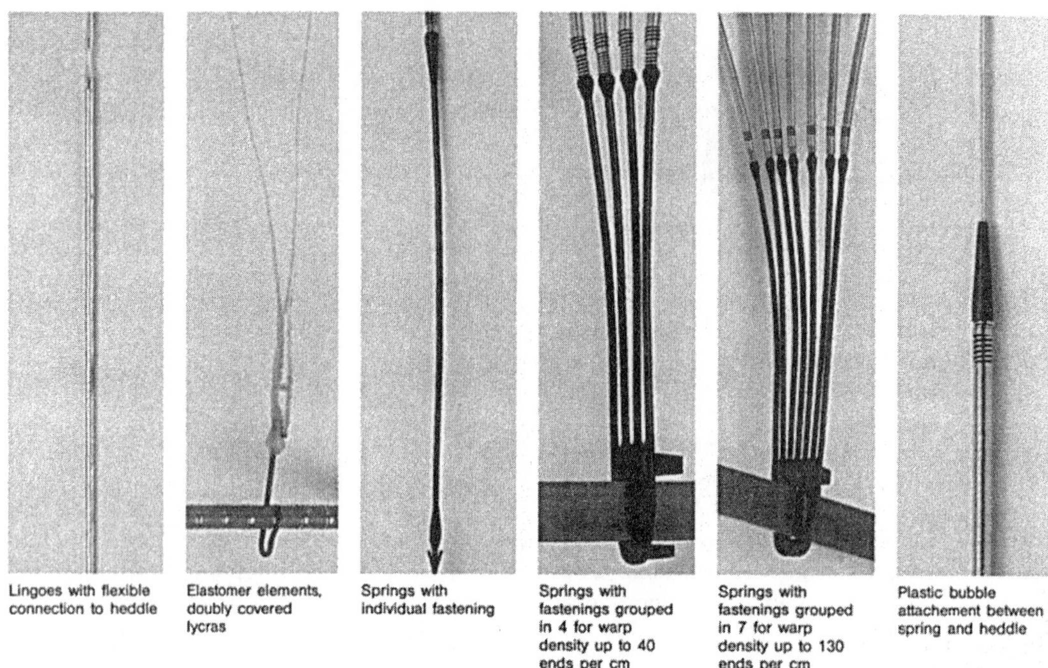

FIGURE 6.63 Downpull elements for harnesses (courtesy of Staubli).

fabrics that are heatset for stability. They also have to have good abrasion resistance. Figure 6.62 shows various connecting elements used in jacquard systems.

Figure 6.63 shows downpull elements for harnesses. Lingoes, elastics or springs with varying downpull forces are used. Table 6.2 shows typical forces for various fabrics. The downpull elements can be fastened to the machine or to the floor. A blower system can be installed for heavy lint fly.

Pick Finding Device

Modern weaving machines are equipped with pick finding devices (Figure 6.64). If a filling yarn is broken, the machine is automatically stopped and the faulty pick is found and removed

TABLE 6.2 Typical harness forces for various fabrics (Staubli).

<table>
<tr><td colspan="3"></td><td colspan="2">Mattress Duck</td><td colspan="4">Furnishing Fabric</td><td colspan="2">Blankets</td></tr>
<tr><td colspan="3">Harness with lingoes</td><td colspan="2">26 g</td><td colspan="4">26–34 g</td><td colspan="2">34–40 g</td></tr>
<tr><td colspan="3"></td><td colspan="2">Silks</td><td colspan="4">Narrow Fabrics</td><td colspan="2">Furnishing Fabric</td></tr>
<tr><td colspan="3">Harness with lycras</td><td colspan="2">30 g</td><td colspan="4">50 g</td><td colspan="2">50 g</td></tr>
<tr><td rowspan="2"></td><td rowspan="2">Silks</td><td rowspan="2">Linings</td><td colspan="2">Upholstery Fabrics</td><td rowspan="2">Matress Duck</td><td rowspan="2">Table Cloths</td><td colspan="2">Terry Cloth</td><td rowspan="2">Blankets</td><td rowspan="2">Velours</td></tr>
<tr><td>Light</td><td>Heavy</td><td>Base</td><td>Pile</td></tr>
<tr><td>Harness with springs</td><td>20–30 g</td><td>30–50 g</td><td>50 g</td><td>50–75 g</td><td>50–75 g</td><td>80–180 g</td><td>80 g</td><td>50–80 g</td><td>180 g</td><td>80 g</td></tr>
</table>

from the fabric. Pick finding devices are suitable for cam motions, dobbies and jacquard machines.

When a filling yarn breaks, a signal from the weaving machine automatically uncouples the shedding machine. Then the pick finder reverses the harness frames of the weaving machine (Figure 6.65).

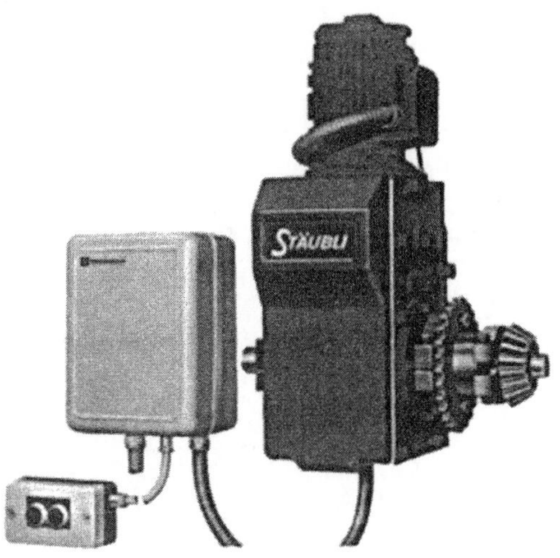

FIGURE 6.64 Pick finding device for dobby shedding (courtesy of Staubli).

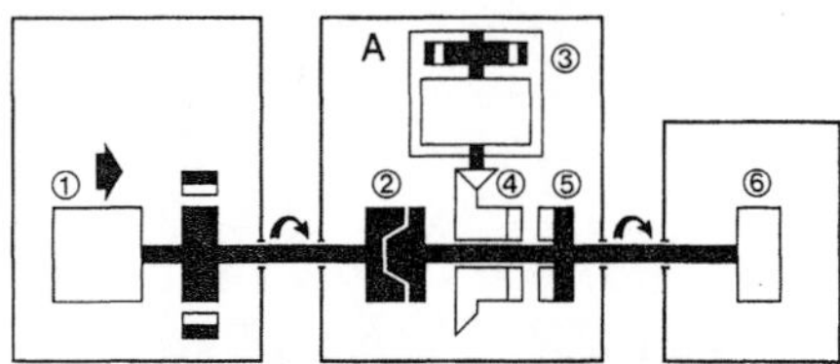

A Pickfinding device
1 Weaving machine
2 Claw coupling
3 Brake motor
4 Bevel gearing
5 Coupling
6 Shedding machine

The dobby is driven directly by the weaving machine via a claw coupling

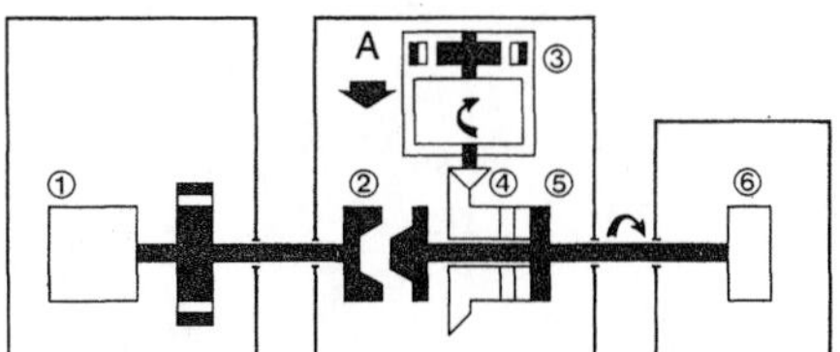

The dobby is disengaged from the weaving machine and reversed pick by pick by the motor of the pickfinding device via a coupling

FIGURE 6.65 Working principle of pick finding device (courtesy of Staubli).

REFERENCES

1. Marks' Standard Handbook for Mechanical Engineers, T. Baumeister, Ed., McGraw Hill Book Company, 1978.
2. Marks, R., and Robinson, A. T. C., Principles of Weaving, The Textile Institute, 1976.

SUGGESTED READING

- Lord, P. and Mohamed, M. H., Weaving: Conversion of Yarn To Fabric, Merrow Technical Library, 1982.
- Bissmann, O., "Shedding Motions: Essential for Every Weaving Machine", ITB International Textile Bulletin, 6/98.

REVIEW QUESTIONS

1. Does the choice of shedding system depend on fabric style? Explain why or why not.
2. What is the reason that limits the number of cams on a weaving machine?
3. What are the differences between negative and positive dobby?
4. Explain the working principle of rotary dobby.
5. What are the advantages and disadvantages of jacquard shedding compared to other shedding systems?
6. How does a pick finding device work?

7

Shuttle Weaving

Filling insertion with shuttle is the oldest filling insertion system. Shuttle weaving was used for centuries beginning with manual looms. In early manual looms, individual weights were being used to tension the warp yarns and the filling was being inserted with a stick. The beat-up was still there but the shedding was not well defined. In 1733, Kay invented the fly shuttle which increased the filling insertion rate and needed to have a clear shed opening. However, Kay's shuttle was still a hand-operated shuttle. Cartwright invented a "power loom" in 1785 which could be operated by "two strong men". With the invention of steam power, the looms of the early 1800s were made of cast iron. Gradual improvements to the power loom resulted in the shuttle looms of today (Figure 7.1), which are about to become extinct due to development of shuttleless weaving machines. The shuttle looms are no longer manufactured for traditional woven fabric manufacturing; however, there are still some 2.5 million shuttle looms in operation throughout the world. The existing shuttle looms are being replaced by shuttleless machines due to disadvantages of shuttle looms such as low speed, low productivity, high power requirements, high noise level and dangers involved with their operation. A detailed description of shuttle weaving is not relevant anymore. Besides, there are some good books specifically dedicated to shuttle weaving; therefore, only a basic description of shuttle weaving is given in this section for three primary reasons:

- For comparison purposes with shuttleless looms. Weaving machines are still categorized as being a "shuttle loom" or a "shuttleless loom".
- For historical reasons, since the shuttle loom was the only loom used for centuries until the widespread use of shuttleless weaving machines.
- Although the shuttle loom has been obsolete, the fundamental motions of a shuttle loom have been transferred to the modern shuttleless weaving machines.

In shuttle weaving, the filling is inserted by a shuttle that traverses back and forth across the loom width. Figure 7.2 shows a shuttle with a quill (pirn) in it. Shuttles can be made of wood, plastic or a combination of both. Figure 7.3 shows various shuttles and Figure 7.4 shows different quill designs. Filling yarn is wound on the quill. As the shuttle moves across the loom, the filling yarn is unwound from the pirn and laid in the shed. The shuttle moves continuously back and forth across the loom. On each side of the loom, there is a picking stick that propels the shuttle by hitting it and causes it to fly across the loom inside the open shed. Picking sticks are

FIGURE 7.1 A shuttle loom.

usually made of special woods that can absorb energy without fatigue. Once in contact, the picking stick and shuttle travel together approximately 20 cm, which is called picking, after which the picking stick stops and the accelerated shuttle continues its flight across the loom. The shuttle reaches a speed of around 50 km/h. When at the other side of the loom, the shuttle has to be decelerated by the other picking stick which is called checking.

The flight time of the shuttle, t, is given by:

$$t = W/V \tag{7.1}$$

where W is the width of the loom and V is the average shuttle speed. The displacement curve of the shuttle is parabolic, which is expressed as:

$$x = c_1 t^2 + c_2 t + c_3 \tag{7.2}$$

where c_1, c_2, c_3 : constants
t : time
x : shuttle displacement

Shuttle velocity is given by:

$$V = dx/dt = 2c_1 t + c_2 \tag{7.3}$$

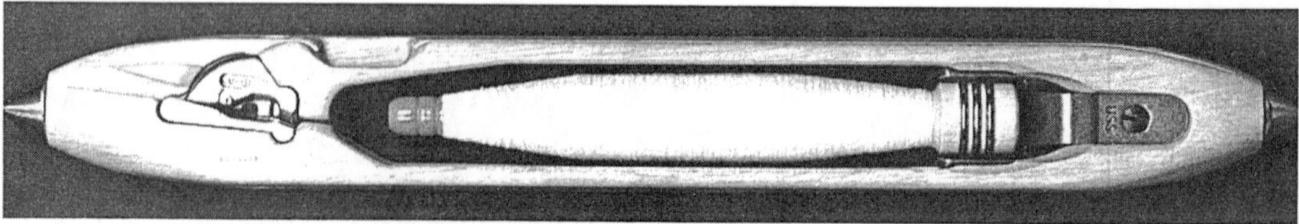

FIGURE 7.2 Shuttle with a quill (courtesy of Vermeutex).

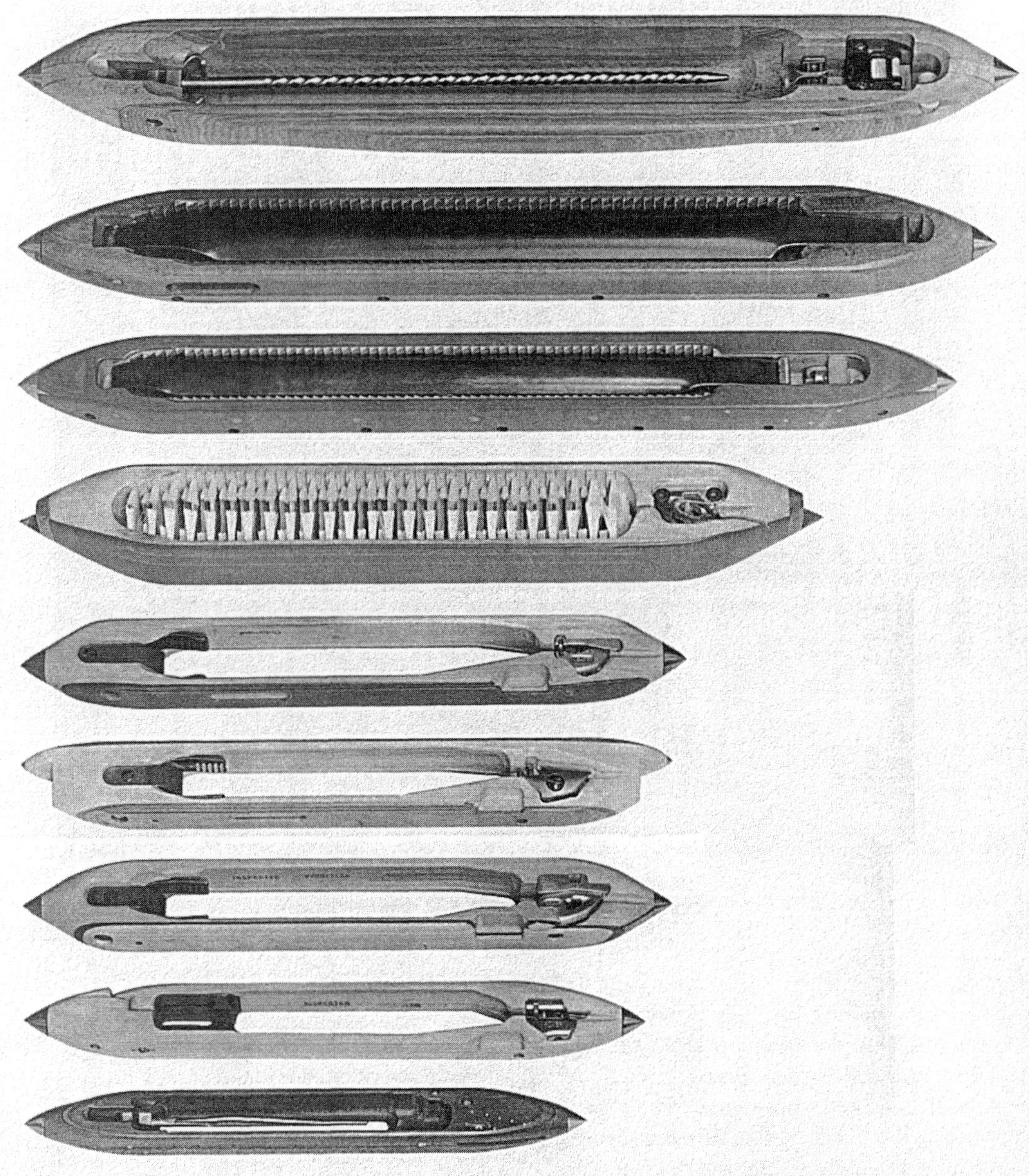

FIGURE 7.3 Various shuttles (courtesy of Vermeutex).

Shuttle acceleration is given by:

$$a = d^2x/dt^2 = 2c_1 \tag{7.4}$$

There are different mechanisms for shuttle picking. Figure 7.5 shows a picking cam mechanism. A continuously rotating picking cam generates energy to displace the picking shaft via a cone. The picking shaft pulls the picking stick with a lug strap, the picking stick accelerates and hits the shuttle. In this mechanism, the shuttle speed changes with the loom speed. Picking can also be done by a spring mechanism which stores energy and releases it suddenly to accelerate the shuttle. For multi-color weaving or for the purpose of mixing filling yarns to prevent

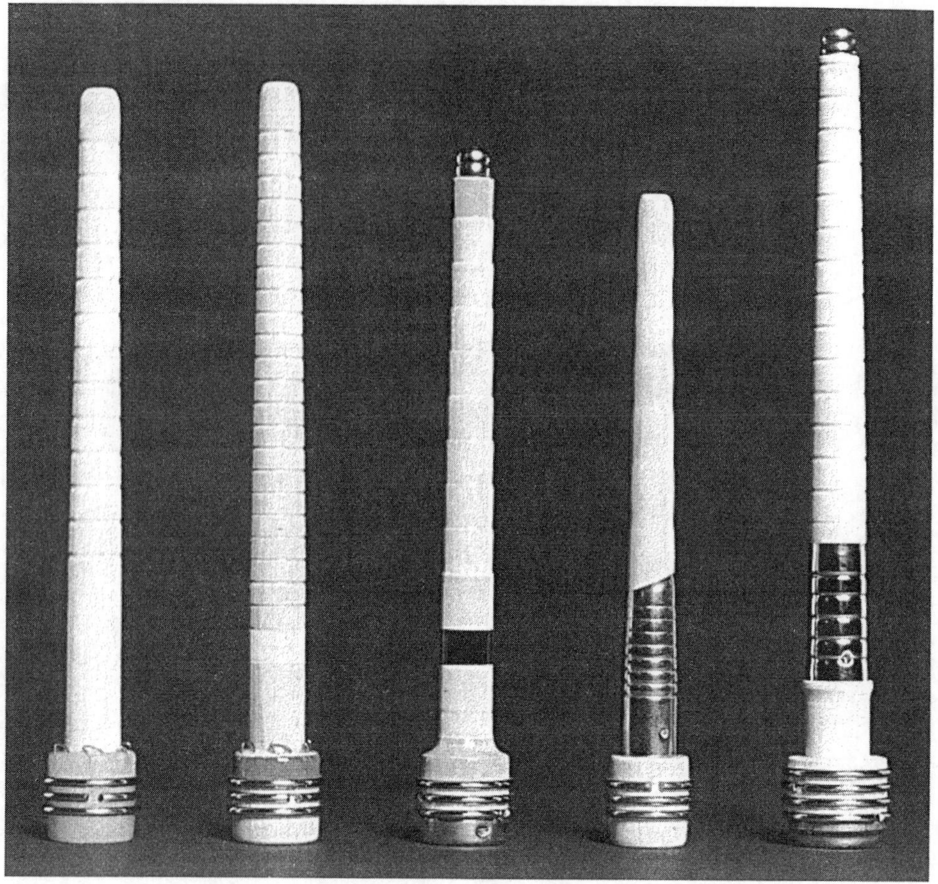

FIGURE 7.4 Quills (courtesy of Vermeutex).

fabric barre, multishuttle looms are used. In these looms, there are usually two or four shuttle boxes on each side of the loom.

After it is picked, the shuttle travels on the race board, the lower portion of the warp yarns being in between the shuttle and the race board (Figure 7.6). The race board supports the shuttle while the reed helps guide the shuttle. The reed and race board are assembled together which is called a lay or sley. The lay oscillates back and forth by two levers that are called lay swords. Lay swords pivot about rocking shafts which are driven by a crank and connecting arm as shown in Figure 7.7. The lay operates once every weaving cycle to beat-up the filling yarn and performs a continuous harmonic motion. Since the shuttle loom has one picking mechanism on each side, each picking mechanism operates every other weaving cycle.

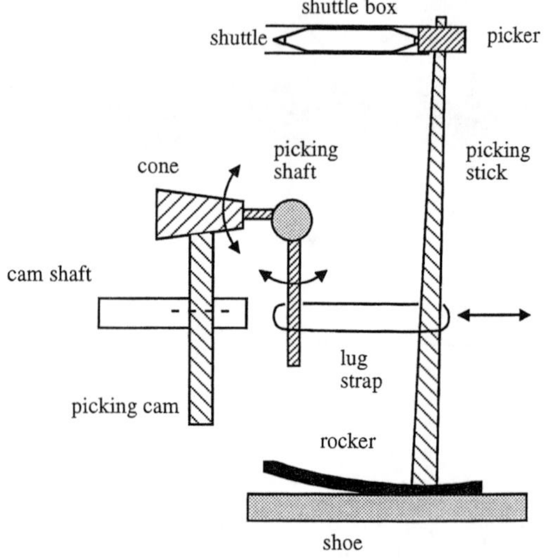

FIGURE 7.5 Shuttle picking with cone underpick mechanism.

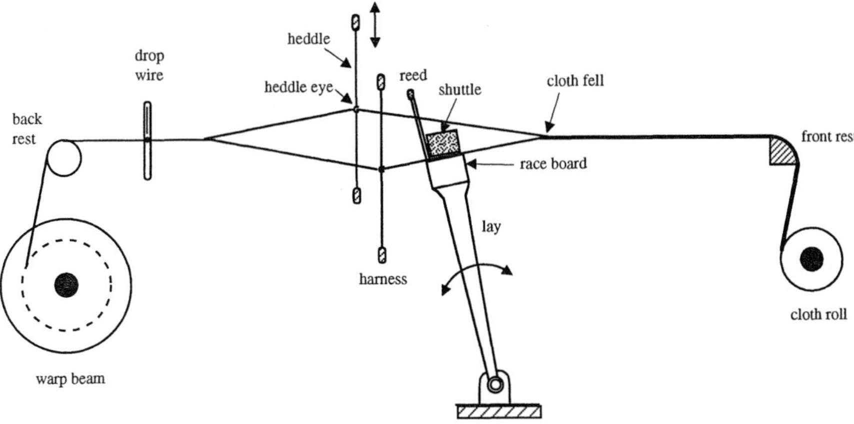

FIGURE 7.6 Cross section of shuttle loom along the warp direction.

The flight of the shuttle has to be carefully controlled as much as possible because the shuttle has to enter the shuttle box properly. If the shuttle does not follow its intended path, it may get out of the shed (called flying shuttle) which is a dangerous situation.

All the events on the loom have to be properly synchronized which can be demonstrated on a loom timing diagram. Figure 7.8 shows a typical loom timing diagram for a shuttle loom. It should be noted that the timing diagram can be different for each loom and fabric design. However, the five basic motions of a loom have to be completed in 360°. In this particular diagram, the beat-up position (where the reed is at the most forward position) is represented as 0°. Note that the movement of the reed takes place over several angles. The shuttle starts moving at around 80° following the beat-up. The contact between the shuttle and picker lasts around 30° which corresponds to approximately 30 cm. At this position the shuttle reaches a speed of around 15 m/sec, the picker stops pushing the shuttle and the shuttle flies across the loom by itself. Once at the other side of the loom, the shuttle is brought to a stop by the checking mechanism which is similar to picking.

Insertion with shuttle is an inefficient process in the sense that the shuttle weighs around 0.5 kg while the weight of the inserted yarn at one pick is less than 1/1000 of the shuttle weight. As a result, other forms of filling insertion without

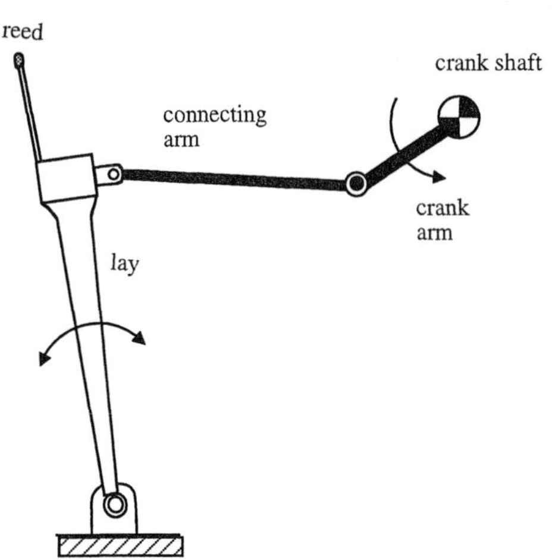

FIGURE 7.7 Schematic of lay mechanism.

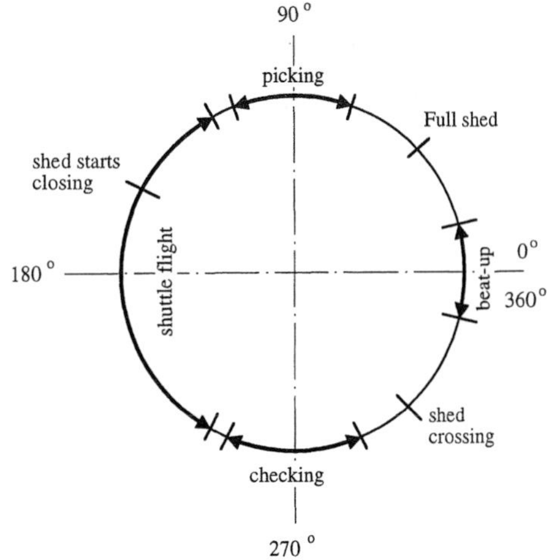

FIGURE 7.8 Typical timing diagram for shuttle loom.

a shuttle have been developed which replaced shuttle weaving.

SUGGESTED READING

- Lord, P. R. and Mohamed, M. H., Weaving: Conversion of Yarn to Fabric, Merrow Technical Library, 1982.
- Marks, R. and Robinson, A. T. C., Principles of Weaving, The Textile Institute, 1976.

REVIEW QUESTIONS

1. What are the reasons for the extinction of shuttle looms?
2. Are there any niche markets that still necessitate the use of shuttle looms?

8

Jet Weaving

8.1 AIR-JET WEAVING

Air-jet weaving is a type of weaving in which the filling yarn is inserted into the warp shed with compressed air. Figure 8.1 shows a schematic of air-jet weaving utilizing a multiple nozzle system and profiled reed which is the most common configuration in the market. Yarn is drawn from a filling supply package by the filling feeder and each pick is measured for the filling insertion by means of a stopper. Upon release of the filling yarn by the stopper, the filling is fed into the reed tunnel via tandem and main nozzles. The tandem and main nozzle combination provides the initial acceleration, where the relay nozzles provide the high air velocity across the weave shed. Profiled reed provides guidance for the air and separates the filling yarn from the warp. A cutter is used to cut the yarn when the insertion is completed. Figure 8.2 shows a modern air-jet weaving machine.

The air-jet weaving machine combines high performance with low manufacturing requirements. It has an extremely high insertion rate. Due to its exceptional performance, air-jet machines are used primarily for the economical production of standard fabrics, covering a wide range of styles. Meanwhile, more and more niches and special fabric segments are covered: heavy cotton fabrics such as denim, terry fabrics, glass fabrics, tire cord, etc.

Air-jet filling insertion is the simplest way of inserting the filling yarn which probably explains why air-jet weaving machines are one of the most popular machines in the market today. The major components of the insertion system are the tandem and main nozzles, ABS brake system and relay nozzles which are relatively simple in design. The insertion medium mass to be accelerated is very small, relative to the shuttle, rapier or projectile machines, which allows high running speeds. Unlike rapier or projectile insertion systems, there are not many mechanically moving parts to control and insert the filling yarn. The advantages of air-jet weaving machines are:

- high productivity
- low initial outlay
- high filling insertion rates
- simple operation and reduced hazard because of few moving parts
- reduced space requirements
- low noise and vibration levels
- low spare parts requirement
- reliability and minimum maintenance

Figure 8.3 shows a typical sequence of events during a complete cycle in air-jet filling insertion. After the tandem and main nozzles are turned on, yarn is released from the clamp (stopper). When all the coils of the particular pick have been pulled off the feeder, the stopper

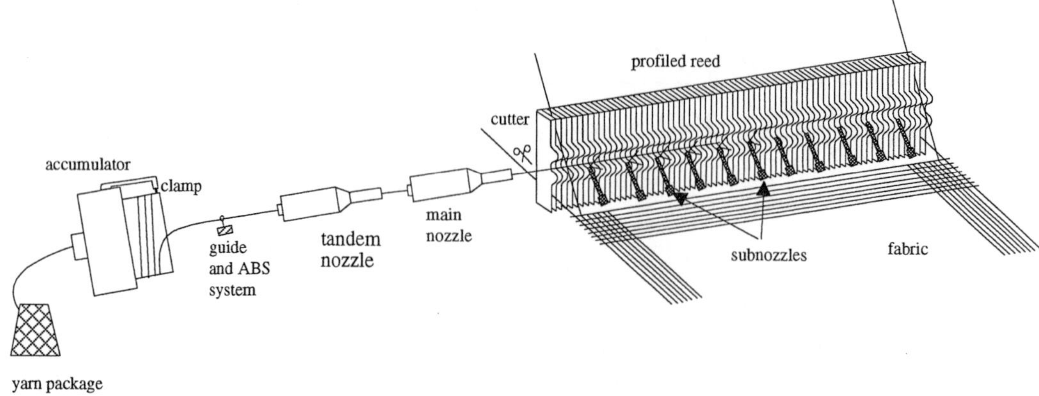

FIGURE 8.1 Schematic of air-jet filling insertion with profiled reed [1].

closes, the yarn decelerates and then will be beaten into the fabric. Thereafter, the air is turned off and the pick is cut to complete the cycle. A typical timing diagram of main and relay nozzles is shown in Figure 8.4. The timed groups of relay nozzles blow air on the tip of the yarn across the machine width. As a result, the yarn is pulled by the air at the tip (rather than pushed from behind) throughout the insertion, minimizing the possibility of buckling which may cause weaving machine stops. This also assures the lowest possible air consumption.

8.1.1 Historical Development of Air-jet Machines

The first application of compressed air as the filling yarn carrier can be traced back to 1914. At that time, Brooks [2] used a projection nozzle and was granted the first U.S. patent on the air-jet weaving concept. He was granted three more patents on air-jet weaving in the following years [3–5]. In 1929, Ballou [6] made a further step by adding a suction nozzle at the receiving side.

The first commercial machine was the Maxbo air-jet weaving machine which was introduced

FIGURE 8.2 Modern air-jet weaving machine.

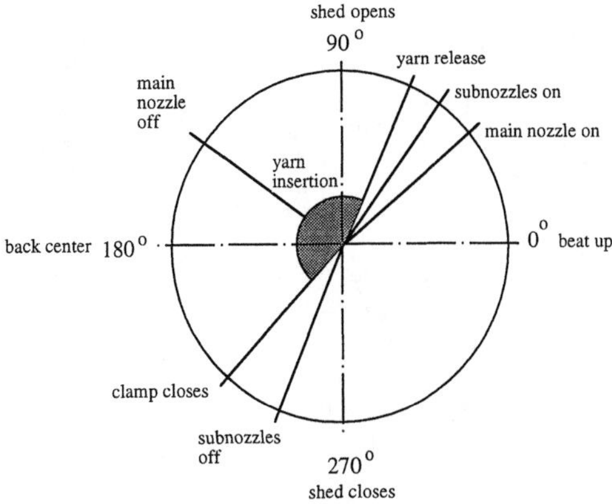

FIGURE 8.3 Typical timing diagram of an air-jet machine with multi-nozzles and profiled reed [1].

by Max Paabo [7,8] and first exhibited in Sweden in 1951, weaving cotton cloth 80 cm wide at a speed of 350 picks per minute (ppm). At about the same time, another air-jet weaving machine was introduced in Czechoslovakia which was developed by Vladimir Svaty [9] and marketed by Kovo trading company. Svaty used the confusor system.

In the early 1960s, use of confusors on the weaving machines as means of constraining the expansion of the air jet extended the weaving machine width to 150 cm [10]. Multi-jet weaving was proposed by Walter Scheffel of Germany [11] and Hifumi Saito of Japan [12]. In 1969, Te Strake developed the filling insertion system with main nozzle, relay nozzles and tunnel reed. After the take-over of Te Strake by Ruti Machinery Works, the system was introduced into the market in 1975 (ITMA 75 in Milan). After these developments, 330 cm weaving machine width and 600 ppm were achieved [13].

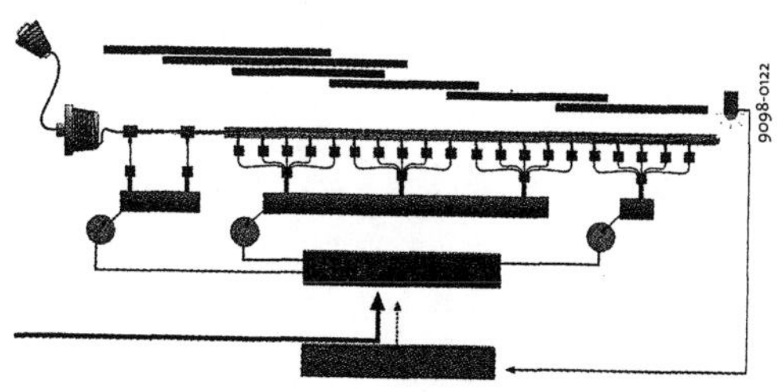

FIGURE 8.4 Programmable air blowing times with automatic correction at speed changes and automatic pressure optimization due to individual air supply.

In 1979, Nissan started using plastic confusors. Toyoda combined subnozzles with plastic confusors. In 1981, Picanol introduced a two color weaving machine and, in 1983, Bonas introduced a four color one. Also to make the guide approaching a closed tunnel, they used flexible plastic pieces that cover the filling exit of the confusor preventing the air from going out. As a result of this, a 390 cm weaving machine width was reached.

Air-jet weaving machines have become very popular in recent years. The air-jet weaving machine was selected as one of the top innovations of the last 25 years by the *Textile World* magazine [14]. The number of air-jet weaving machine manufacturers has increased considerably over the past several years. Intensive research and development on air-jet weaving machines have continued. As a result, air-jet weaving machines are getting wider, faster and more economical than before. Today, a running speed of over 1800 ppm (at 190 cm width) is achieved although it is not commercially feasible yet. Weaving widths of around 400 cm are available. At ITMA '99 in Paris, Dornier exhibited a 430 cm air-jet weaving machine with a jacquard head. The filling insertion rate has exceeded 2400 m/min in commercial running and up to eight color pick-at-will is available. Increased width and speed of air-jet weaving machines created the demand for improved filling yarn characteristics in terms of strength and evenness. Figure 8.5 shows a modern weaving plant with air-jet weaving machines.

Air-jet weaving machines are under constant development. Current research is mainly focused on the air and yarn interaction as well as the guide system to increase the yarn velocity and reduce the compressed air consumption. Widening of the application range is another topic of the current developments.

8.1.2 Yarn Insertion Systems

Yarn Feeders

The condition of the yarn feed into the nozzle exerts a great influence on the movement of the yarn through the shed. Because of the high yarn

FIGURE 8.5 A modern weaving plant with air-jet weaving machines.

velocity during insertion, it is not possible to unwind yarn intermittently from the filling package. Therefore, yarn storage and feed systems are used between the nozzle and the yarn package. There have been two main systems to store the yarn behind the nozzle before insertion: loop storage and drum storage (Figure 8.6). In loop storage, which has become obsolete, air was used to form a loop and keep the yarn straight without getting entangled. This storage system is simple in construction, however, it does not provide positive control on the yarn. The length of the loop depends on the diameter and speed of the measuring drum and may change due to slippage between the measuring drum and the feeding roller. The principle of drum storage is based on the storage of a sufficient yarn length (more than one pick) on a highly polished metal cylindrical body, "spider legs" or ceramic bands. Drum storage and yarn package may or may not be arranged in one straight line. A stopper pin or hook controlled by a timing control unit releases the required number of wraps of filling yarn to form each pick. Today, drum storage is the only yarn storage system in air-jet machines (Chapter 5).

There are two basic types of drum feeders: wire feeder and roller feeder. In a wire feeder, radially positioned wires allow minimum yarn contact for low unwinding tension, less air consumption and faster insertion. A laser beam is used to check the yarn release to assure stability of filling insertion. A drum diameter adjustment facilitates weaving width changes. In a roller feeder (also called "drum feeder with winding

1. Loop storage

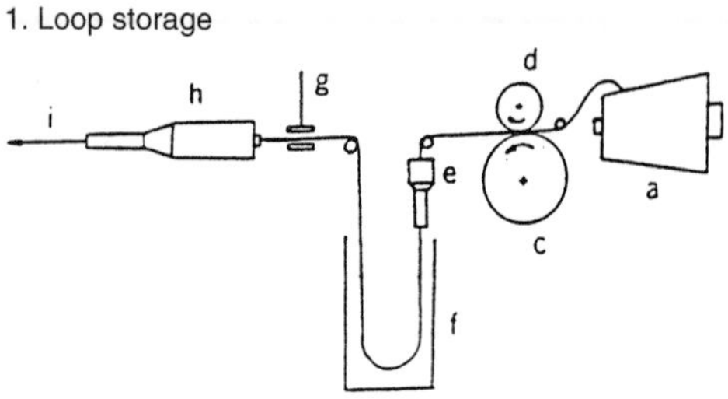

2. Drum storage

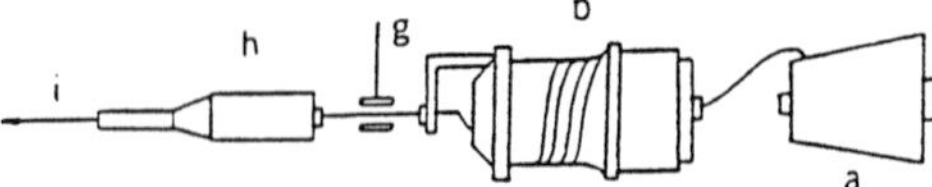

a. Yarn supply package
b. Drum feeder
c. Measuring drum
d. Feeding roller
e. Auxiliary nozzle
f. Storage pipe
g. Clamp
h. Main nozzle
i. Yarn

FIGURE 8.6 Schematic of yarn storage systems [15].

separation") yarn is spaced on the drum without overlapping which is well suited for hairy filling yarns.

The loop storage system gives higher yarn velocity and different insertion behavior than the drum storage system. On average, drum storage gives higher yarn tension than the loop storage which hinders the yarn movement [15]. This means that the mechanical friction forces are higher in the case of drum storage wherein the yarn mass is gradually included in the motion, depending on the distance traveled inside the guide channel. The initial yarn mass in drum storage is small compared to the loop case. Therefore, the initial acceleration is higher than the loop storage. In loop storage, the effect of the yarn mass is significant in affecting the motion. The yarn length in the reserve loop can be roughly considered as a compact body which is available for flight without much resistance. Once it is accelerated, it gains higher velocities than in the case of drum storage and the velocity continues to be high during the rest of the insertion because of its inertia. In drum storage, the yarn is drawn from the drum surface uniformly.

In spite of the velocity advantage of loop storage, drum storage is used on almost all of the new air-jet weaving machines. One reason for this is that it allows better control on the yarn than in the case of loop storage. It is difficult to form a loop without getting the yarn entangled for some yarns, such as high twist or stiff yarns. The space requirements also favor drum storage, especially for wide weaving machines.

Insertion Configurations

Three different systems have been used mainly on commercial air-jet weaving machines:

1. Single nozzle, confusor guides and suction on the other side
2. Multiple nozzles with guides
3. Multiple nozzles with profiled reed

Figure 8.7 shows schematics of these three yarn insertion systems. In System 1, a single nozzle is used to insert the yarn. Diffusor lamellae are placed across the entire width to guide the air stream which is injected into the shed. The disadvantage of this system is the weaving width limitation due to lack of additional nozzles to keep the air (and yarn) velocity high enough across the fabric width. Therefore, weaving machines with this configuration are limited in width. In System 2, in addition to the main nozzle, auxiliary nozzles are used. They are arranged across the warp width at certain intervals and inject air sequentially and in groups in the direction of yarn movement. The main nozzle consumes only a small fraction of the compressed air used in air-jet weaving machines as compared to relay nozzles. System 3 has the lamellae built in the reed and auxiliary nozzles (also called relay or subnozzles) across the warp. In this system, the entrance and exit of the lamellae in and out of the shed are eliminated. Thus, abrasion on the warp ends is reduced and misplacement of the warp ends between lamellae, which may cause fabric defects, is prevented. With the profiled reed, the restriction on warp density is also less severe than the case of the confusor guide system. Figure 8.8 shows the typical dimensions of profiled reed lamella.

Although all the three systems have been used on air-jet weaving machines, System 3, multiple

1. Single nozzle, confusor guides and suction

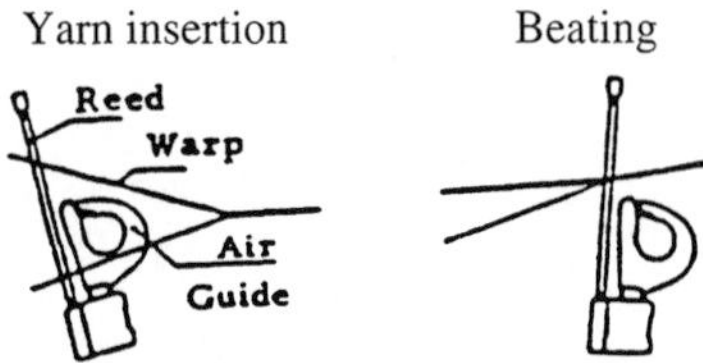

2. Multiple nozzles with guides

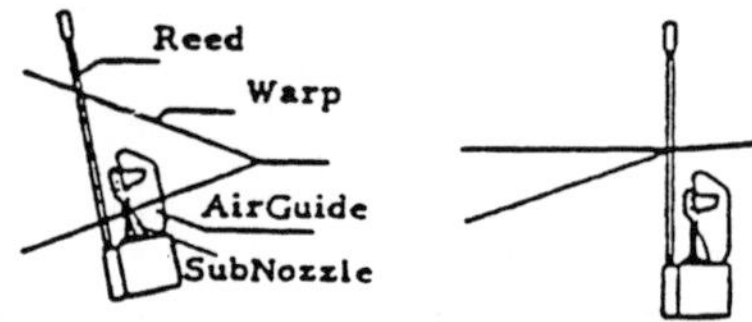

3. Multiple nozzles with profile reed

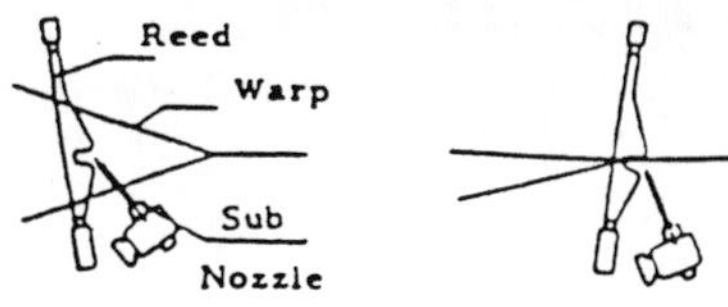

FIGURE 8.7 Yarn and air guiding systems in air-jet weaving [16].

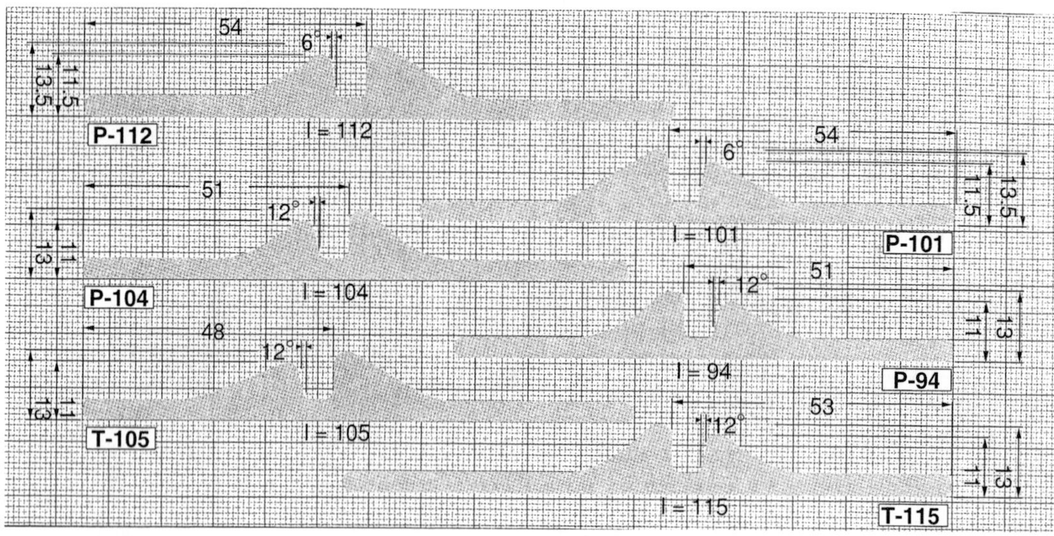

FIGURE 8.8 Dimensions of profiled reed lamella (courtesy of Laminage de Precision N.V.).

FIGURE 8.9 Relay nozzles and profiled reed (courtesy of Te Strake).

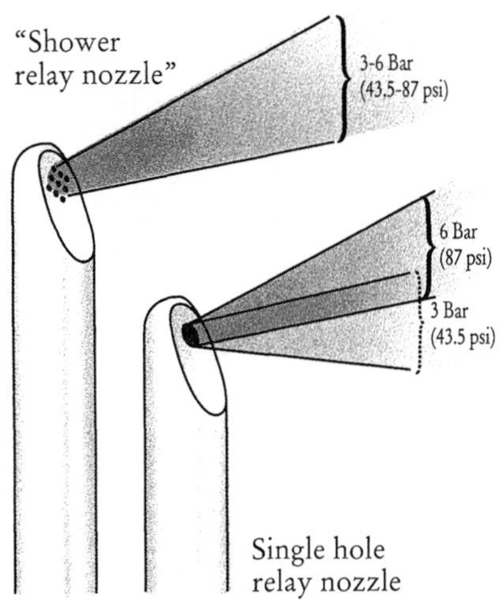

FIGURE 8.10 Multihole (shower) relay nozzle and single hole relay nozzle (courtesy of Te Strake).

nozzles with profiled reed, is the standard in the market today (Figure 8.9). Auxiliary nozzles are placed across the machine width on the lay. A relay nozzle can have a single hole or multiple holes arranged in the direction of yarn flight (Figure 8.10). A multihole relay nozzle is also called a shower nozzle.

It is reported that the convex shape of the relay nozzle top may prevent warp damages in special cases such as in filament warp weaving by avoiding splitting of the yarn (Figure 8.11). Table 8.1 shows the typical application areas of the two types of relay nozzles.

There have been other configurations of insertion systems with air-jet. Some of these methods are still in the development stage. In a method of air-jet weaving, called pneumatic rapier method [17,18], the filling yarn is placed into the shed by an air stream confined in hollow tubes. The yarn is projected by the compressed air to the middle of the shed where it is seized by the suction flow in the other rapier and is inserted to the end of the shed. In 1966, Elitex introduced a weaving machine in which a single missile, which grips the filling yarn at the leading end, is propelled by compressed air. The missile was inserted from both sides alternatively.

The patent literature indicates that the future generation of air-jet weaving machines may be using tubes as the guide system. A patent application by Ruti Machinery Works which was published in 1981 is an example of this trend [19]. Figure 8.12 shows this invention. Two lamellae combs (b) and (c) consist of plate-like elements (d) and (e) which possess an opening (g) for guiding the filling yarn and a yarn outlet or exit opening (h). During the operation, the lamellae combs (b) and (c) are periodically interleaved or shoved into one another and retracted away from one another. Their yarn exit slots (h) are closed by contact with the reed dents (a), as in Figure 8.12(C), where the position during insertion is shown. The openings (g) of the lamellae (d) and (e) form a continuous guide channel.

With this invention, it is claimed that since there is no depression or sinks at the guide channel walls, the yarn is not deflected in the direction of such sinks and cannot become

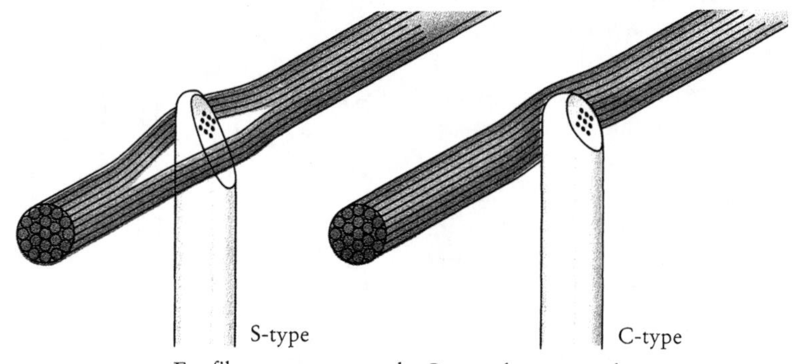

FIGURE 8.11 Relay nozzle tops (courtesy of Te Strake).

TABLE 8.1 Typical applications of relay nozzles (courtesy of Te Strake).

Te Strake Shower Nozzles Selection Table		Shower Nozzle Type			
		Coated		Uncoated	
Warp		S-type	C-type	S-type	C-type
Staple Yarn					
Cotton		O	O	X	O
Rayon	1)	X	O	X	O
Polyester	1)			X	O
Wool		X	O	—	—
Polyester/Cotton	1)	X	O	X	O
Polyester/Rayon	1)	X	O	X	O
Polyester/Wool		X	O	—	—
Other blends	1)	X	O	X	O
Filament Yarn					
Rayon	1) twisted	X	O	X	O
Rayon	air entangled	X	XX	X	XX
Acetate	1) twisted	X	O	X	O
Acetate	air entangled	X	XX	X	XX
Polyester/Polyamide	1) twisted	X	O	X	O
Polyester/Polyamide	zero twist	X	XX	X	XX
Polyester/Polyamide	air entangled	X	XX	X	XX
Polyester/Polyamide	texturized	O	X	O	X
Glass	zero twist	X	O	—	—
Glass	texturized	X	O	—	—
Polypropylene		—	XX	—	—

Note:
1) With dull fibers or filaments always use coated nozzles

Advice: XX strongly recommended for warps over 80 ends/inch
X advised
O alternative
— not advisable

caught or entrapped at the walls. Also, there is a reduction in the consumption of air since it cannot escape because of the closed wall of the channel. Further, it is reported that, with appropriate design of the airtight guide channel, the filling can be transported across by a suction action rather than a blowing action. In this case, the yarn can be controlled considerably better since there is applied a pure traction or tension force from the front instead of pressure at the rear. The described closed guide channel may also be applied to a flat weaving machine or to a rotary weaving machine having simultaneous sheds extending in the warp direction. Although this invention sounds promising, to date, there has not been any commercial or prototype machine displayed based on this concept.

In another invention by McGinley [20,21], helical annular pieces (a) are used to form a closed tube as shown in Figure 8.13. These pieces are mounted on a base (b) and can rotate around their vertical axis. They have slots for the yarn release. They enter the shed parallel to the warp. Then, upon rotation around their axis, they form a closed tube (position A). After the filling insertion, which takes place at position B, the rotation is reversed and the yarn (c) is pulled out from the guide channel when the pieces are withdrawn from the shed (position C). This invention is a multiphase warp-wave weaving machine in which sheds are opened, retained and moved towards the fabric with the use of simultaneous multiple air-jet insertion.

8.1.3 Principles of Air-jet Filling Insertion

Air Flow

The transporting medium, air, has a complicated motion during the insertion. The air flow

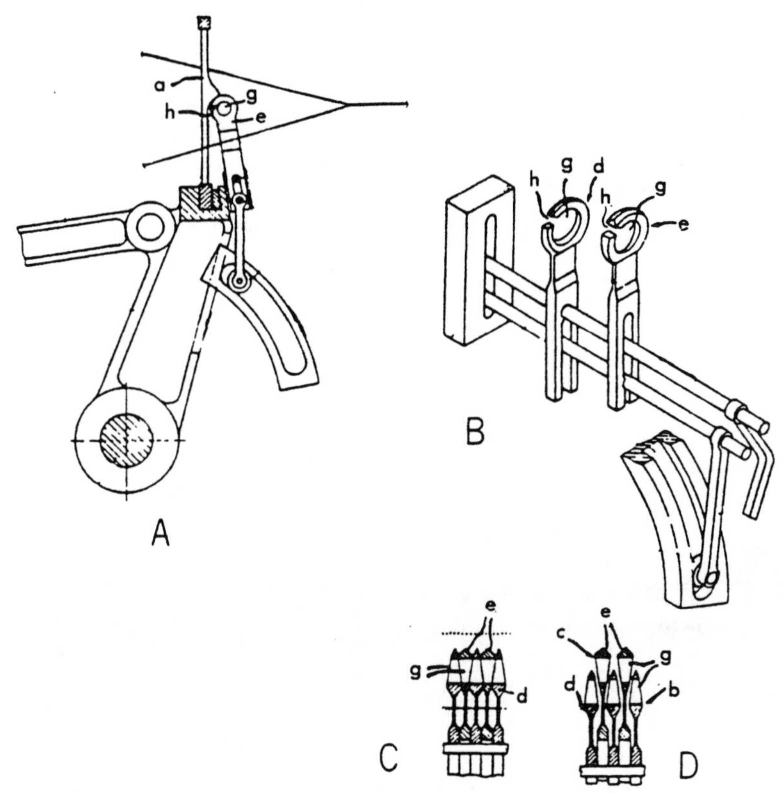

FIGURE 8.12 Air and yarn guiding system by Ruti Machinery Works [19].

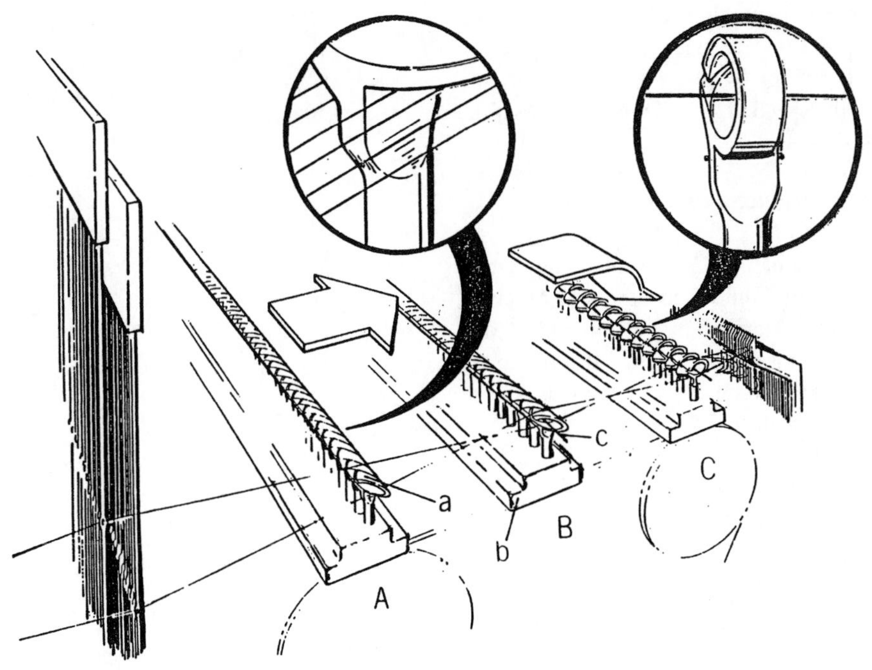

FIGURE 8.13 Air and yarn guiding system by McGinley [20,21].

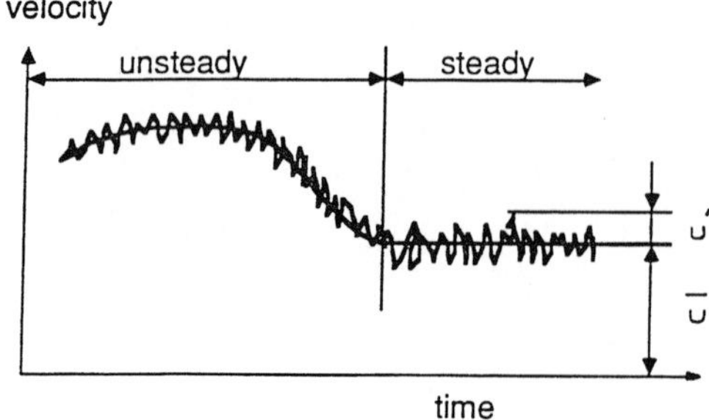

FIGURE 8.14 Type of air flow in air-jet insertion [16].

in air-jet insertion is unsteady, turbulent and can be either compressible or incompressible depending on the velocity (Figure 8.14). It is turbulent because Reynolds number is normally high. It is unsteady, because, in today's modern weaving machines, the air is turned on and off several hundred times per minute, which does not allow a steady flow to form. In addition to this, the guide channels are partly open and the walls are not very smooth which also contributes to the formation of turbulence. It is generally accepted that, for air velocities above 100 m/s, the compressibility should be taken into account. In air-jet weaving systems, it is possible to have air velocities above and below 100 m/s during one insertion cycle, depending on the position and time along the insertion length. In addition to all these, the velocity near the nozzle exit can be close to the speed of sound, making it possible for shock waves to occur inside the nozzle.

Flow Through Nozzles

The function of a nozzle is to increase the kinetic energy of a fluid in an adiabatic process. This involves a decrease in pressure and is accomplished by the proper change in flow area. The nozzle theory given in the literature is generally based on isentropic flow [22–24].

A typical shape of the nozzles used on air-jet weaving machines is shown in Figure 8.15 [25]. Due to the existence of the yarn tube, the performance of this nozzle differs from a conventional nozzle. The yarn tube separates the flow and develops another boundary layer which reduces the flow energy, due to friction. The flow inside the nozzle is complicated by the air that flows through the yarn tube. The direction and the amount of this flow depend on the pressure ratio at both ends of the yarn tube. The pressure at one end is atmospheric, but at the other end varies according to the pressure distribution inside the nozzle and the position of the yarn tube. The distance between the yarn tube tip and the air tube entrance L is used to describe the effect of the yarn tube position on the air velocity at the nozzle. Assuming that the yarn tube is blocked (i.e., no flow passing through), the nozzle can be approximated to the nozzle shown in Figure 8.15(b), which represents the actual area of the flow. The performance of this nozzle is similar to the performance of the converging-diverging nozzle, as shown in Figure 8.15(c). Depending on the pressure, various flow conditions may occur. In case 1, the stagnation pressure is equal to the atmospheric pressure and there is no flow. In case 2, isentropic flow occurs throughout the nozzle and the flow is entirely subsonic. In case 3, the Mach number is unity at the throat; a normal shock appears down-

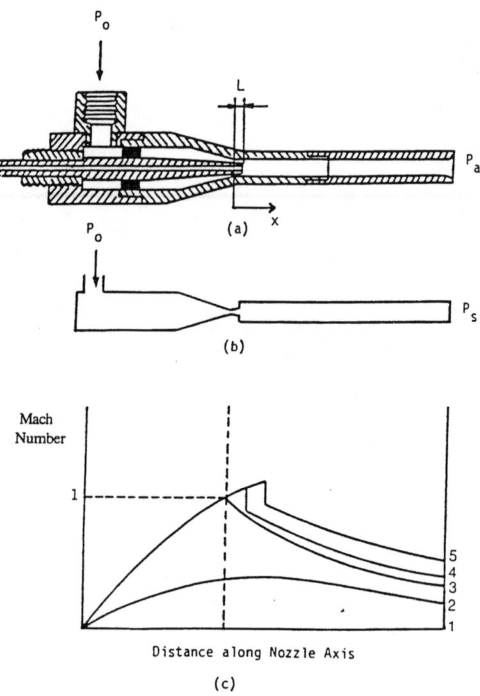

FIGURE 8.15 Schematic of a typical main nozzle used in air-jet weaving machines [25].

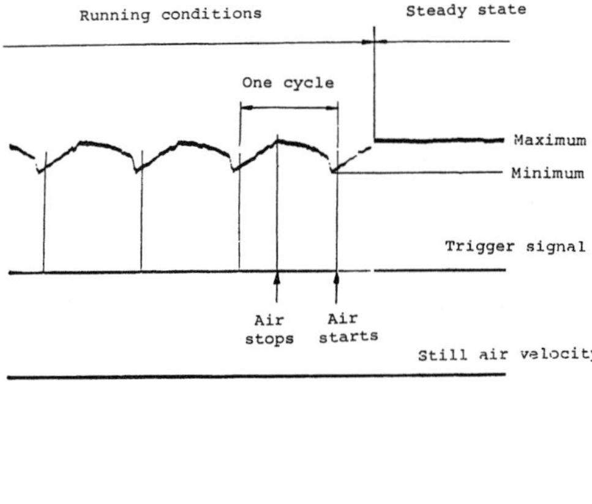

FIGURE 8.16 Air velocity chart during insertion [15].

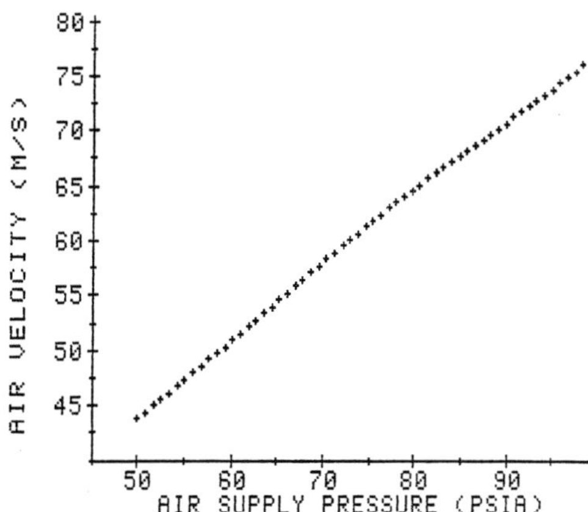

FIGURE 8.17 Effect of air supply pressure on air velocity [15].

stream of the throat and the process after the shock is subsonic deceleration. In cases 4 and 5, the shock happens to the flow downstream of the throat at a high Mach number. The actual air velocity at the nozzle exit will be affected by the length and diameter of the air tube and the frictional effects.

The axial air velocity at the nozzle exit decreases as the air tube length increases, due to the frictional effects in the air tube [26]. As the air tube length is increased, the turbulence level at the nozzle exit decreases.

Because of the discontinuous insertion of the air at the main nozzle, the air velocity at the nozzle exit changes between a maximum and a minimum value during the insertion cycle as shown in Figure 8.16. However, it does not drop to the velocity of the still air. If the main nozzle moves with the reed assembly, which is usually the case, then the flow situation becomes more complex.

With increasing air supply pressure, the average air velocity at the main nozzle exit also increases as shown in Figure 8.17.

Jet Flows

Turbulent flows are termed free if they are not confined by solid walls. A free jet is formed when a fluid is discharged from a nozzle or orifice into the atmosphere as shown in Figure 8.18 [27,28]. Except for very small velocities of flow, the jet becomes completely turbulent at a short distance from the point of discharge. Because of the turbulence, the emerging jet becomes partly mixed with the surrounding fluid at rest. Particles of fluid from the surroundings are carried away by the jet so that the mass-flow increases in the downstream direction. Empirically it was shown that a turbulent jet grows from its source at approximately 13° half angle, regardless of its actual Reynolds number. The relation between U_{max} (velocity at the center line) and x (distance along the axis) is given by:

$$U_{max} = \text{constant } (1/x) \qquad (8.1)$$

Flow in Air Guiding Systems

The air velocity in the shed of an air-jet weaving machine can be expressed as [29]:

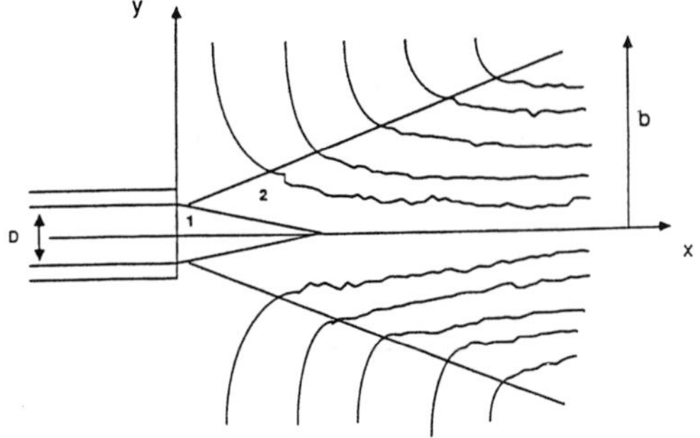

FIGURE 8.18 Free jet.

$$U = U_o \exp(-c\,x),\ (15D < x) \qquad (8.2)$$

where U_o : air speed at the entrance of the shed
c : loss coefficient in the shed
D : nozzle diameter

With a steady flow in the shed, the pressure drop, ΔP, due to friction and divergence of the air, is given by [29]

$$\Delta P = \frac{\lambda\, L_1\, L_2}{8\, A\, \rho\, U_m^2} \qquad (8.3)$$

where λ : pressure-loss coefficient which depends on Reynolds number
L_1 : periphery of the shed
L_2 : length of the shed,
A : area of the shed
ρ : air density
U_m : mean velocity

The pressure loss increases with an increase in the length of the nozzle. However, the nozzle must be long enough to make the air flow parallel after it has reached a throttled condition.

The air in the channel of a pneumatic weaving machine moves by inertia as in a jet, i.e., the loss in energy is made up from the kinetic energy only [30]. The unavoidable drop in air velocity along the channel is caused by turbulent exchange through the gaps and the longitudinal slots in the channel, as well as by the resistance arising from the configuration of the channel elements. Compared with a free jet, the channel provides improved conditions for filling insertion.

When air is discharged into a free space, its velocity drops to 4% of its original value at a distance of 1.5 m from the nozzle. When air is discharged from the nozzle into a confusor, the velocity drop becomes smaller, diminishing the air velocity at a distance of 1.5 m from the nozzle to 14–23% of its original value, depending on the confusor diameter [31]. The percentage drop in air velocity depends on the ratio of confusor diameter to jet diameter.

The velocity drop of the air-jet is enormous because of the nature of air, i.e., it is compressible, and the air ejected is very easy to expand. The velocity from the nozzle is a result of the pressure difference between the outer pressure and the inner pressure at the nozzle. To prevent expansion and to control the air velocity, several devices are used by weaving machinery manufacturers such as profiled reed, confusor guides, etc. A small shed opening may also be helpful for this purpose.

The turbulence of the air flow slows the yarn movement because it causes the yarn to deviate from the center of the flow. Therefore, the turbulence level should be minimal. However, it is impossible to insert the yarn in laminar flow because the air velocity would be very low.

For single nozzle insertion, air velocity distribution during an insertion cycle depends on both distance and time. It increases with time (at a fixed point on the insertion length during the yarn insertion cycle) but decreases with the distance along the insertion length. Figure 8.19 shows the air velocity distribution with respect to distance and time for a complete cycle which was measured on an air-jet weaving machine simulator that utilizes a single nozzle [16]. The measurement was done without yarn insertion. High insertion frequency, combined with the inertia effects of the air, develop an unsteady continuous flow inside the guide channel during the whole cycle. The air velocity fluctuations within a cycle are more obvious at the initial section of the channel, decreasing towards the end.

Yarn Tension

Yarn tension affects the filling insertion by hindering the yarn movement. Yarn velocity depends to a large extent on its tension. During the insertion, high tension causes longer insertion times, hence lower yarn velocities. Resulting from the high air pressure, high tensions also cause weak yarns to break. Therefore, it is desirable to even out the tension fluctuations during insertion. Several factors determine the yarn tension on the filling yarn: friction between the yarn and still air (before the nozzle), mechanical friction between the yarn and guides, and air-friction force on the yarn inside the insertion channel.

Figure 8.20 shows the yarn tension diagrams for drum storage for one pick and the average of 40 picks in single nozzle air-jet filling insertion [16]. Yarn tension was measured at the front

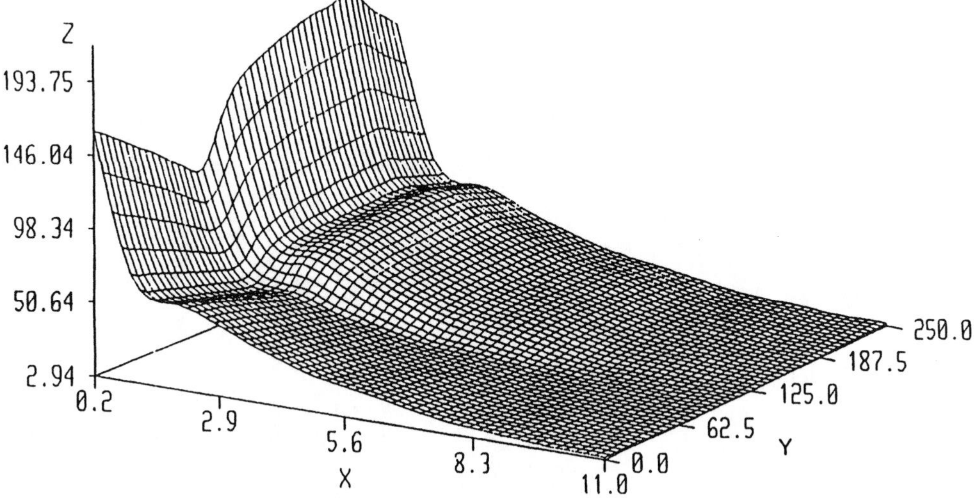

FIGURE 8.19 Air velocity distribution depending on time and distance along the insertion length [16]. X: distance along the insertion length (ft), Y: time (millisecond), Z: air velocity (m/s).

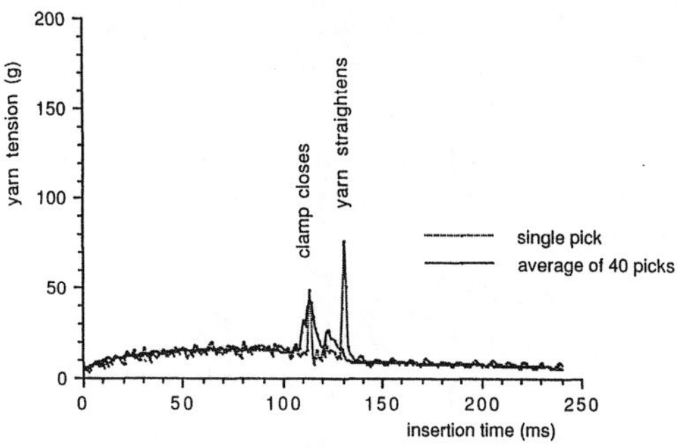

FIGURE 8.20 Tension diagrams for drum storage [16].

of the nozzle using a tensometer. Because of the nature of the air-jet filling insertion (i.e. the yarn motion is not positively controlled), the tension peaks take place at different times. Therefore, although the single picks show greater tension peaks and fluctuations, when the average of 40 picks is taken, the tension peaks are evened out and hence their magnitude decreases. This is shown more clearly in Table 8.2.

Analysis of Yarn Tension during Unwinding of Yarn from Drum Feeders

The overhead unwinding of yarn from a cylindrical surface by air-jet is a boundary value problem. Consider an inertial reference frame XOR, whose origin is fixed at the center of the drum surface and an X-axis which coincides with the drum axis (Figure 8.21). Consider, also, a moving coordinate system XOR′ that rotates about its X axis at a constant rotational velocity ω, which is the unwinding velocity of the contact point between the yarn and drum surface. The unit vectors of the reference frame that are associated with the yarn consist of tangent $\mathbf{e}_t$, principal normal $\mathbf{e}_n$ and binormal $\mathbf{e}_b$ (bold letters indicate a vector). The unit vectors of cylindrical coordinate system are $\mathbf{e}_r$ (radial outward), $\mathbf{e}_\phi$ (circumferential), and $\mathbf{e}_x$ (parallel to the drum axis). The moving coordinate system rotates

TABLE 8.2 Yarn tension values for drum storage [16].

	Yarn Tension (g)	
	Loop	Drum
Single pick:		
Average	10.898	11.907
At the end of the loop	186.572	
At clamp closing	80.957	48.974
When yarn straightens	115.136	77.636
Average of 40 picks:		
Average	10.396	11.958
At the end of the loop	55.810	
At clamp closing	40.380	42.48
When yarn straightens	46.093	26.513

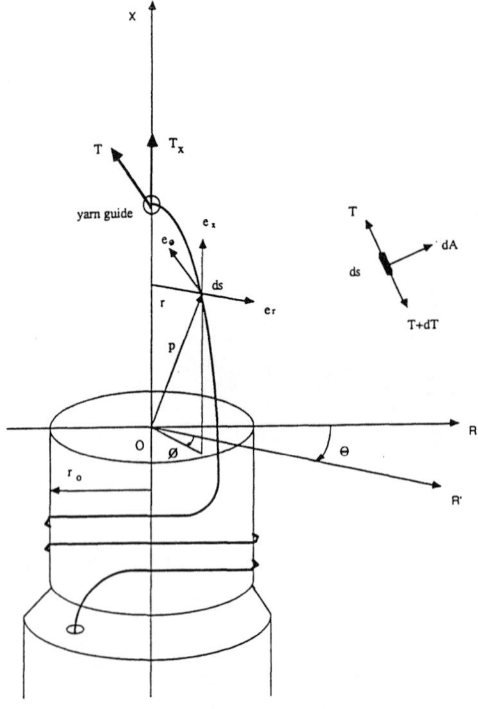

FIGURE 8.21 Theoretical model of yarn tension for drum storage [16].

about its X-axis at the constant (assumed) rotational speed ω (= $\omega\, \mathbf{e_x}$).

The yarn is assumed to be inextensible and perfectly flexible. Consider an infinitesimally small yarn segment of length ds with a position vector $\mathbf{p}$. From Figure 8.21,

$$\mathbf{p} = r\, \mathbf{e_r} + x\, \mathbf{e_x} \tag{8.4}$$

By definition,

$$\mathbf{e}_t = \frac{d\mathbf{p}}{ds}$$

Since,

$$\frac{d\mathbf{e_r}}{ds} = \frac{d\phi}{ds}\mathbf{e}_\phi, \text{ then} \tag{8.5}$$

$$\mathbf{e_t} = \frac{dr}{ds}\mathbf{e_r} + r\frac{d\phi}{ds}\mathbf{e}_\phi + \frac{dx}{ds}\mathbf{e_x}$$

The mass of the yarn segment ds is $dM = m\, ds$, where m is the mass linear density of the yarn. Then, the equations of motion of the yarn segment, by Newton's second law, are

$$\mathbf{a}\, dM = \Sigma\, d\mathbf{F} \tag{8.6}$$

Differentiating the position vector $\mathbf{p}$ with respect to time, the absolute velocity of the yarn is obtained (the derivative is calculated at a fixed point),

$$\mathbf{v} = \frac{d\mathbf{p}}{dt}$$

Expressing $d\mathbf{p}/dt$ in terms of its value relative to the rotating XOR′ system,

$$\mathbf{v} = \left[\frac{d\mathbf{p}}{dt}\right]_r + \boldsymbol{\omega} \times \mathbf{p} \tag{8.7}$$

where $(d\mathbf{p}/dt)_r$ is the rate of change of $\mathbf{p}$ as viewed from the rotating system and $\boldsymbol{\omega}$ is the absolute angular velocity of the rotating system.

To obtain the absolute acceleration of ds, the rate of change of each term in Equation (8.7) should be found at a fixed point. Thus,

$$\frac{d}{dt}\left[\frac{d\mathbf{p}}{dt}\right]_r = \left[\frac{d^2\mathbf{p}}{dt^2}\right]_r + \boldsymbol{\omega} \times \left[\frac{d\mathbf{p}}{dt}\right]_r \tag{8.8}$$

and

$$\frac{d}{dt}[\boldsymbol{\omega} \times \mathbf{p}] = \frac{d\boldsymbol{\omega}}{dt} \times \mathbf{p} + \boldsymbol{\omega} \times \left[\frac{d\mathbf{p}}{dt}\right]_r + \boldsymbol{\omega} \times (\boldsymbol{\omega} \times \mathbf{p}) \tag{8.9}$$

In Equation (8.8), $(d^2\mathbf{p}/dt^2)_r$ is the acceleration of the yarn segment relative to the XOR′ system, which is zero. In Equation (8.9), the first term on the right-hand side is also zero because $\boldsymbol{\omega}$ is constant. Adding Equations (8.8) and (8.9) together, the absolute acceleration of the yarn segment is obtained,

$$\mathbf{a} = 2\boldsymbol{\omega} \times \left[\frac{d\mathbf{p}}{dt}\right]_r + \boldsymbol{\omega} \times (\boldsymbol{\omega} \times \mathbf{p}) \tag{8.10}$$

In this expression, $(d\mathbf{p}/dt)_r = \mathbf{v_r}$ is the translational velocity of the yarn segment relative to the moving system and can be expressed as $v\mathbf{e_t}$.

The first term on the right-hand side is the Coriolis acceleration which arises from two sources. One is due to the changing direction of the velocity of yarn segment in space relative to the moving system and the other is due to the rate of change of the velocity $\boldsymbol{\omega} \times \mathbf{p}$ because of the changing magnitude or direction of the position vector **p** relative to the moving frame. The second term on the right side of Equation (8.10) represents the centripetal acceleration which is directed toward the axis of rotation perpendicularly to the yarn segment.

External forces acting on the yarn segment are:

1) Yarn tension, $\mathbf{T} = T\mathbf{e_t}$
2) Air resistance (drag force), **A**

Air drag force, **A**, is caused by the friction between the yarn and the air and has two components (Figure 8.22): along the yarn axis ($\mathbf{e_t}$) and perpendicular to the yarn path. Even at high rotational velocities, the tangential component of air drag is negligible and gravitational force on the yarn can also be neglected. Thus, Equation (8.6) reduces to

$$\mathbf{a}\, dM = d\mathbf{T} + d\mathbf{A} \tag{8.11}$$

The Coriolis acceleration is

$$2\,\boldsymbol{\omega} \times \mathbf{v_r} = 2\,\omega\, v\, \mathbf{e_N} \tag{8.12}$$

which is normal to the yarn path ($\mathbf{e_N}$: unit vector normal to the yarn but not the principal normal) [32].

The centripetal acceleration is given by

$$\boldsymbol{\omega} \times (\boldsymbol{\omega} \times \mathbf{p}) = -\omega^2\, r\, \mathbf{e_r} \tag{8.13}$$

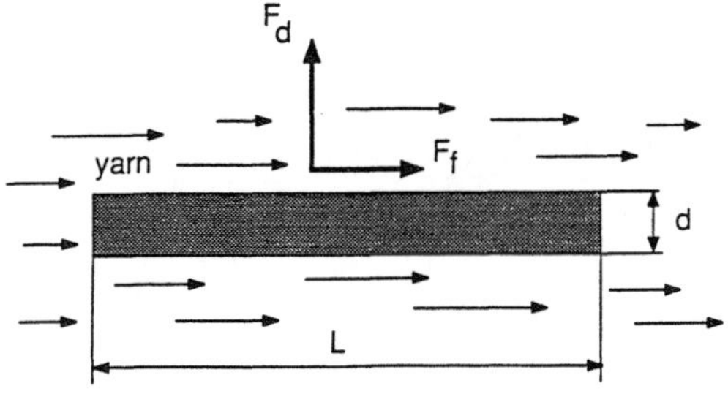

FIGURE 8.22 Forces acting on a moving yarn in air flow [16].

The component of Equation (8.13) along the yarn axis is $-\omega^2 r(dr/ds)$. As a result, the component of the equation of motion (8.11) along $\mathbf{e_t}$ results in

$$-\omega^2 m\, r\, dr = dT \tag{8.14}$$

Upon integration,

$$T = -\frac{\omega^2 m\, r^2}{2} + c \tag{8.15}$$

From the boundary condition, at $r = r_o$, $T = T_o$ which is the frictional force between the drum surface and the yarn,

$$c = T_o + \frac{\omega^2 m\, r_o^2}{2}$$

Thus, the tension distribution along the yarn path is

$$T = T_o + \frac{\omega^2\, m\, r_o^2}{2} - \frac{\omega^2\, m\, r^2}{2} \tag{8.16}$$

The maximum tension occurs at the yarn guide, i.e., when $r = 0$,

$$T = T_o + \frac{\omega^2\, m\, r_o^2}{2} \tag{8.17}$$

But, $\omega = V/r_o$ where V is the withdrawal speed of the yarn. Then,

$$T = T_o + 0.5\, m\, V^2 \tag{8.18}$$

Neglecting the friction force T_o, the tension at the yarn guide is given by:

$$T = 0.5\, m\, V^2 \text{ (Newton)} \tag{8.19}$$

where m : mass linear density of the yarn (kg/m)
V : withdrawal speed of the yarn (m/s)

This force is in the direction of yarn axis. The force in the drum axis direction is given by (considering the Capstan effect on yarn guide)

$$T_x = T\, e^{\mu\alpha} \tag{8.20}$$

where μ : coefficient of friction between the yarn and guide
α : wrap angle at the guide

Dynamics of Yarn Insertion with Air

Propelling Force by Air-jet

The total force on a body placed in a stream of fluid consists of skin friction (equal to the integral of all shear stresses taken over the surface of the body) and of pressure drag (integral of normal forces). The sum of the two is called total or profile drag [27]. The component of the resultant force parallel to the undisturbed initial velocity is referred to as the friction force (F_f) and the component perpendicular to that direction is called drag (F_d) as shown in Figure 8.22. The propelling force to move the yarn in air-jet insertion is provided by friction between the air and yarn surface and is given by the following formula:

$$F_f = 0.5\ C_f\ \rho\ (U - V)^2(\pi\ d\ L) \qquad (8.21)$$

where C_f : skin friction coefficient
ρ : air density
U : air velocity
V : yarn velocity
d : yarn diameter
L : yarn length subject to air

This force is proportional to the square of the relative velocity between the air and yarn. The propulsive force increases with an increase in the air velocity and the yarn diameter. This is because with increasing diameter, the yarn surface area that is in contact with the air becomes larger. The dimensionless coefficient C_f is a function of Reynolds number. In the case of compressible fluids, they also depend on Mach number. Skin friction coefficient for spun yarns and thick yarns (with a certain hairiness) is higher than that for fine and smooth yarns. For untreated cotton yarns the C_f is twice that of singed cotton yarns. For textured yarns, C_f varies depending on the openness of the yarn structure [33].

C_f can be determined from Equation (8.21) using a set-up as shown in Figure 8.23. The tension on a certain length of yarn is measured in an air flow with constant velocity. For this purpose, a plastic tube with constant inner diameter (D) is used. When using the tubes with larger diameters, approximately constant velocities can be obtained along a tube length of L_2. As the tube diameter is increased to obtain higher air velocities, the wall friction losses increase inside the tube causing a decrease in air velocity along the tube. For all cases, air velocity can be measured at several points along the tube and the average is taken to be used in Equation (8.21).

The tube is bent approximately 15° at one end as shown in the figure. The yarn is clamped on the left side in order to prevent relative movement against air. The yarn enters into the tube through a small opening. It is essential that the yarn does not touch the tube during measurements. After an initial tension is given to the yarn, it is clamped at the other side, as well. First, this initial tension (T_1) is measured. Then,

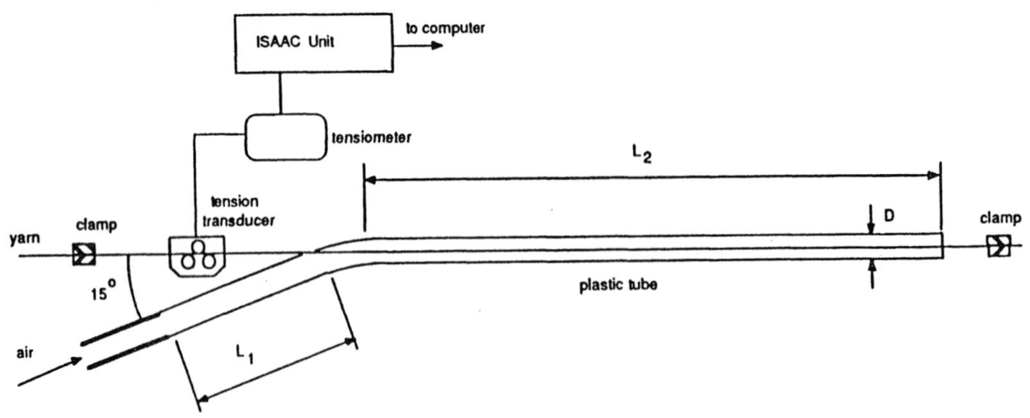

FIGURE 8.23 Arrangement for air friction coefficient measurement [16].

air is turned on and the tension on the yarn is measured (T_2). The tension on the yarn due to air friction is obtained as

$$T = T_2 - T_1 \qquad (8.22)$$

The surface area (A) of the yarn subjected to air flow is given by

$$A = \pi \, d \, L_2 \qquad (8.23)$$

where d : diameter of the yarn (m)
L_2 : length of the yarn under the influence of air (m)

Table 8.3 shows the air-friction coefficient results for a two-ply ring spun cotton yarn (50/2 Ne) with $d = 2.115 \times 10^{-4}$ m and $L_2 = 1$ m. Figure 8.24 shows the relation between the air-friction coefficient and air velocity for the air velocity range of 20–200 m/s. The relation is geometrical in the form of

$$C_f = c_1 \, U^{-c_2} \qquad (8.24)$$

where c_1 and c_2 are constants and U is the air velocity (m/s).

Yarn Flight in Air-jet

The movement of yarn in air-jet insertion is a complex motion. It is not a positively controlled process. Once the yarn is released from the clamp, there is little, if any, that can be done to control its movement along the insertion length.

The yarn is subjected to sudden acceleration and deceleration during its flight. Uniform acceleration is better than sudden acceleration for air-jet weaving because it reduces yarn breaks. Drum storage provides more uniform motion for the yarn than the loop storage.

When a stationary yarn is left in an air stream from a nozzle, the yarn accelerates as a result of the air-yarn friction. The part of the yarn flight where the yarn velocity is lower than air velocity is called the acceleration part as shown in Figure 8.25. The acceleration part can also be divided into two portions. The portion of constant air velocity (core of the nozzle) is called

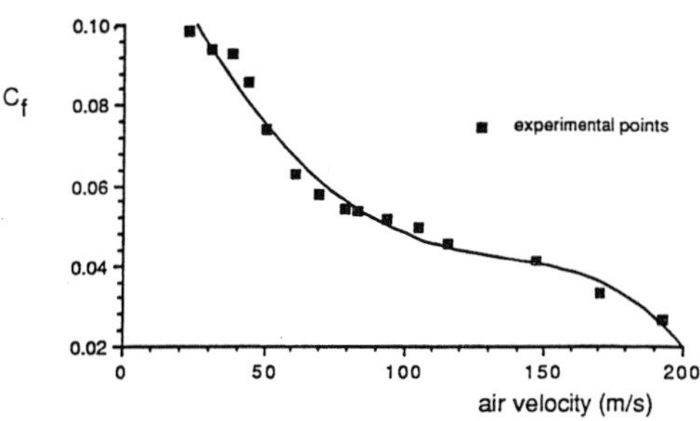

FIGURE 8.24 Air friction coefficient versus air velocity for a two ply ring spun cotton yarn (50/2 Ne) [16].

TABLE 8.3 Data for air-friction coefficient measurement [16].

Air Velocity (m/s)	Tension (cN)	C_f
23.311	2.214	0.0986
31.145	3.767	0.0939
37.976	5.536	0.0929
43.904	6.815	0.0855
50.128	7.594	0.0739
60.979	9.653	0.0628
70.083	11.734	0.0578
78.821	13.894	0.0541
83.440	15.381	0.0534
93.871	19.904	0.0515
104.821	22.491	0.0495
115.581	24.996	0.0452
147.000	36.953	0.0413
170.490	39.819	0.0331
192.666	40.829	0.0266

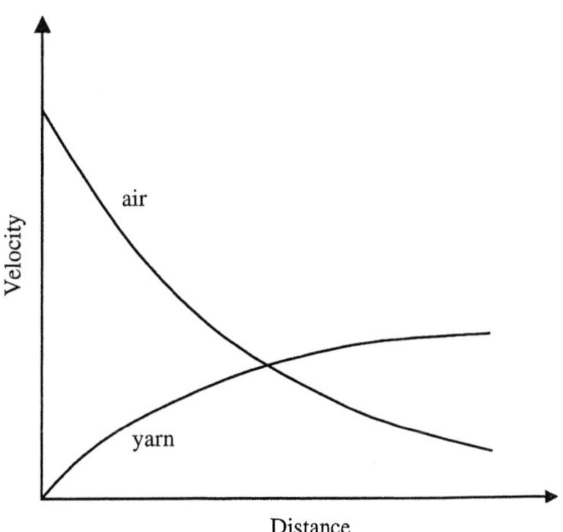

FIGURE 8.25 Air and yarn velocities with single nozzle [1].

the intense-acceleration section. The second portion is called the weak acceleration section, in which the yarn velocity increases slowly until it is the same as that of the air [34]. Then, yarn acceleration ceases; friction changes from a moving to a breaking force. The yarn begins to move by inertia and loses velocity as a result of air resistance. In other words, air acts as a brake on the yarn rather than as a propulsive force. This is the main cause of the divergence of filling yarn from the centerline of the shed. Therefore, in an ideal situation on the machine, yarn velocity should never exceed the air velocity. In this case, the buckling of the yarn tip can be reduced as a result of the reduction in the deviation of the yarn from a straight path. To achieve this, relay nozzles (subnozzles, auxiliary nozzles) are used to maintain high air velocity (as well as high yarn velocity) across the machine width.

Analysis of Yarn Motion

In addition to the complexity of air flow discussed earlier, the transported object, yarn, also has its complications. It is highly flexible, it is not uniform in structure and its diameter is variable along the length. The surface characteristics, hairiness, manufacturing method, twist level, whether single or ply, texturing, etc., all affect the yarn movement in the air flow, making the analysis of yarn motion a challenging task.

In contrast to the complications in air flow and yarn structure, there is a single physical law which governs the yarn motion throughout the course of insertion: Newton's second law, which states that, for a body, the rate of change of momentum is equal to the total force acting upon it. However, the application of this principle to the yarn motion is not so straightforward.

The literature has several models to describe the yarn motion during insertion [16,26,29,34–46]. An exact theoretical model is hardly possible for the insertion with air-jet, therefore the models given in the literature are partly theoretical and partly experimental. Many simplifying assumptions had to be made because of the complexity of the task. The lack of an exact model in the literature is in part due to the extreme complexity of the mathematical analysis of compressible or incompressible turbulent flow which is the transporting medium for the yarn during insertion with air-jet. In addition, yarns that are used for insertion are extremely flexible and not uniform in structure which makes their movement much more difficult to determine in a turbulent flow. The yarn movement in the air stream depends on the interaction between the yarn surface and air. In other words, air-jet filling insertion is not a positively driven process. The channels that are used to guide the air and yarn are partly open, which makes the continuity equation difficult to apply. Besides, due to the pulsing nature of the air from the nozzle, the air flow along the insertion width is never steady.

In general, the existing models for yarn motion assume that the yarn moves along a straight line inside the guide channel (center line). Fluctuations in the movement of yarn due to turbulent and unsteady flow are not taken into account. Therefore, the buckling of the filling yarn is neglected. It is also assumed that the filling yarn has uniform properties such as constant diameter and linear density and it is inextensible. The effect of gravitational force on the yarn motion is neglected since its value is small.

Considering Figure 8.26, the guide channel entrance is taken as the starting point for the yarn motion. The schematic in the figure shows a typical arrangement of elements in air-jet weaving. The arrangements of the elements may slightly vary from machine to machine. The equation of motion for the yarn in air flow, based on Newton's second law, can be written as:

$$\frac{d}{dt}[M\ V] = \Sigma\ F \qquad (8.25)$$

where M : total yarn mass involved in the motion (kg-mass)
V : yarn velocity (m/s)
F : force acting on the yarn (Newton)
t : time (s)

The tip of the yarn is considered to represent the whole yarn. For example, the "yarn velocity" means the velocity of the tip of the yarn during insertion. However, the motion of the yarn tip does not necessarily represent the motion of the

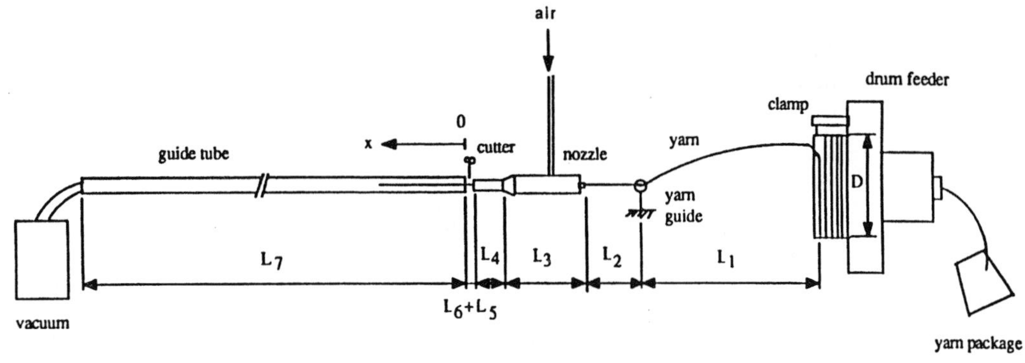

FIGURE 8.26 Schematic of the drum storage model [16].

whole yarn. The motion of the yarn tip determines the insertion time of the yarn.

The total force acting on the yarn is,

$$\sum F = F_1 - F_2 \tag{8.26}$$

where F_1 : air-friction force on the yarn
F_2 : yarn tension at the yarn guide

F_1 is given by,

$$\begin{aligned} F_1 &= 0.5\ \pi\ d\ \rho\ C_f(U_o - V)^2(L_4 + L_5 + L_6) \\ &+ 0.5\ \pi\ d\ \rho \int_{x=0}^{x=L_7} C_f(U - V)^2\, dx \\ &- 0.5\ \pi\ d\ \rho\ C_f V^2 L_2 \end{aligned} \tag{8.27}$$

where d : yarn diameter (m)
ρ : air density (kg/m^3)
C_f : air-friction coefficient
U_o : air velocity at the core of nozzle (m/s)
V : yarn velocity (m/s)
L_2 : yarn length between the guide and nozzle
L_4 : length of the air tube of the nozzle
L_5 : distance between the nozzle exit and cutter
L_6 : distance between the cutter and guide tube
L_7 : tube length
x : distance from the tube entrance (m)

The first term on the right side of Equation (8.27) is the force exerted on the yarn by the nozzle. Inside the yarn tube, there is air flow in the direction of yarn movement due to the low pressure generated at the nozzle throat. Since this velocity is relatively small, its effect on the yarn movement is compensated for by the friction between the yarn and yarn tube walls. The second term on the right side of Equation (8.27) is the air friction force on the yarn length x, created by the air flow inside the guide tube. The third term, which has a minus sign, represents the air resistance on the yarn along the length L_2.

The yarn tension due to unwinding from the drum in the X direction is given by Equation (8.20).

For simplicity, let

$$A = 0.5\ \pi\ d\ \rho$$

$$K_1 = L_1 + L_2 + L_3 + K_2$$

$$K_2 = L_4 + L_5 + L_6$$

The total mass involved in the motion is given by,

$$M = m\ (x + K_1)$$

From equation (8.25),

$$\frac{d}{dt}[M\ V] = m\ V^2 + m\ (x + K_1)V\frac{dV}{dx} \tag{8.28}$$

Substituting Equations (8.20), (8.27), and

(8.28) into Equation (8.25) and after rearranging, the following formula is obtained for the equation of the yarn motion.

$$\frac{dV}{dx} = A\left\{\frac{C_f(U_o - V)^2 K_2}{m\ V\ (x + K_1)} + \frac{1}{m\ V\ (x + K_1)} \int_{x=0}^{x=L_7} C_f(U - V)^2\ dx - \frac{C_f\ V\ L_2}{m\ (x + K_1)}\right\} - \frac{V}{(x + K_1)}\{0.5\ e^{(\mu\ \alpha)} + 1\} \quad (8.29)$$

This equation shows that the yarn velocity can be increased by increasing the air friction force (i.e. increasing the air velocity) and by decreasing the yarn tension. This equation, which is a first order, non-linear, ordinary differential equation, can be solved using numerical methods such as Fehlberg Fourth-Fifth Order Runge-Kutta method [16]. To solve Equation (8.29), the air velocity should be known which can be measured using an anemometer.

Initial Conditions

To solve Equation (8.29), initial conditions must be determined. Referring to Figure 8.27(b), the air valve is turned on but the yarn is still clamped. At this moment, the length of the yarn that is under the influence of air-jet is equal to $L_4 + L_5$. Equation (8.25) is still valid for this situation. The total force on the yarn is given by,

$$\Sigma\ F = 0.5\ \pi\ d\ \rho\ C_f\ U_o^2\ (L_4 + L_5) \quad (8.30)$$

and the yarn mass (considering the moment when the clamp opens but yarn is still stagnant) is:

$$M = m\ (L_1 + L_2 + L_3 + L_4 + L_5) \quad (8.31)$$

The rate of change of momentum,

$$\frac{d}{dt}[M\ V] = m\ (L_1 + L_2 + L_3 + L_4 + L_5)\frac{dV}{dt} \quad (8.32)$$

It is worth noting that, in Equation (8.32), although $V = 0$, the acceleration (dV/dt) is not zero, since there is force on the yarn. Substituting

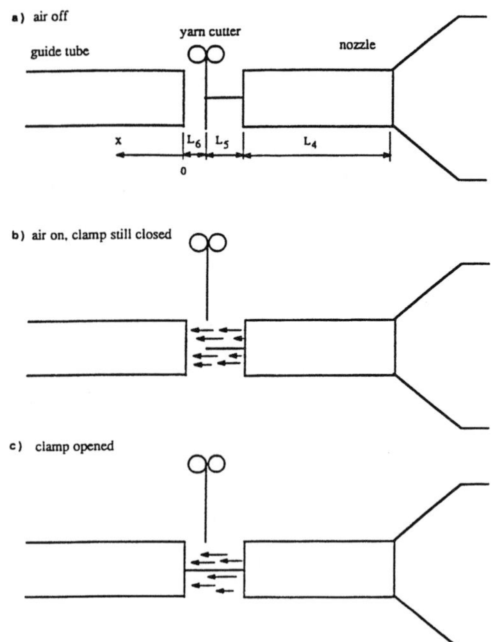

FIGURE 8.27 Initial conditions.

Equations (8.30) and (8.32) into Equation (8.25), the acceleration of the yarn is obtained:

$$\frac{dV}{dt} = a = \frac{0.5\ \pi\ d\ \rho\ C_f\ U_o^2\ (L_4 + L_5)}{m\ (L_1 + L_2 + L_3 + L_4 + L_5)} \quad (8.33)$$

Since the air velocity is constant along $L_5 + L_6$ (potential core of the jet), and the distance L_6 is small, the acceleration can be taken as constant along the distance L_6. Therefore, yarn velocity at the end of distance L_6 can be written as

$$V_o = \sqrt{2\ a\ L_6} \quad (8.34)$$

This is the yarn velocity at $x = 0$ and $t = 0$.

The factors which affect the yarn insertion are the air velocity at the main nozzle, the air velocity distribution along the guide channel, the yarn structure and the conditions behind the main nozzle [26]. Insertion time depends on the following parameters:

- yarn structure
- yarn surface characteristics (lubrication, finishing, etc.)
- yarn linear density
- yarn twist

- yarn surface area subject to air
- relative velocity of the air and yarn

To increase the yarn velocity, the air friction force should be increased and the tension which hinders the yarn motion should be decreased. To increase the air friction force, air velocity should be increased. For this, closed tubes will be ideal which will also reduce the air consumption.

8.1.4 Performance of Yarns in Air-jet Insertion

Air-jet weaving machines are ideal for cost-effective production of bulk fabrics with a wide range of styles. Air-jet machines can handle both spun (natural, synthetic or blended) yarns and continuous filament yarns. Textured yarns are especially suitable for air-jet weaving due to high propelling force. However, monofilament yarns are not suitable for air-jet weaving because of low friction between air and yarn which is due to smooth surface of the monofilament yarn. A wide range of fabrics from gauze fabrics to dense, heavy cotton fabrics, from patterned dress fabrics to ribbon fabrics can be woven on air-jet weaving machines. Air-jet weaving is also ideal for fine glass fabric production. Specially designed air-jet weaving machines are used for tire cord manufacturing with tuck-in selvage in plain weave.

Since the force required to move the yarn mass is provided exclusively by air friction against the yarn surface, it is largely dependent on the yarn structure, the yarn and fiber surface, and relative motion of air and yarn. The propulsive force is largely independent of the fiber material [15]. Minute disturbances in the flow field can also lead to undesirable aberrations of the yarn tip that result in faults or machine stops. The air consumption of the main jet depends on the yarn type and denier. Spun yarns and coarse yarns (with a certain hairiness) have higher air resistance coefficients than fine and smooth materials. This explains why monofilament yarns cannot be inserted with air-jet. The factors that essentially determine whether a yarn is suitable for pneumatic insertion are its count, structure and twist [33].

Effect of Yarn Structure

High twist, large denier, long staple, high fibril cohesion increase the standability of spun yarns to air-jet, giving longer yarn breaking time. More air is needed to weave continuous filament fabrics than spun fabrics due to less frictional force between yarn surface and air flow. Yarn velocity in the insertion channel increases with the number of filaments due to the larger yarn surface that is in contact with the air [47,48].

Yarns having a larger diameter require increased air pressure for filling insertion. This is because the mass of the yarn increases in proportion with the square of the yarn diameter, whereas the yarn surface area increases linearly with the diameter.

Several researchers reported that open-end (OE) yarns give higher mean yarn velocities than ring spun (RS) yarns but ring spun yarns have higher initial acceleration [15,49]. OE yarns are composed of concentric structures with an inner core which contains most of the fibers in a compact and highly twisted assembly (Figure 4.92). The outer layers are wrapped around the core and are less packed. OE yarns are 15% bulkier and less hairy than RS yarns [50]. The bulkier structure of OE yarns, which increases the yarn surface area, causes the increased air friction. Although OE yarns have higher average velocity, RS yarns have higher velocity at the beginning of the insertion. The initial acceleration of the RS yarns is slightly higher than that of OE yarns which must be due to the higher hairiness of the RS yarns. Table 8.4 shows the acceleration values of some OE and RS yarn pairs at 10 milliseconds into the insertion. Murata air-jet

TABLE 8.4 Acceleration of open-end (OE) and ring spun (RS) yarns 10 milliseconds after the yarn insertion [15].

Yarn Count (Ne)	Twist Multiple	Structure	Acceleration (m/s^2)
10/1	4	RS	968
10/1	4	OE	865
13/1	4	RS	1189
13/1	4	OE	1083
16/1	4	RS	1258
16/1	4	OE	1235

spun (MJS) yarns have higher yarn velocity compared to ring spun yarns due to the fact that MJS yarns are bulkier than RS yarns. In fact, MJS yarns have a similar structure to OE yarns, i.e. the yarn is composed of a main fiber bundle wrapped by a small group of fibers.

Effect of Yarn Count

High linear density causes longer insertion times. The significance of this effect increases with increasing range of linear density. Tests with single nozzle and slotted tubes showed that in the case of small changes, linear density does not have significant effect on the average velocity [15]. However, it has an effect on the acceleration and instantaneous velocity. Coarse yarns have a high linear density, hence their initial acceleration is normally low as shown in Figure 8.28. This is because yarn acceleration is inversely proportional to the yarn mass. Towards the end of the insertion, the velocity of the coarse yarns increases due to the high inertia. The effect of inertia on yarn velocity compensates for the low initial acceleration resulting in approximately the same velocity irrespective of the yarn count when the change in count is within a small range. However, the effect of linear density becomes important if the range is too large. Finer yarn has considerably higher average velocity compared to coarse yarn. As shown in Figure 8.29, fine yarn has higher acceleration and velocity at the first 2 m. After that, again because of the inertia, the velocity of the coarse yarn stays above that of the fine yarn. However, this inertia effect is not as significant as the mass effect and

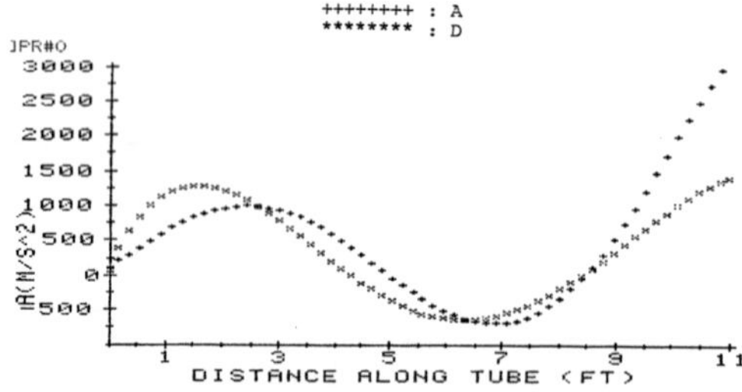

FIGURE 8.28 Acceleration curves of ring spun yarns with smaller difference in linear density [15]. Twist multiple: 4. Yarn A: Ne 10/1, Yarn D: Ne 16/1.

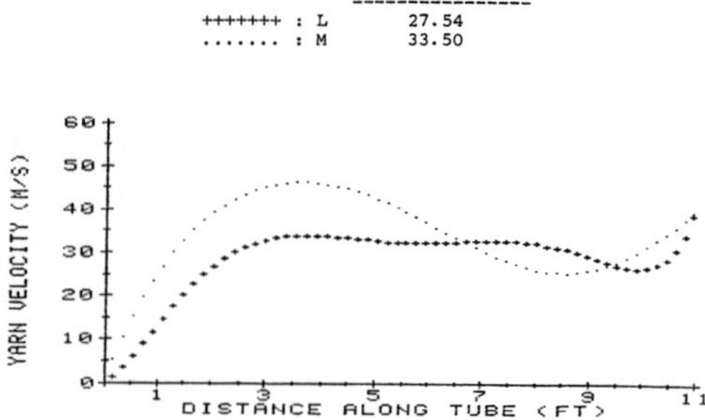

FIGURE 8.29 Velocity distributions of ring spun yarns with large differences in linear density [15]. Twist multiple: 4. Yarn L: Ne 6/1, Yarn M: Ne 50/1.

fine yarn shows higher average velocity along the insertion length.

Effect of Twist and Ply

Twist level plays an important role in the behavior of yarn in air-jet insertion. Twist increases the strength of the yarns by creating lateral forces which prevent the fibers in the yarn from slipping over one another. These forces bring the fibers closer which makes the yarn more compact. High twist level increases the insertion time since twist reduces the diameter of the yarn and makes the yarn surface smoother. The propulsive force decreases with a decrease in the diameter and the smoother surface reduces the friction between the yarn surface and the air. As a result, yarns with low twist have higher velocities. There is no significant difference between S and Z twist for air-jet insertion [51].

Freedom of the filling to untwist during insertion results in twist loss during weaving which affects the strength of the fabric, its dye uptake, and possibly other properties. There is a minimum twist multiple below which the filling yarn disintegrates because of the air pressure [52,53].

Plying is done by twisting several yarns together to obtain more durable yarns. The ply twist is applied in the opposite direction to the twist direction of component strands. Plied yarns give longer insertion times in air-jet filling insertion than one-ply yarns with the same count. The reason is that additional twist makes the yarn surface smoother and reduces air friction.

Effect of Texturing

False-twist and air-jet texturing increase the friction force on the yarn leading to an increase in yarn velocity compared to straight filament yarns (Figure 8.30). Because of the bulky structure of the textured yarn, air penetrates into the textured yarn better, causing higher propelling force.

Statistical Variations

Since the yarn movement in air-jet filling insertion is not positively controlled, there is some variation associated with it. The air velocity distribution is different for each pick. The filling yarn does not have uniform structure and properties. Its diameter is not constant and its surface hairiness is not the same throughout its length. Padfield [54] observed up to 10% velocity fluctuations due to the variation in the yarn diameter alone. It was also reported [55] that the pressure fluctuations in the compressed air circuit cause variations in the insertion time, these must be prevented by sufficient dimensioning/lay out of the compressed air installation. These variations affect the air friction force on the yarn and therefore cause variations in the filling yarn velocity from pick to pick. Since the timing of air and clamp operations is fixed, this situation may cause variations of the yarn lengths during the insertion. This results in waste of filling yarn in air-jet weaving. To minimize this waste, new air-jet weaving machines use microprocessors to adjust the air pressure on the basis of the average velocity over a certain number of picks and to maintain that average. The yarn velocity and insertion time show a normal distribution in air-jet filling insertion [16].

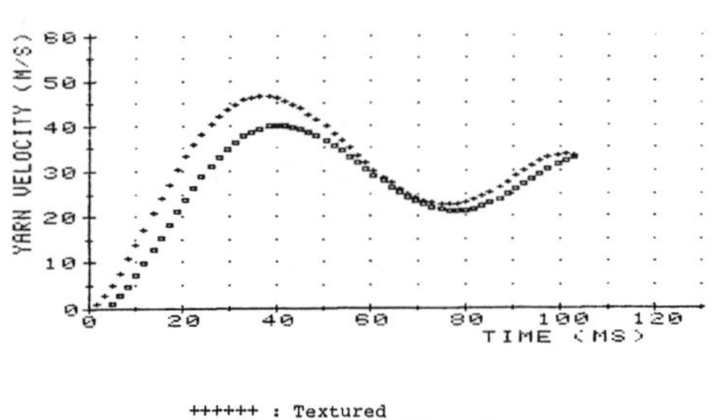

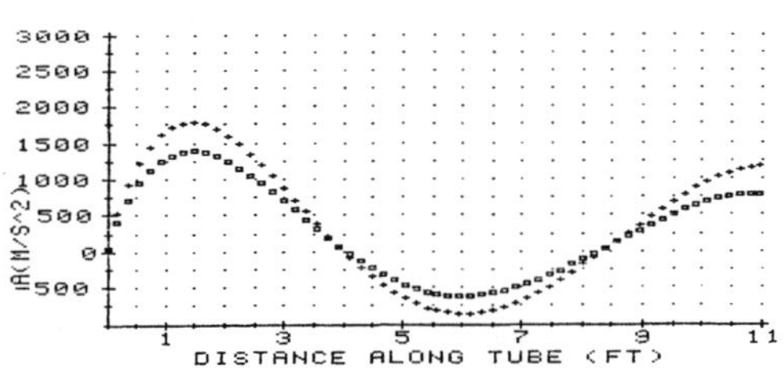

FIGURE 8.30 Yarn velocity and acceleration distributions for textured and straight filament yarns with a single nozzle [15].

8.1.5 Functional Characteristics of Modern Air-jet Weaving Machines

Air-jet weaving has become one of the most common methods of fabric formation in recent years. The absence of mechanical devices for picking and checking provides air-jet machines a quieter operation than the other types of machines. A flexible program of styles, short style changing times and a minimum of attendance and maintenance requirements guarantee high economical efficiency with air-jet weaving machines. The robust and compact construction of today's air-jet weaving machines allows for weaving the most heavy fabrics. Gentle handling of warp and filling yarns permits the highest speeds. In these machines, due to optimized air consumption, power requirement is modest and there is little filling yarn waste. The amount of air supply in an air-jet weaving mill should be based on the number of weaving machines. The air used must be free of impurities such as oil, water and dust. The pressure pipes must be corrosion resistant. Typical pressure at the machine entrance is 0.6 MPa.

Profiled reed and relay nozzles, which are the most common, assure safe filling insertion even with larger weaving widths. Single nozzle machines normally do not exceed 165 cm in width. In theory, there is no width limitation of multi-jet machines; however, the practical limit is slightly over 400 cm. The air pressure in the nozzles adjusts itself automatically to the insertion conditions. Programmable blowing times with automatic correction at speed changes and automatic pressure optimization due to individual air supply provides air-saving and gentle filling insertion (Figure 8.4). The air-jet timing is controlled through a solenoid valve by computer for good weaving quality, maintenance and energy conservation. The filling insertion time is digitally displayed for easy inspection and verification. During the filling insertion process, yarn velocities of 50 to 80 m/s are attained. Operation speed depends on type of cloth, conditions in weaving mill, yarn quality and fabric cover factor.

Typical yarn characteristics for a universal air-jet weaving machine are as follows:

warp yarn count	: virtually unlimited
warp density	: 3–111 ends/cm
filling yarn counts	: staple fiber yarns: 6–110 tex
continuous filament yarns	: 45–500 dtex
filling density	: 3.5–140 picks/cm

Typical characteristics of modern air-jet weaving machines are given below. These characteristics may change from machine to machine and from manufacturer to manufacturer. Figure 8.31 shows the overview of the subunits in a typical air-jet weaving machine.

Touch Screen Terminal and Memory Card (Figure 8.32)

The operator can activate the desired function at the touch of a finger. Guided by a menu, the operator enters pattern data and machine settings, optimizes settings, retrieves operating data or receives information concerning causes of and remedies for malfunctions. The entire system routine is microprocessor controlled.

Setting data can be transferred from one machine to the other with a memory card. It can also be used for feeding in machine and style data records. If no setting data are available for a new style, the microprocessor calculates standard settings from the fabric specification and automatically adjusts them on the electronically controlled functional units. The weaving machine is thus very quickly set and ready for operation. This program, called ICS (Initial Condition Setting), is based on empirical values.

Drive (Figure 8.33)

Machine main motor 1 drives main shaft 3 via two multi-V-belts. The right- and left-hand weaving machine gearboxes are connected by the main shaft. All mechanical units are directly or indirectly controlled from the main shaft. The weaving machine can be rotated in creep speed by the frequency control.

The brake stopping position of the weaving machine can be adjusted to suit the style. Automatic reversing during pick finding is controlled from the main motor. The weaving machine speed is controlled by using V-belt pulleys on the main motor axle with different diameters. A speed control can be added to the weaving machine.

Warp Let-off (Figure 8.34)

The power warp let-off can handle a wide weft density range. The weft density range depends on the warp beam flange and tube diameters as well as on the machine speed.

Force sensor 1 measures the warp tension of the entire warp. The difference between the measured warp tension value and the parameterized reference warp tension is corrected by the speed of AC servo motor 2. The speed of the servo motor is transmitted to warp beam gear 4 via reduction gear 3, thus driving warp beam 5. The warp tension remains uniform from full to empty warp beam. If necessary, a warp tension correction can be parameterized depending on machine stoppage duration.

With the so-called "kick-back" device, the warp beam is briefly reversed by one parameterized value before the machine starts up. When the machine starts, the system regains the reference warp tension. As a further option, the warp let-off speed can be altered for a pre-set time when weaving with varying weft densities, which assures a smooth transition from one weft density section to the next.

In terry weaving, it is extremely important that a certain pile height be kept constant. For this reason, a constant warp tension is required. Electronic warp let-off motions provide uniform, low tension for constant pile height.

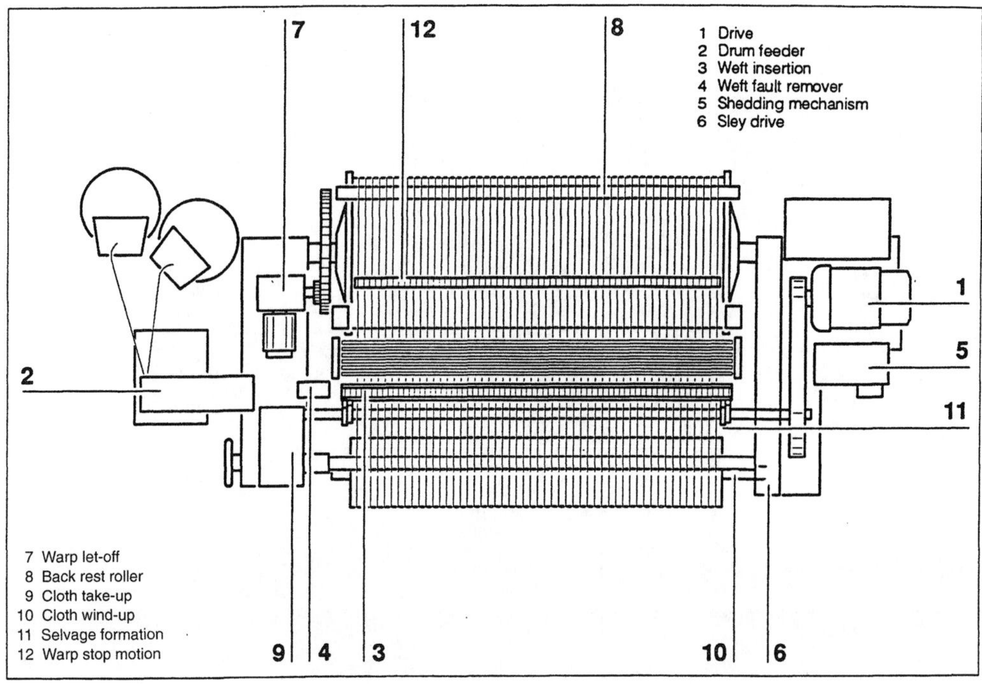

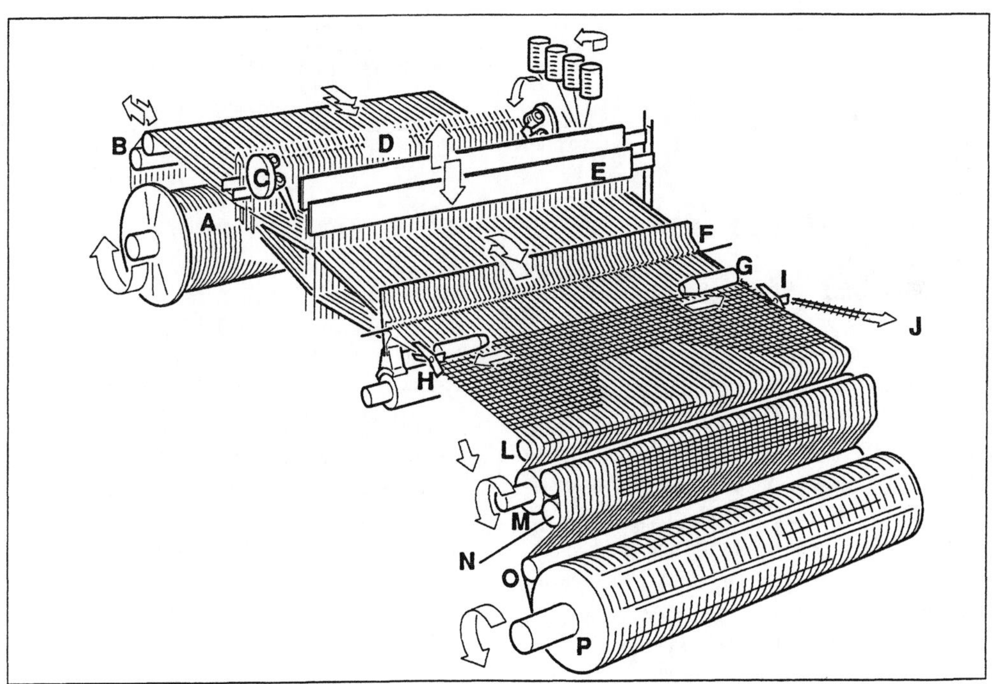

Overview of sub-units

A Warp
B Back rest roller
C Leno device
D Warp stop motion
E Shafts
F Reed
G Temples
H Weft cutter
I Auxiliary selvage cutter
J Auxiliary selvage
L Spreading roller
M Take-up roller
N Press roller
O Press roller
P Cloth beam

FIGURE 8.31 Schematic of a typical air-jet weaving machine.

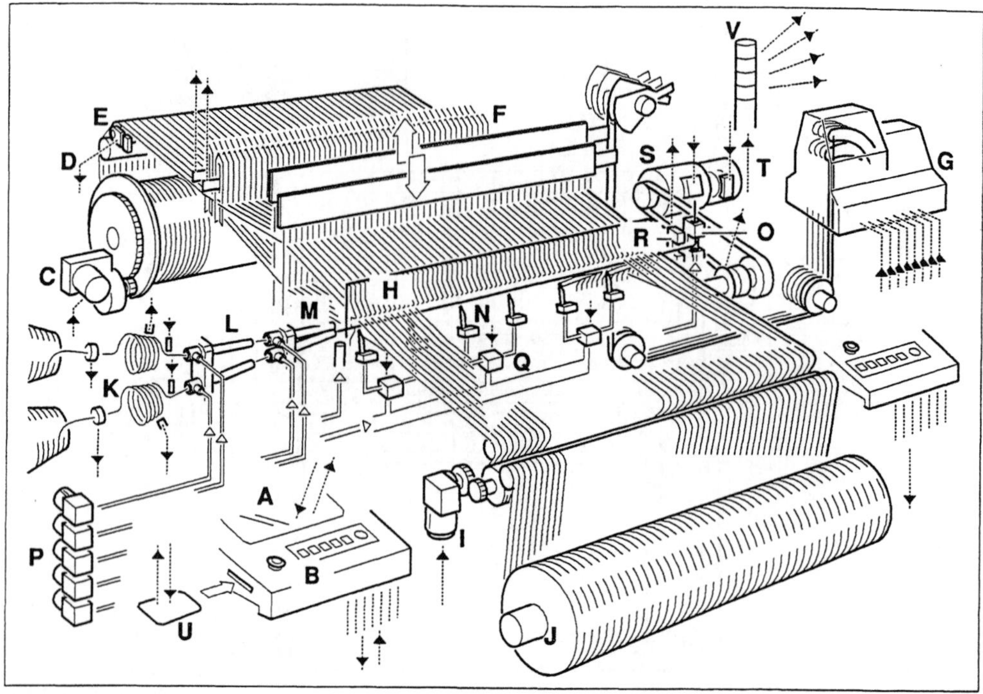

Overview of sub-units

A	Operating and display terminal	I	Cloth take-up	P	Pressure control valves
B	Keypad	J	Cloth wind-up	Q	Relay nozzle valve
C	Warp let-off	K	Drum feeder	R	Weft monitoring
D	Warp tension sensor	L	Tandem nozzle	S	Main motor
E	Back rest roller	M	Main nozzle	T	Electromagnatic machine brake
F	Warp stop motion	N	Relay nozzle	U	Memory card for setting data
G	Shedding mechanism	O	Stretching nozzle	V	Signal lamp
H	Reed				

FIGURE 8.31 (continued) Schematic of a typical air-jet weaving machine.

Back Rest Roller (Figure 8.35)

A positively controlled back rest roller 1 is used for weak, low twist yarns. The differences in warp tension at shed change are thus largely compensated for.

The positive control of the back rest roller is affected via crank gear 2 from the right- and left-hand machine gear boxes. Crank stroke 3 is adjusted to the shed opening and warp yarn. When weaving synthetic yarns, the back rest roller is negatively controlled. The oscillating motion of the back rest roller is set up by the shed change, i.e. the differences in tension in the warp yarn.

Fabric Take-up (Figure 8.36)

For driving the fabric take-up roller, there are two variants:

a. Mechanical fabric take-up—The weft density is determined by the change and standard wheels in the fabric take-up gear. The resulting rotational speed (rpm) of the fabric take-up roller gives the weft density in the fabric. In mechanical continuous indirect take-up, the fabric can be doffed without stopping the weaving machine. A multi-disc brake system prevents slippage and ensures smooth and wrinkle-free take-up.

b. Fabric take-up by means of servo motor—The weft density is determined by the frequency controlled servo motor. Various weft densities can be programmed, for instance for change of weave, satin stripes, etc. Quick warp pull-through can be activated either by key operation or via the servo motor. With both variants the fabric is drawn from fabric take-up roller 2 over spreading roller 1. Press roller 3 prevents the fabric from slipping back. Depending on the type

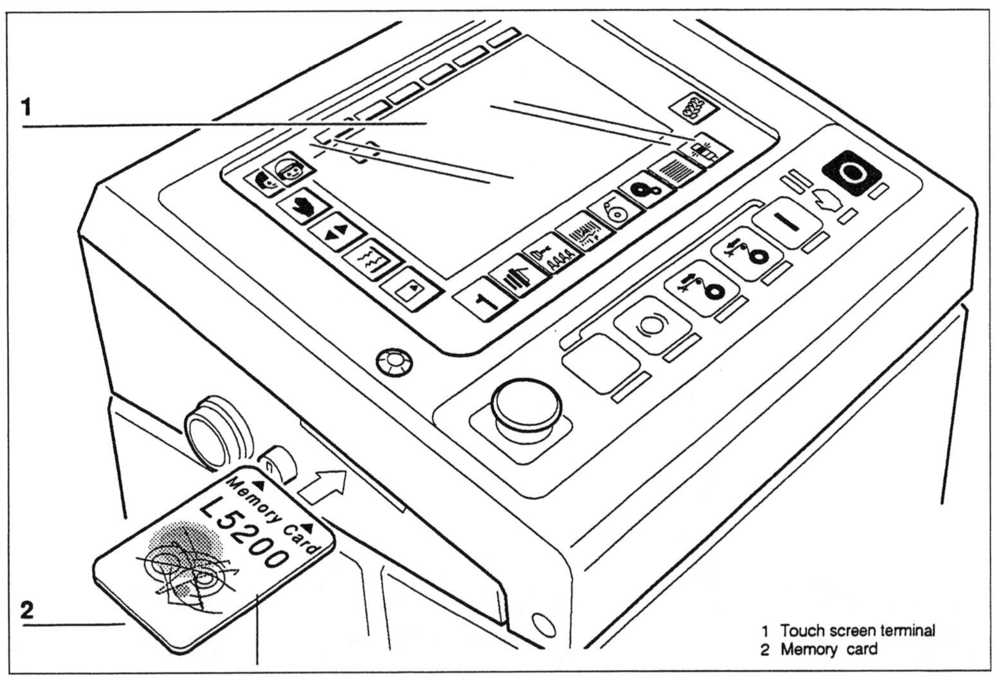

FIGURE 8.32 Touch screen terminal and memory card.

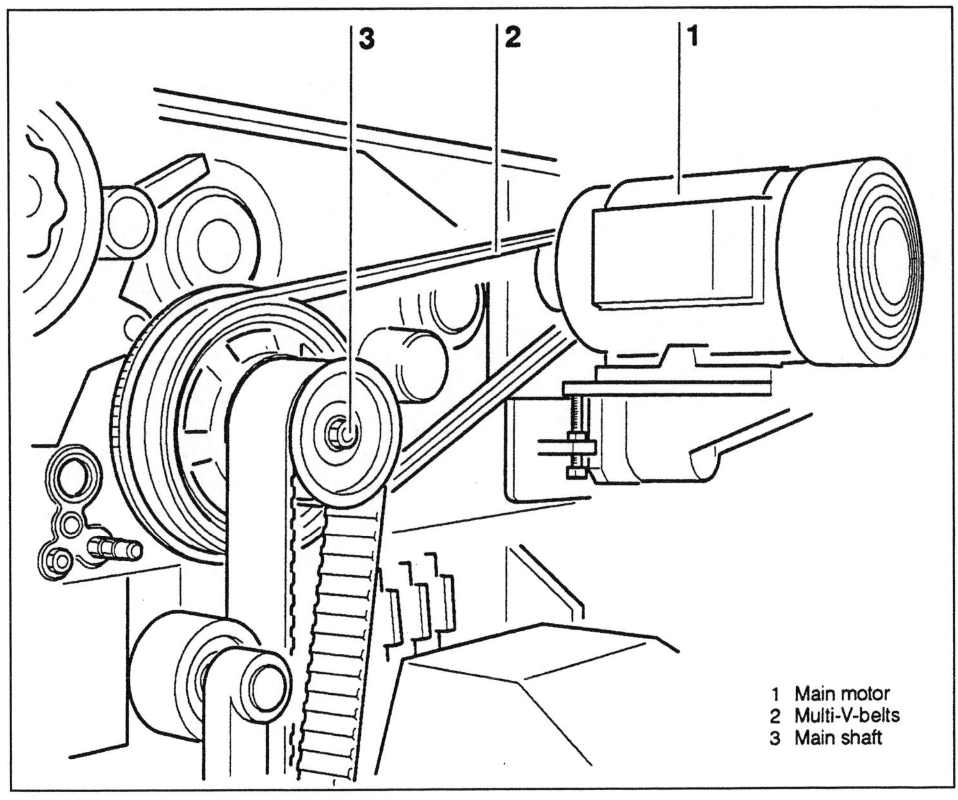

FIGURE 8.33 Schematic of drive mechanism of an air-jet weaving machine.

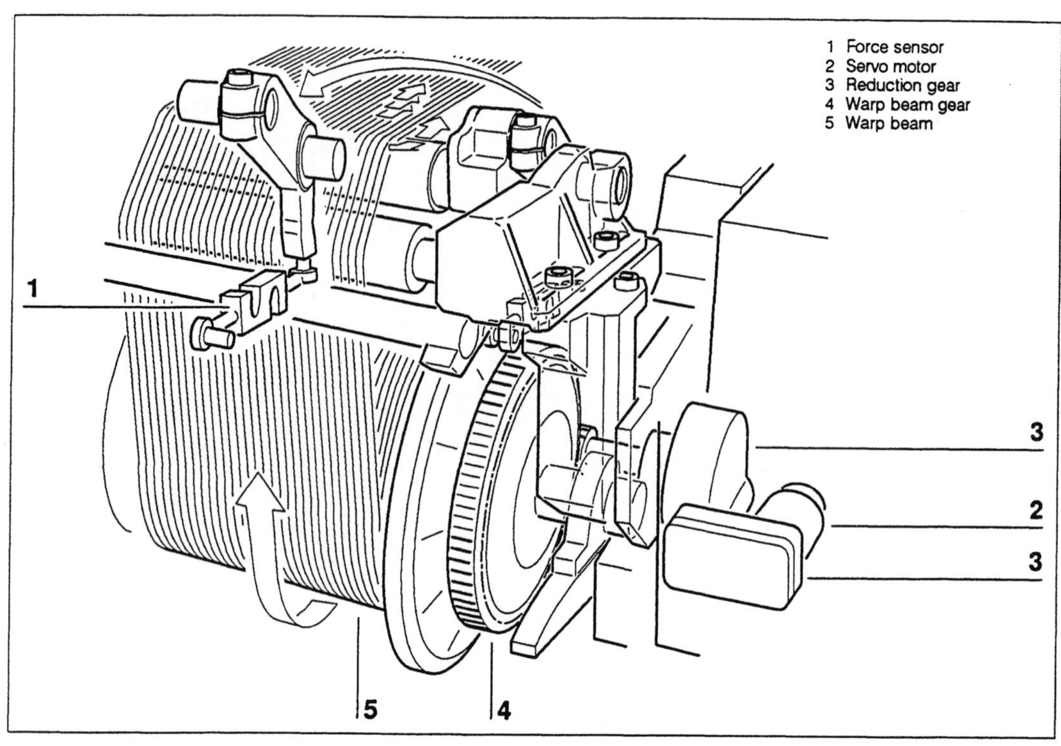

FIGURE 8.34 Schematic of warp let-off mechanism on a typical air-jet weaving machine.

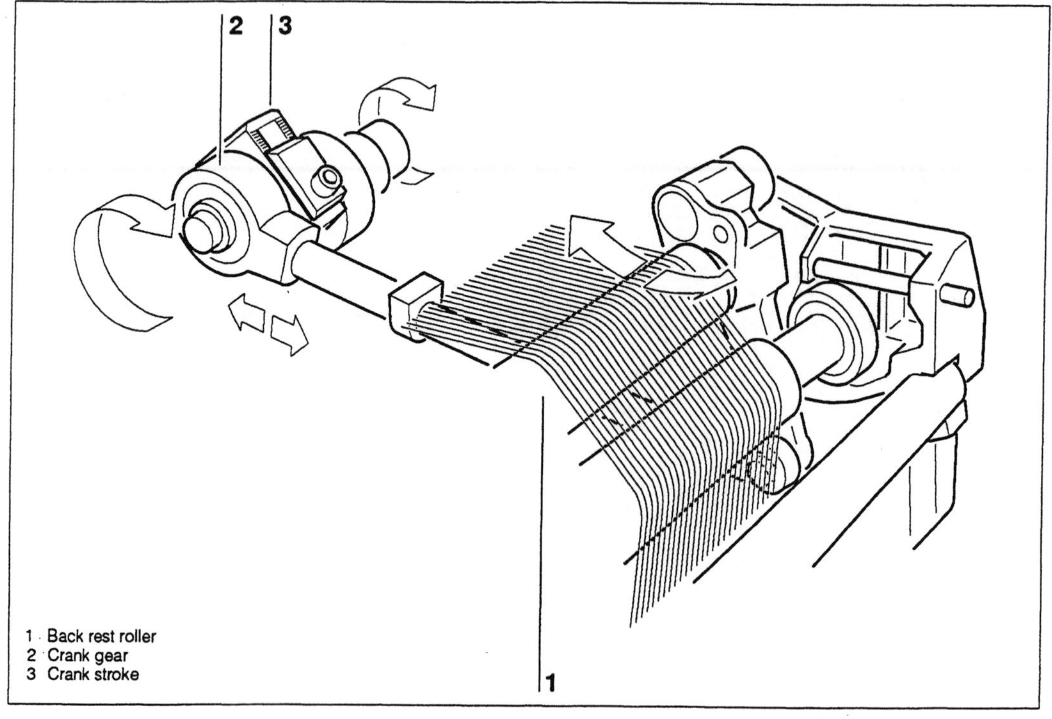

FIGURE 8.35 Back rest roller on air-jet weaving machine.

FIGURE 8.36 Schematic of fabric take-up.

of fabric, one or two press rollers 3 are used. One press roller is generally sufficient for light weight fabrics with low warp tension. A radio controlled device could be used to forward or reverse the take-up and let-off.

Fabric Wind-up (Figure 8.37)

The fabric wind-up drive is controlled by the cloth take-up gear. The fabric winding hardness is governed by friction clutch 1. This can be set with higher or lower tension. Press roller 2 ensures crease free winding of the fabric. For loosely woven, low density fabrics, a rotating press roller is used instead of the fixed press roller.

In automatic doffing systems, the cloth is promptly doffed by a signal from the weaving machine without stopping the machine. The fabric data from the weaving machine can be marked onto the final product.

Shedding Mechanisms

Air-jet weaving machines generally utilize symmetrical or asymmetrical inside treadle crank motion (up to 6 harnesses), positive or negative cam (up to 8 harnesses), negative or positive dobby (up to 16 harnesses) and jacquard shedding systems (Chapter 6).

As an example, negative cam (tappet) is suitable for very high speed operation (Figure 8.38). Drive belt pulley 1 on the main drive drives cam motion 3 via a synchronous belt 2. Up to eight shafts can be controlled. The non-positive cam motion brings the shaft into low shed position via cable traction 4. The shaft is moved into high shed position by means of spring motion 5. Roll 6 of the roller lever is pressed by spring force against the motion controlling contour of cam 7.

Up to six harnesses can be controlled by the inside treadle crank motion (Figure 8.39). Pendulum levers 1 are driven by both drive units via crank gear 2. The individual shafts are brought into high or low shed position by the up and down motion of the pendulum levers. With the inside treadle motion, it is only possible to produce plain weave.

Figure 8.40 shows a negative dobby for an air-jet weaving machine. Drive pulley 1 on the

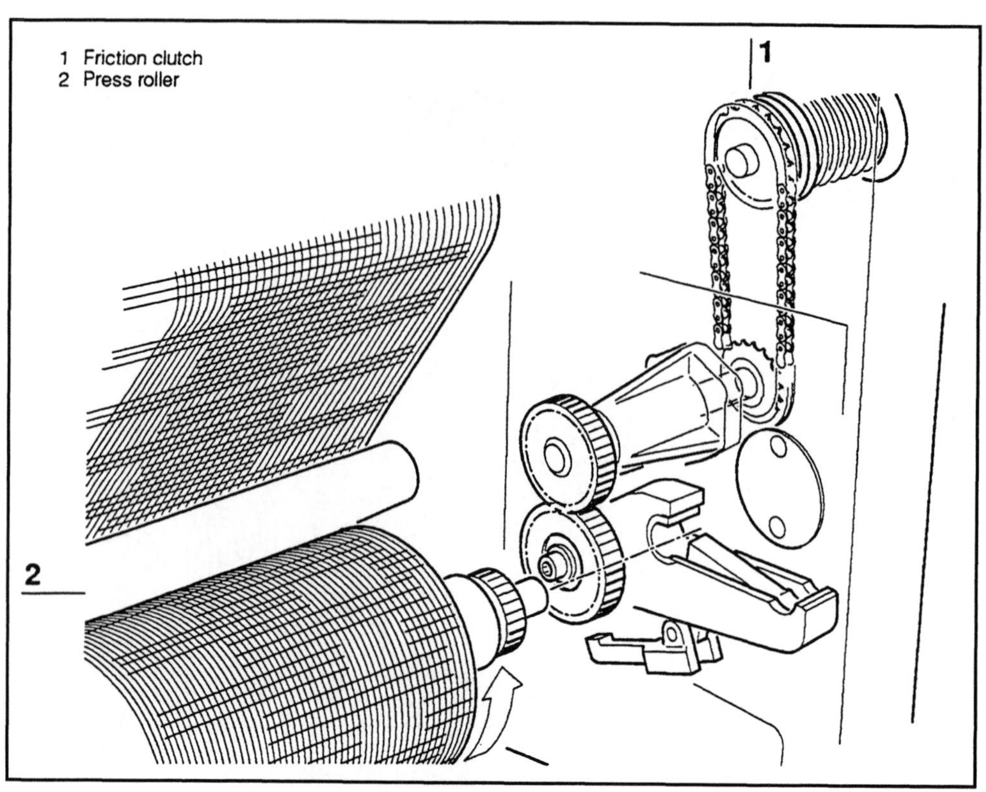

FIGURE 8.37 Fabric wind up.

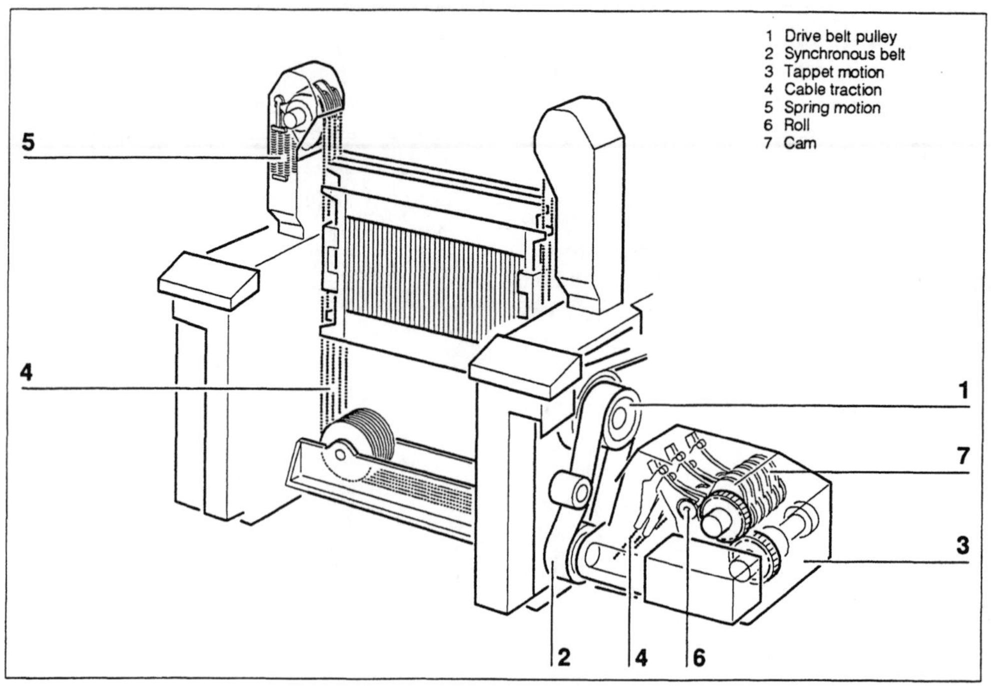

FIGURE 8.38 Schematic of negative cam motion.

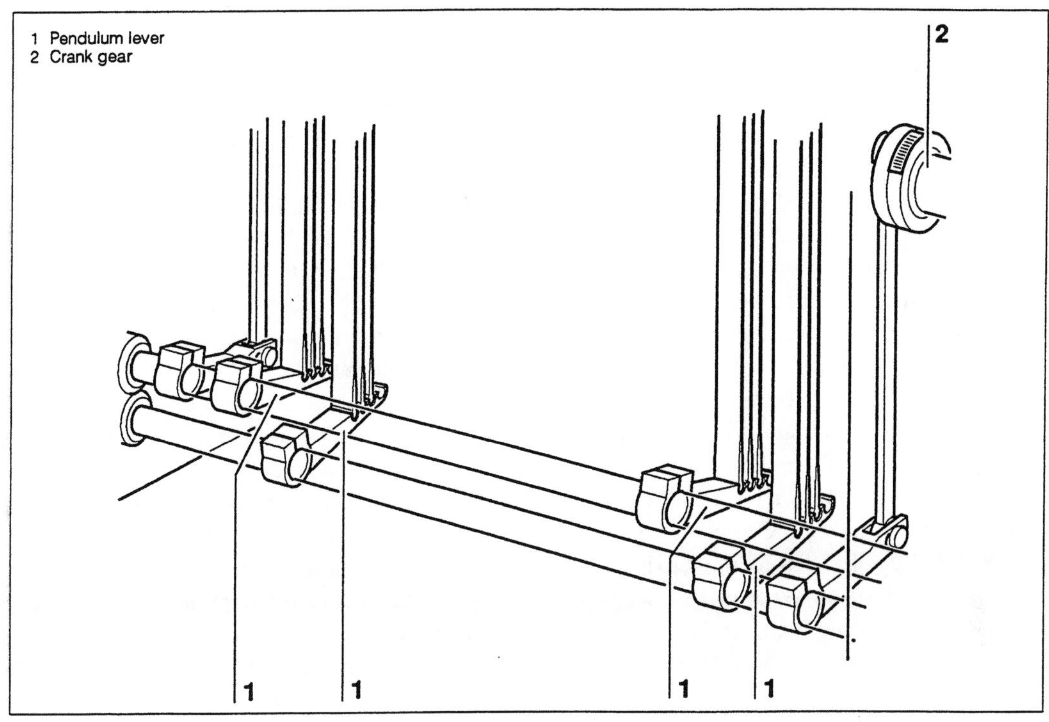

FIGURE 8.39 Schematic of inside treadle motion.

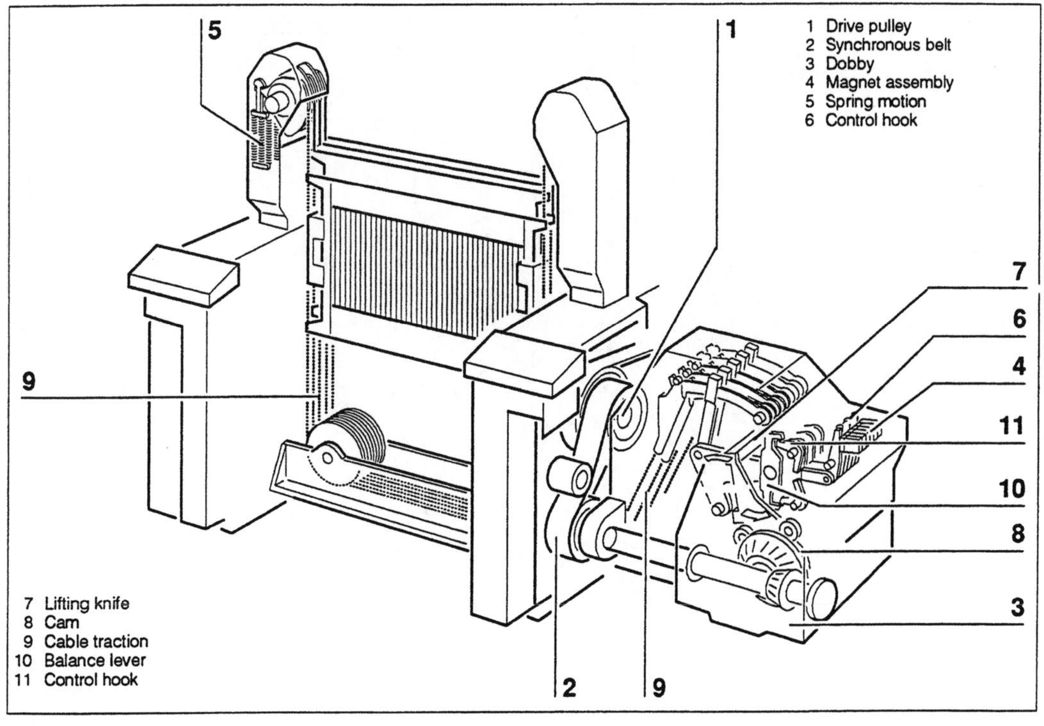

FIGURE 8.40 Schematic of negative dobby.

main drive drives the dobby 3 via a synchronous belt 2. Up to 16 shafts can be inserted. The shafts are controlled by programming magnet assembly 4 controlling the hooks. The shafts are held in high shed position by spring motion 5. When a control hook 6 is released by the magnet assembly, the corresponding shaft moves into the lower shed. The actual lifting and lowering motion of the shafts is generated by the reciprocating motion of the lifting knives 7. These lifting knives are driven by cams 8 and move in the opposite direction. The transmission of energy to the shaft is affected by cable traction 9. When the shaft is lowered from high to low shed position, the hooks of balance lever 10 are held by control hook 11 of the magnetic assembly. The balance lever is held by the control hook in this position until the shaft has returned to high shed position, when the control hooks are disengaged.

Electronically controlled jacquard machines enable high economy production of fabrics of extremely complex weaves. Modern jacquard machines permit speeds of up to 800 ppm.

Drum Feeder

Figure 8.41 shows the schematic of a drum feeder for air-jet weaving machines. The weft yarn is drawn off the package and wound onto measuring bands 2 by the rotating motion of yarn guiding tube 1. The pick length depends on the fabric width. The pick length is set by adjusting the measuring bands and the number of coils. The electromagnetically controlled stopper pin 3 releases the weft yarn at the machine angle set.

A sensor 4 controls the number of coils to be unwound. As soon as the parameterized number of coils is drawn from the drum feeder, the stopper pin blocks the following coils. The pick length is fine tuned by adjusting the measuring bands on the drum feeder.

The electronically controlled filling feeder guarantees exact metering of the required filling yarn length, thus ensuring a minimum of filling yarn waste. High filling insertion rates inevitably entail high peak values of tensile stress when the

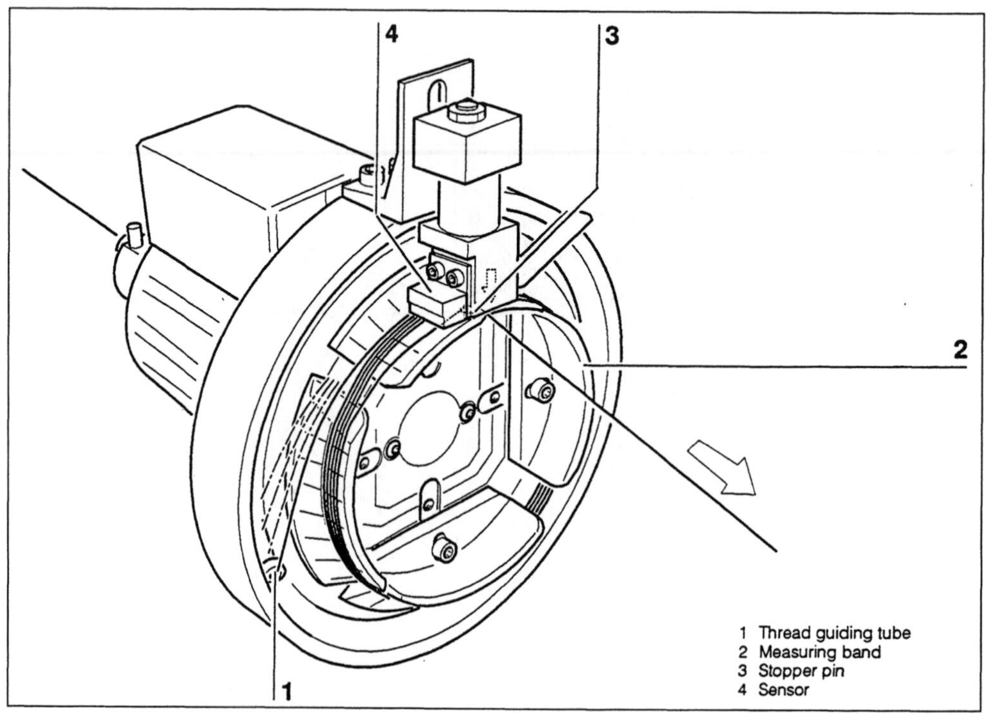

FIGURE 8.41 Schematic of a drum feeder for air-jet weaving machines.

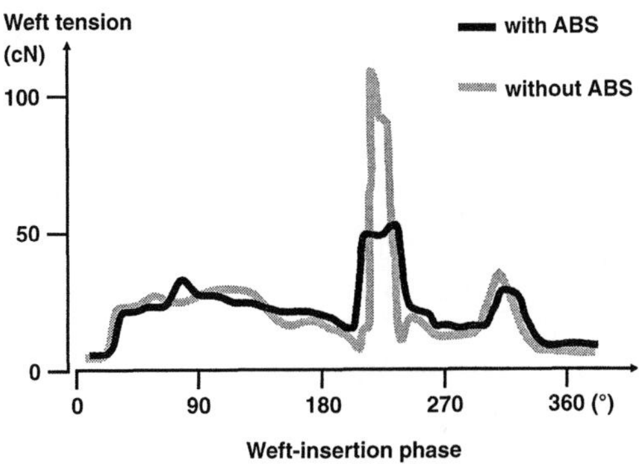

FIGURE 8.42 Filling yarn tension with (solid line) and without controlled brake.

filling yarn is slowed down. The electronically controlled filling yarn break system reduces peak values of the yarn tension forces considerably as shown in Figure 8.42. As a result, there is fewer machine stops, particularly when weaving coarse and delicate yarns.

In some machines, the rotation of the yarn guide in the drum feeder is synchronized with the rotation of the weaving machine. The solenoid pin for filling release is controlled by computer. Therefore, there is no need to adjust the drum when changing the weaving machine revolution.

Main Nozzle (Figure 8.43)

The filling yarn threaded into nozzle tubes 1 is accelerated by the concentrated air-jet and inserted into the shed. The amount of air needed depends on the yarn structure and the yarn count. It is controlled by means of pressure control valve 3. To keep the motionless pick tensioned, a weak air stream is blown onto the yarn lying in the main nozzle. The air volume needed for this purpose can be individually controlled.

To reduce the weft insertion time, a tandem nozzle 4 is used, which is the equivalent of two main nozzles positioned one behind the other. The tandem nozzle has the job of unwinding the weft yarn from the drum feeder. The winding resistance is thus compensated for by the tandem nozzle. The pressure can be reduced at the main nozzle, assuring very gentle yarn insertion.

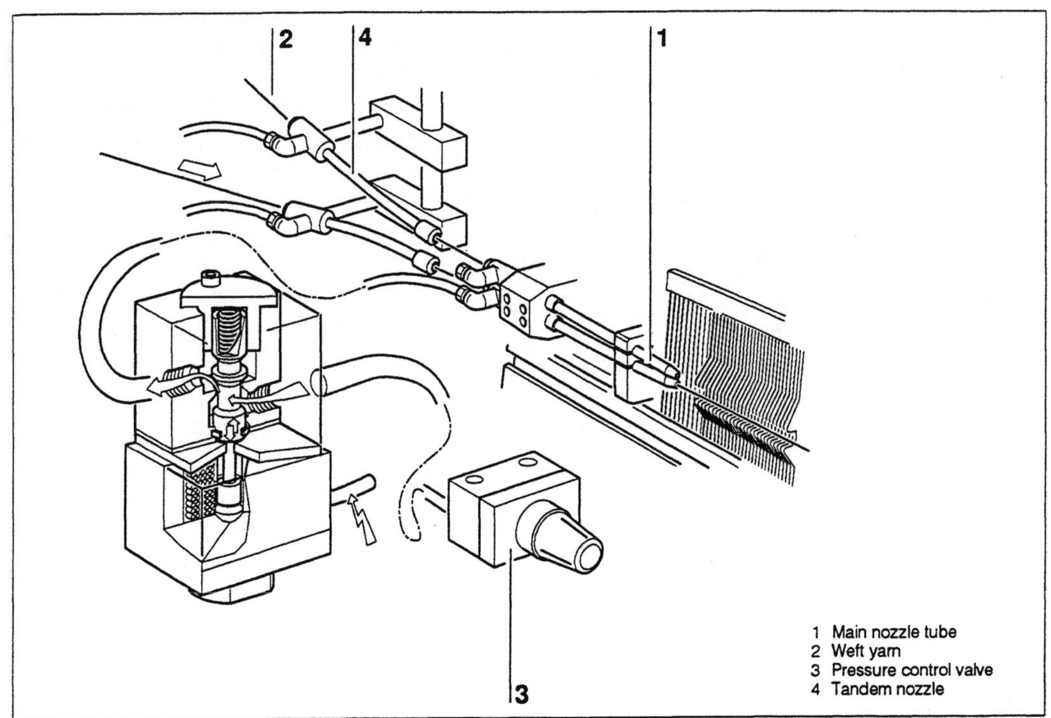

FIGURE 8.43 Main nozzle.

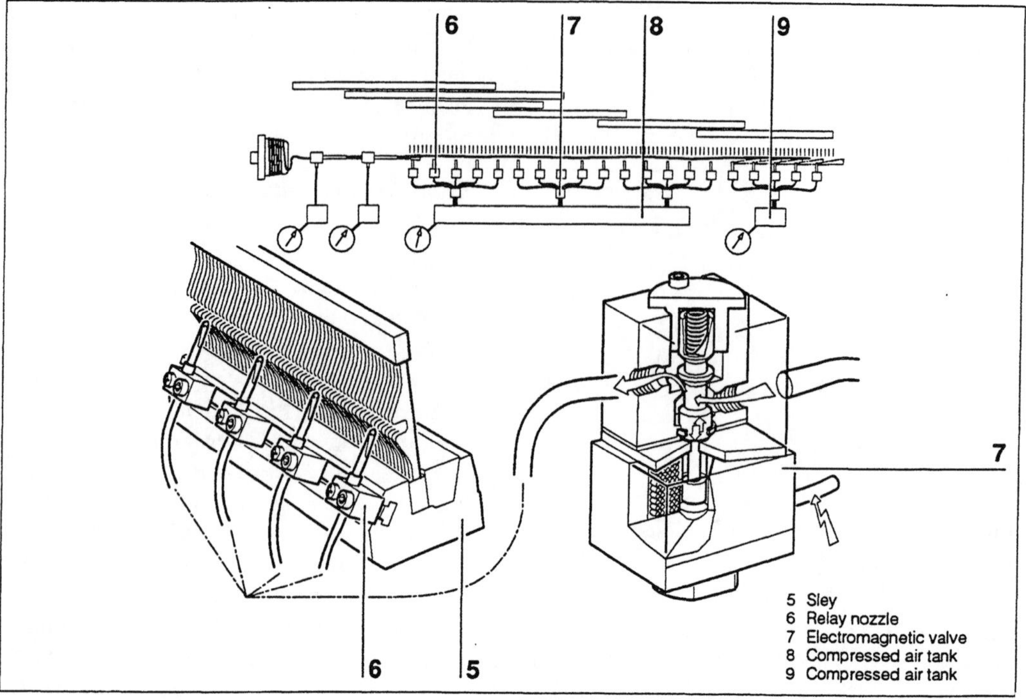

FIGURE 8.44 Relay nozzle unit.

Relay Nozzle Unit (Figure 8.44)

Relay nozzles 6 mounted in sley 5 are connected in groups to electromagnetic valves 7. The air-jet is started by the electromagnetic relay nozzle valve. The length of time the valve is opened depends on the reed width and relay valve spacing as well as on the yarn. The compressed air is distributed from compressed air tank 8 via the valves to the nozzles.

If the stretching nozzle is not used, the last relay nozzles can be supplied from a separate compressed air tank 9, thus keeping the weft yarn tensioned until the shed is closed. This version is used in the spun fiber yarn sector.

Reed, Filling Detector and Stretching Nozzle (Figure 8.45)

The reed 10 is tunnel shaped, owing to the profile of the dents. With this configuration, the dynamic pressure set up by the nozzles is kept in the weft channel. The reed beats up the inserted pick at the fell of the fabric.

When the pick arrives at the right-hand end of the reed, it is monitored by weft detector 11. The weft detector, like the reed, is mounted in the sley. If the pick is too late or does not arrive at the weft detector, the optical control reacts and the weaving machine shuts off.

For some weft yarns, especially filament yarns, a stretching nozzle 12 is used. The stretching nozzle has the task of keeping the weft yarn tensioned after its arrival in weft detector 11. The air pressure and timing can be set individually for this function.

Automatic Mis-Pick Remover (Figure 8.46)

If a pick is not properly inserted, the weft fault remover extracts it from the shed. The process is monitored until the weaving machine is restarted.

Procedure: An incorrectly inserted pick 1 is registered by the weft detector.

- Activation of the electromagnetic weft cutter 2 is suppressed, to prevent the faulty pick from being cut.

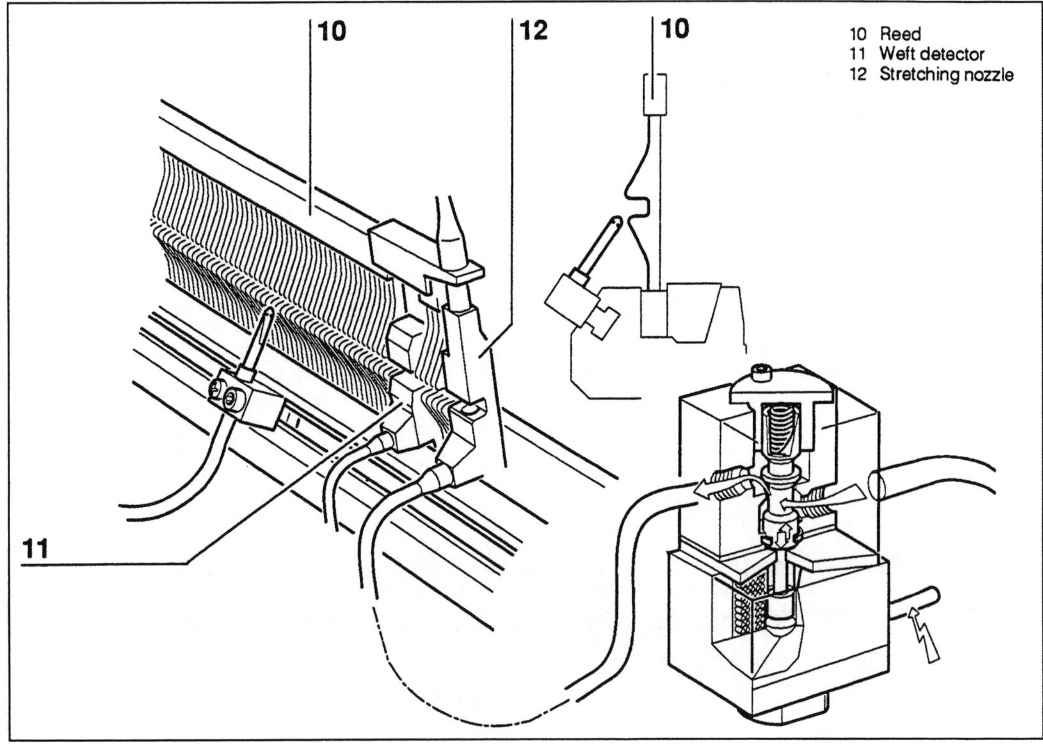

FIGURE 8.45 Schematic of reed, weft detector and stretching nozzle.

- The weaving machine makes one more revolution before stopping.
- The weaving machine reverses into open shed position.
- Weft is once more released from the drum feeder, forming a loop.
- The loop is blown by the vertical blowing nozzle not into the shed, but into blowing duct 4. The end of the loop then passes through suction nozzle 5 into suction duct 6.
- The pick is nipped as press roll 7 is lowered onto winding roll 8. The pick is removed from the shed by turning the winding roller. The pick is severed by a cutter from main nozzle 9.
- The pick length removed is compared with the parameterized pick length setpoint.
- The weaving machine automatically restarts once the pick is correctly removed.

There are also devices that detect filling breakages before the main nozzle, automatically repair them and restart the weaving machine. If the first pick after a stop is inserted at the normal timing, problems such as long pick or tip trouble may occur. Therefore, the filling insertion timing of the first pick is controlled by the computer. Start marks are prevented by the swift acceleration of the main motor of the weaving machine to full speed.

Sley (Figure 8.47)

Main shaft 1 of the weaving machine drives sley crank gear 2 on both sides of the machine. Driving arm 3 transmits the crank motion to sley tube 4. Sley 5, which moves back and forth, is mounted on sley tube 4.

The reed 6 is secured in aluminum profile 8 by means of a wedge type reed holder 7. The sley stroke is approximately 74 mm. At machine position 0° or beat-up, the reed is at an angle of 6° to the fabric plane.

Warp Stop Motion (Figure 8.48)

The warp stop motion stops the machine the instant a warp yarn breakage occurs. The yarn

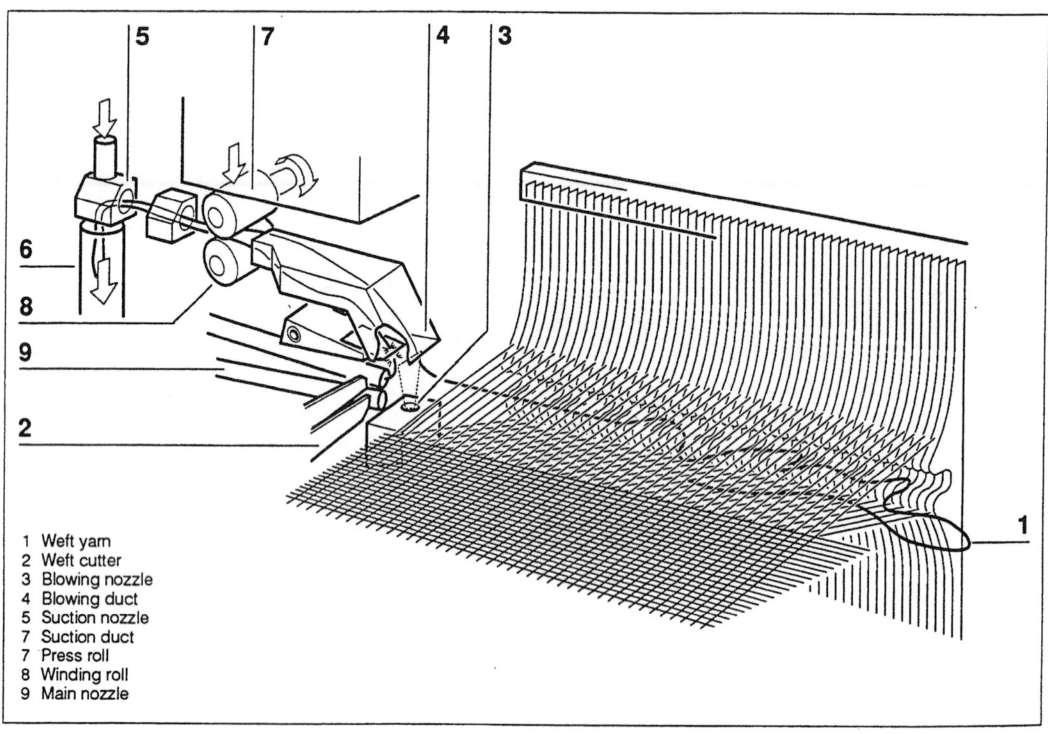

FIGURE 8.46 Automatic mis-pick remover.

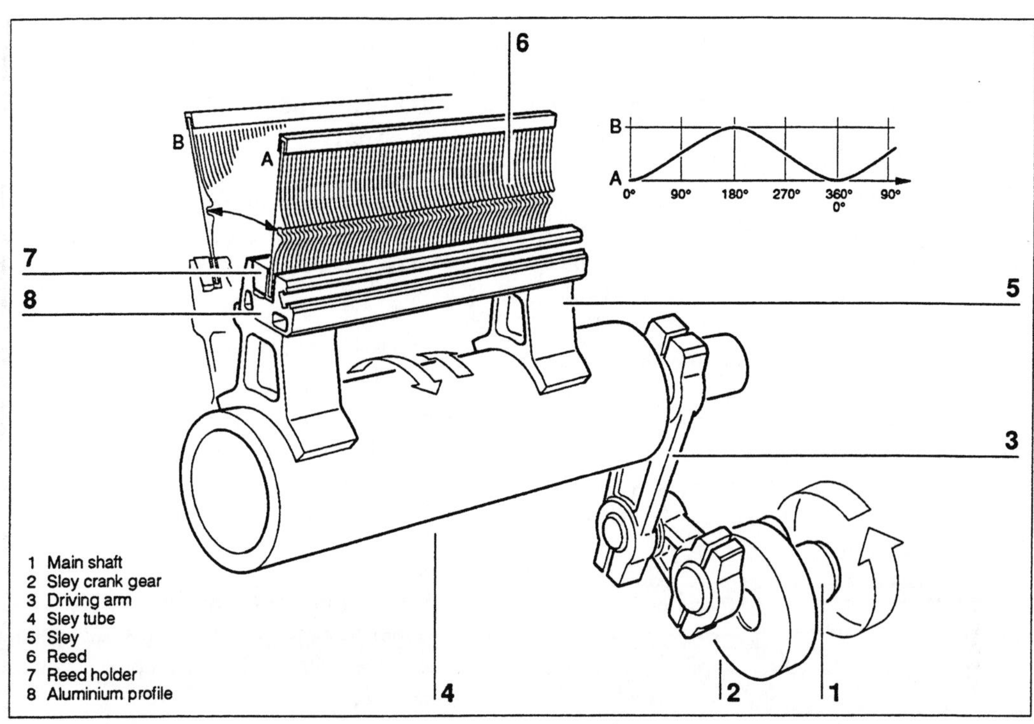

FIGURE 8.47 Sley movement.

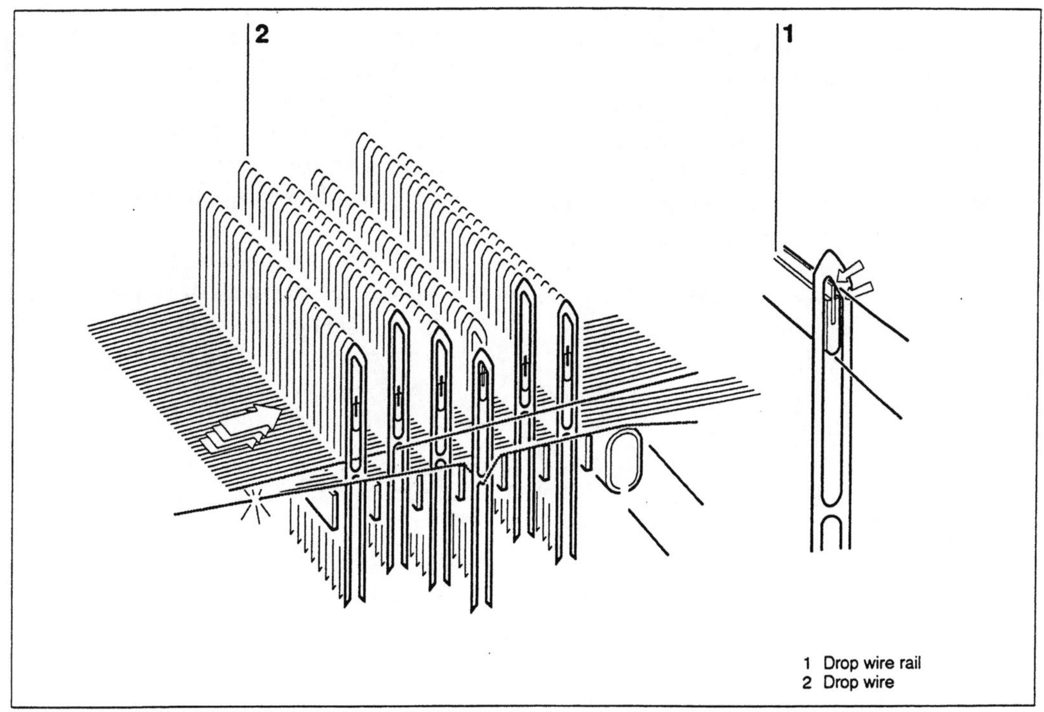

FIGURE 8.48 Warp stop motion.

breakage position is indicated automatically for quick repair work.

The configuration of the warp stop motion depends on warp density, yarn count and type of yarn. Up to six drop wire rails 1 can be fitted. The weight of the drop wires 2 depends on yarn type and count. For fine fabrics, a warp stop motion with a low drop wire dropping height can be fitted, to take the strain off the warp yarn.

Open or closed drop wires can be used, depending on whether the drop wires are pinned on the weaving machine, on the drawing-in machine or by hand. The signal column can be controlled so as to indicate in which area of the warp the broken warp end is located. As a drop wire drops, a circuit is closed and the weaving machine is stopped.

Selvages

Air-jet weaving machines typically produce cut selvages reinforced with leno. The trimmed ends of filling yarn on the receiving side may be fixed with additional yarns fed from a special creel and the cut off edge is fed into a waste box. High quality of selvages is assured by various selvage forming units such as full and half leno selvage devices and mechanical-pneumatic tucking units.

Figure 8.49 shows a full cross leno device. This device binds the picks on both sides of the fabric with two leno yarns 1 to produce a firm selvage. It is driven via gears; the leno device and the stationary gear being attached to the same shaft. The stationary gear drives the intermediate gears with the planetary gears. Spool holders 2 thereby rotate. The leno yarns unwound from the spools move up and down to produce a full cross leno selvage.

Tuck-in devices are used on air-jet weaving machines for fringeless selvages (Figure 8.50). A tuck-in device can be mechanic/pneumatic. Intermediate tucker devices are used when weaving multiwidth fabrics on wide weaving machines.

Phase A:

- Pick 1 has been inserted and is lying ready for cutting in front of tucking unit cutter 2.

Phase B:

- Pick 1 is beaten up by the reed.
- Cutter support 3 moves forward with tucking unit cutter 2 and cuts pick 1.
- The weft end is lying ready below the catch hole.
- Rotary valve 4 opens.
- The air streams through lower nozzle 5 and blows the tip of pick 1 into the catch hole.
- Tucking needle 6 dips into the shed.

Phase C:

- The eye of tucking needle 6 is exactly below the catch hole.
- Rotary valve 4 opens.
- The air current of upper nozzle 7 blows the weft end into the eye of tucking needle 6.
- Tucking needle 6 moves back.

Phase D:

- The weft end is tucked in.
- Cutter support 3 returns to its basic position and the weft end is beaten up with the next pick.
- Selvage formed by tucking unit.

Fusible fabrics can be severed and sealed with the sealed selvage device (Figure 8.51). A varying number of fusing elements 1 can be connected. When the auxiliary selvage 2 is cut off, the selvage is sealed.

The low temperature of the fusing elements prevents burning of the fabric. The sealing temperature can be adapted to the fabric being woven.

Safety devices are mounted on machines to guard against hand injuries. A light beam detects intrusion of hand or foreign objects into the lay area, triggering immediate weaving machine stoppage. Two hand starting motion is also available to increase operator safety.

8.1.6 Economics of Air-jet Weaving

Since the introduction of filling insertion by means of air-nozzles, air and energy consumption have been a concern.

Apart from investment and labor costs, energy costs are an important economical factor in weaving. A certain amount of air is required to transport the filling yarn into the shed. The volume of this air depends on the type of yarn used and is in proportion to the inserted volume of

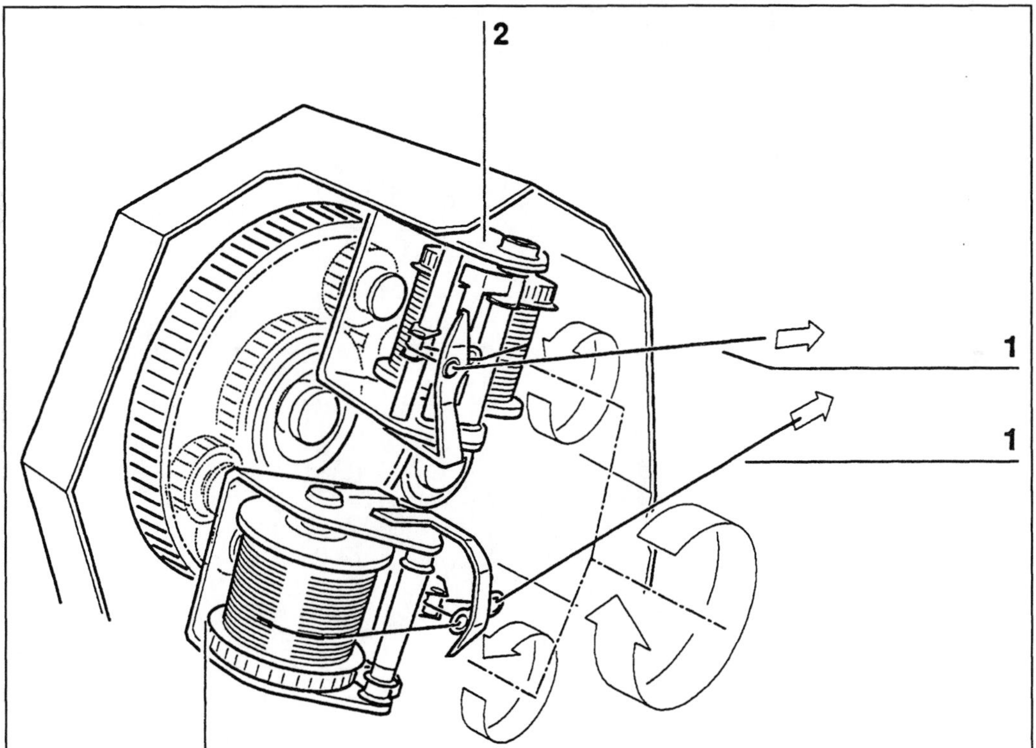

2

1

1

1 Leno thread
2 Spool holder

2

FIGURE 8.49 Full cross leno device.

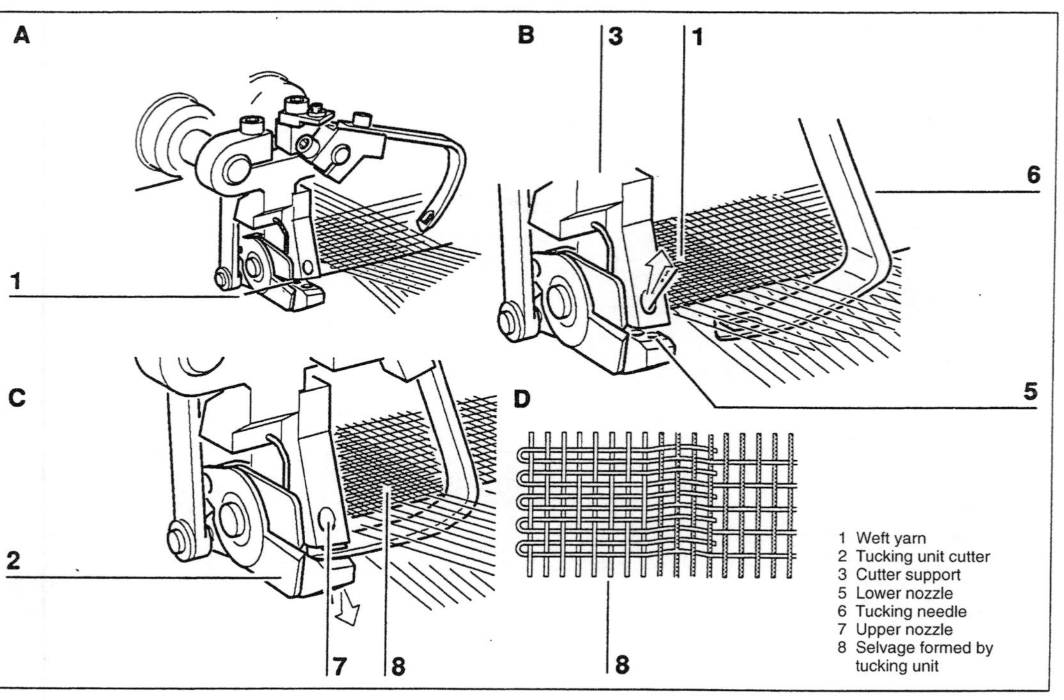

FIGURE 8.50 Tuck-in and intermediate tuck-in units.

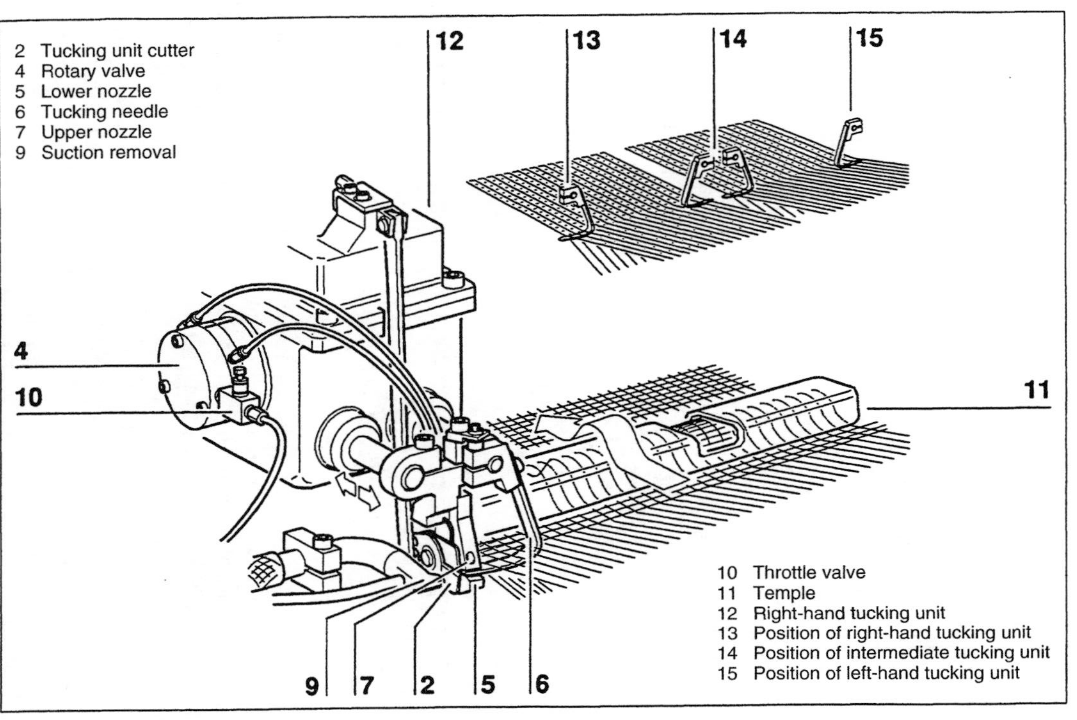

FIGURE 8.50 (continued) Tuck-in and intermediate tuck-in units.

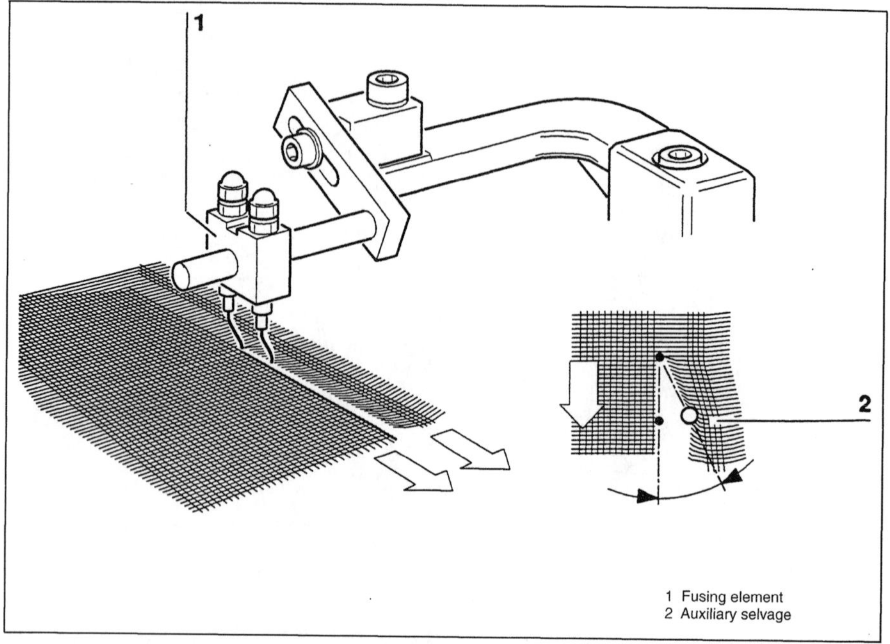

FIGURE 8.51 Sealed selvage.

Figure 8.4 shows the filling insertion system of a typical air-jet weaving machine. Automatic pick control regulates the pressure of the tandem, main and relay nozzles automatically without the need of human intervention. With spun yarns the control of the main nozzle pressure only is usually sufficient.

Reduction of the yarn load is affected by employing balloon breakers. In general, it is possible to work with reduced air pressures which not only reduces stoppages but also lowers air consumption. Tandem nozzle, i.e., arrangement of two main nozzles one behind another, allows filling insertion with a reduced pressure.

For a more even filling insertion, the opening of the relay nozzles is tapered as shown in Figure 8.52. This tapered outlet has a bundling effect on the air-jet and thus, at a distance of 50 mm from the nozzle, an increase of the air pressure of 30% (Figure 8.53). It was shown that the

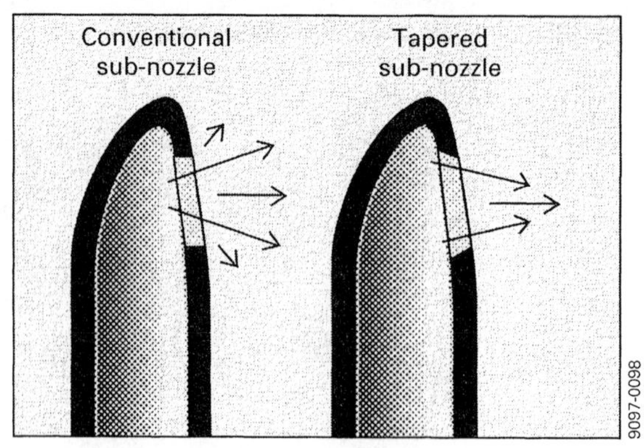

FIGURE 8.52 Relay nozzles with tapered opening.

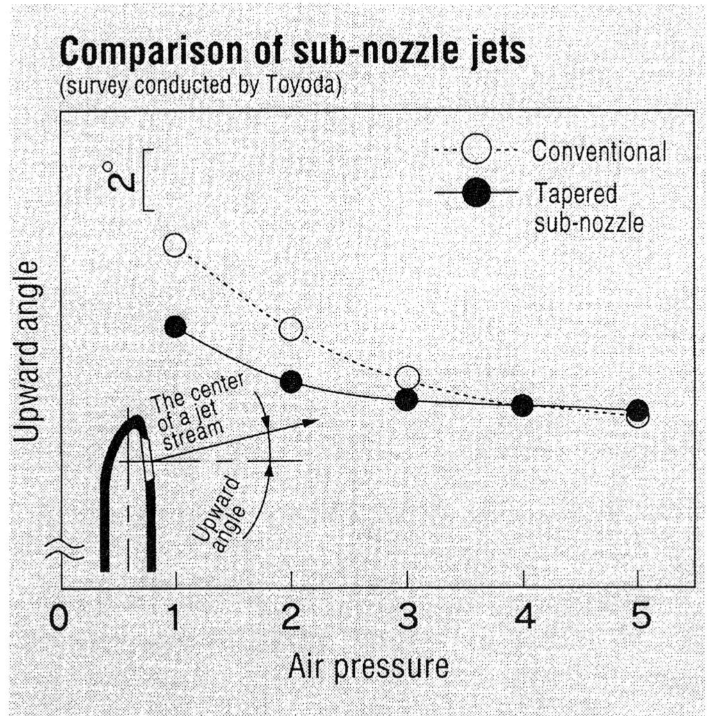

FIGURE 8.53 Reduced deflection of the blow angle with tapered opening on relay nozzles.

blow angle with tapered relay nozzles remains far more constant. With conventional nozzles the blowing direction varies with the pressure of the relay nozzles within a range of 4° to 6°, while this area can be cut in half by using tapered nozzles as shown in the figure.

Approximately 80% of the compressed air used in an air-jet weaving machine is used by the relay nozzles. Optimization of the blowing times of the relay nozzles can be achieved by increasing the number of valves of the relay nozzles. The total quantity of relay nozzles remains unchanged; however, only 3 instead of 4 or 5 relay nozzles are connected per valve. This allows a reduction of the blowing time of the individual valve (Figure 8.54). To guarantee complete stretching of the filling yarn, with a blowing time of the relay nozzles as short as possible, the stretching nozzle must be used. Increasing the cross-sectional area of the supply tube enables insertion of the necessary quantity of air at reduced pressure. Reduction of the pressure level results in power saving at the compressor.

Filling insertion rate and machine speed are influenced by several factors:

- weaving machine (nominal weaving width, shedding, number of harnesses, selvage formation)
- fabric style (fabric density, warp yarn tensile force, weft yarn count, weave)
- style dependent weaving machine settings (shed movement, shed angle, harness frame and harness weight)
- yarn material

As the nominal weaving width increases, the filling insertion rate increases. However, the machine speed has to decrease due to longer insertion time for the filling yarn. Figure 8.55 shows these relationships for air-jet weaving machines. Air jet weaving machines offer space and labor savings in production of fabrics. Figure 8.56

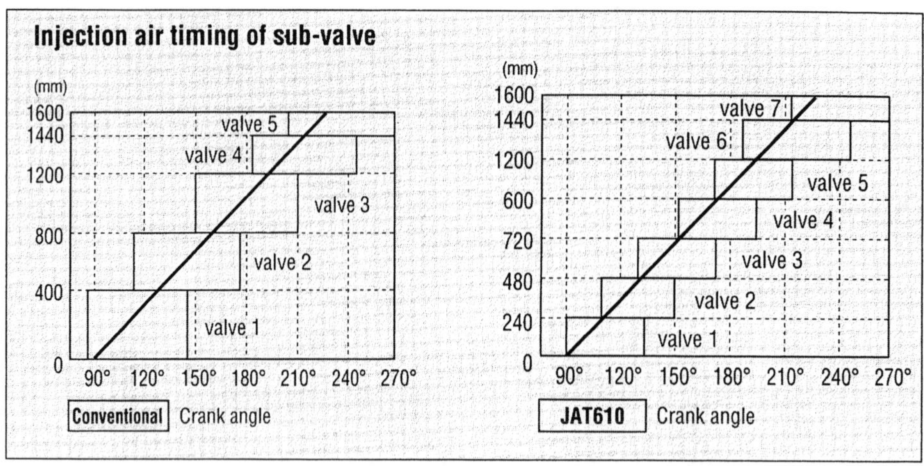

FIGURE 8.54 Reduced blowing times due to increased quantity of relay nozzle valves.

shows a typical plant layout of air-jet weaving machines.

Compressed air is an ideal means of filling insertion. Air is gentle to the yarn and allows high quality weaving of almost all current staple and filament yarns. Compressed air gives high speed weaving at low noise level.

The quality and reliability of the compressed air itself is critical for successful air-jet weaving. If the air filtering system fails, oil can reach the weaving machine and clog nozzles, disturb the operation of the machine and stain the fabric. This results in expensive production slow down, downtime and spoilage.

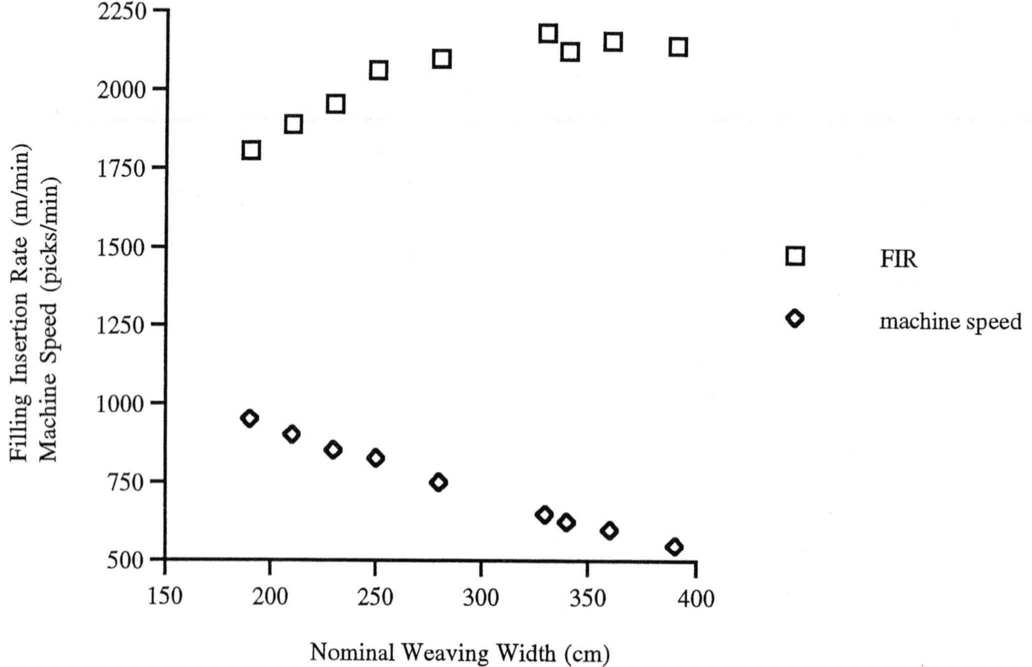

FIGURE 8.55 Effect of weaving width on filling insertion rate (FIR) and machine speed.

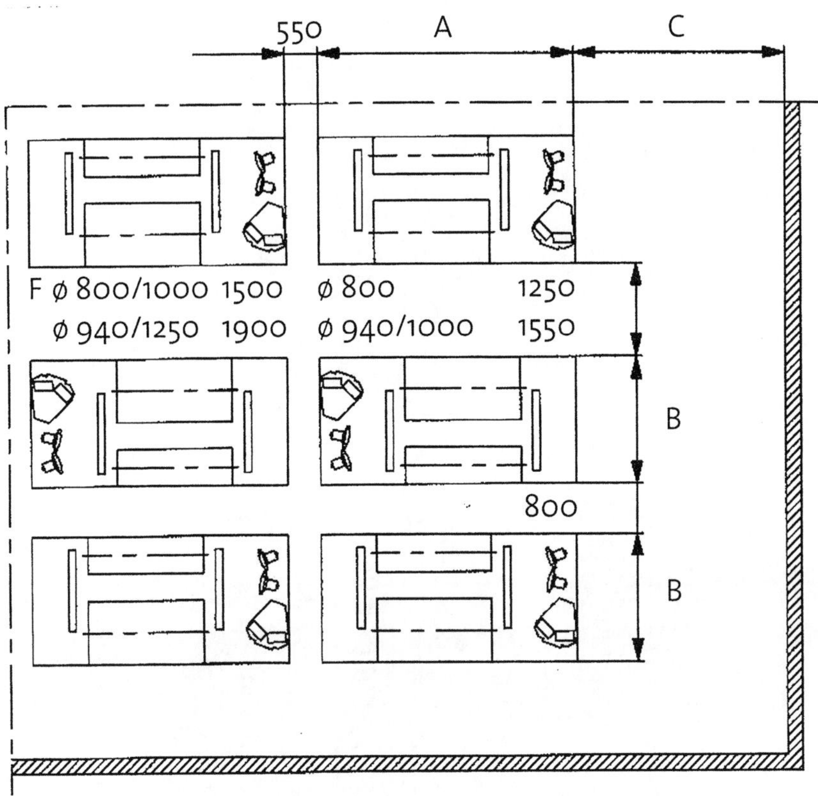

FIGURE 8.56 Air-jet weaving machine layout.

Although modern oil removal filters are very efficient, they cannot guarantee 100% clean air. Their efficiency is dependent on the lubricating oil's temperature. For example, up to 10 times more oil passes through a filter at 40°C than at 10°C. The filter systems require close monitoring and in large installations, can add significantly to operating costs through frequent filter replacement and the man-hours required to replace them. That is why the use of oil free compressors is recommended.

Humidity in the air causes corrosion in the air net. It may even cause corrosion in the weaving machine itself. This leads to reduced efficiency, high power requirement and high maintenance costs. In the weaving machine, rust particles can clog nozzles and spoil fabrics. With dry air, a simpler and lower cost distribution system is possible as no condensate drains are required. In the textile industry, the dew point of the compressed air must be kept between +2° to +5°C to avoid condensation. Residual solid contents of compressed air in the air jet weaving machine should not exceed 0.1 mg/m^3, and a maximum particle size is 5 μm in order to avoid disturbance in the weaving machines and spoilage of the end product. The best way to produce the pure, oil-free compressed air for air-jet weaving machines is to ensure that it never comes into contact with oil in the first place.

8.2 WATER-JET WEAVING

A water-jet weaving machine inserts the filling yarn by highly pressurized water. The tractive force is provided by the relative velocity between the filling yarn and the water jet. If there is no velocity difference between the water and yarn, then there would be no tension on the yarn which would result in curling and snarling of the yarn. The tractive force can be affected by the viscosity of the water and the roughness and length of the filling yarn; higher viscosities

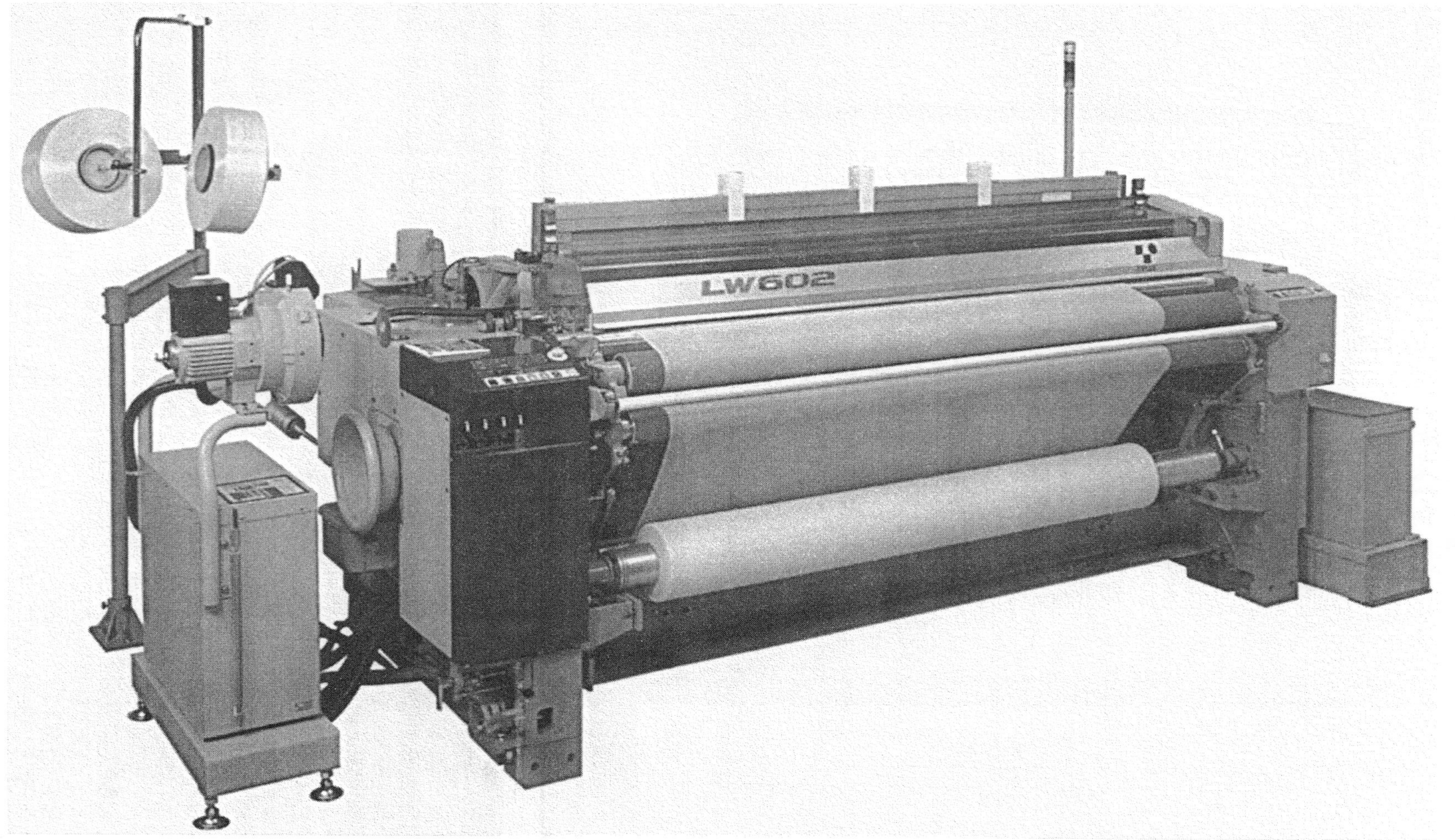

FIGURE 8.57 Water-jet weaving machine (courtesy of Toyoda).

cause higher tractive forces. The viscosity of water depends on the temperature.

The first water-jet weaving machine was KOVO, which was developed by Vladimir Svaty at the Research Institute for Textile Technology in Czechoslovakia and shown at the Brussels Textile Machinery Exhibition in 1955. This machine had a speed of 600 ppm with a reed width of 40 inches. Up to 12 harnesses were possible. The Prince Jet loom was another earlier water-jet weaving machine. It had a reed width of 65 inches at 400 ppm speed. The width and speed of the water-jet looms gradually increased and the modern water-jet weaving machines can have a speed of around 1,500 ppm while the maximum reed width is 3 m and the filling insertion rate is 1800 mpm. Figure 8.57 shows a modern water-jet weaving machine. Table 8.5 shows the typical characteristics of water-jet machines.

TABLE 8.5 Major characteristics of water-jet machines (courtesy of Toyoda).

LW600 Specification

			601 : Versatile machine	603 : Heavy fabric machine	602 : Single-color crank machine
Nominal Reed Space(cm)			150,170,180,190,210,230	190,210,230	150,170,180,190,210
Drive	Adjustable Space		0~-600mm		
	Main Motor	Crank	2.6kw		
		Dobby	2.6kw 3.7kw(opt)		-
	Main Pulley	Crank	195mm.dia , 3 Belts		
		Dobby	258mm.dia , 5 Belts		-
	Main Brake		Electromagnetic bake(70kgm)		
	Inching		Inverter-type inching, Contactor-type inching(opt)	Inverter-type inching	Inverter-type inching, contactor-type inching(opt)
Warp Line			900mm height (800mm dia, let-off)		
Frame			Box type frame with 4 stays		
Weft Insertion	Measuring		Feeder measuring system (Single-color: Wire type , 2-color : Roller type)		
	Picking		Single color, 2 color	2 color	Single color
	Brake/Pull-back		Single color :Weft brake(opt)	-	Single color :Weft brake(opt)
			2 color :Weft brake & pull-back (R & B)		-
	Pump		Plunger type, compression spring activated		
			Plunger dia(mm) :16,17,18,20,22,24,26		
			Single pump, Double pump (opt)		Single pump
	Nozzle		Adjustable rectifier type nozzle		
	Package Stand		2-package(single color),4-package(2color)off-loom type	4 package,off-loom type	2 package, off-loom type
Shedding	Crank Type		4 heald frames (std)/14mm pitch 6 heald frames (opt)/14mm pitch	-	4 heald frames (std)/14mm pitch 6 heald frames (opt)/14mm pitch
	Upper Mount Dobby		16 heald frames /12mm pitch		-
	Positive Cam		10 heald frames /12mm pitch		-
Beating	Drive		Crank type, both sides drive		
	Sleysword		Multi-sleysword	Multi-sleysword with intermediate support	Multi-sleysword
	Reed Stroke		65,75,85,95mm	85,95mm	65,75mm
	Reed Protecting Device		Cam drive(*1)	-	Cam drive(*1)
Let-Off	Mechanism		Electronic-controlled automatic tension adjusting positive continuous-drive let-off motion		
	Effective Tension		40kg~400kg	250kg~800kg	40kg~250kg
	Beam flange dia		800(std),914(opt),1000(opt)		φ800(std), φ914(opt)
	Easing		Negative easing(std), Positive easing(opt)		Negative Easing
Take-Up	Mechanism		Continuous positive take-up system		
			Mechanical take-up motion(std), Electronic take-up motion(opt)	Mechanical take-up motion	Mechanical take-up motion(std), Electronic take-up motion(opt)
	Cloth Line		Flat cloth line(std)、Slanted cloth line(opt)	Slanted cloth line	Flat cloth line
	Max. roll dia		520mm		
	Counter		Electronic type built into left side operation panel		
	Drive		Gear driven with automatic tension adjusting mechanism		
Leno			Leno interweaving system by planetary gear motion(closed gear motion)		
Catch Cord			Waste filling end gripped.false-twisted and carried into waste yarn bucket		
	CC shedding		Heald frame shedding(std)、separate cc shedding(opt)		Heald frame shedding
Water Extraction	Mechanism		Separater system		
	Blower		Energy-saving high pressure blower 400W(std) , 750W(opt)		
Cutter			Mechanical cutter with ceramic and hard alloy blades		
Electric Spec	Control Panel		Microprocessor control (Compact unit type)		
	Operation Panel		Sheet key type operation panel with emergency button		
	Presetter		Built-in type presetter at left side operation panel for Electronic let-off,Electronic take-up,Single-color settings		
	Memory Card		Input & output function built-in for let-off motion,electronic take-up,Single color feeder, 2 color PAW and EDU(*2)		
	Electronic Dobby control		EDU(opt)		-
	Dropper Monitor		LED light at left side operation panel		
	Feeler		IR feeler		
Lubrication			Oil bath system for main drive parts , greasing system for bearings		

* 1 This device is not available on a slanted cloth line machine
* 2 Memory Card is used by inserting card into the presetter built into the left hand operation panel. Memory cards are sold as separate items.
* All drawings, data, and photographs which appear in this catalogue are subject to change without prior notice"

FIGURE 8.58 Double pumps for weaving two different fillings (courtesy of Toyoda).

Water-jet weaving machines have the same basic functions of any other type of weaving machines. The principle of filling insertion with a water-jet is similar to the filling insertion with an air-jet: they both use a fluid to carry the yarn. However there are some differences that affect the performance and acceptance of water-jet weaving machines. For example, the yarn must be wettable in order to develop enough tractive force.

The flow of water has three phases: 1) acceleration inside the pump prior to injection into the nozzle, 2) jet outlet from the nozzle, 3) flow inside the shed. The water flow inside the shed has a conical shape with three regions: compact, split and atomized. Compact and split portions are better for yarn insertion. Due to water weight, the jet axis forms a parabola which necessitates adjusting the axis of the nozzle upward by some angle. The flow of water then follows the motion of angular projection.

Unlike the air-jet weaving machines, the pump and picking system is fixed firmly to the machine frame to ensure that the beat-up mechanism moves the reed only. Due to the viscosity of water and its surface tension, a water-jet is more coherent than an air-jet. As a result, the water-jet does not break up that easily and has a longer propulsive zone. There are no varying lateral forces in a water-jet to cause the filling yarn to contort. Besides, since the wet moving element is more massive, there is less chance for the filling yarn to entanglewith the warp. The braking of the filling yarn is provided by the reed.

The width of a water-jet weaving machine depends on the water pressure and diameter of the jet. Since water is not compressible, it is relatively easy to give enough pressure to the water-jet for insertion. The diameter of the jet is around 0.1 cm and the amount of water used for one pick is usually less than 2 cc. Double pump system, with two nozzle at will filling insertion, is suitable for weaving fabrics with two different fillings (Figure 8.58).

At ITMA-99 in Paris, Toyoda exhibited a 4 color pick-at-will water jet weaving machine (Figure 8.59). The nozzles are positioned by a

FIGURE 8.59 Four color pick-at-will (courtesy of Toyoda).

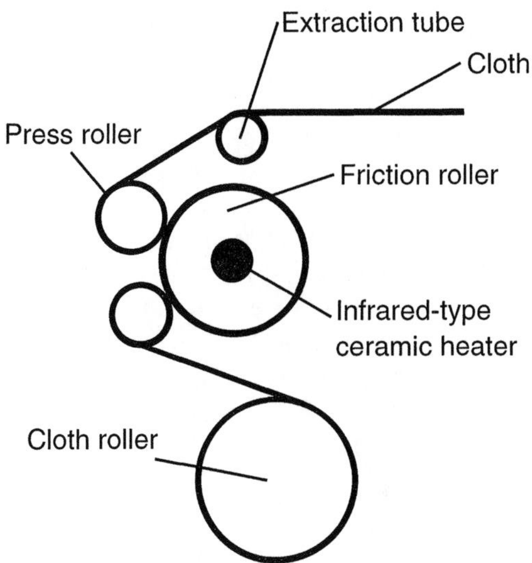

FIGURE 8.60 Schematic of evaporator (courtesy of Toyoda).

servo motor. Filling insertion occurs from one nozzle position.

In some machines, an evaporator is used to dry the fabric on-loom (Figure 8.60). In the figure, the drying system is built into the take-up friction roller.

The wastewater after insertion is usually removed into a drainage system.

REFERENCES

1. Adanur, S., Lecture Notes, Auburn University, 1995.
2. Brooks, J. C., U.S. Patent 1,096,283.
3. U.S. Patent No. 1,386,550 (1921).
4. U.S. Patent No. 1,368,691 (1921).
5. U.S. Patent No. 1,405,096 (1922).
6. Ballou, E. H., U.S. Patent No. 1,721,940 (1929).
7. Swedish Patent No. 253, 930 (1945).
8. British Patent No. 616,323 (1949).
9. British Patent No. 693,751 (1953).
10. Svaty, V., British Patent No. 860,970 (1961).
11. British Patent No. 942,737 (1963).
12. British Patent No. 1,004,850 (1963).
13. Mohamed, M. H., "The Current State of Weaving", Textile Fundamentals Short Course, North Carolina State University, 1980.
14. Textile World, Sept. 1993.
15. Adanur, S., Air-Jet Filling Insertion: Velocity Measurement and Influence of Yarn Structure, M.S. Thesis, North Carolina State University, Raleigh, NC, 1985.
16. Adanur, S., Dynamic Analysis of Single Nozzle Air-Jet Filling Insertion, Ph.D. Thesis, North Carolina State University, Raleigh, NC, 1989.
17. British Patent No. 862,093 (1961).
18. Buakev, P. T. and Vlasov, P. V., "The Pneumatic-Rapier Method of Weft Insertion", Tech. of the Text. Ind. USSR, No. 2, 1969.
19. GB Patent 2,072,719 A, Ruti.
20. McGinley, T. F., USP 4,122,871, October 1978.
21. McGinley, T. F., USP 4,122,872, October 1978.
22. Holman, J. P., Thermodynamics, McGraw-Hill, 1980.
23. Shapiro, A. H., The Dynamics and Thermodynamics of Compressible Fluid Flow, The Ronald Press Company, 1953.
24. Van Wylen, J. G., Thermodynamics, John Wiley & Sons, 1960.
25. Mohamed, M. H., and Salama, M., "Mechanics of a Single Nozzle Air-Jet Filling Insertion System, Part I: Nozzle Design and Performance", TRJ, Vol. 56, No. 11, November 1986.
26. Salama, M., Mechanics of Air-Jet Filling Insertion, Ph.D. Thesis, North Carolina State University, 1984.
27. Schlichting, H., Boundary-Layer Theory, McGraw-Hill, 1979.
28. White, F. M., Viscous Fluid Flow, McGraw-Hill, 1974.
29. Uno, M., et al, "A Study on Air-Jet Looms", J. Text. Mach. Soc. of Japan, 7(1), 28A, 1961.
30. Pilipenko, V. A., "Analysis of the Air Flow in the Channel of Pneumatic Looms", Tech. of the Text. Ind. USSR, No. 2, 1965.
31. Kohlhaas, O., "Investigations of Weft Storage Systems", Melliand Textilberichte (Eng. Ed.), November 1979.
32. Batra, S. K., et al, "Dynamic Analysis of Yarn in Ring Spinning: An Integrated Approach", NCSU, March 1987.

33. Krause, H. W., "The Air-Jet Weaving Machine in Practical Use", Melliand Textilberichte (Eng. Ed.), September 1980.
34. Pilipenko, V. A., "Contribution to the Analysis of the Weft Yarn Speed on Pneumatic Loom", Tech. of the Text. Ind. USSR, No. 1, 1964.
35. Duxbury, V., et al, "A Study of Some Factors Involved in Pneumatic Weft Propulsion", JTI, Vol. 50, No. 10, 1959.
36. Lyubovitskii, V. P., "Analysis of the Acceleration and Flight of the Shuttle in a Gaseous Jet", Tech. of the Text. Ind. USSR, No. 5, 1965.
37. Lyubovitskii, V. P., "The Weft Yarn Flight in the P-105 Loom", Tech. of the Text. Ind. USSR, No. 3, 1966.
38. Pilipenko, V. A., "Analysis of the Weft Yarn Acceleration of the Pneumatic Loom", Tech. of the Text. Ind. USSR, No. 3, 1968.
39. Klockhova, G. M., "Weft Insertion by Air-Jet", Tech. of the Text. Ind. USSR, No. 6, 1969.
40. Lyubovitskii, V. P. and Akaev, A. A., "The Dynamics of a Yarn End of Finite Dimensions in the Air Flow on the P-105 Air-Jet Loom", Tech. of the Text. Ind. USSR, No. 5, 1970.
41. Lyubovitskii, V. P. and Akaev, A. A., "The Yarn Dynamics in the Non-Stationary Air Flow of the P-105 Loom", Tech. of the Text. Ind. USSR, No. 6, 1970.
42. Klockhova, G. M., "The Movement of the Pick on the Pneumatic Loom", Tech. of the Text. Ind. USSR, No. 1, 1970.
43. Uno, M., "A Study on Air-Jet Loom with Substreams Added, Part 1: Deriving the Equation of Motion for Weft", J. Text. Mach. Soc. of Japan, Vol. 18, No. 2, 1972.
44. Uno, M. et al, "A Study on an Air-Jet Loom with Substreams Added, Part 2: Analysis of Various Weaving Factors by the Equation of Motion of Weft", J. Text. Mach. Soc. of Japan, Vol. 18, No. 3, 1972.
45. Balagushkina-Pilipenko, V. A., "Mispicks in Pneumatic Weft Propulsion", Tech. of the Text. Ind. USSR, No. 6, 1972.
46. Lunenschloss, J. and Wahhoud, A., "Investigation into the Behavior of Yarn in Picking with Airjet Systems", Melliand Textilberichte (Eng. Ed.), April 1984.
47. Kissling, U., "Experimental and Theoretical Analysis of Weft Insertion by Air-Jet", Melliand Textilberichte (Eng. Ed.), February 1985.
48. Ishida, T., "Air-Jet Loom, Present and Future, Part 5: Technical Problems Caused by Air-Jet", JTN, November 1982.
49. Hasegawa, J., et al., "A Study of Weft Insertion System on Air-Jet Loom", paper presented at ASME Textile Engineering Conference, Raleigh, NC, 1981.
50. Mohamed, M. H. and Lord, P. R., "Comparison of Physical Properties of Fabrics Woven From Open-End and Ring Spun Yarns", Text. Research J., March 1973.
51. Wahhoud, A., "Investigations Into the Behavior of Yarns in Pneumatic Weft Insertion", Melliand Textilberichte (Eng. Ed.), April 1983.
52. Yao-Qi, Q., "The Twist Loss of the Weft During Air-Jet Weaving", JTI, No. 1, 1984.
53. Mohamed, M. H., et al., "Influence of Filling Yarn Characteristics on the Properties of Curdory Fabrics Woven on an Air-Jet Loom", TRJ, Vol. 57, No. 11, November 1987.
54. Padfield, D. G., "A Note on the Fluctuations of Tension During Unwinding", JTI, Vol. 47, No. 6, T301, 1956.
55. Lunenschloss, J., et al, "Optimization of Air Consumption and Weft Insertion Characteristics of Spun Yarns in Industrial Air-Jet Weaving", Melliand Textilberichte, 67, 1986.

SUGGESTED READING

- Melling, K. G., "The Best Selvage: Necessity, Not Choice", Textile World, September 1998.
- Lord, P. R. and Mohamed, M. H., Weaving: Conversion of Yarn to Fabric, Merrow Technical Library, 1982.

REVIEW QUESTIONS

1. Explain the major characteristics of air flow in air-jet weaving.
2. What are the factors that affect the efficiency of an air-jet nozzle?

3. What are the forces acting on the yarn in an air-jet? Which forces are useful to propel the yarn?
4. Why does air-jet weaving require high quality filling yarns?
5. Compare the performance of the following yarns in air-jet weaving. Assume that the yarns have the same count.
   - ring spun yarns
   - open-end yarns
   - air-jet spun yarns
6. Compare the air-jet weaving with the other weaving systems economically.
7. What are the main physical differences between an air jet and a water jet? How does this affect the flight of the yarn? Explain.
8. What are the factors limiting the wider use of water-jet machines.

9

Projectile Weaving

The projectile weaving machine was introduced to the market in 1952 by Sulzer as the first successful, shuttleless weaving machine. It was exhibited at the International Exhibition of Textile Machinery in Brussels in 1955. This filling insertion system produces good fabric quality with high economical efficiency and low energy consumption. At ITMA-99 in Paris, improved performance of these machines was exhibited. Use of new electronics optimized smooth running and made the machine easier to operate. A modified filling insertion system further reduced yarn loading.

Projectile weaving machines use a projectile equipped with a gripper to insert the filling yarn across the machine. The unique principle of projectile filling insertion allows the insertion of practically any yarn: cotton, wool, mono- and multifilament yarns, polypropylene ribbon, and even hard fibers like jute and linen. This is because all yarns, fine or coarse, are securely gripped and inserted by the projectile (Figure 9.1), resulting in a wide variety of fabrics, from simple staple goods through superior fashion cloth and from wide heavy industrial fabrics to complex jacquard cloths. Figure 9.2 shows the schematic principle of projectile filling insertion. The gripper projectile draws the filling yarn into the shed. Energy required for picking is built up by twisting a torsion rod. On release, the rod immediately returns to its initial position, smoothly accelerating the projectile by means of a picking lever. The projectile glides through the shed in a rake shaped guide. Braked in the receiving unit, the projectile is then conveyed to its original position by a transport device installed under the shed.

Figure 9.3 shows a projectile weaving machine. The projectile weaving machines offer the following advantages:

- low power consumption
- reduced waste of filling material due to unique clean, tucked-in selvages
- quick warp and style change
- mechanical and operational reliability and ease of use
- low spare parts requirement and easy maintenance
- long machine life

Another major advantage of projectile weaving machine is that more than one width of fabric can be woven at a time. Different widths, from 33 cm to 540 cm, make the projectile weaving machine even more economic, saving energy and space. Yet, because of the favorable speed at such weaving widths, wear and tear on the machine are also much reduced. Projectile weaving machines can be a single color or multicolor machine for any sequence of up to four or six different filling yarns.

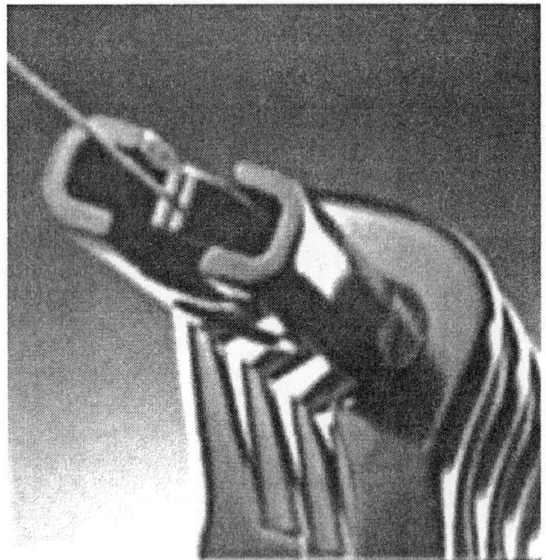

FIGURE 9.1 Filling grab by projectile.

9.1 FUNCTIONAL DESCRIPTION OF PROJECTILE WEAVING MACHINES

There are two major models of projectile weaving machines in production today: P7150 and P7250. Another version, called P-Lean, is a low cost alternative to these machines. The following functional descriptions are based on P7150 and P7250 weaving machines. P7150 is a further development of the P7100. Figure 9.4 shows the drive sequence in the projectile weaving machine.

Machine Drive

Figure 9.5 shows the schematic of the machine drive for a Sulzer P7250 weaving machine. The machine is driven by main motor 1 via four V-belts 2 to two flywheels 3. When the weaving machine is switched on, the machine brake is released by the eccentric crank motion of the starting mechanism with motor 6, and the two flywheels are pressed against clutch plate 4. The clutch plate is attached rigidly to main shaft 5 so that the weaving machine starts up.

When the weaving machine is switched off, electric power is no longer supplied to holding magnet 7. The tension exerted by spring 8 engages the machine brake and releases clutch pressure. At the same time, main motor 1 is switched off via limit switch 9.

Two modes of operation are possible with the starting gear:

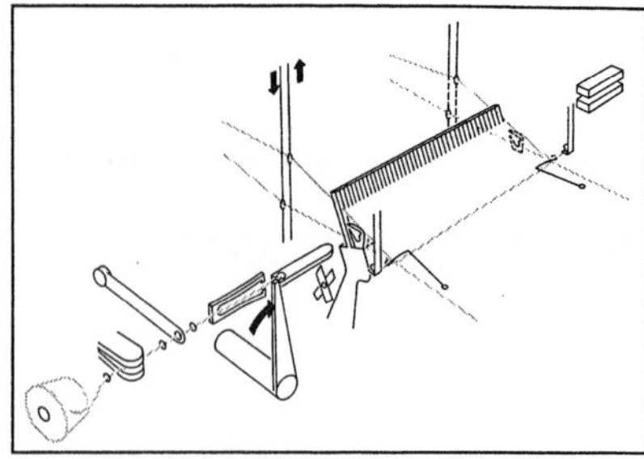

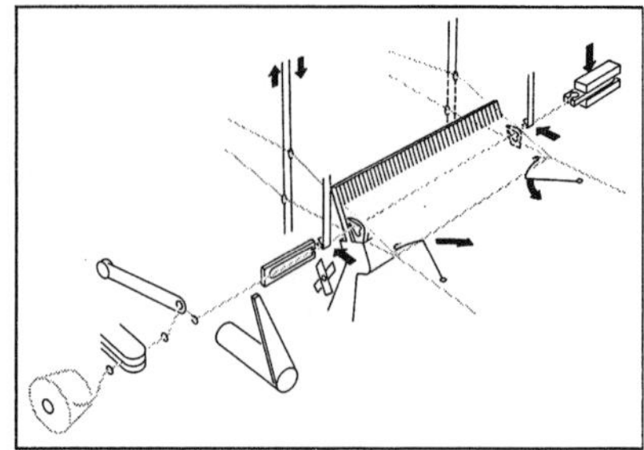

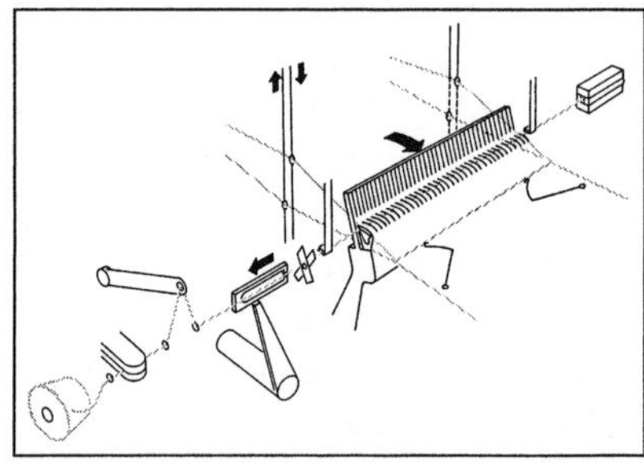

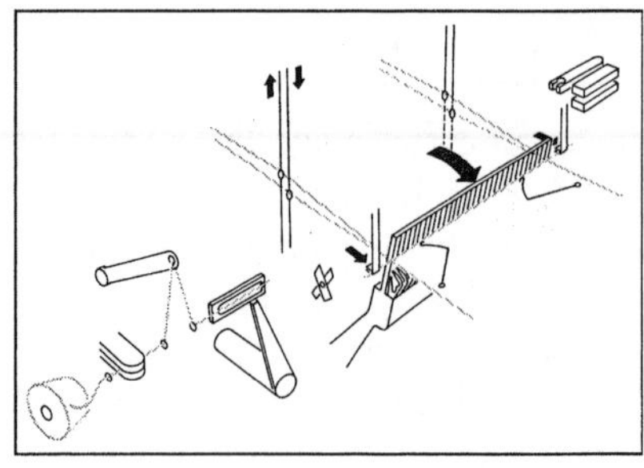

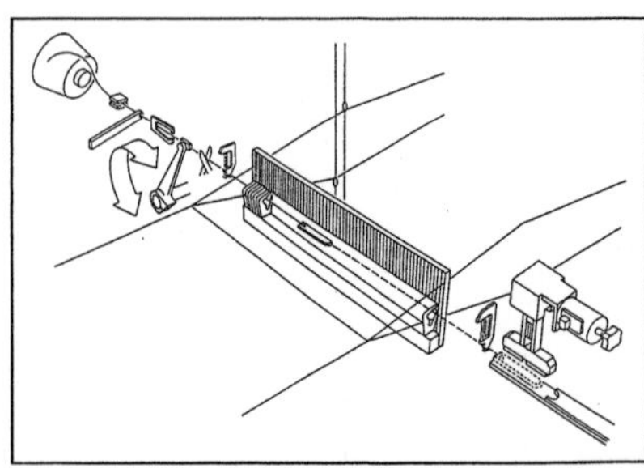

FIGURE 9.2 Schematic principle of filling insertion with projectile.

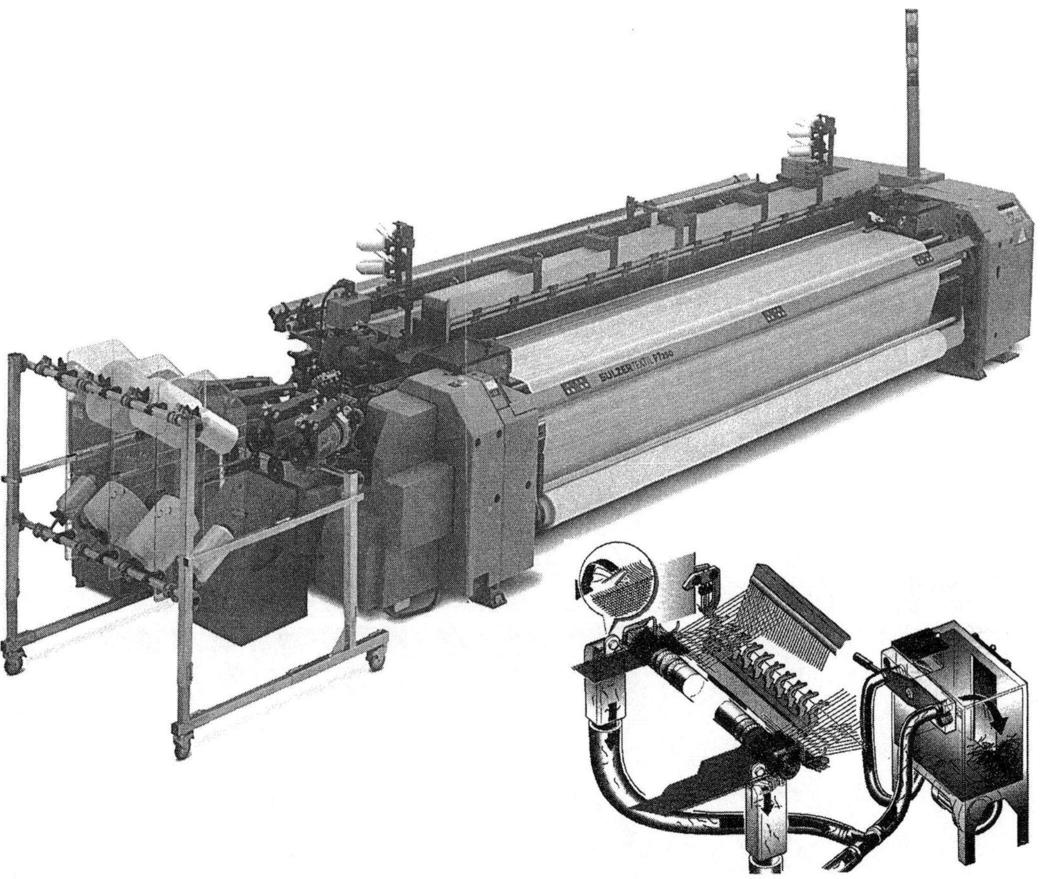

FIGURE 9.3 Projectile weaving machine.

1. Switch on machine. Full revolution of the crank, upward movement to release the machine brake, downward movement to engage the machine clutch.
2. Creep speed operation. Reduced crank rotation, only to release the machine brake. Creep speed motion is then generated by the creep speed motor via the electromagnetic clutch in the creep speed gear to the driving shaft of the weaving machine.

Figure 9.6 shows the machine drive for P7150 machine. At the upward movement of the engaging device 1 the machine brake 2 is raised and the lock 7 slides under the link 8. At the same time the motor 6 starts. When the machine is ready for switching on, a solenoid holds the link 8 secure and the lock 7 cannot escape.

During the downward movement of the engaging device 1, the disc flywheels 4 are pressed together and entrain the clutch disc 3 with them. If a fault occurs, the solenoid is deenergized and the lock 7 and machine brake 2 are pulled down by the spring 9. At the same time the motor is switched off via limit switch 10.

If the fault is not put right, the solenoid cannot hold the lock 7 and the machine does not start when the engaging device 1 is operated. To turn the machine with the handwheel 5, the motor may be switched off. All other mechanisms function normally, so that also in this case the machine cuts out in the event of trouble.

Stop Motions

The stop motions in P7250 operate partly mechanically, partly electronically (Figure 9.7). They protect the individual units of the weaving machine against damage and switch off the weaving machine if motion is obstructed.

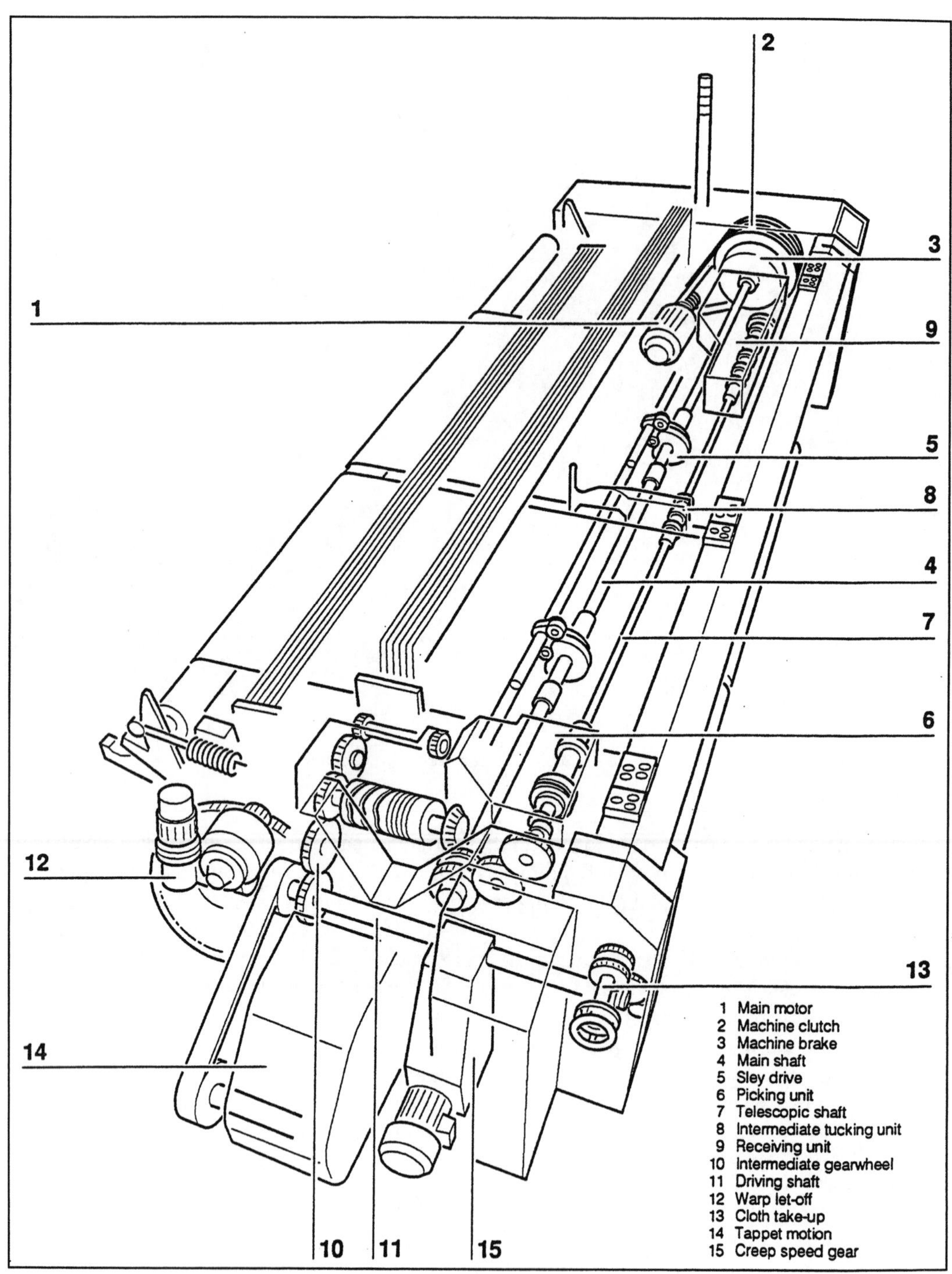

FIGURE 9.4 Drive sequence of projectile weaving machine.

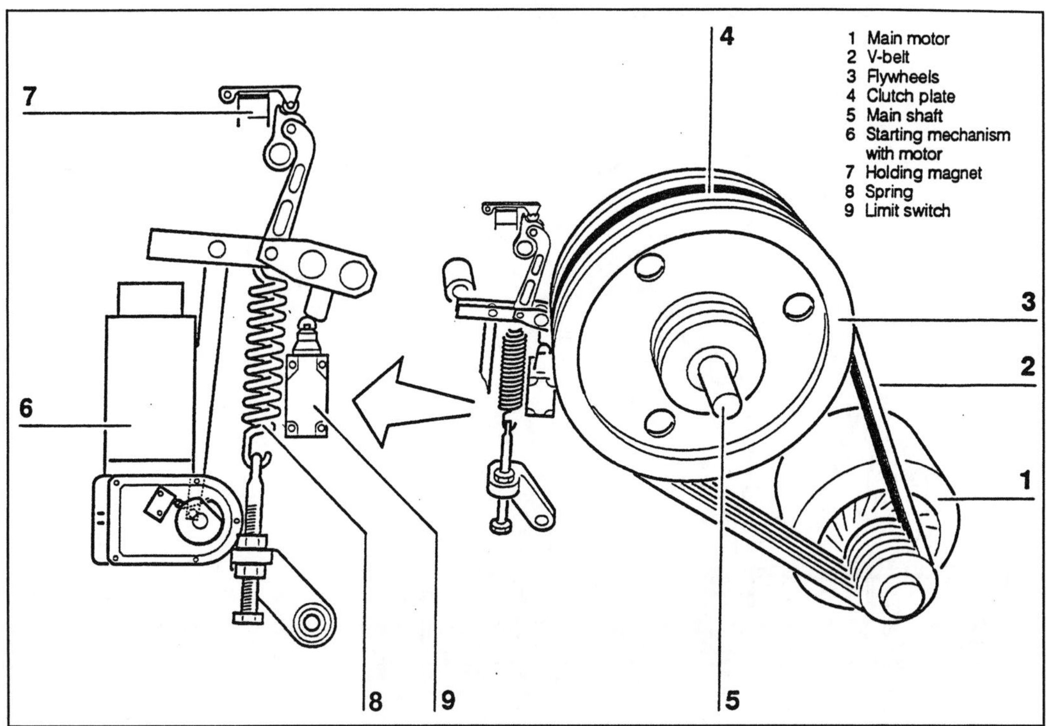

FIGURE 9.5 Schematic of machine drive for P7250.

Filling detectors 5, warp stop motions 8 and projectile monitor 10 operate electronically. Depending on the type of malfunction, the machine is switched off directly by cutting off the electric power supply to holding magnet 15, or via sensors 6, 7 and 12, which actuate an electrical signal in the event of mechanical malfunctions; the power supply to the holding magnet is then interrupted and the weaving machine switched off.

Warp Let-off Motions

Warp let-off in the projectile weaving machine is mechanically or electronically controlled. The warp let-off provides constant warp tensions regardless of the warp beam being full or empty (Figure 9.8).

Figure 9.9 shows the schematic of power warp let-off mechanism for P7200. Back rest roller 1 serves as an automatic control device for the warp let-off. The position of the back rest roller is scanned continuously by the sensor and switch flag 2 while the machine is running, and is transmitted to the control electronics in the switch cabinet.

Whenever back rest roller (whip roller) 1 is about to leave its reference position, the control electronics respond by accelerating or slowing down warp let-off motor 3.

When operations are interrupted, the current speed of the warp let-off is stored for re-starting. In order to prevent starting marks after stoppages, a start delay and a slight tensioning or slackening of the warp, depending on the length of the stoppage, can be programmed at the terminal.

Warp let-off motion is transmitted from the motor to the warp beam via a worm gear and warp beam gear wheel. If two half-width warp beams are being used, differential gear 4 ensures uniform tension on both warps.

Back rest roller 1 serves as a yarn guide and control for the warp let-off (Figure 9.10). Warp tension is adjusted by means of springs 2 on both sides of the machine.

For heavy fabrics, deflecting roller 3 can be used in addition to back rest roller 1, especially

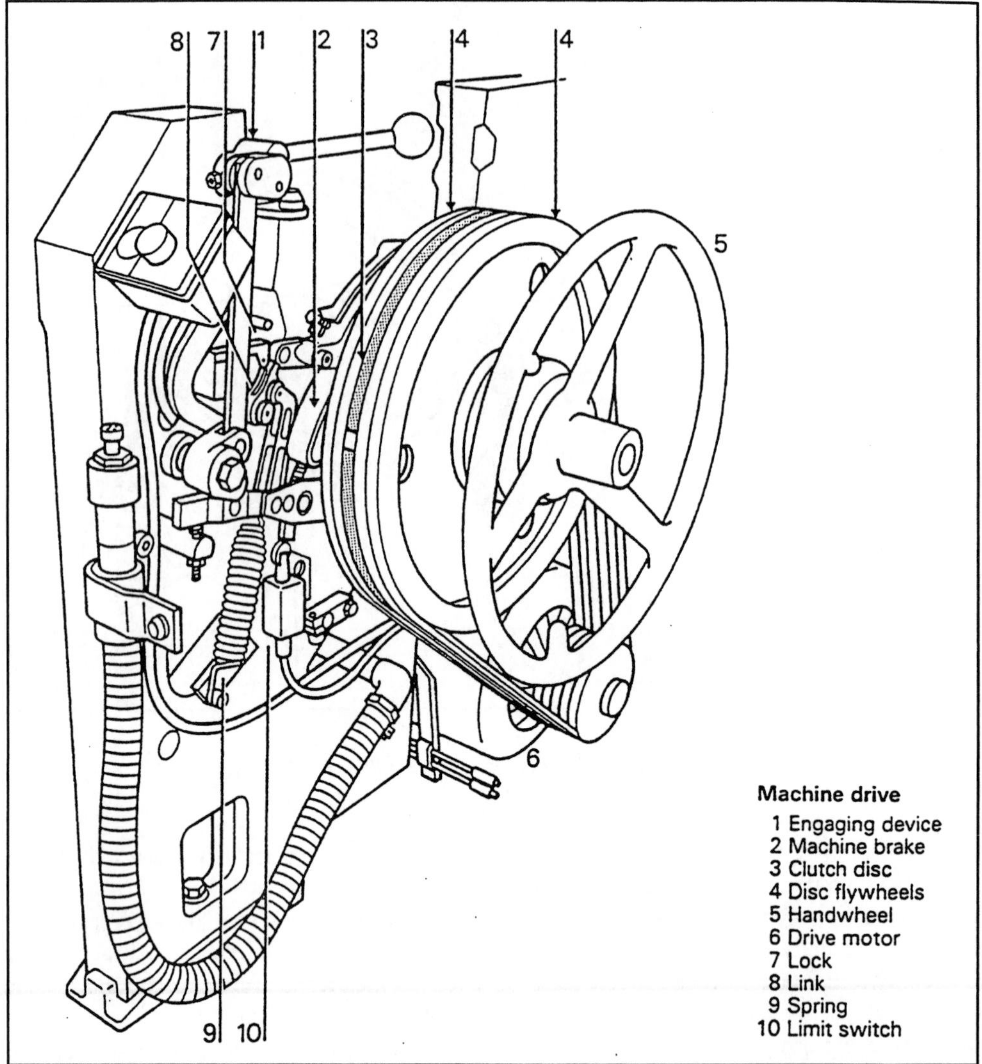

FIGURE 9.6 Schematic of machine drive for P7150.

on wide weaving machines (Figure 9.11). This reduces the load on the back rest roller, which serves mainly as an automatic control for the warp let-off. Depending on the nominal width, this allows up to 10% higher filling density or allows reduction of the recoiling fell at unchanged filling density.

P7100 has a mechanical let-off mechanism. In mechanical warp let-off (Figure 9.12), the warp tension is adjusted with the springs 5. Actuated by a cam, the roller 6 presses the cam disc 4 briefly against an actuating disc, letting off the warp beam via the worm drive 3. As the warp beam diameter diminishes, the warp tension tends to rise, and the whip roller 1 moves down a little. Via the regulating linkage 2, this sets the roller 6 closer to the cam disc so that the let-off increases again. When the cam disc is unloaded, the actuating disc contacts a brake under spring pressure.

If two half width beams are woven off, the worm drive 3 includes a differential to compensate for the differences in tension between the two half width warp beams.

P7150 has a power let-off mechanism which is shown in Figure 9.13. Via a sensor and switch

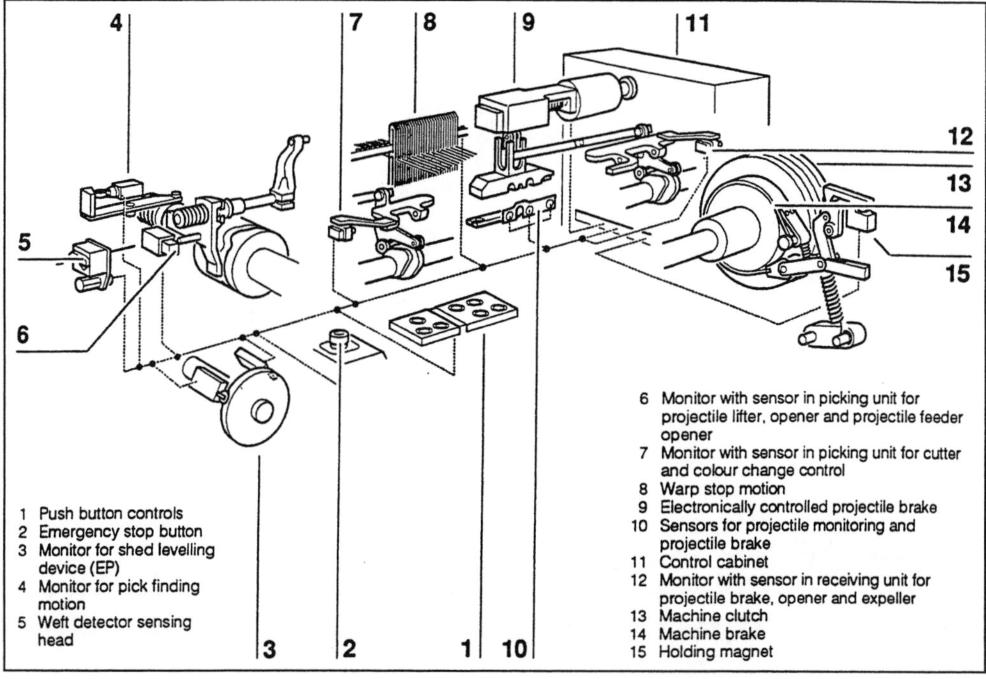

FIGURE 9.7 Stop motions on P7250 projectile weaving machine.

vane 2, the speed of the warp let-off motor is controlled so that the position of the whip roller 1 remains constant and with it the warp tension. When operation is interrupted, the last speed of the warp let-off motor 3 is stored. Upon restarting the machine, the warp let-off motor 3 starts again at the right speed.

Warp Tensioner

In P7150 and P7250, the warp tensioner is used as an alternative to the standard back rest roller, especially with heavy fabrics (Figure 9.14).

Torsion bar 3 in tension tube 2 presses back rest roller 1 against the warp ends. Depending upon the style being woven, torsion bar 3 is pretensioned to the required torsion, thus setting the warp tension. The warp tensioner may be used in conjunction with the power warp let-off.

Color Change

Figure 9.15 shows different versions of color change controls.

a) 1-1 filling mixer. This is done with shackle gear and electronic switch-off control. This system can easily be changed over to single color weaving.

FIGURE 9.8 Force sensing unit for constant warp tension.

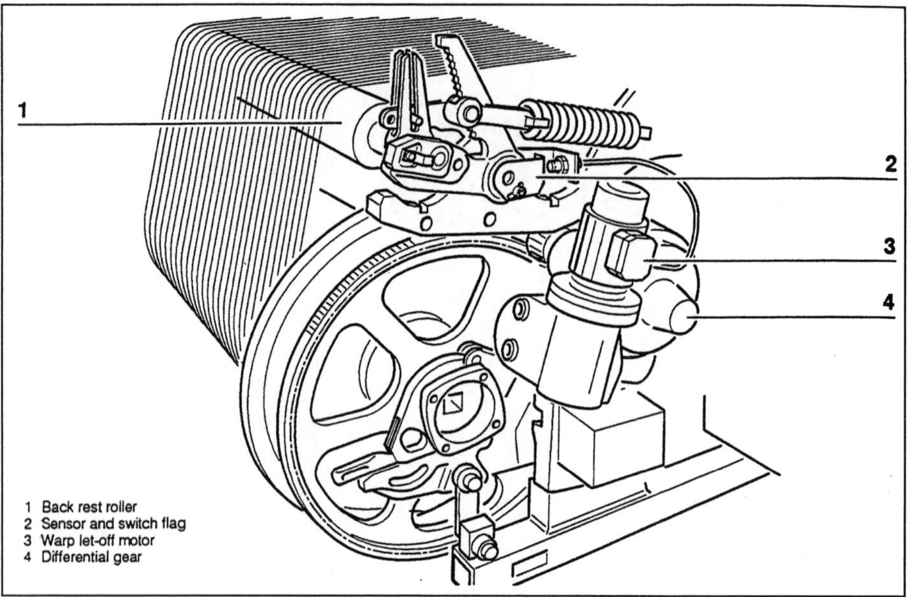

FIGURE 9.9 Schematic of warp let-off mechanism for P7200.

b) Two color unit ZSQ. This color change system is electronically programmable for two colors. This system is designed for weaving machines with cam motions, certain dobbies and jacquard machines.
c) Four color unit MSA (not shown). This is an electronically programmable color change system for four colors. It is designed for weaving machines with cam (tappet) motion, dobby and jacquard machines.

Shedding Mechanisms

Projectile weaving machines can have cam (tappet) or electronically controlled rotary dobby systems. Rotary dobby can be controlled

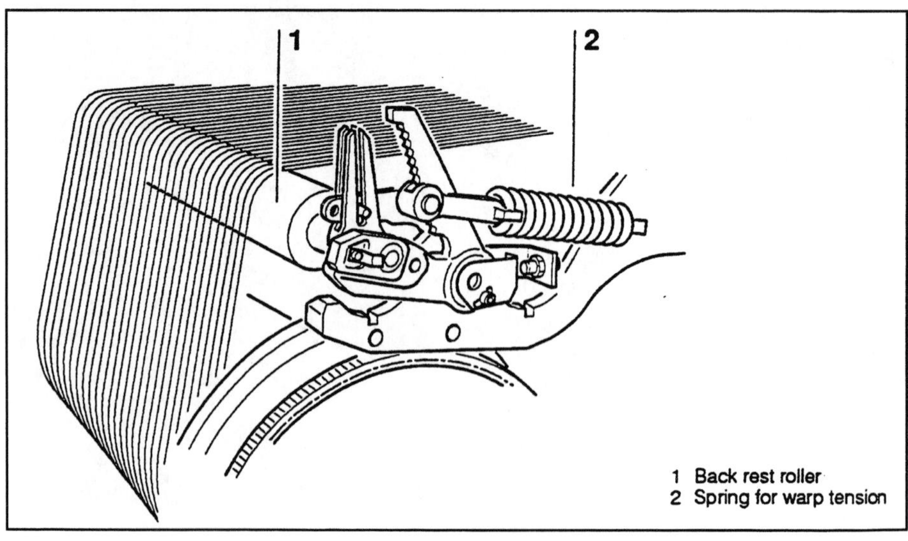

FIGURE 9.10 Back rest (whip) roller.

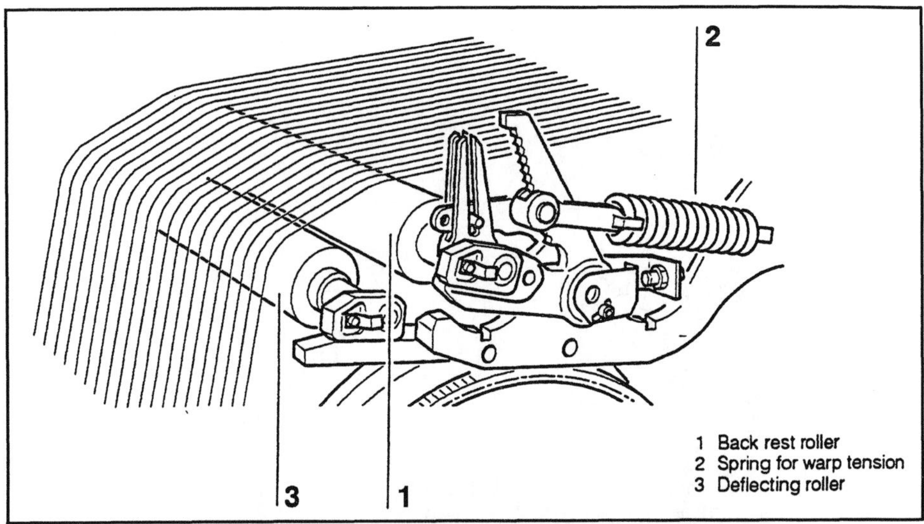

FIGURE 9.11 Back rest roller with deflection roller.

mechanically or electronically. Jacquard system can be mechanical or electronic. For P7150/P7250 only electronically controlled dobby and jacquard are available. Mechanical or electronic shed leveling equipment is used to level the harnesses.

a. Positive Cam (Tappet) Motion (Figure 9.16)

Cam motion 1 is designed for 6, 10 or 14 harnesses and can be used for weave repeats of up to 10 picks (P7150 and P7250 have up to 8 pick repeat). The shedding motion is positively operated via double cams 2.

The shed leveling device enables roller levers 3 to be lifted off the double cams and all harnesses 4 to be brought to the same height.

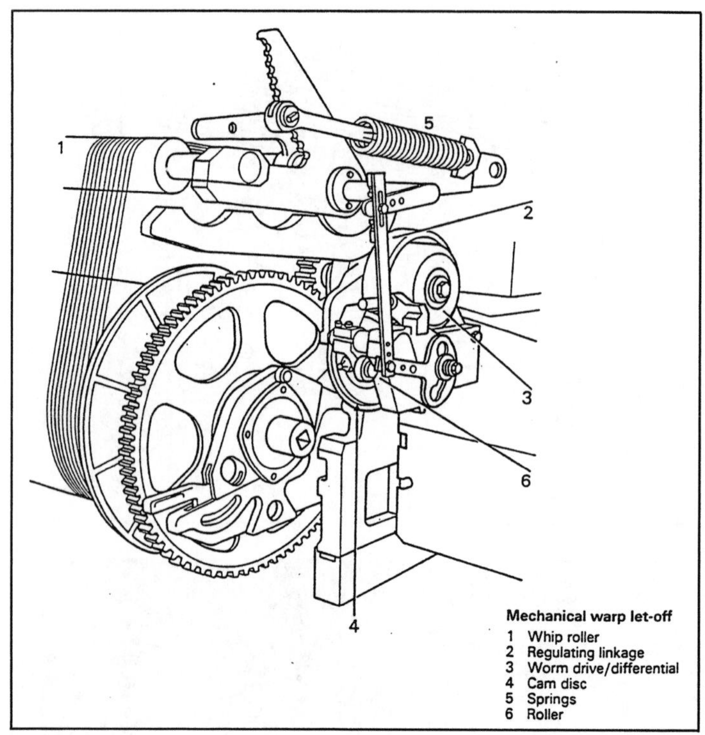

FIGURE 9.12 Mechanical warp let-off for P7100.

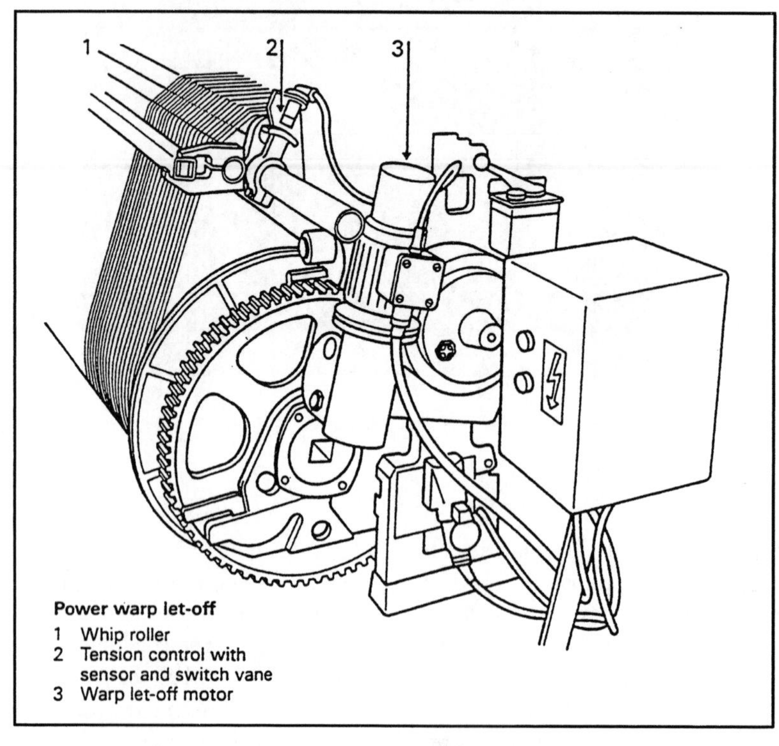

FIGURE 9.13 Power warp let-off for P7150.

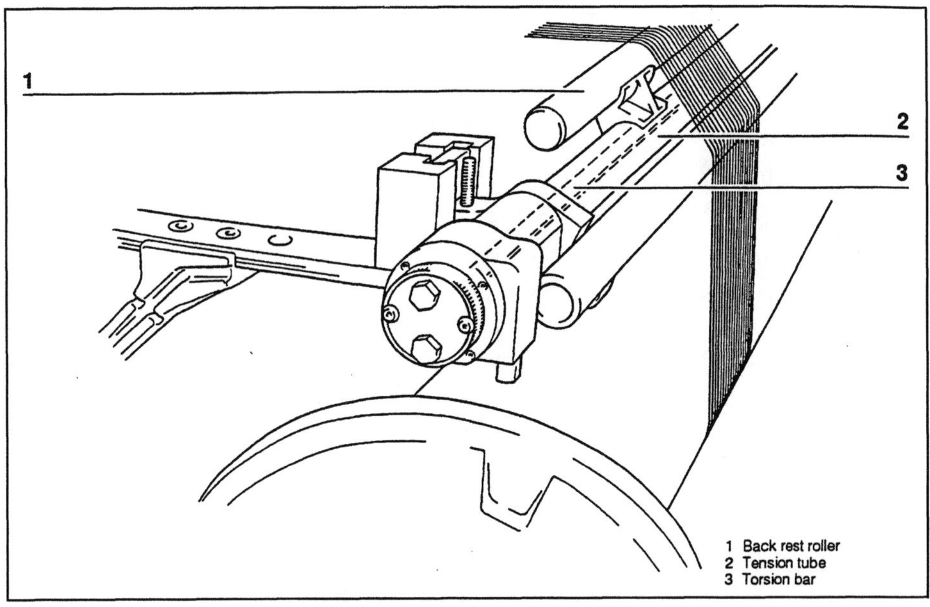

FIGURE 9.14 Warp tensioner.

b. Staubli Electronically Controlled Rotary Dobby (Figure 9.17)

This dobby permits weaving with up to 18 harnesses. Weaving machines equipped with microprocessor control are controlled by entering the fabric and machine specific parameters at the terminal. Harness movement is defined by moving jack runners 1.

c. Electronically Controlled Jacquard Machine (Figure 9.18)

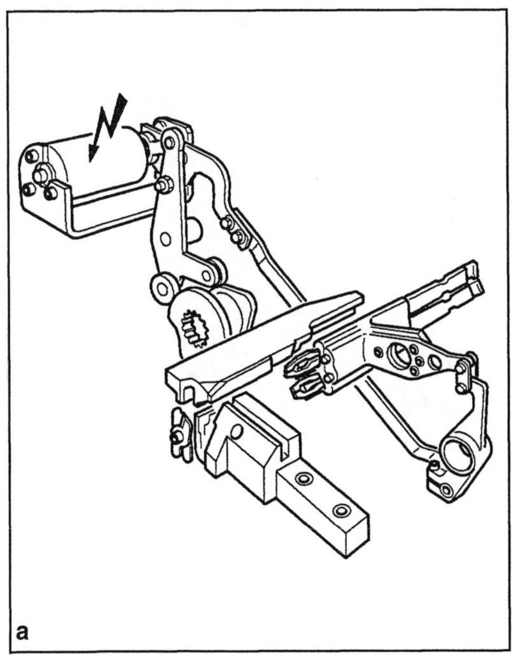

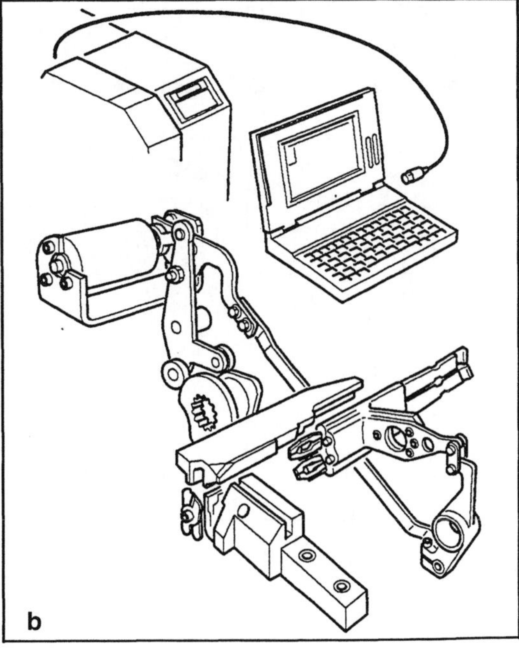

FIGURE 9.15 Different versions of color change controls.

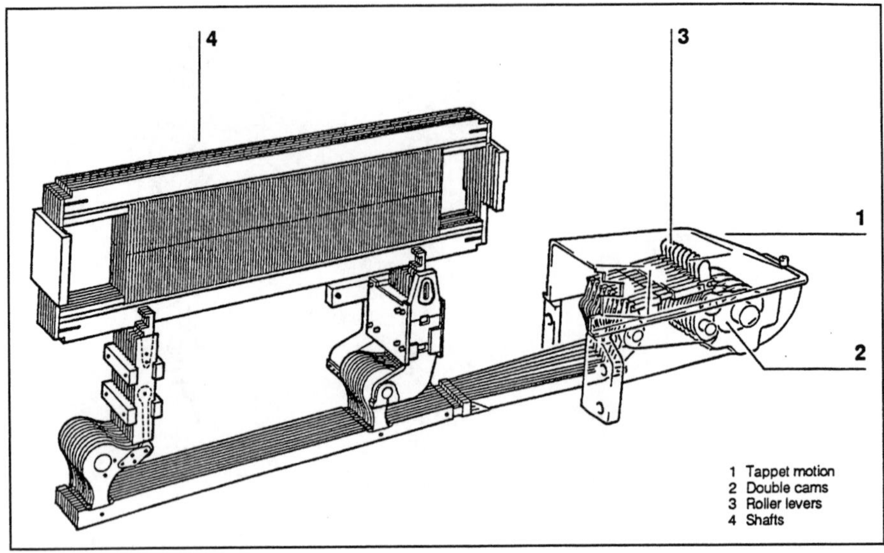

FIGURE 9.16 Positive cam motion.

Jacquard machines enable a large number of warp ends (hooks) to be controlled simultaneously and individually, i.e., extensive weave repeats are possible.

Lifting knives 1 move harness cords 2 up via hook 3 and pulley 4 by means of electronic control 5, depending on the weave, down movement is provided by a spring.

An automatic shed leveling device can be fitted to P7150 and P7250. By the automatic leveling of the harness frames during downtimes, the load acting on the warp yarns is kept even. This obviates the start-up marks. At the same time, the drawing-in of broken warp yarns is facilitated, since the eyelets of the harnesses are all in one and the same plane.

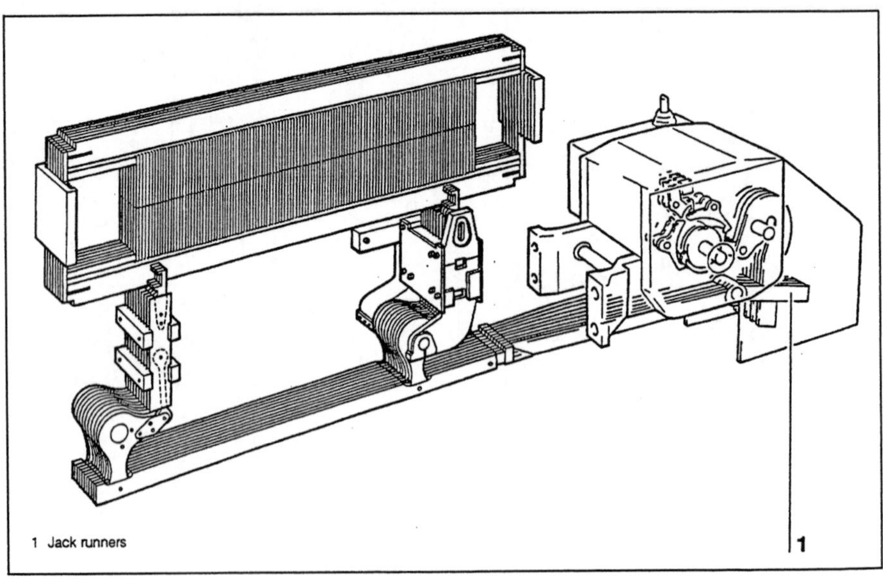

FIGURE 9.17 Electronically controlled rotary dobby.

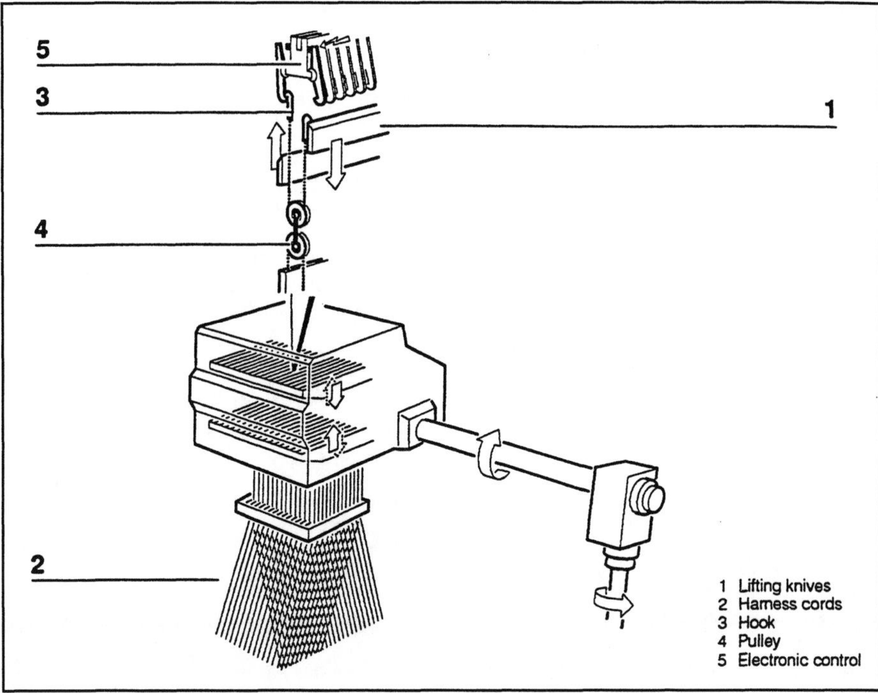

FIGURE 9.18 Electronically controlled jacquard machine.

Filling Tensioner (Figure 9.19)

The filling tensioner consists of an electronically programmable filling brake 1 and a cam controlled filling tensioner lever 2.

Shortly before the projectile is picked, the filling brake opens and the filling tensioner lever moves upwards so that the filling yarn can be inserted without deflection.

When the projectile approaches the receiving unit, the filling brake closes and prevents too much filling yarn from being inserted. The filling tensioner lever then moves downwards and stretches the slack yarn in the shed.

Filling Insertion Sequence (Figure 9.20)

A. Projectile 1 moves into the picking position.
B. Projectile feeder 2 opens after projectile 1 has gripped the filling end presented to it.
C. The projectile has drawn the yarn through the shed, while filling tensioner lever 3 and adjustable filling brake 4 act to minimize the strain placed on the yarn during picking.
D. The projectile is braked by projectile brake 8 in the receiving unit and pushed back, while the filling tensioner lever and the filling brake hold the filling yarn lightly stretched. At the same time, the projectile feeder moves close to the selvage.
E. Projectile feeder 2 takes over the filling yarn, while filling end grippers 5 hold it at both sides of the fabric.
F. The yarn is released by the projectile on the receiving side. Filling cutter 6 cuts the filling yarn. A conveyor takes expelled projectile 1 and carries it outside the shed back to the picking position.
G. The reed has beaten up the pick. Tucking needles 7 tuck the filling tails into the next shed (tucked selvage). Filling tensioner lever 3 has taken up the length of yarn remaining when projectile feeder 2 returns. The next projectile 1 is brought into the picking position.

Air Supported Filling Insertion

Pre-acceleration of the filling with compressed air reduces the strain on the filling (Figure 9.21). Two part, staggered guiding teeth reduce the strain on warp and filling. Therefore,

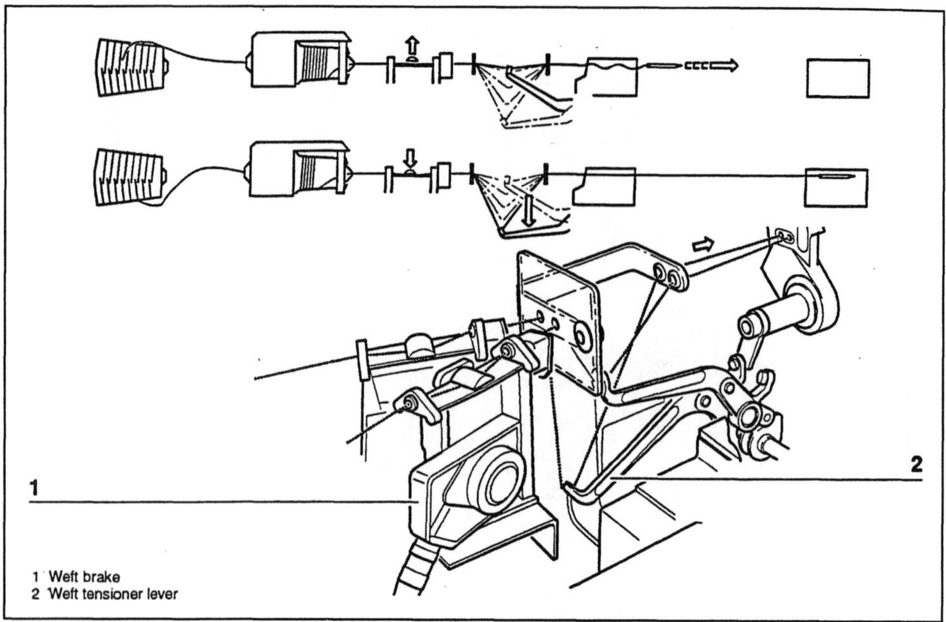

FIGURE 9.19 Filling tensioner.

even delicate yarns, such as those without twist or air-intermingled, can be used in the warp to make fabrics of good quality.

With the air supported filling insertion, the peak values of the tensile stress, which the filling yarns are subjected to, are reduced to a degree that permits insertion of low tenacity yarns even at the highest speeds. Due to the support provided by the compressed air, the flow of the tensile load phase is markedly softer and the peak values of the tensile stress are reduced by approximately 50% (Figure 9.22). This allows weaving of fine yarns.

Picking Mechanism

a. Before Picking (Figure 9.23)

Torsion rod 9 is twisted via the cam 8, roller lever 6 and picking shaft 10 till the knee joint 5 tilts slightly beyond dead point.

The front end of the torsion rod 9 fits in the picking shaft 10, on which picking lever 11 is clamped so that it performs the rotary movement of the picking shaft 10 and torsion rod 9.

When projectile 2 is ready for picking, the rollers 12 left and right beside the cam 8 run up onto the ribs of the roller lever 6, causing the knee joint 5 to yield.

b. After Picking (Figure 9.24)

The picking lever 11 jerks forward quickly, accelerating the projectile 2 through the picking shoe 1. The unloading movement of the torsion rod 9 is cushioned by an oil brake 7.

Projectile Circulation (Figure 9.25)

Conveyor chain links 2 push a projectile 1 into projectile lifter 3 in picking unit 4, pick by pick. Projectile opener 5 rises to open the projectile gripper and moves parallel with the projectile lifter into the picking position. During this process the open projectile gripper grips behind the clamping surfaces of the projectile feeder 6. The projectile opener leaves the projectile, which seizes filling yarn 7.

The projectile feeder gripper is opened and the projectile accelerated. It passes through the guide teeth and is stopped in projectile brake 11. Sensors monitor the arrival time and position of the projectile in receiving unit 12.

The projectile returner 13 pushes the projectile into the correct position to open the projectile

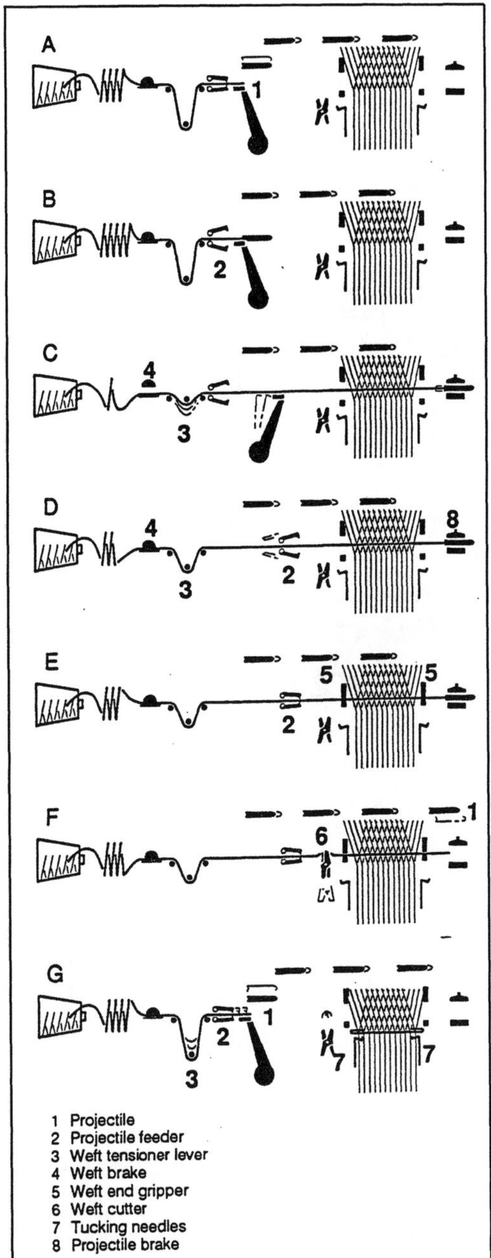

FIGURE 9.20 Filling insertion sequence with projectile.

gripper. During this process, filling tensioner lever 14 stretches the filling yarn and takes up the length of yarn corresponding to the distance the projectile is pushed back.

The filling yarn thus inserted is seized by filling end grippers 15 and the projectile feeder at cutter 16, and cut. It is then beaten up by the reed.

The package end of the filling yarn is seized by the projectile feeder at the cutter and taken back for transfer to the next projectile. Expeller 17 presses the projectile between two links in conveyor chain 18 at the appropriate moment.

FIGURE 9.21 Pre-acceleration of the filling yarn with compressed air.

Projectile Types and Dimensions

Projectiles have been made of steel or carbon composites (Figure 9.26) which have different variations. Figure 9.27 shows the projectile types and Table 9.1 gives the characteristics of projectiles. D1 is the standard steel projectile for the vast majority of commercial yarns. D12 is the same as D1 with a larger yarn clamping surface to ensure more reliable gripping, even of delicate yarns. D2 has a big cross section and large clamping surface and is used for extremely coarse yarns. K3 is the synthetic (carbon composite) projectile which was intended to economically produce very delicate fabrics. Figure 9.28 shows the various types of projectile jaws to hold the yarn. Table 9.2 lists the applications of projectile grippers.

The popularity of carbon composite projectiles has been declining due to their poor fatigue performance compared to steel projectiles.

Weft thread tensile force (cN)
without preacceleration
with
preacceleration (LUS)
Weft insertion cycle

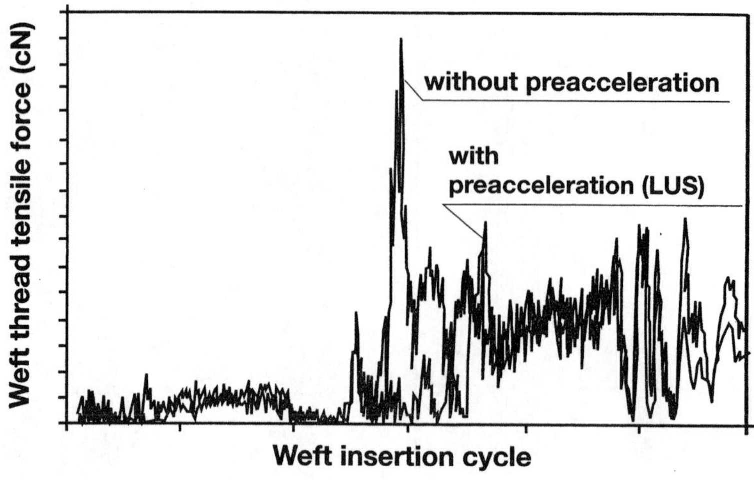

FIGURE 9.22 Reduced tension on the filling yarn by air support.

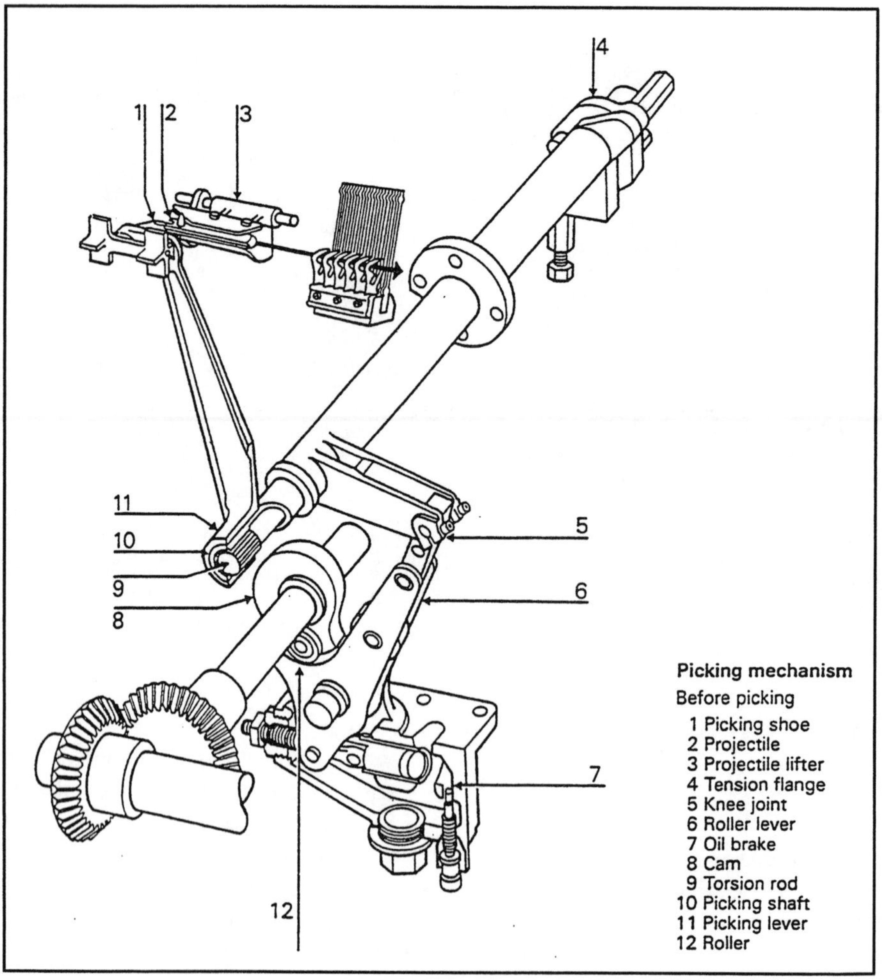

1
2
3
4
5
6
7
8
9
10
11
12
Picking mechanism
Before picking
1 Picking shoe
2 Projectile
3 Projectile lifter
4 Tension flange
5 Knee joint
6 Roller lever
7 Oil brake
8 Cam
9 Torsion rod
10 Picking shaft
11 Picking lever
12 Roller

FIGURE 9.23 Schematic of mechanism before picking.

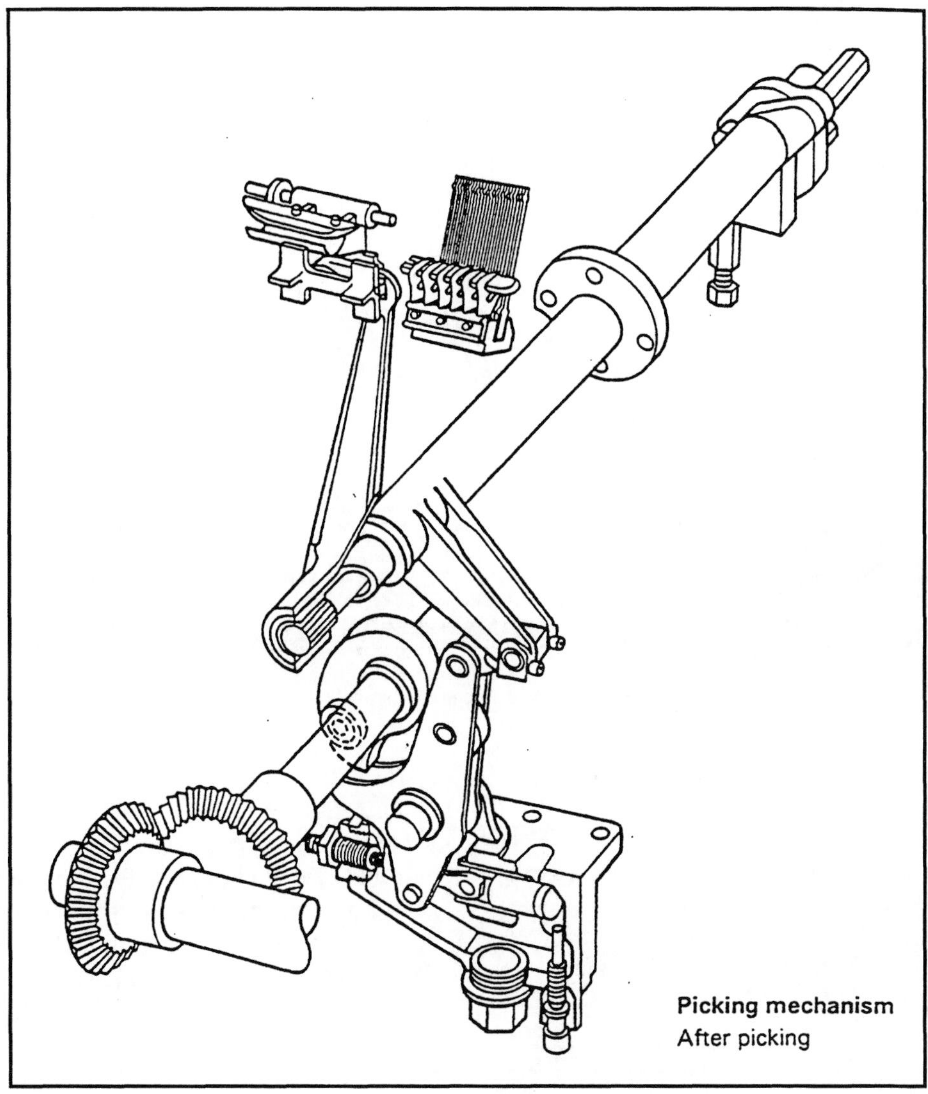

FIGURE 9.24 Schematic of mechanism after picking.

Yarn Centering and Cutting (Figure 9.29)

Filling yarn 1 is aligned by centering blade 2 so that it can be seized by the projectile feeder, i.e. projectile feeder gripper 3, and filling end gripper 4.

After cutter 5 has severed the filling yarn, the projectile feeder moves back with the filling end to the starting position to transfer the yarn to the next projectile. The remaining length of yarn is taken up by the filling tensioner lever of the filling tensioner.

Projectile Guides

Figure 9.30 shows two types of projectile guides: aligned guide teeth and staggered guide teeth. Guide teeth D1/D12-D2 on both systems are molded in a block. Aligned guide teeth are used for applications such as bolting cloth, silk, fly screen, coated glass, rain proof fabric, voile, linings and tape yarns. Staggered teeth are used for applications such as multifilament yarns, monofilament yarns, staple yarns including terry, woolen yarns, hard fiber yarns and leno fabrics.

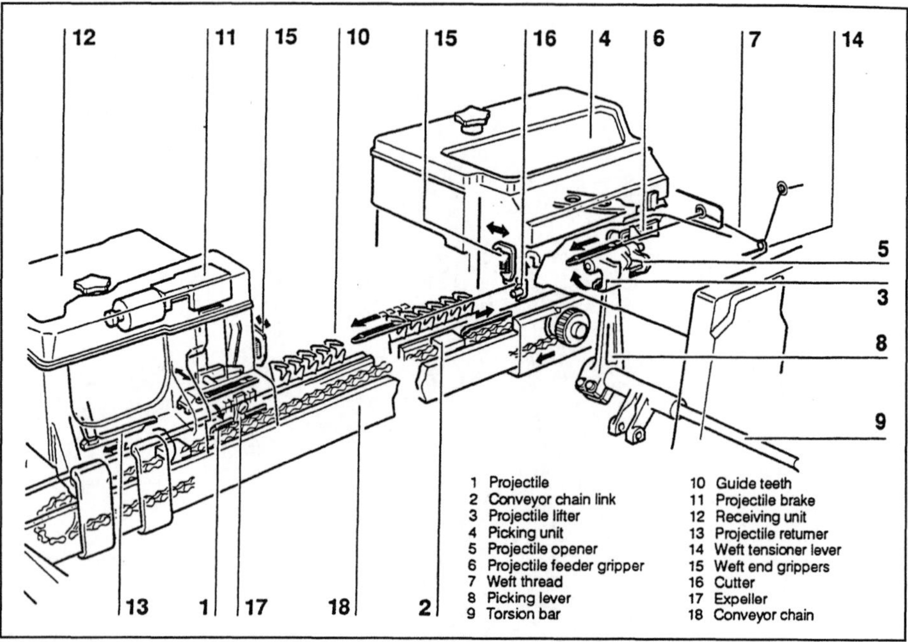

FIGURE 9.25 Projectile circulation.

Electronically Controlled Projectile Brake (Figure 9.31)

The purpose of electronically controlled projectile brake 5 is always to bring the projectiles to a stand-still in the correct position in the receiving unit without manual intervention.

FIGURE 9.26 Composite (top) and steel projectile.

The position of each projectile in the receiving unit is detected by sensors 1 and 3 and transmitted to a microprocessor. This processes the signals into control commands for stepping motor 6, which opens or closes the projectile brake in steps. Sensor 2 monitors the timely arrival of the projectiles in the receiving unit.

Sley Drive (Figure 9.32)

Sley 2 carries the reed 1 and guide teeth 3, which form the flight channel of the projectiles. The dwell of the sley during filling insertion varies with the machine width (Table 9.3). The picking lock, projectile guide and receiving lock must then be in line.

TABLE 9.1 Characteristics of projectiles.

	D1 *steel*	D12 *steel*	D2 *steel*	K3 *composite*
Length (mm)	89	89	89	96
Width (mm)	14.3	14.3	15.8	18
Height (mm)	6.35	6.35	8.5	8

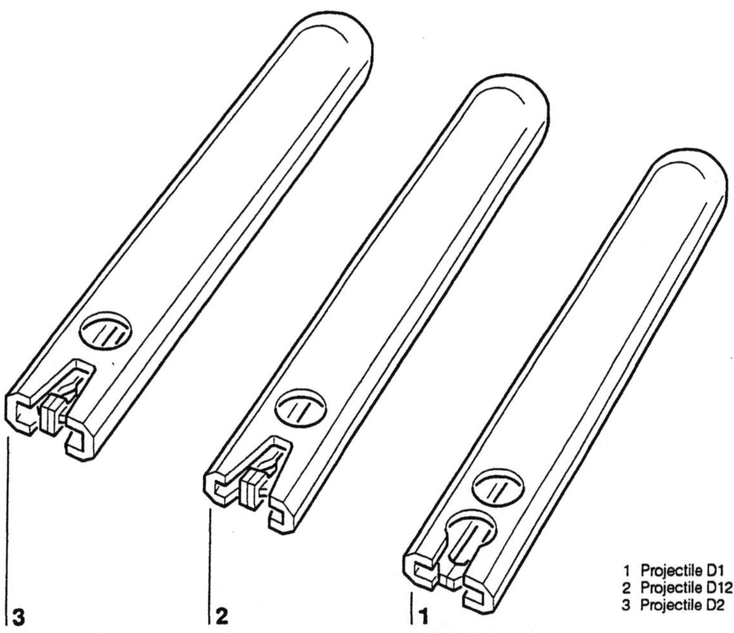

FIGURE 9.27 Projectile types.

The sley motion is symmetrical and is positively controlled via double cams 4. During reed beat-up, the guide teeth emerge below the shed and release the filling yarn through the upward-facing opening.

It is reported that the 4° inclination of the reed with the wedge sley reduces the recoiling fell when weaving heavy fabrics (Figure 9.33), as a result, higher filling densities can be woven. Filling yarn beat-up is performed more gently.

TABLE 9.2 Applications of projectile grippers.

Projectile Grippers			D1	D1	D12	D2
Clamping Force	*Colour*	*Clamping Area (mm)*	*2,2 x 3*	*2,2 x 4*	*3,8 x 5*	*4 x 5*
600 g	violet	smooth	X			
1000 g	yellow	smooth	X			
1900 g	orange	smooth				
		with friction coating			X	X
2200 g	white	smooth	X	X	X	X
		with friction coating		X	X	X
		with groove	X	X	X	
		crisscross knurls		X	X	X
2500 g	blue	smooth			X	
		with friction coating			X	X
		with groove	X	X	X	
		crisscross knurls	X	X	X	X

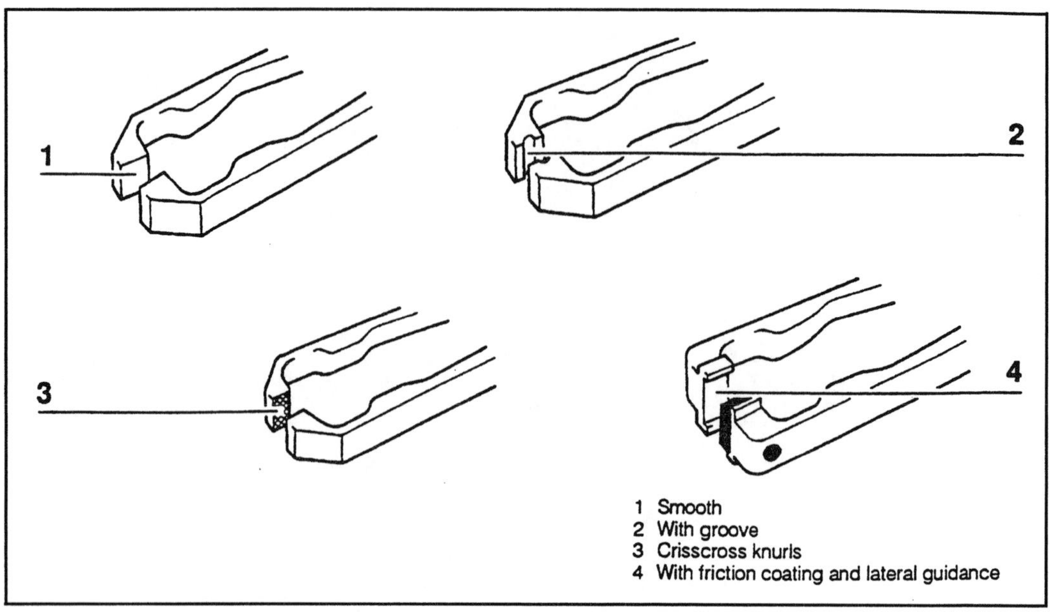

FIGURE 9.28 Projectile jaws to hold the yarn.

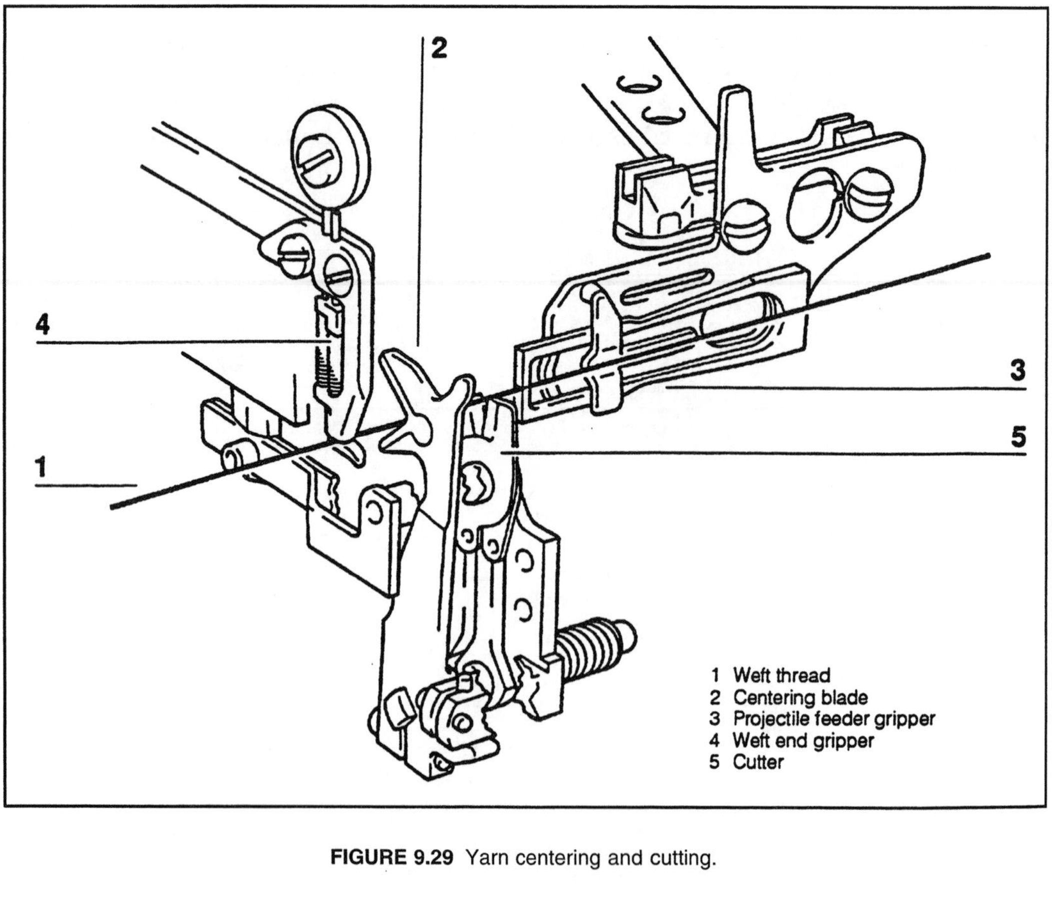

1
2
3
4
5
1 Weft thread
2 Centering blade
3 Projectile feeder gripper
4 Weft end gripper
5 Cutter

FIGURE 9.29 Yarn centering and cutting.

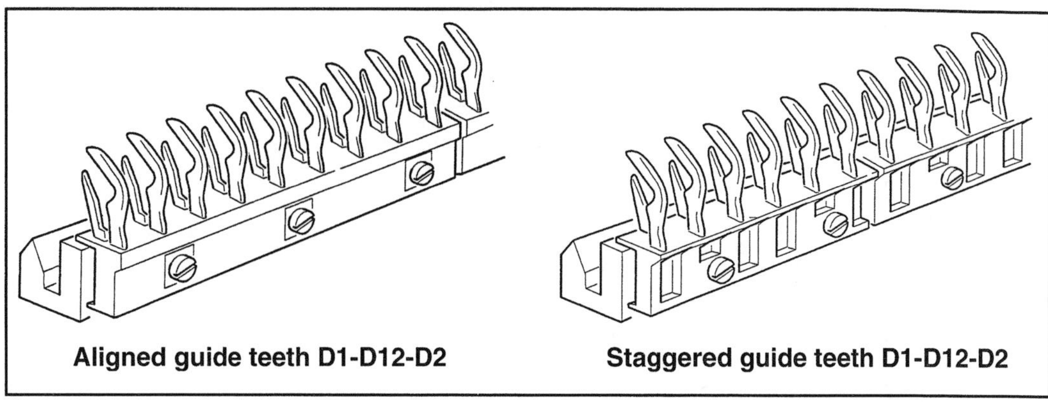

FIGURE 9.30 Aligned and staggered guide teeth.

Selvage Formation

The tucking unit or intermediate tucking unit of the projectile weaving machine forms clean tucked-in selvages (Figure 9.34). Because of the filling insertion system, the filling yarn is of exactly the right length and need not be shortened, as a result, there is no yarn wasted.

The projectile weaving machine can also be fitted with units to make leno or sealed selvages.

The yarn end is tucked in by the interaction of filling end gripper 1 and tucking needle 2. After filling insertion, the filling end grippers seize the pick on the left and right of the reed.

After the projectile gripper has opened in the receiving unit and the filling yarn has been cut

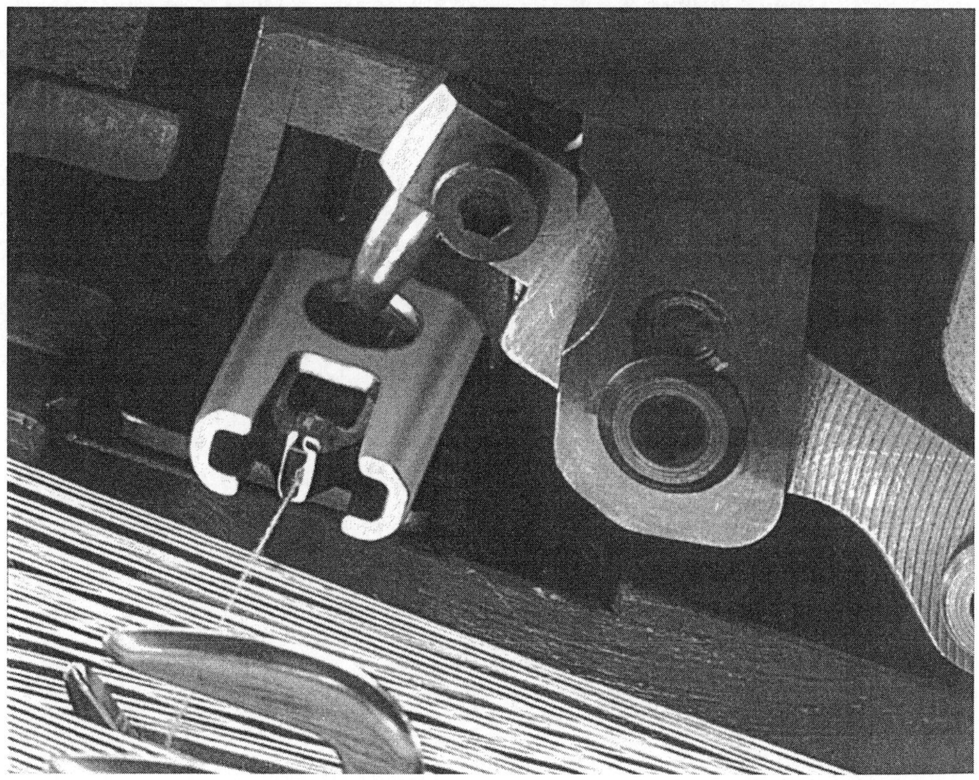

FIGURE 9.31a. Projectile arriving on the receiving side.

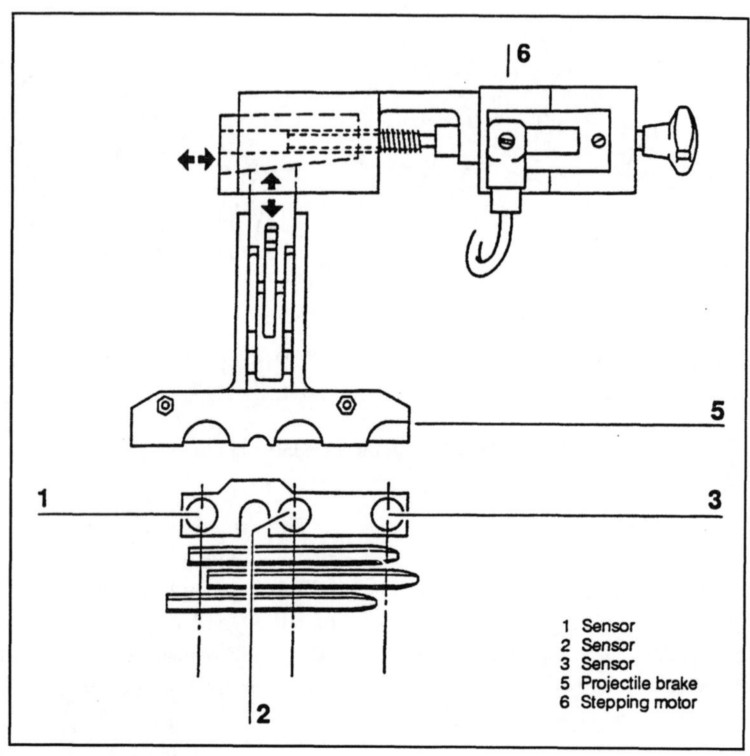

FIGURE 9.31b. Schematic of electronically controlled projectile brake.

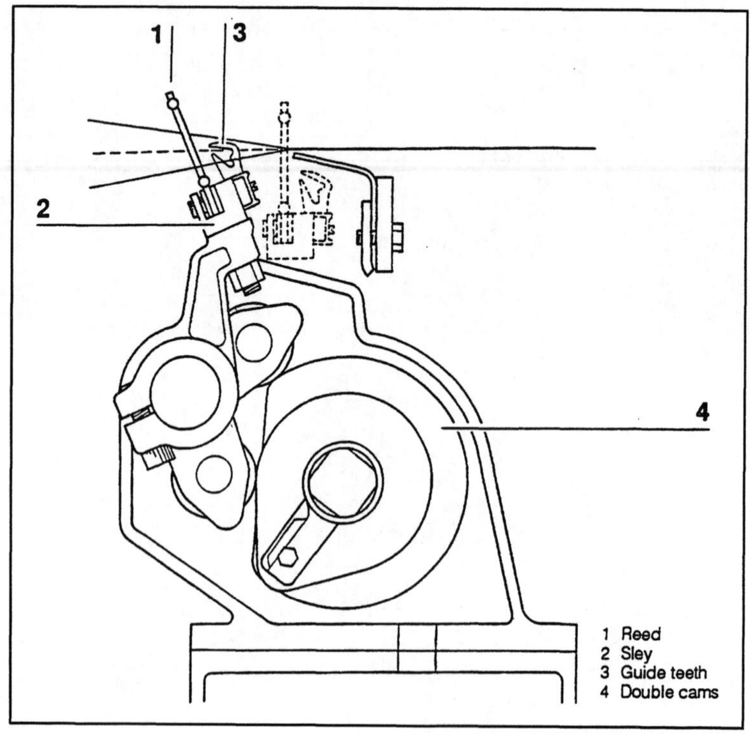

FIGURE 9.32 Sley drive.

TABLE 9.3 Sley dwell during filling insertion.

Machine Width	Sley Dwell
360 cm to 540 cm	110° 340°
330 cm	120° 340°
280 cm	135° 340°
190 cm and 220 cm	150° 340°

on the picking side, the sley moves towards reed beat-up. The filling end grippers with the clamped yarn ends move forward with it at the same time. While the next pick is inserted, the tucking needles pull the filling ends out of the filling end grippers and place them in the newly formed shed.

In the case of very filling-intensive styles or when weaving with coarse filling yarn, a split selvage with half cross leno interlacing can be used. For synthetic fabrics, a selvage sealing device can be used for severing the selvage.

Intermediate tuck-in units (Figure 9.35), used between the fabric widths during multi-width weaving, form the tucked selvages at the cloth width dividing lanes. The tucked ends are obtained by a 30 mm wide gap between the warp ends of the adjacent fabric widths.

At reed beat-up, filling end grippers 1 seize the free length of filling yarn, which is severed in the middle by cutter 2. Tucking needles 3 then grasp the filling tails and insert them in the shed which has been newly formed in the meantime.

Fabric Take-up

Fabric take up systems are similar for both P7150 and P7250 weaving machines.

a. With Take-up Roller Mounted on Lateral Bearings (Figure 9.36)

Take-up roller 1 is driven by driving shaft 3 via worm gear 2. Filling density is adjusted with a combination of four change wheels A to D. The entire range of filling densities is covered with different change wheels and worm gears. During pick finding, the driving shaft is reversed

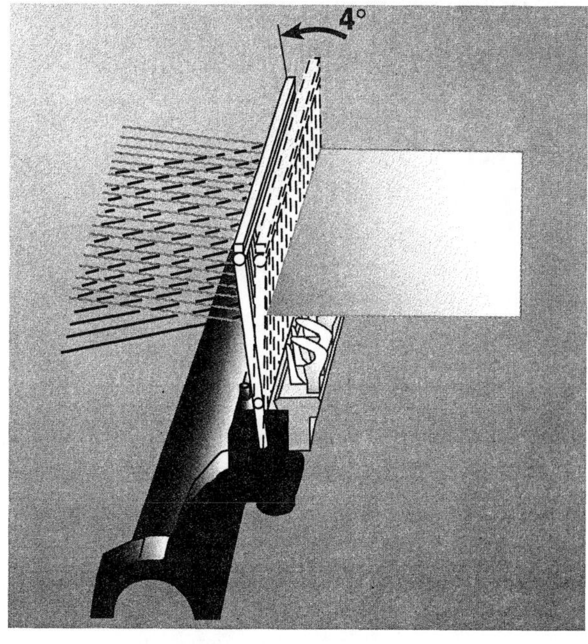

FIGURE 9.33 Inclination of reed.

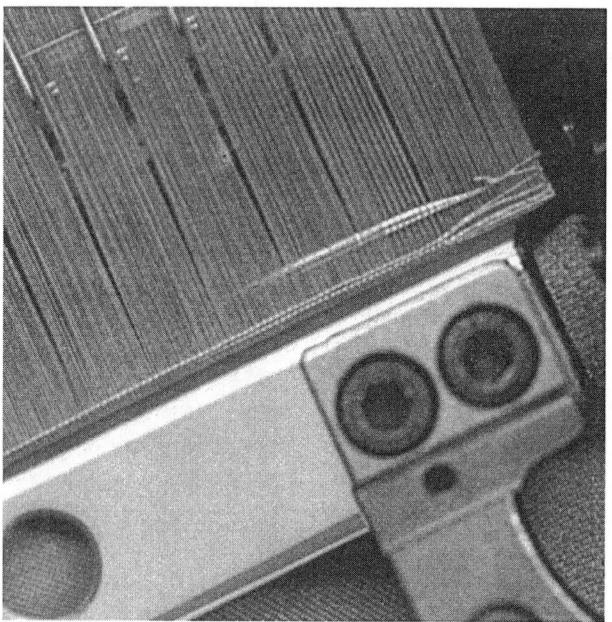

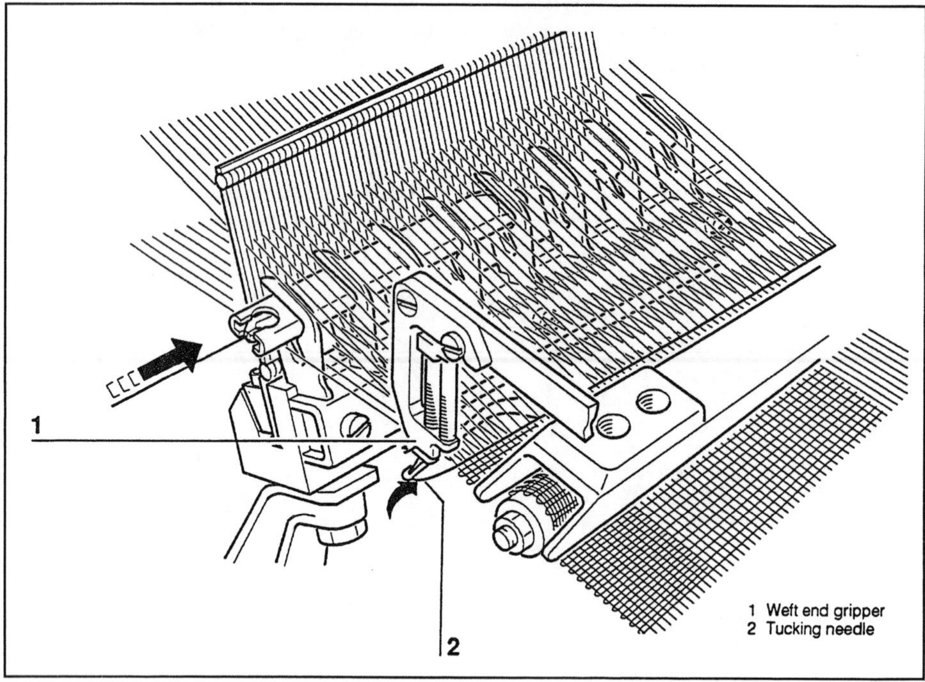

FIGURE 9.34 Selvage formation on the projectile machine with the unique tucked-in selvage.

and with it the cloth take-up. Cloth beam 4 is driven via adjustable friction clutch 5.

b. With Floating Take-up Roller (Figure 9.37)

Floating take-up roller 1 without fixed lateral bearings is especially suitable for heavy and plain fabrics. The warp tension pulls the take-up roller against plastic rollers 2, so that the cloth cannot slip back.

Pressing device 3 can be lifted off the take-up roller, so that the latter descends when warp tension is released and can be taken out after the plastic rollers have been removed. A spreading

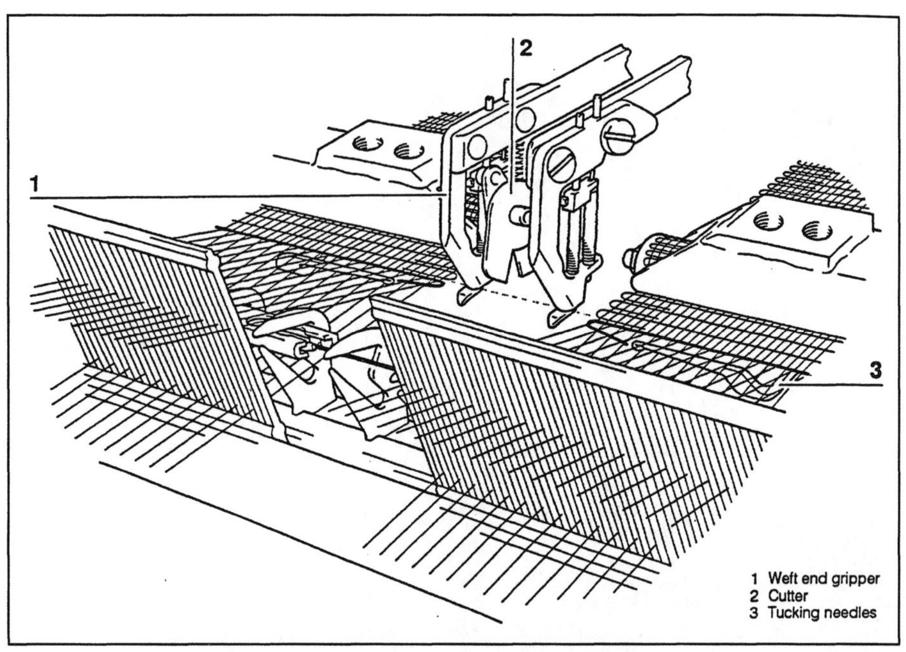

FIGURE 9.35 Intermediate tuck-in unit.

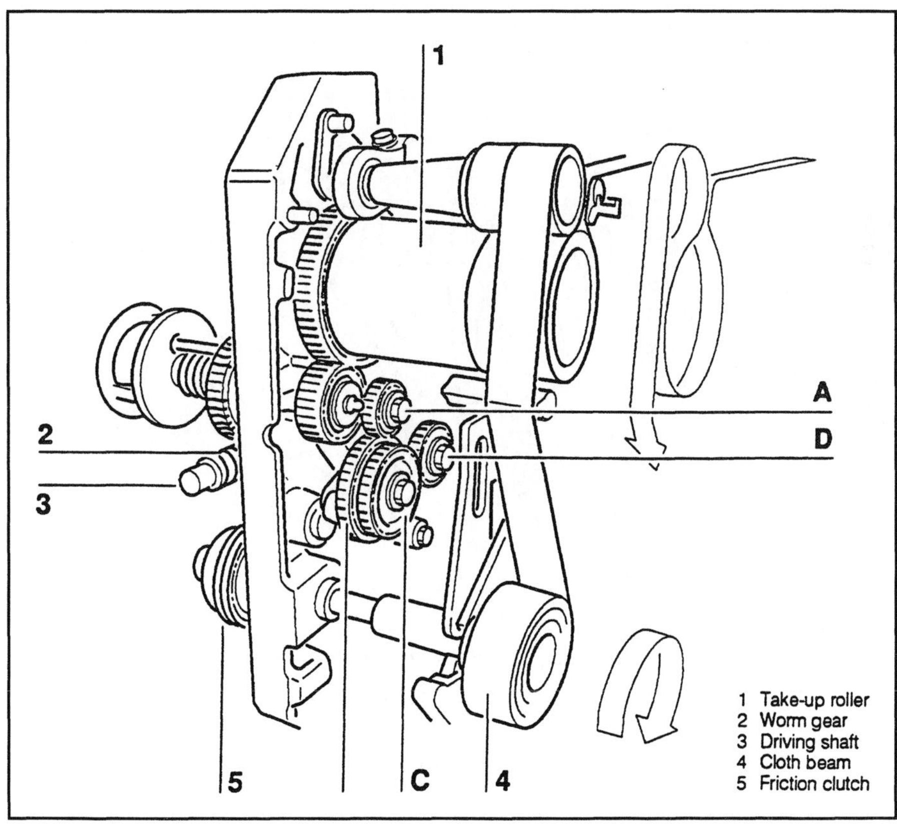

FIGURE 9.36 Fabric take-up with take-up roller mounted on lateral bearings.

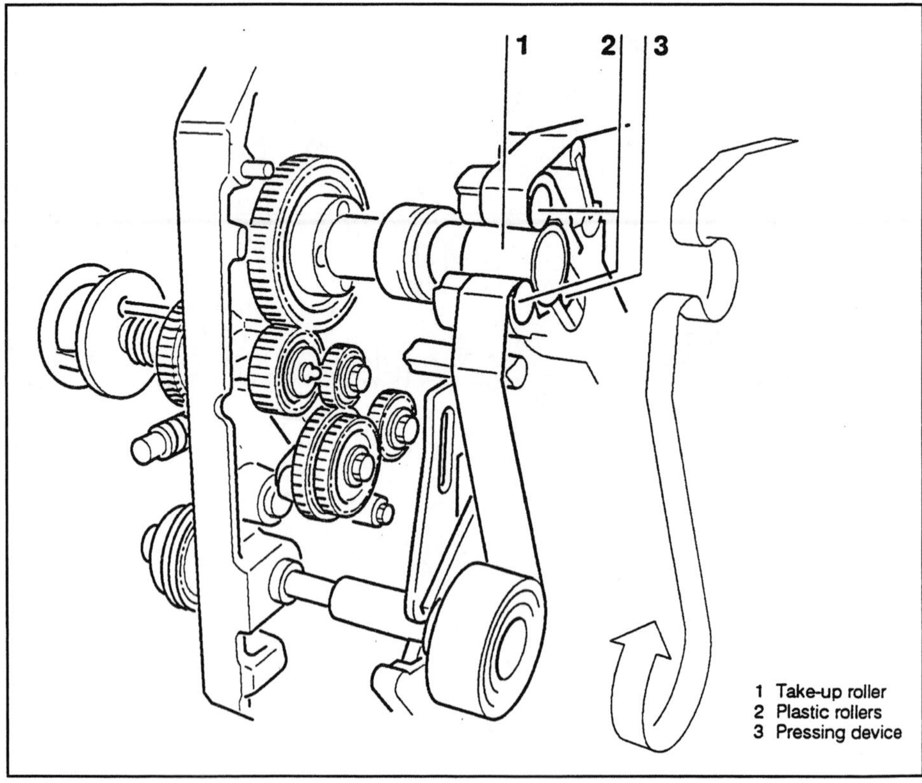

FIGURE 9.37 Fabric take-up with floating take-up roller.

roller can be fitted instead of the upper plastic roller. The drive, change wheels and cloth beam operate in the same way as for cloth take-up with lateral bearings.

The floating take-up roller increases stability and improves perfect cloth take-up. With the self-regulating contact pressure, sliding back of the fabric is eliminated even at high warp tensions. With heavy fabrics, the recoiling fell is reduced.

9.2 PROJECTILE TERRY WEAVING MACHINE

Towels are finished in the warp direction with tucked-in selvages and in the filling direction with borders and fringing. The tucked-in selvage reduces the finishing costs of towels.

The pile warp beam in the high position has a larger flange diameter. Both the ground and pile warps have warp tensioners to control warp let-off. The warp tensioner for the pile warp reacts rapidly and sensitively to the consumption of pile warp according to the pattern being produced which results in uniform pile (Figure 9.38). A constant pile height is a key criterion in terry cloth manufacturing.

Figure 9.39 shows the cross section of a projectile terry weaving machine. Loop formation on the projectile terry weaving machine is obtained by the cloth control principle.

While the reed motion always has the same stroke, the cloth fell is periodically pulled away from the reed beat-up position by the joint horizontal motion of the breast beam (rocking arm) and the temple. Thus, two or three partial beat-ups are produced for the subsequent loop formation of the pile warp after each full beat-up of the reed (three or four filling terry). Loop height is determined mainly by the distance between the pick group and the withdrawn cloth fell.

Figure 9.40 shows the schematic diagram of pile formation. The standard projectile terry weaving machine is equipped for weaving two different pile heights.

A draw hook 3 is mounted on each side of roller lever 2 for terry cam 1 as a link to the rocking arm. Depending upon the terry pattern, one of the draw hooks is disengaged upwards, while the other is engaged. Both draw hooks are

FIGURE 9.38 Low-mass warp tensioner for constant pile height.

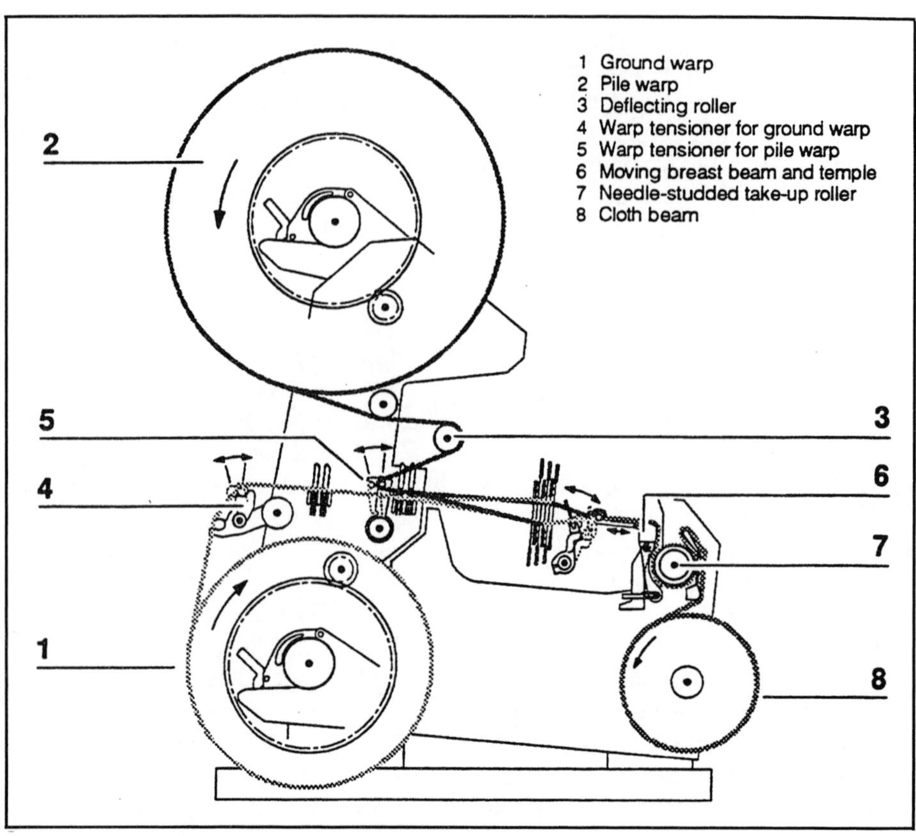

FIGURE 9.39 Cross section of projectile terry weaving machine.

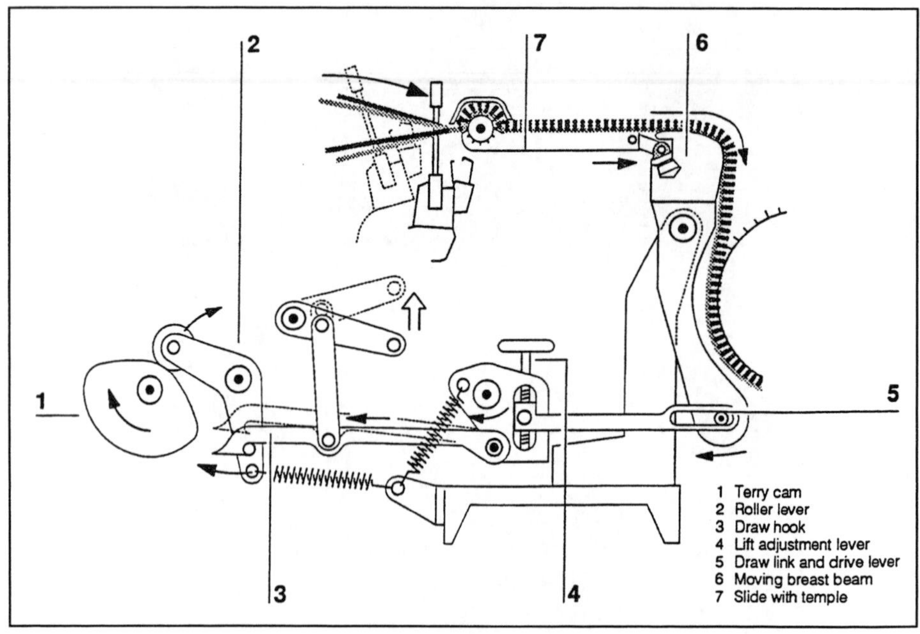

FIGURE 9.40 Schematic diagram of pile formation.

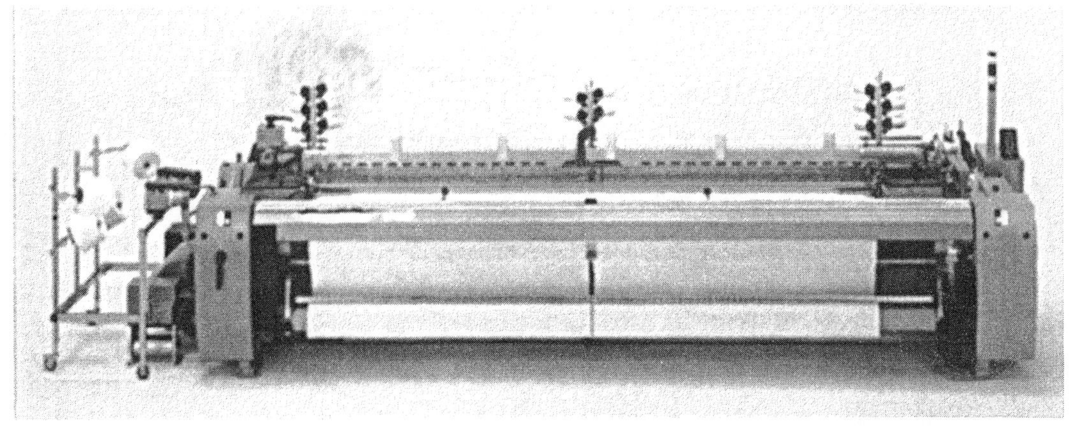

FIGURE 9.41 The projectile weaving machine P-lean.

disengaged for weaving without pile loops. The pile height can be adjusted individually by means of lift adjustment lever 4.

Projectile terry weaving machines are supplied in two color and four color versions with shed control by cam motions, dobbies and jacquard machines.

9.3 P-LEAN MACHINE

Standard fabrics (such as shirtings, sheetings, poplin and print clothes) and medium weight industrial fabrics constitute about two-thirds of the textile surfaces produced all over the world. Due to fierce competition, cost effective production is a must in these markets. For this reason, some weaving machine builders developed special machines strictly aimed at the economic production of standard fabrics. One of these machines is SulzerTextil's "P-lean" projectile machine that has been made for the weaving of medium weight fabrics of spun fiber yarns of count 15–60 tex (Ne 10–40), with 10–40 yarns/cm (25.5–103 ends/inch) in warp and filling (Figure 9.41).

Figure 9.42 shows the design details in the area of the warp. Standardized warp beam bearings permit usage of all full size or half warp

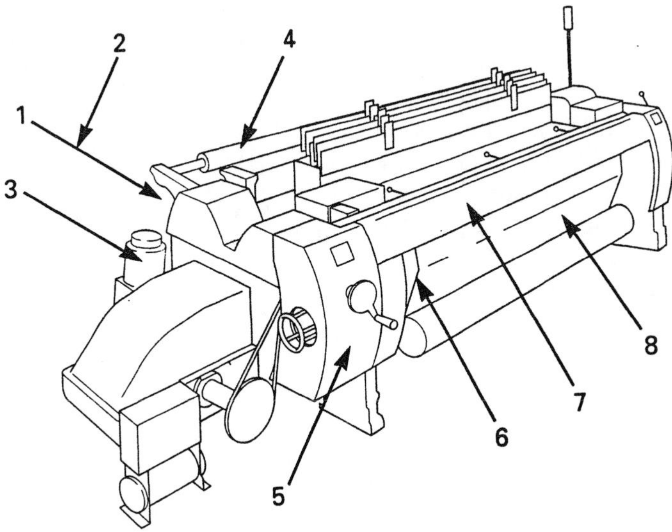

FIGURE 9.42 Major components of P-lean machine.

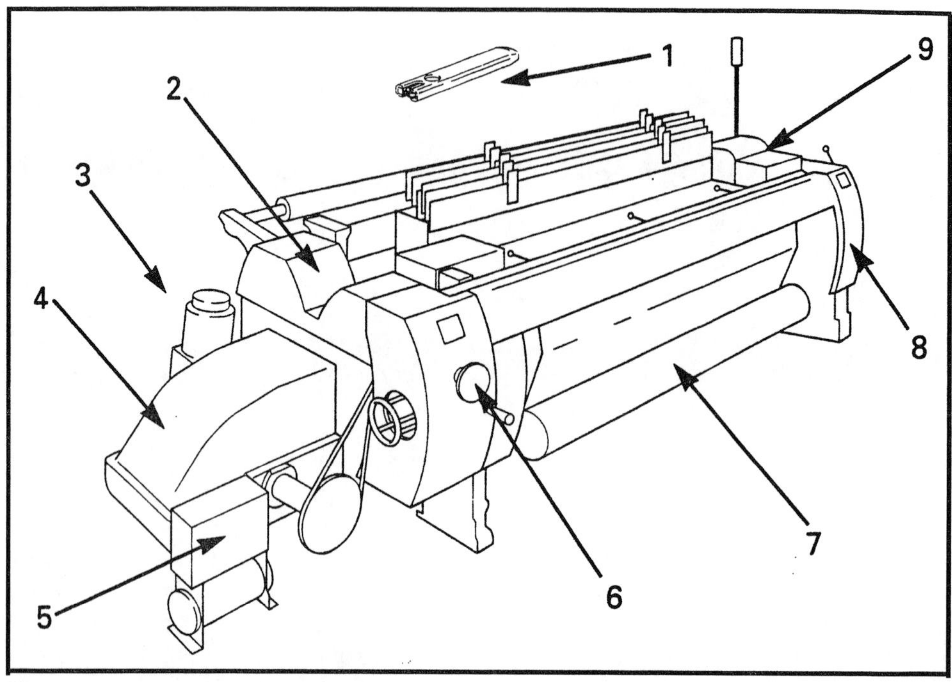

FIGURE 9.43 Filling insertion components of P-lean machine.

beams 1,2 that are compatible with any other projectile weaving machine. Depending on requirements, mechanically or electronically controlled warp let-off systems may be employed 3. Deflection of the warp can be affected with the aid of back rest and deflection rollers 4. Filling density can be altered by replacing the change wheels 5. In terms of selvage formation, the weaver has the choice between half leno and tucking unit 6. Take up and press rollers enable high precision fabric control 7. The "P lean" allows for single or multiwidth weaving 8.

Figure 9.43 shows an overview of the technological features of filling insertion. The projectiles grip yarns of a wide count range 1. The filling yarn brake can be optimally adjusted to the requirements of the yarn type 2. The machine is equipped with electronically controlled filling feeders for either filling mixer, or two color filling insertion 3. The cam motion, part of the standard equipment, is designed for up to ten harness frames 4. An electronically controlled lubricating device with controlled dosage provides a continuous supply of lubricant to all lubricating points 5. For pick finding, the weaver can choose between a mechanical and a self-acting device. To facilitate the repair of warp yarn breaks, the harness frames are automatically brought to a common level 6. The fabric width is infinitely variable in the range of 287–360 cm for 360 cm wide machine and 312–390 cm for 390 cm wide machine 7. The machine speed can be adjusted to suit the fabric width 8. For a filling insertion rate of 1200 m/min, the wattage is 4.25 kW.

TABLE 9.4 Fabric data.

Fabric:	Cretonne
Warp yarn	
Material:	Cotton 100%
Count (tex(Ne)):	29.5 (20)
Warp density (ends/cm (ends/″)):	23.6 (60)
Weft yarn	
Material:	Cotton 100%
Count (tex(Ne)):	29.5 (20)
Warp density (ends/cm (ends/″)):	23.6 (60)
Reed width (cm (″)):	170 (66.9)
grey (cm (″)):	162 (63.9)
Weight (g/m_2 (ounces/sq yd)):	146.8 (4.4)
Working hours (h/a):	8400
Production (running m/a):	13.45 million

TABLE 9.5 Requirements in weaving machines.

	Weaving Machines:			
	Projectile Weaving Machine P-Lean	Air-jet Weaving Machine	Rapier Weaving Machine	Rapier Weaving Machine
Filling insertion system:	Projectile	Air-jet	Rapier	Rapier
Weaving width (cm):	360	190	190	360
Speed (1/min):	350	750	500	300
Efficiency (%):	92	90	92	89
Wattage (kW):	4.25	3.0 + 9.0*	6.0	7.0

*: For generation of compressed air.

In the range of standard fabrics, economic manufacturing is of decisive significance for the competitiveness of an enterprise. A comparison of identical fabrics and identical productions yields expressive figures concerning the advantages and disadvantages of the selected system. As an example, Table 9.4 shows the data of a cretonne fabric. The quantity of weaving machines required for an annual production of 13.45 million meters, is calculated in Table 9.5. According to this calculation, 100 projectile weaving machines of type "P-lean", 140 or 120 rapier weaving machines with working widths of 190 or 360 cm or 95 air-jet weaving machines, with working width of 190 cm, are needed to produce the mentioned 13.45 million running meters of fabric. As far as production volume and quantity of weaving machines are concerned, the air-jet weaving machines are the winners. The economy calculation, however, calls for further factors to be considered, such as, filling yarn waste and energy consumption. Including these two characteristics in the calculation will change the result in favor of the filling insertion with projectiles.

Filling Yarn Waste

In designing the P-lean machine, reduction of filling waste was a central concern. Therefore, with cut selvage, the filling yarn waste produced by this machine is only 3 cm per filling insertion for two fabric panels, with selvage saver. When using the tucking device, no filling yarn waste is generated. If the filling parameters given in Table 9.4 (yarn count 29.5 tex and filling density 23.6 yarns/cm or 60 yarns/inch) are used as a basis of the calculation, the variables shown in Table 9.6 are obtained. The projectile weaving machine produces some 24 tons of filling yarn

TABLE 9.6 Filling yarn waste with cut selvage.

	Weaving Machines:			
	Projectile Weaving Machine P-Lean	Air-jet Weaving Machine	Rapier Weaving Machine	Rapier Weaving Machine
Filling insertion system:	Projectile	Air-jet	Rapier	Rapier
Weaving width (cm):	360	190	190	360
Filling yarn waste per pick (cm):	5	6	10	12
Filling yarn waste (kg/year):	23960	57270	95860	57230
Difference (kg/year):	—	33310	71900	33270
Additional requirement in cotton cultivation area* (ha):	—	58	126	58

*: Average harvest of cotton = 570 kg/ha.

TABLE 9.7 Energy consumption of various weaving machines.

	Weaving Machines:			
	Projectile Weaving Machine P-Lean	Air-jet Weaving Machine	Rapier Weaving Machine	Rapier Weaving Machine
Weft insertion system:	Projectile	Air-jet	Rapier	Rapier
Weaving width (cm):	360	190	190	360
Wattage (kW):	4.25	3.0 + 9.0*	6.0	7.0
Energy consumption (MWh/a):	3280	8620	6490	6280
Difference (MWh/a):	—	5340	3210	3000
Additional oil consumption for energy generation (mill. l/a):**	—	+1.06	+0.64	+0.59

*: For the generation of compressed air.
**: 1 kWh = 0.2 l oil, inclusive of 45% efficiency in energy generation.

waste per year, which puts it into the leading position regarding waste saving properties, followed by the 360 cm wide rapier weaving machine, the air-jet machine with approximately 57 tons/year, and the 190 cm wide rapier weaving machine with almost 96 tons/year. The difference in filling yarn waste between the various insertion systems, converted into cotton plantation areas, is also included in the table.

Energy Consumption

As can be seen in Table 9.7, the P-lean projectile weaving machine features a low energy requirement compared with the other filling insertion systems. The reason is the inherent advantage of the projectile insertion system. Compared to the insertion systems of rapier and air-jet weaving machines, filling insertion by

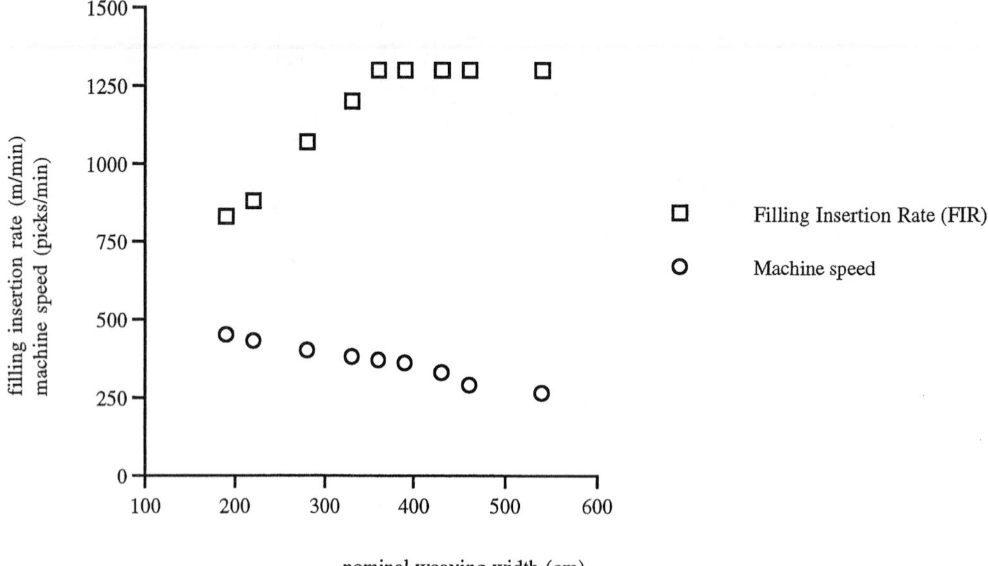

FIGURE 9.44 Relations between nominal weaving width, filling insertion rate and machine speed for projectile weaving machines.

FIGURE 9.45 Projectile weaving plant.

means of projectiles features the least energy requirement. The masses to be moved for insertion by projectiles are substantially smaller than those required for rapier insertion, and they are far more efficient than filling insertion with the aid of compressed air. Table 9.7 shows the differences in power consumption. The P-lean machine is the leader, followed by the 360 and 190 cm wide rapier and air-jet weaving machines, respectively. When converting the additional requirement of electric power into the equivalent quantities of oil, the advantage of economizing on natural resources is clear.

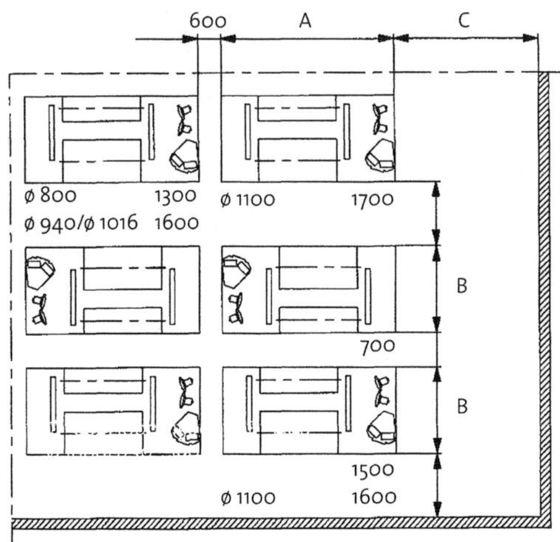

FIGURE 9.46 Typical layout for projectile weaving machines.

9.4 OPERATIONS

Filling insertion rate and machine speed are influenced by yarn quality, yarn count and style as well as the actual weaving width. Figure 9.44 shows the relationship between nominal weaving width, filling insertion rate and machine speed for projectile weaving machines.

Figure 9.45 shows a modern weaving plant with projectile machines. Figure 9.46 shows a typical layout drawing for projectile machines. Table 9.8 shows the recommended projectile weaving machines for industrial textiles.

TABLE 9.8 Projectile weaving machines for industrial textiles.

Machine Type	Application	Material	Weaving Width (cm)	Weave	No. of Harnesses	No. of Wefts	Warp Beams	Speed (rpm)	Insertion Rate (m/min)
P7M FO 846 N6 SMK D2	Forming fabrics	Monofilament PES and PA 0.10–0.50 mm diameter	846	Single layer Multi layer	Up to 20	6	Single Twin	120	1,000
P7 D220 N1 EP D12	Wire fabrics	Al wire 0.4 mm Steel wire 0.3 mm	220	Standard	Up to 10	1	Single	350	770
P7M RSP S655 N2 EP D2	Agrotextiles	Monofilament, tape PP, PES, PA, PET	655	Standard	Up to 10	1, 2 or 4	Up to 4 half beam	185	1,200
P7M RSP S846 N1 EP D2	Agrotextiles Geotextiles	Monofilament, tape PP, PES, PA, PET	846	Standard	Up to 10	1, 2 or 4	Up to 4 half beam	130	1,100
P7M R3 B/S330 N2 SP D12	Heavy fabrics Canvas Conveyor belts	Cotton	334	Standard	Up to 18	2	Single	330	1,000
P7M R3 S540 N2 SP D12	Filter fabrics	Monofilament PES, PA and PET	540	Standard	Up to 18	2 or 4	Single Twin	200	1,100
P7M R3 B/S280 N2 SP Q D12	Heavy fabrics Canvas Conveyor belts Sailcloth	Cotton Multifilament	283	Standard	Up to 18	2 or 4	Single Twin	320	900
P7M FO 540 N6 SMK D2	Forming fabrics	Monofilament PES and PA 0.10–0.50 mm diam.	540	Single Multilayer	Up to 20	6	Single Twin	200	1,000

REVIEW QUESTIONS

1. Explain the working principle of the torsion bar mechanism.
2. What is air supported filling insertion?
3. How many grippers are used in a projectile weaving machine? Why?
4. How is the selvage formed in a projectile machine?
5. Why are projectile weaving machines very suitable for industrial textiles?

10

Rapier Weaving

In this type of weaving, a flexible or rigid solid element, called rapier, is used to insert the filling yarn across the shed. The rapier head picks up the filling yarn and carries it through the shed. After reaching the destination, the rapier head returns empty to pick up the next filling yarn, which completes a cycle. A rapier performs a reciprocating motion.

Rapier weaving machines can be two types:

1. Single rapier machines: A single, rigid rapier is used in these machines. The rigid rapier is a metal or composite bar usually with a circular cross section. The rapier enters the shed from one side, picks up the tip of the filling yarn on the other side and passes it across the weaving machine while retracting (Figure 10.1). Therefore, a single rapier carries the yarn in one way only and half of the rapier movement is wasted. Also there is no yarn transfer since there is only one rapier. The single rapier's length is equal to the width of the weaving machine; this requires relatively high mass and rigidity of the rapier to ensure straight movement of the rapier head. For these reasons, single rapier machines are not popular. However, since there is no yarn transfer from rapier to rapier, they are suitable for filling yarns that are difficult to control.
2. Double rapier machines: Two rapiers are used in these machines. One rapier, called the giver, takes the filling yarn from the yarn accumulator on one side of the weaving machine, brings it to the center of the machine and transfers it to the second rapier which is called the taker. The taker retracts and brings the filling yarn to the other side. Similar to the single rapier machines, only half of the rapier movements is used for filling insertion.

Double rapier machines can be rigid or flexible. There are two types of double rigid rapiers: Dewas system and Gabler system. In Dewas system, the giver grips the tip of the yarn, brings it to the center and transfers it to the taker which retracts and carries the yarn to the other side of the weaving machine (Figure 10.2).

In the Gabler system, the yarn is not gripped. The giver extends the yarn in the form of a "U" shape to the center of the weaving machine. The yarn is then transferred to the taker, which extends the yarn to the other side of the weaving machine by straightening it (Figure 10.3). Since both rapiers extend to the outside of the weaving machine, the space requirement for double rigid rapier machines is high.

In flexible rapier machines, the rapier has a tape-like structure that can be wound on a drum (Figure 10.4). This saves space and allows narrower machine widths compared to double rigid machines. The yarn is gripped both by the giver and taker. Double flexible rapier machines are

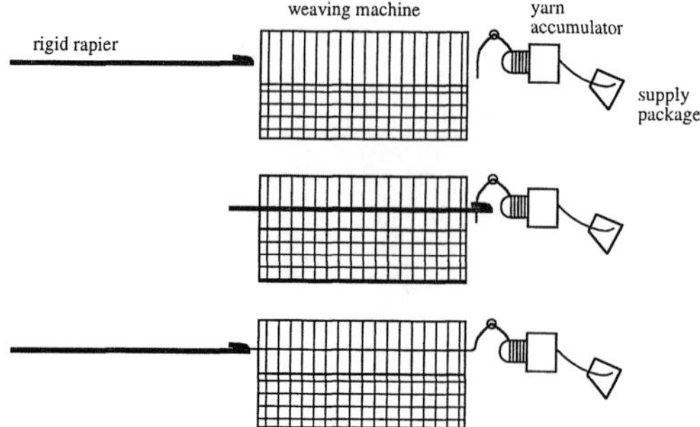

FIGURE 10.1 Schematic of single rapier insertion.

more common than the rigid rapier machines. Figure 10.5 shows a double flexible rapier machine.

Rapier weaving machines are known for their reliability and performance. Since 1972, the rapier weaving machine has evolved into a successful, versatile and flexible weaving machine. A very wide range of fabrics can be woven on a rapier weaving machine which is typically from very light fabrics with 20 g/m^2 to heavy fabrics with around 850 g/m^2. Rapier machines are widely used for household textiles and industrial fabrics. Designed for universal use, the rapier weaving machine can weave not only the classic wool, cotton and man-made fibers, but also the most technically demanding filament yarns, finest silk and fancy yarns. Figure 6.43 shows a

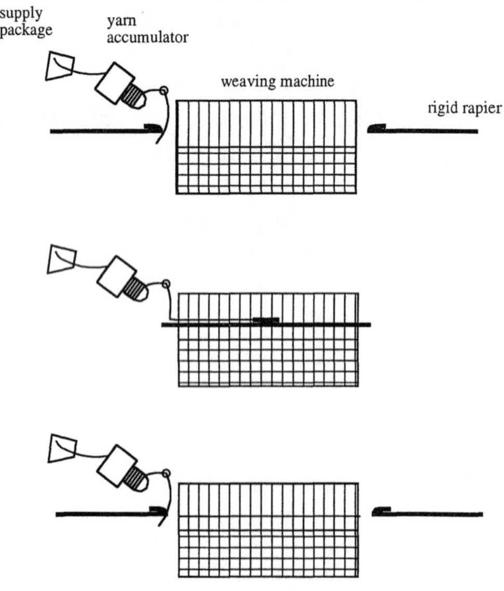

FIGURE 10.2 Schematic of Dewas system.

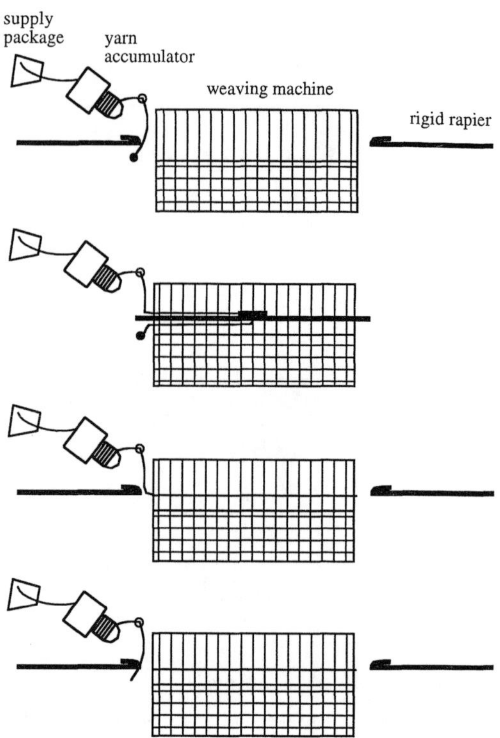

FIGURE 10.3 Schematic of Gabler system.

rapier machine weaving fine silk yarns. Lightweight grippers with a balanced center of gravity, tape bracings and extended tape guides, combined with support teeth in the lower shed, provide smooth, yarn friendly gripper run. An advantage of rapier machines is that they impart minimal stress on the filling yarn. The gripper heads take a wide range of yarn counts ranging from 5 to 1000 tex. Up to 16 different filling yarns can be inserted. Rapier machines can have up to 28 harnesses. Today, modern rapier machines achieve weft insertion rates of around 1400 m/min. Rapier machines can have a cam (tappet), dobby or electronically controlled jacquard shedding systems. With the rapier system, insertion of double picks is possible. Rapier machines have low yarn breakage rates.

In the latest machines, the rapiers are made of composite materials and the rapier guide is eliminated. As a result, there is no contact of the warp yarns with the metallic rapier guide. Since there is no friction with the rapier guide, the stress on the tape and rapier assembly is reduced. At ITMA-99 in Paris, more user

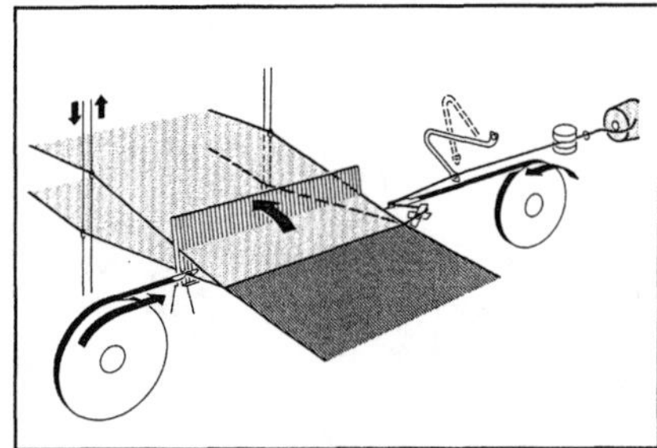

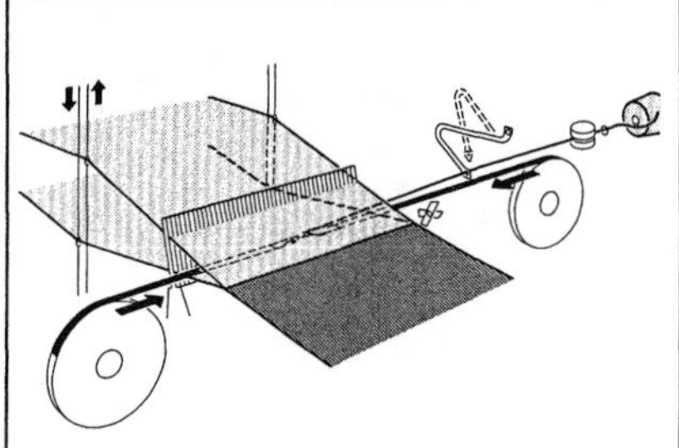

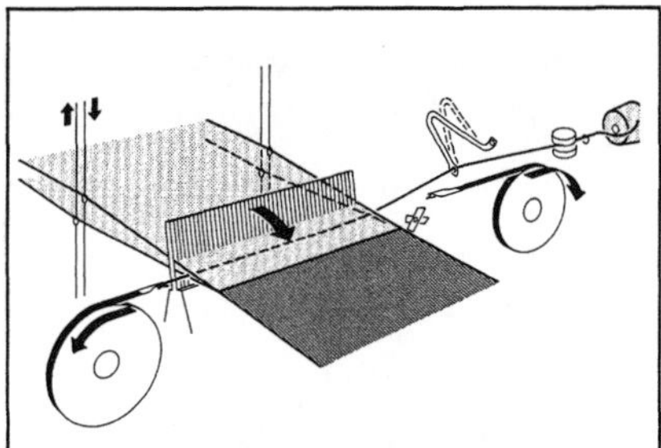

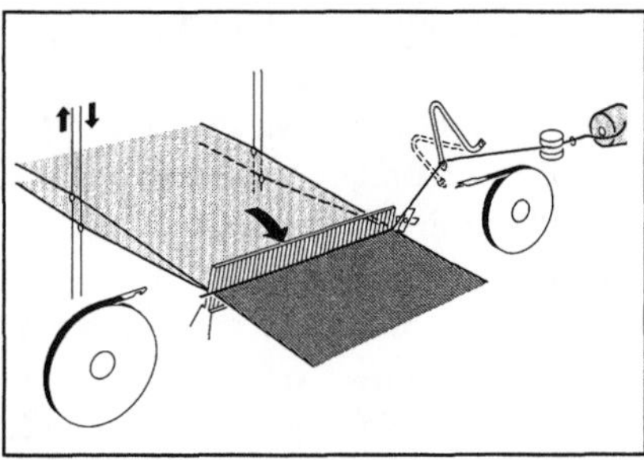

FIGURE 10.4 Schematic of double flexible rapier system.

friendly rapier machines were exhibited. Sulzer-Textil exhibited G6300 with industrial speed of up to 700 ppm. Their performance increased, the range of applications extended, the weaving widths increased and filling insertion was optimized for better fabric quality. The choice of shedding mechanisms also increased. Rapier machines allow quick style change which increases flexibility.

10.1 FUNCTIONAL DESCRIPTION OF DOUBLE FLEXIBLE RAPIER WEAVING MACHINE

Since double flexible rapier machines are the standard in the industry, it is appropriate to analyze this type of machine in more detail. Figure 10.6 shows the major subunits of a typical double flexible rapier machine.

Weaving Machine Drive (Figure 10.7)

Main motor 1 drives disc flywheel 3 via three V-belts 2. The electromagnetic clutch transmits the torque to the driving gear wheel and the main shaft. Four speeds are possible with the built-in discs.

The weaving machine or the shedding mechanism is slowly turned forward and backward by creep speed motor 4. A separate fan 5 is fitted to generate the required vacuum for cleaning the two rapiers.

The disc flywheel increases the moving mass and contributes to the smooth running of the weaving machine. Figure 10.8 shows the drive diagram of the machine.

Warp Let-Off (Figure 10.9)

Typically, the power warp let-off covers a pick count range from 2 to 250 picks/cm. It can be programmed for two different weft densities. The warp tension is kept constant to a given set point.

The configuration of warp let-off gear 3 depends on the required number of picks per cm, the warp beam flange diameter and weaving machine speed. There is a choice of one warp beam tube diameter, two warp beam winding diameters and two warp let-off gear configurations for these. The warp let-off gear is controlled via a sensor which continuously monitors the position of the warp tensioner (Figure 10.10). The link with the warp let-off gear is disconnected by releasing clamping bar 2 on the warp beam adapter, whereupon warp beam can be turned manually. For borders with higher weft density the warp let-off slows down in synchronism with the cloth take-up. A different warp tension in the border is also possible. A warp tensioner provides the proper tension for the

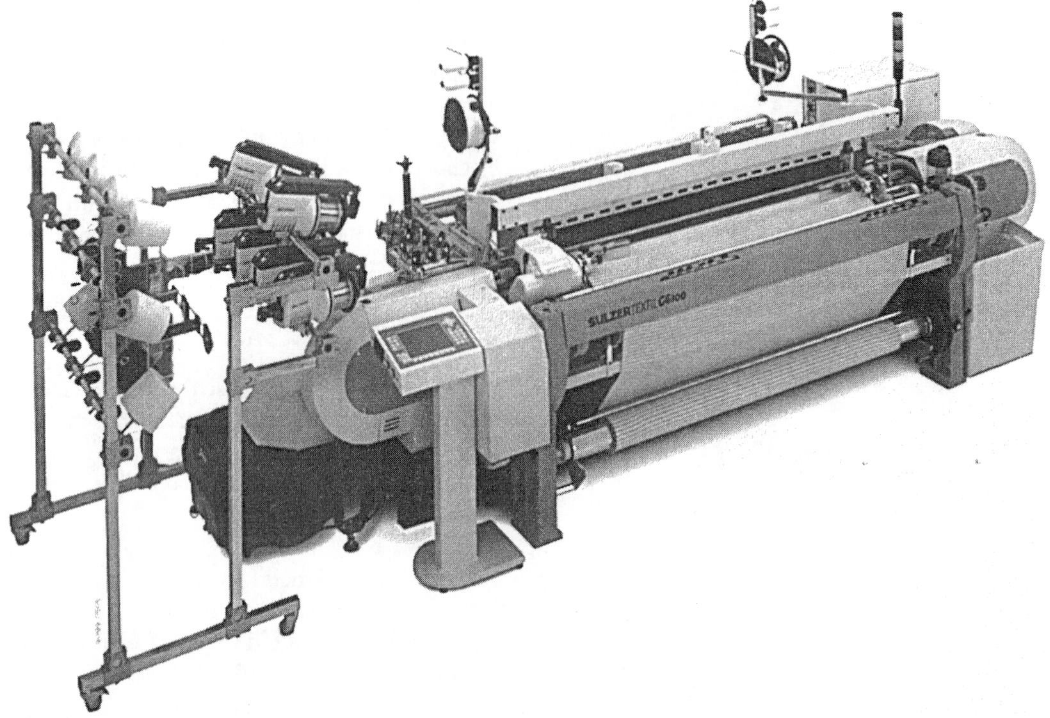

FIGURE 10.5 Double flexible rapier machine.

warp. The warp tensioner can be adjusted horizontally and vertically. It can be positioned to suit the style currently being woven. Minimal stress on the filling yarn during processing results in good fabric quality and high efficiency. The lower shed does not make contact with the gripper head. The precise shed separation gives good warp run and low warp breakage as a result of optimum warp tensioning.

In another type of rapier machine, the tensioner builds up the warp tension by means of springs which also transmit the whole load to the load cell via the tie rod. Warp tension can be entered at control terminal. Preload of springs can be adjusted by means of knobs. A safety switch prevents the warp tension from rising beyond the maximum permissible level. The whip roller can be locked or set free to rotate depending on warp characteristics.

Warp Stop Motion (Figure 10.11)

The configuration of the warp stop motion depends on warp density, warp yarn count and type of yarn. Up to 8 drop wire rails 2 can be fitted. The weight of drop wires 1 depends on yarn type and yarn count.

Open or closed drop wires can be used, depending on whether they are pinned on the weaving machine or drawn in on the drawing-in machine or by hand.

The signal column signals a warp break. The terminal display shows in which sector (1 or 2) of the warp the broken warp end can be found. As a drop wire 3 drops, an electrical circuit is closed and the weaving machine stops (Figure 4.105).

Shedding Mechanisms

The common shedding systems used for rapier machines are positive cam, electronically controlled positive dobby and jacquard systems.

Figure 10.12 shows an electronically controlled rotary dobby mechanism on the rapier machine. This dobby permits weaving with up to 24 harnesses with a speed of 630 rpm. It is controlled by entering the fabric and machine

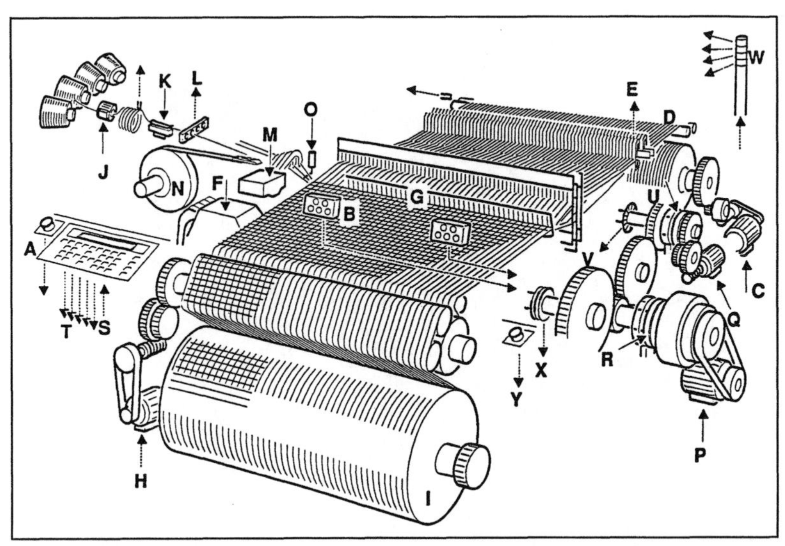

Overview of subunits

A	Operator and display terminal	N	Tape wheel with rapier tape
B	Operating keyboard	O	Weft yarn lifter
C	Warp let-off	P	Main motor
D	Warp tensioner	Q	Creep speed motor
E	Warp stop motion	R	Clutch and brake
F	Shedding mechanism	S	Input of fabric and machine-specific information
G	Sley drive and reed	T	Communication between terminal and weaving machine
H	Cloth take-up	U	Pick finder clutch
I	Cloth wind-up	V	Crank angle transmitter
J	Weft feeder	W	Signal column
K	Weft brake	X	Synchronization monitoring
L	Weft monitoring	Y	Emergency stop button
M	Pick sequence control		

FIGURE 10.6 Schematic of the major components of a double flexible rapier machine.

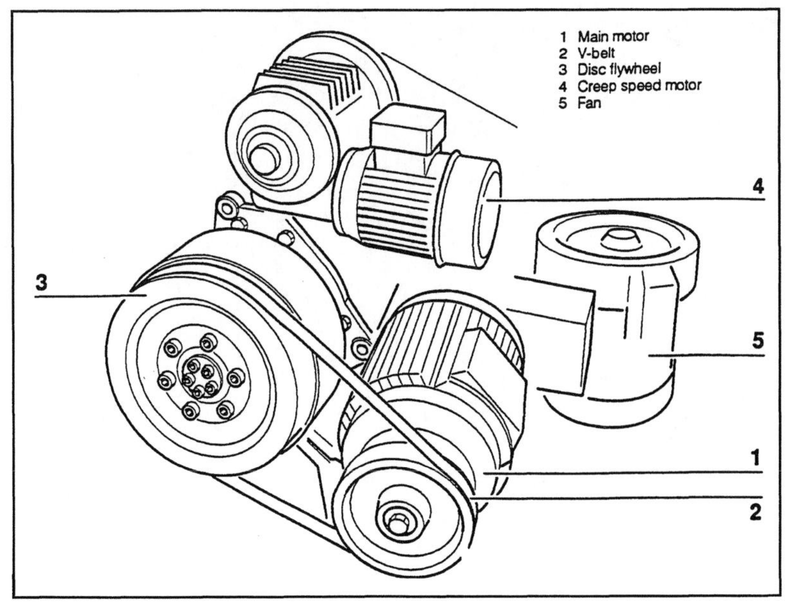

FIGURE 10.7 Weaving machine drive.

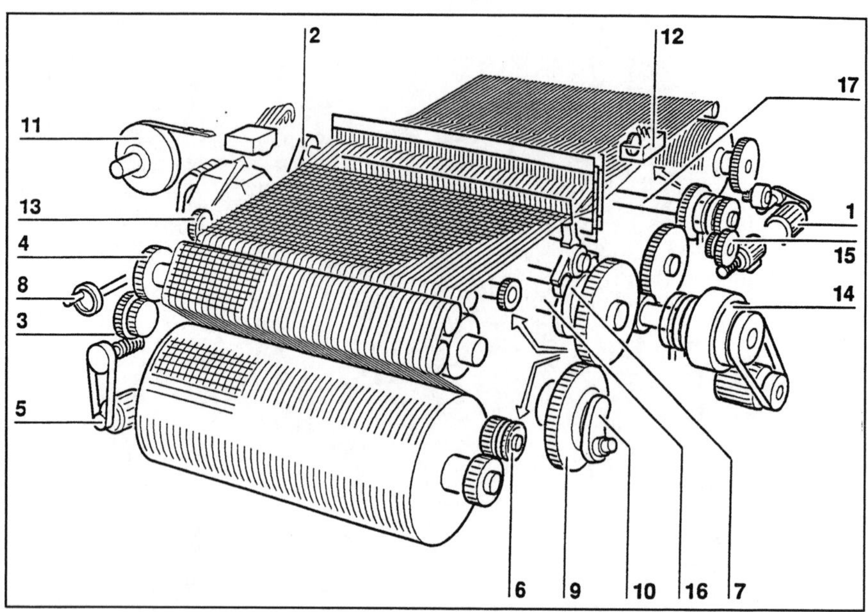

1 Power warp let-off, ground warp
2 Shedding mechanism drive
3 Cloth transport gear
4 Take-up roller drive
5 Cloth take-up drive
6 Cloth wind-up drive
7 Sley drive
8 Handwheel
9 Tape drive gearwheel
10 Tape drive stroke crank arm
11 Tape wheel
12 Full cross leno drive
13 Electronic pick sequence control
14 Main drive and brake
15 Creep speed gear
16 Main driving shaft
17 Rear driving shaft

FIGURE 10.8 Drive diagram.

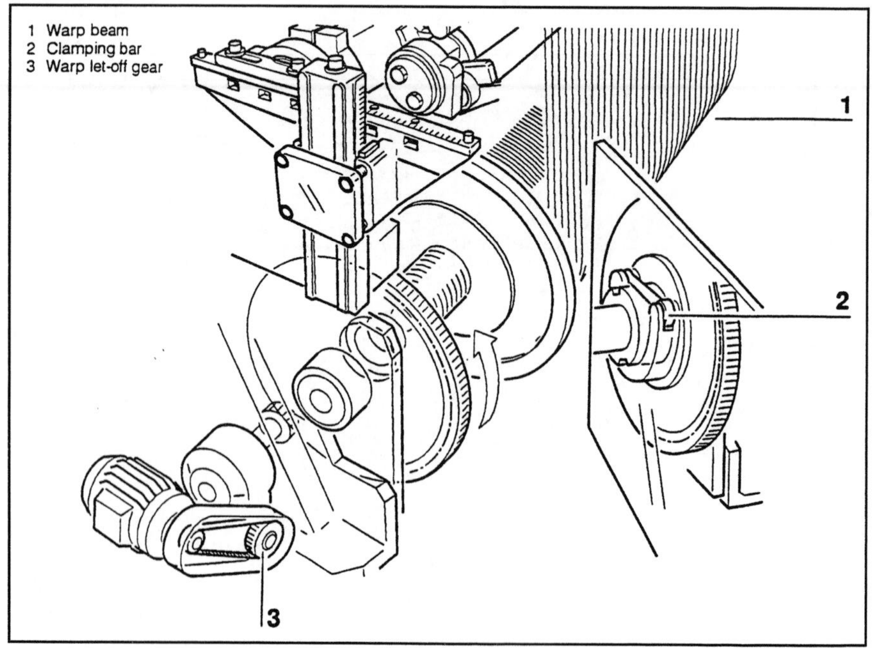

FIGURE 10.9 Schematic of warp let-off mechanism.

FIGURE 10.10 Warp tensioner.

FIGURE 10.12 Electronic rotary dobby on the rapier machine.

specific parameters at the terminal. The shed operating angle is modified between 19–28° with special gauges.

Figure 10.13 shows the schematic of the electronically controlled Jacquard mechanism. Jacquard machines enable a large number of warp yarns (number of hooks 2) to be controlled individually; i.e., long weave repeats are possible (e.g. damask fabrics).

Knives 1 move pulley 4 up and down under electronic control, according to the weave. Magnetic elements 3 are grouped in units and actuated by the central electronic unit.

Sley Motion

Figure 10.14 shows the shed geometry and sley movement of a rapier weaving machine. Figure 10.15 shows the schematic of sley mechanism for the same machine. The sley motion is controlled via sley shaft 1 with clamped reed 2

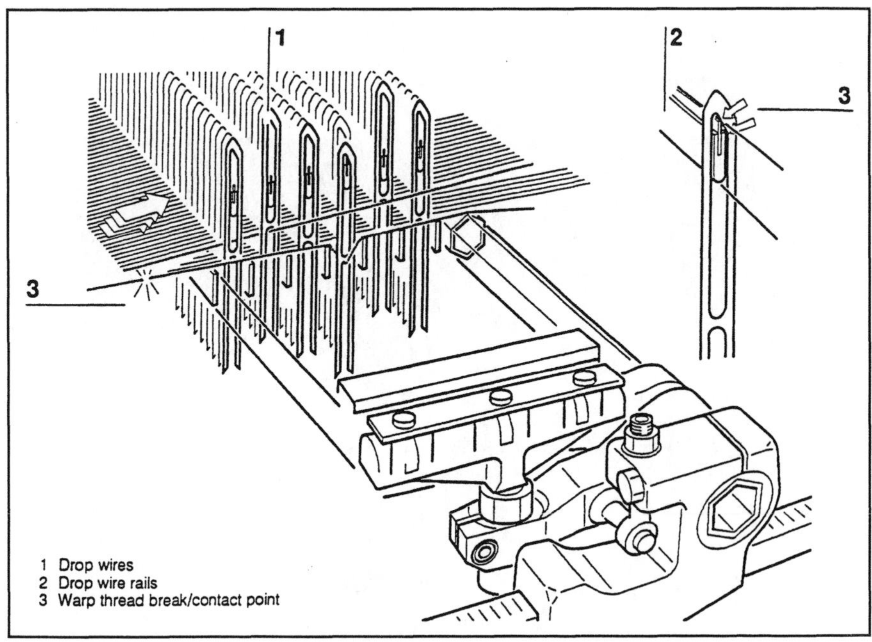

FIGURE 10.11 Warp stop motion.

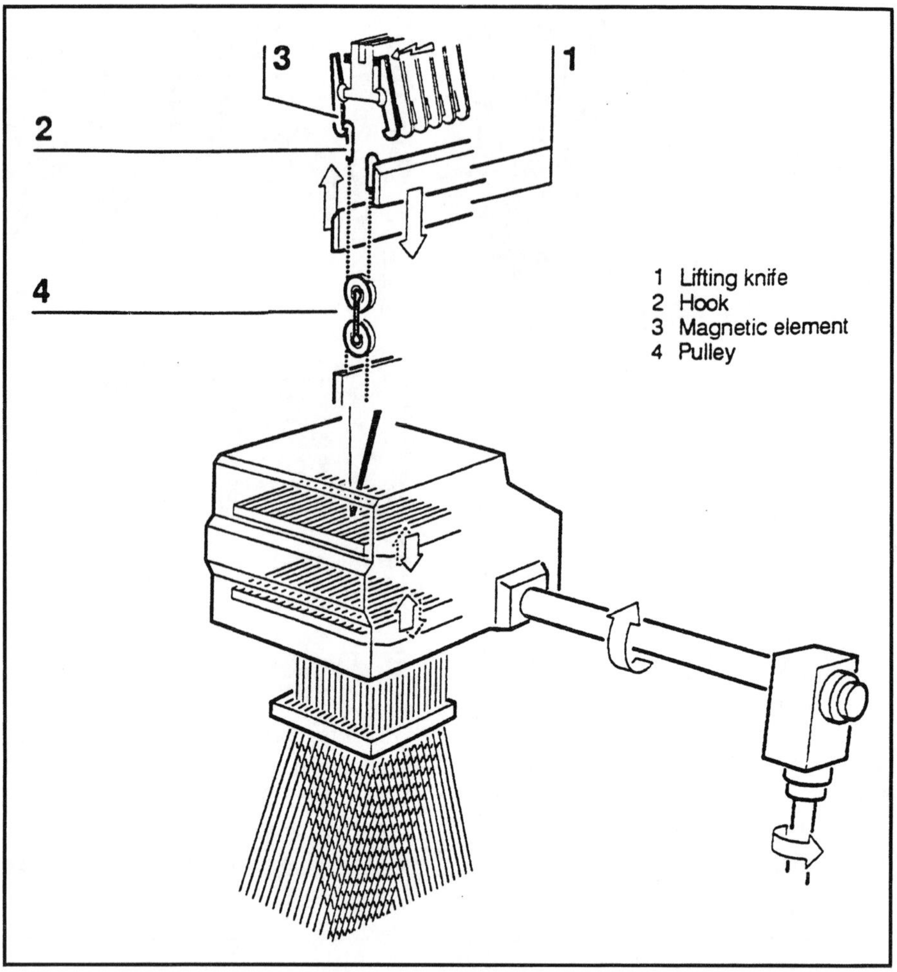

3
1
2
4
1 Lifting knife
2 Hook
3 Magnetic element
4 Pulley

FIGURE 10.13 Electronically controlled jacquard machine.

Temple

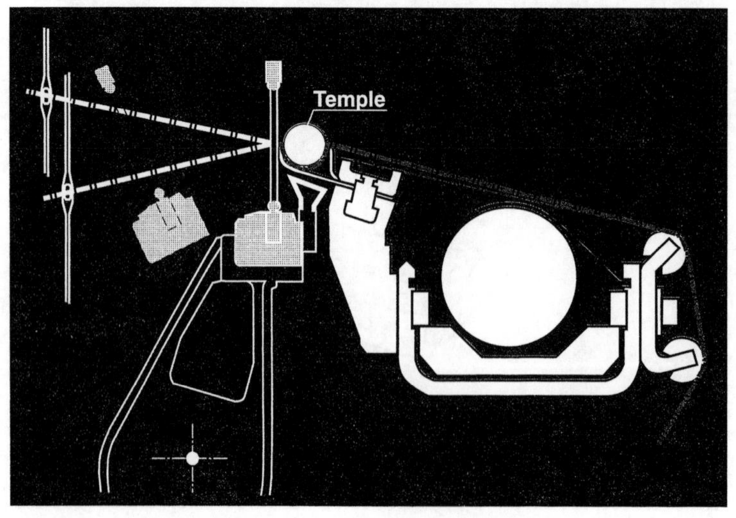

FIGURE 10.14 Sley motion of rapier machine.

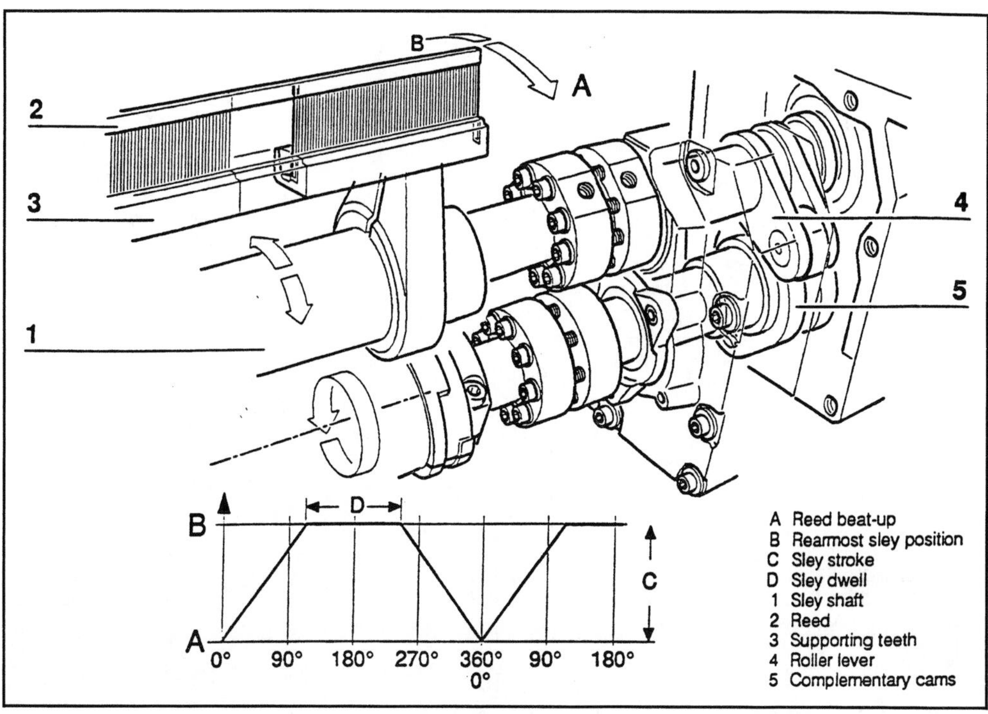

FIGURE 10.15 Schematic of sley mechanism.

and supporting teeth 3. The supporting teeth pass through the lower warp ends when the shed is almost open and protrude by 6–8 mm over those ends. Thus the gripper and tape run in the middle of the shed resting over the teeth. Characteristics of this system are no scraping between tape and warp ends and a wide opening angle for warp ends which allows a good separation for yarns.

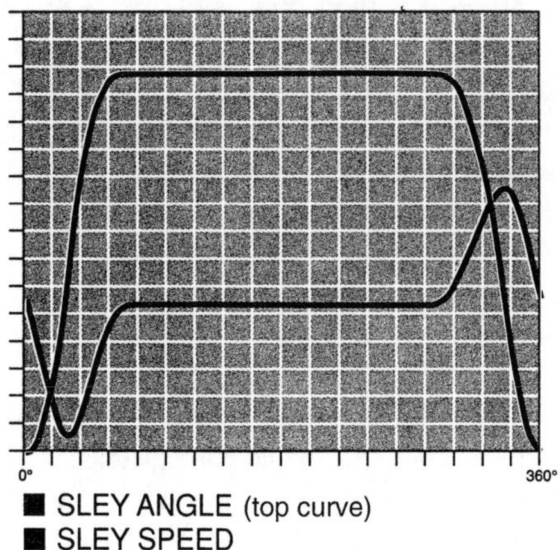

FIGURE 10.16 Sley angle and sley speed.

The motion is transmitted by complementary cams 5 on the left and right to roller levers 4. Complementary cams with different sley dwells are used, depending on the width of the weaving machine. In another variation, the sley has a continuous raceway (coated with velvet or rubber), which lays under lower warp ends that are kept lifted by a steel beam. Thus the gripper and tape run close to lower ends. Characteristic of this system is the absence of any element passing through the warp ends. Figure 10.16 shows the typical diagrams of sley angle and sley speed in a rapier machine.

Weft Insertion

Weft feeder 2 draws the weft yarn off cross-wound package 1 and coils it on to accumulator drum 3 at a preset speed (Figure 10.17). As soon as a sufficiently large weft yarn supply has been wound on, weft reserve monitor 4 interrupts the coiling process until this weft yarn has been inserted. The weft yarn is therefore always unwound from the same drum diameter at weft insertion and thus maintains a defined weft tension.

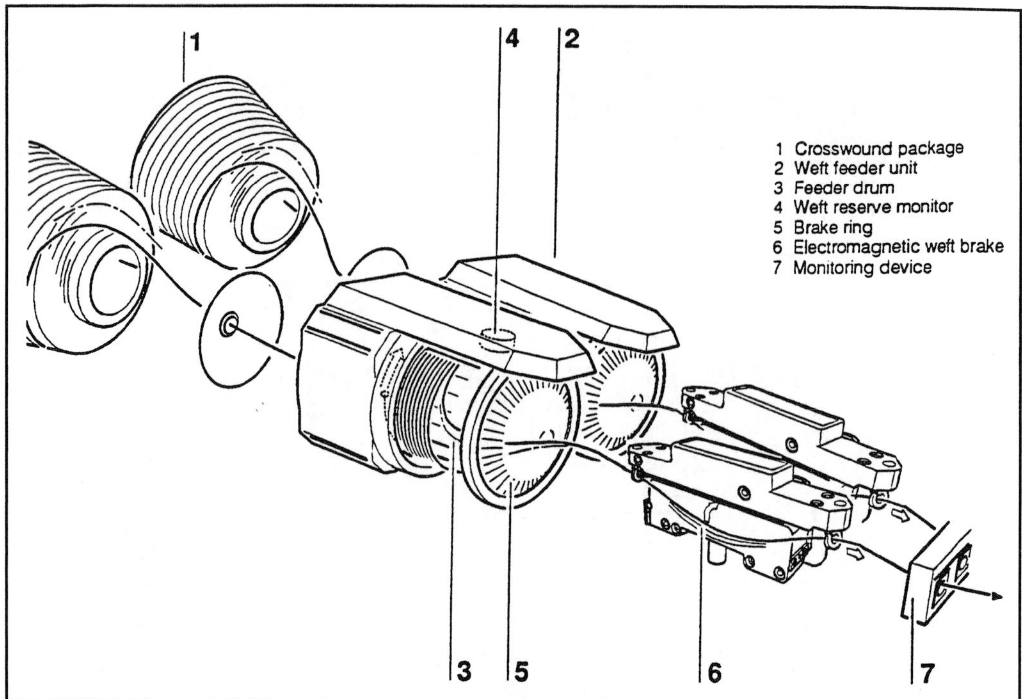

FIGURE 10.17 Schematic of weft yarn supply and weft feeder unit.

Brake ring 5 prevents ballooning during drawing-off.

The weft yarn is tensioned by electromagnetic weft brake 6 depending on the weft insertion conditions. The weft brake responds to the commands entered at the terminal. Braking force and timing can be adjusted individually. Monitoring device 7 monitors the pick during a defined insertion period.

Figure 10.18 shows the schematic of weft insertion diagram with double flexible rapier. In Phase A, weft finger 1 places the pick in the ready position. Left-hand rapier 2 seizes the pick and clamps it, and weft cutter 4 cuts it off.

In Phase B, the left-hand rapier transfers the pick to right-hand rapier 3. In Phase C, the right hand rapier withdraws the pick and releases it. The reed beats up the inserted pick to the cloth fell.

Figure 10.19 shows the schematic of color control. Electronically controlled color selector with optimized filling yarn feeding unit allows yarn-friendly filling insertion of up to eight colors. The pick sequence is determined by the fabric style and is controlled electronically. The required pick sequence is stored in the terminal. During pick finding, the color control is synchronized electronically, so that the weave and the weft color always correspond.

Magnet 2 of the required weft yarn is activated, thus drawing tongue 3 to the magnet in the read-in position of carriage 4. The carriage pushes the tongue forward and weft finger 1 lowers the weft yarn into the zone of the left-hand gripper. If the same weft yarn is inserted several times in succession, the weft finger remains in the insertion position (below).

Figure 10.20 shows the left-hand and right-hand rapier grippers which are all-purpose rapiers. Lightweight grippers with few moving parts can handle a wide range of yarns from 5 to 1000 tex. Gripper head 1,2 is bolted on to rapier tape 5, made of carbon fiber reinforced plastic.

In the initial position, yarn clamp 3 of the left-hand rapier is opened and cleaned by the suction device. Tongue 4 in the right-hand rapier remains open before and after reed beat-up. Due to optimized positioning of the gripper head's center of gravity, optimized tape stiffening and

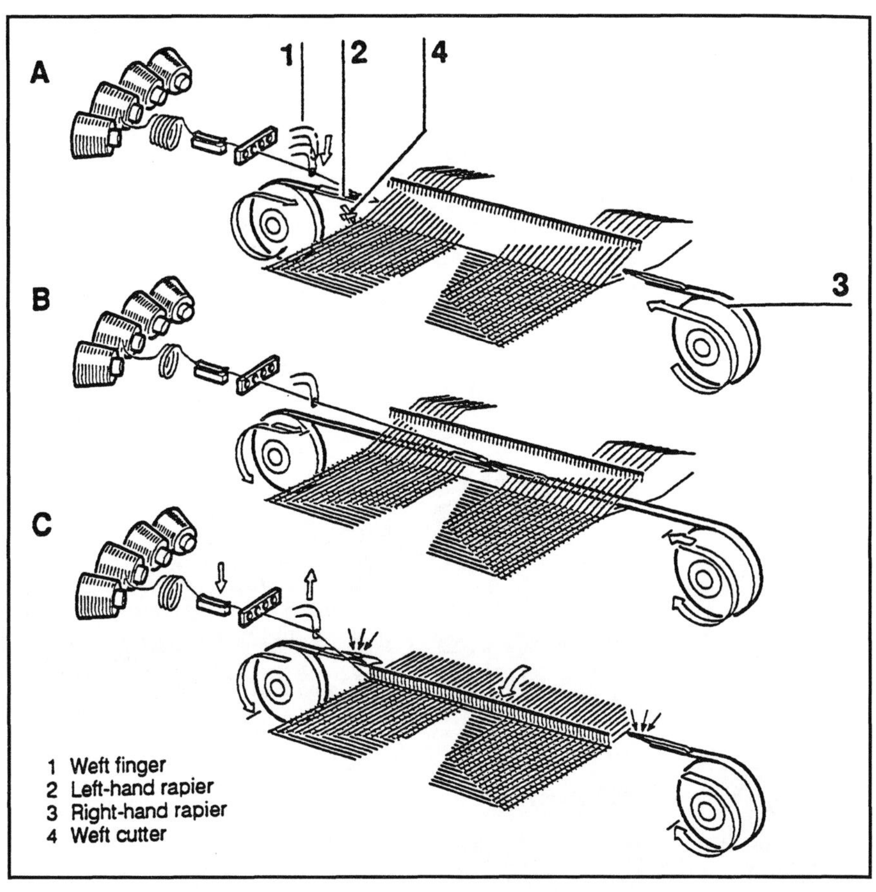

FIGURE 10.18 Weft insertion diagram of double flexible rapier machine.

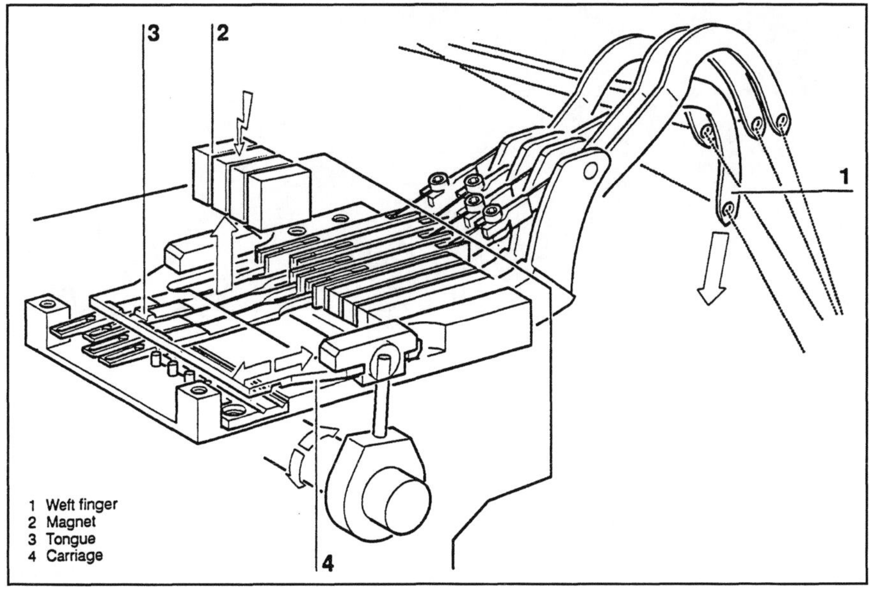

FIGURE 10.19 Schematic of color control.

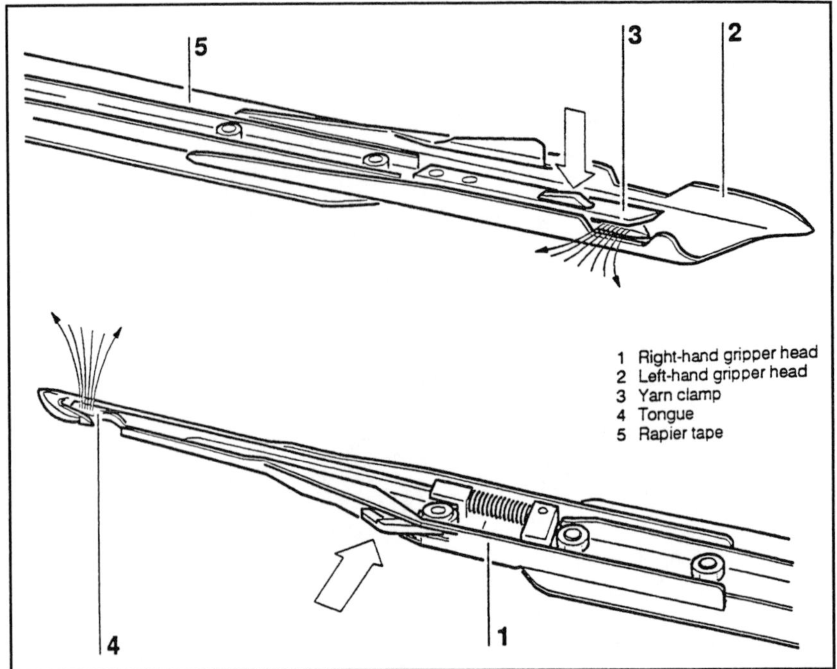

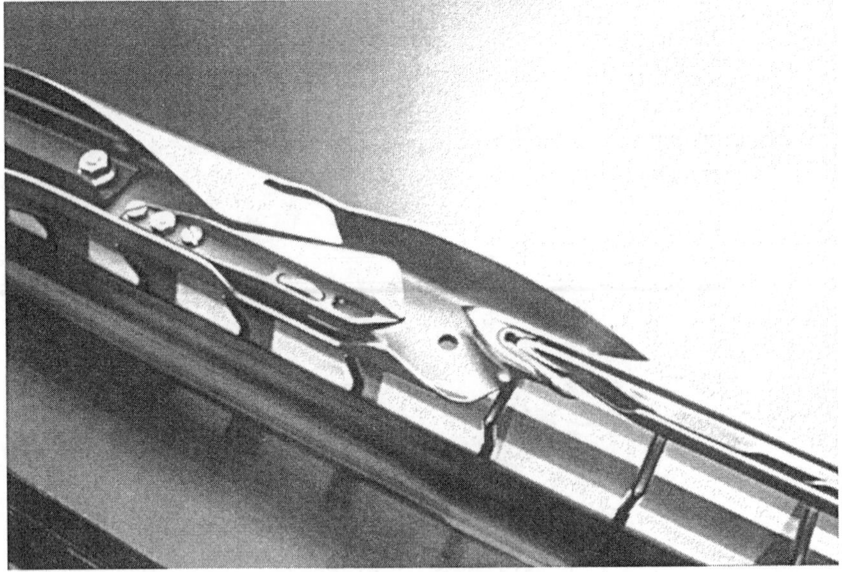

FIGURE 10.20 Grippers.

supporting teeth or race board, lower shed does not contact with gripper which decreases wear. As a result, there is no obstructing guiding elements inside the shed.

Figure 10.21 shows the operating sequence during insertion of the filling yarn. Filling yarn 1 is presented to rapier 2 and seized. When the weft yarn is held in the gripper clamp it is cut off by cutter 3 and inserted by the rapier into the shed.

Figure 10.22 shows the working principle of the filling cutter. Filling yarn in the waiting position is held by retaining hook 1. The weft yarn being inserted moves into the opening of the

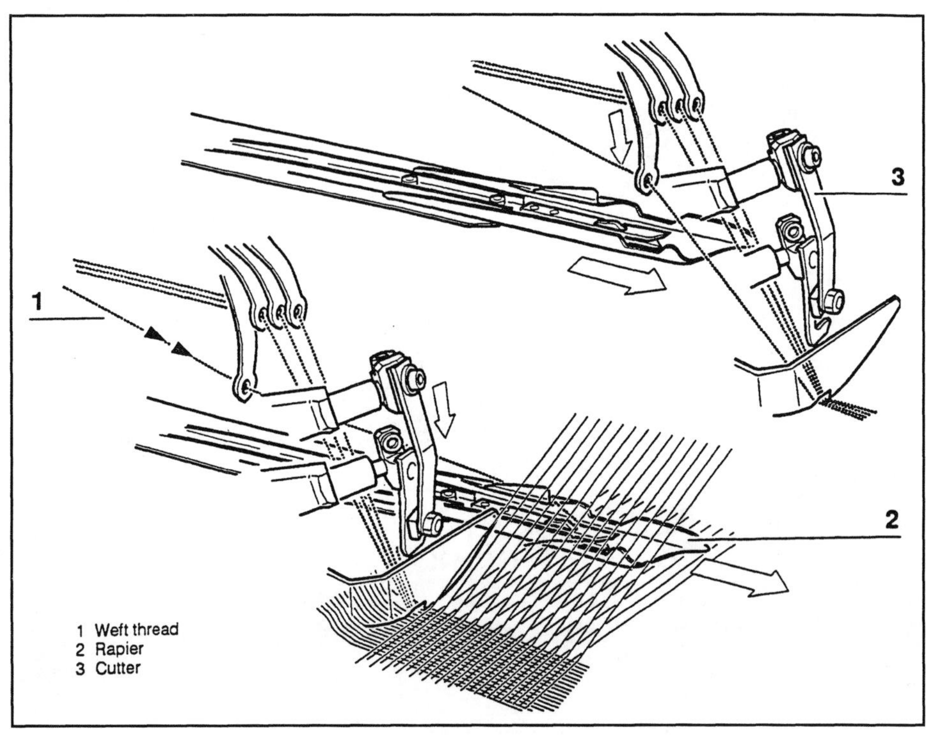

FIGURE 10.21 Operating sequence of insertion.

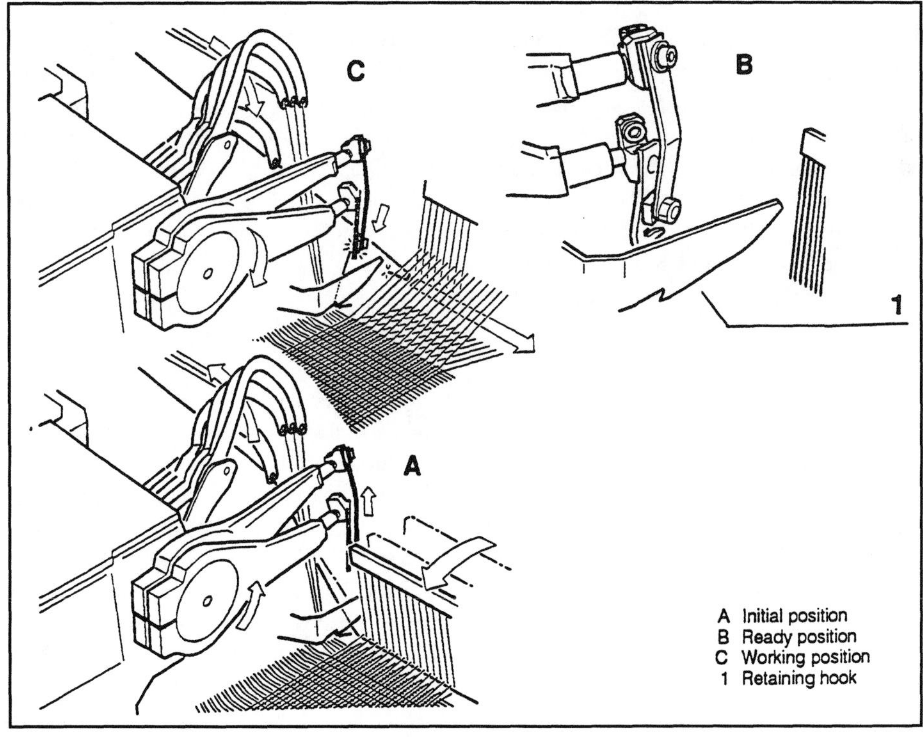

FIGURE 10.22 Filling cutter.

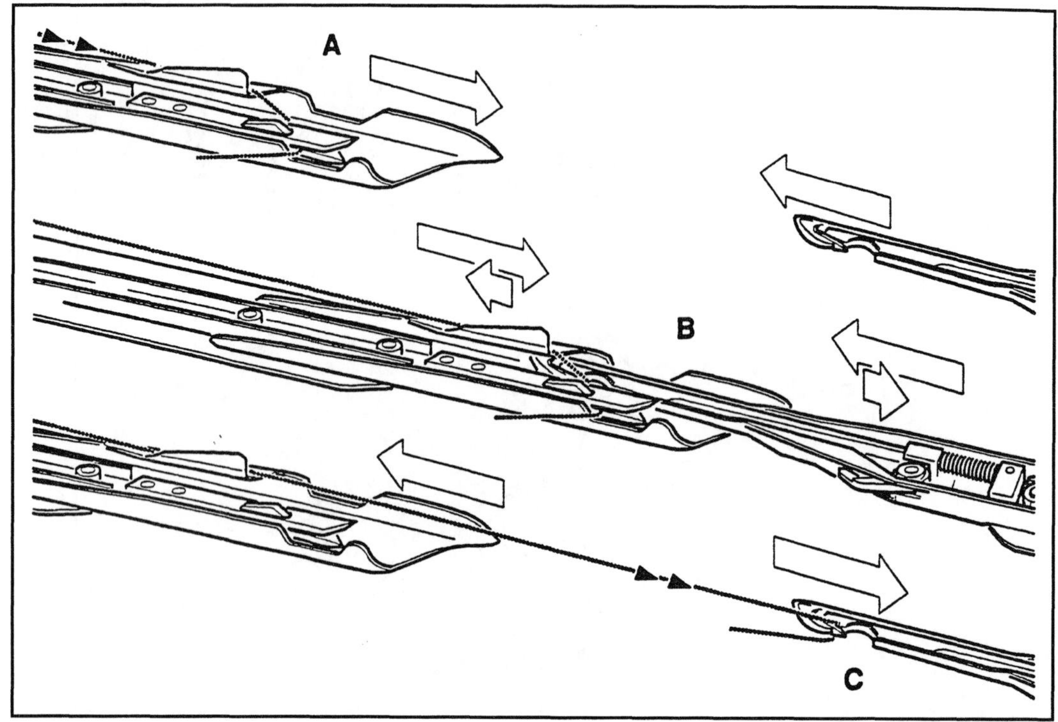

FIGURE 10.23 Schematic of filling yarn transfer.

cutter blades and is cut as soon as the left-hand rapier enters the shed.

Figure 10.23 shows the transfer of yarn from one rapier head to the other. In Phase A, the left-hand rapier takes the pick with accelerated and decelerated speed into the middle of the weaving machine. In Phase B, the yarn is taken over and gripped by right-hand rapier 2. In Phase C, the right-hand rapier then pulls the pick to the right-hand edge of the fabric and releases it.

Rapier Motions

Figure 10.24 shows the schematic of the rapier tape drive which is actuated by a drive gear. The continuous rotary motion of the main drive shaft is transformed into the alternating rotary motion of the rapier drive shaft. The drive gear rotates at the speed of the weaving machine. The circular motion is transmitted to the transmission lever. An oscillating motion results on toothed segment, and is transmitted to tape wheel shaft by gearwheel. A guide presses rapier tape on to tape wheel, thus ensuring smooth rapier action.

Figure 10.25 shows the diagram of rapier movements. The rapier movements are synchronous, i.e., both grippers start to move at the same crank position. They meet in the middle of the weaving machine and their movements overlap, so that pick 6 is transferred when both rapiers are on their way back. Figure 10.26 shows the diagrams of gripper speed.

Filling Yarn Braking

The electronically controlled weft brake makes it possible to process yarns of low tenacity (Figure 10.27). Electronically controlled, highly dynamic yarn brake with programmable braking force and braking duration reduces filling yarn waste and reduces yarn breakage frequency. Automatic feeder control provides switch-over at filling breakage without interrupting the weaving process. If a weft yarn breaks before the feeder, the microprocessor switches over to a feeder with the same yarn. When the operator repairs the broken weft, the first feeder is started up again.

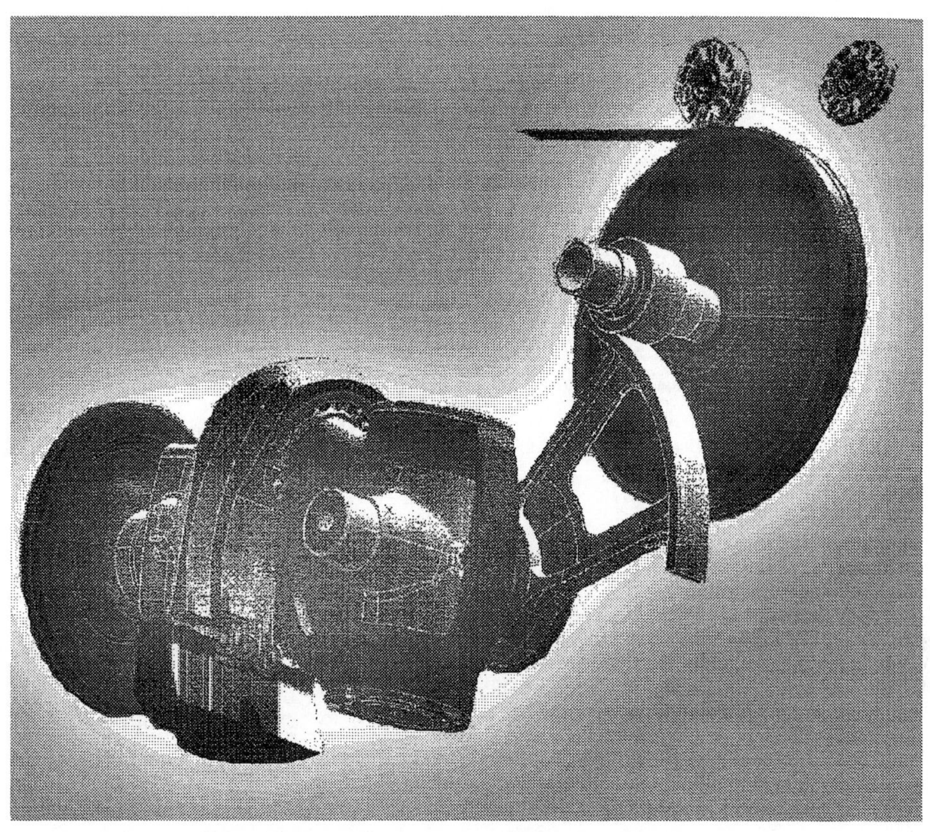

FIGURE 10.24 Single phase tape drive for a rapier machine.

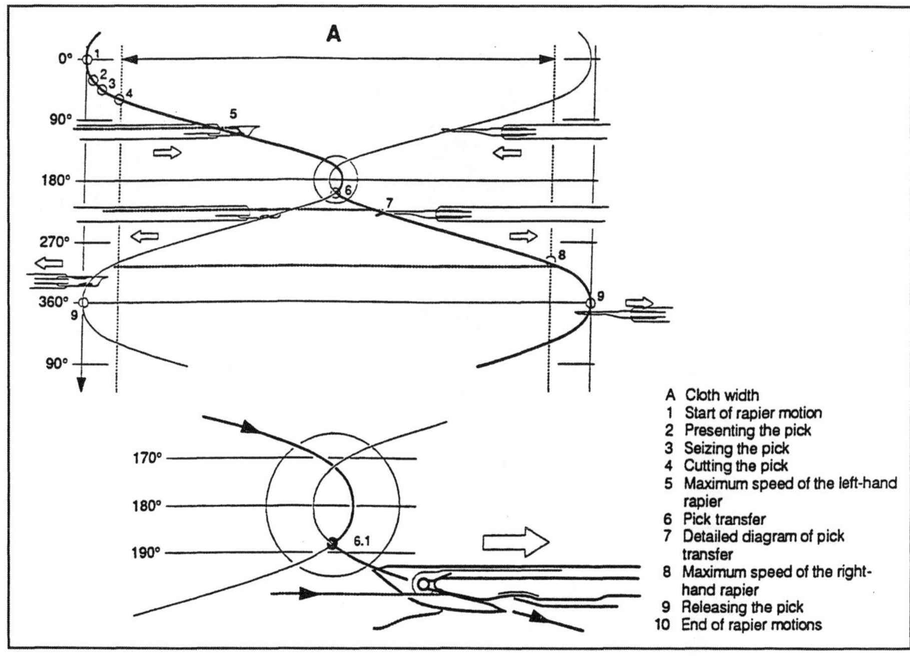

FIGURE 10.25 Diagram of rapier movements during symmetrical rapier travel.

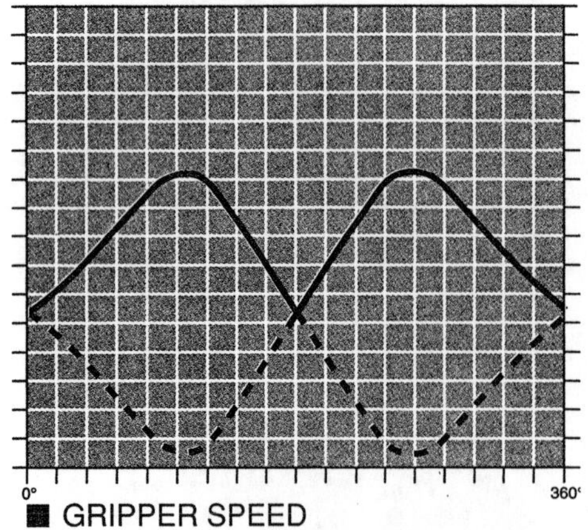

FIGURE 10.26 Gripper speed diagram.

The weft yarn is braked as needed for the weft insertion process by electromagnetically controlled weft brake 1. The yarn is lightly held by the weft brake before and after weft insertion (standby phase). The weft brake is closed before pick transfer in the shed, before it is drawn out to the right. Braking force and timing can be adjusted. Figure 10.28 shows the range of adjustment for the four braking phases in one weft insertion cycle.

Braking phase A

Making angle: automatically at 0°, when the weft yarn is selected.
Braking angle: from 20° to 70°.

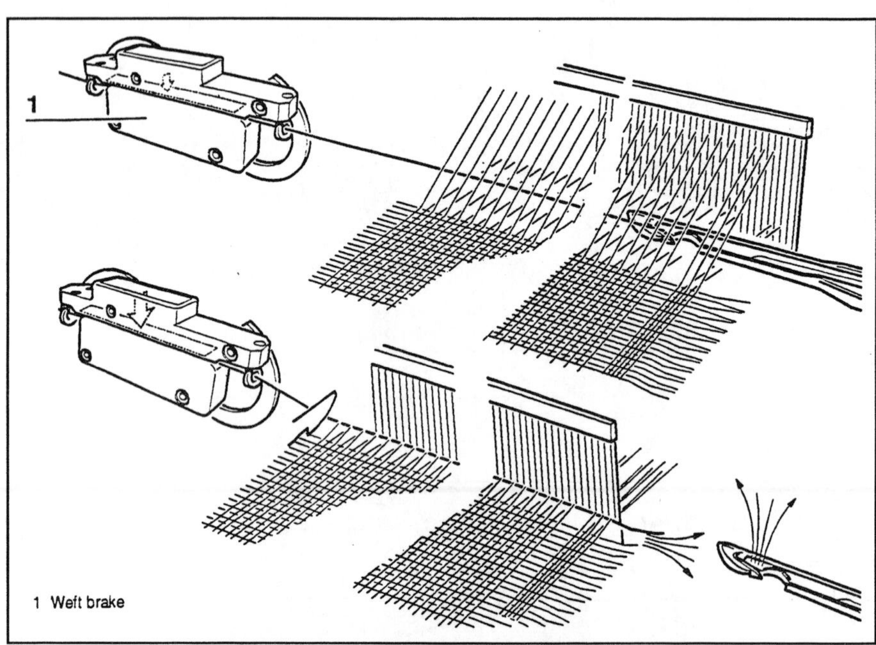

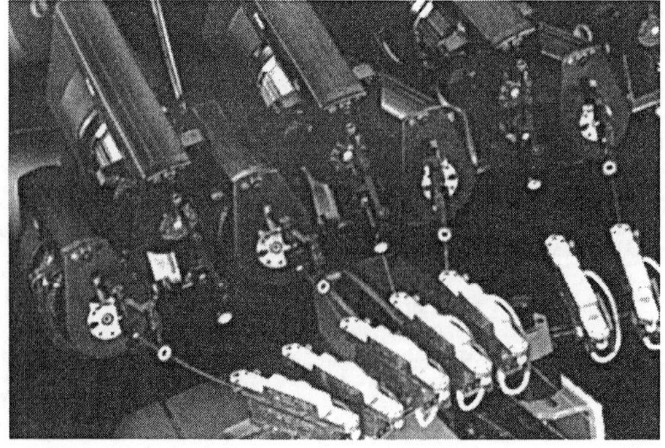

FIGURE 10.27 Microprocessor controlled weft brake on rapier machines.

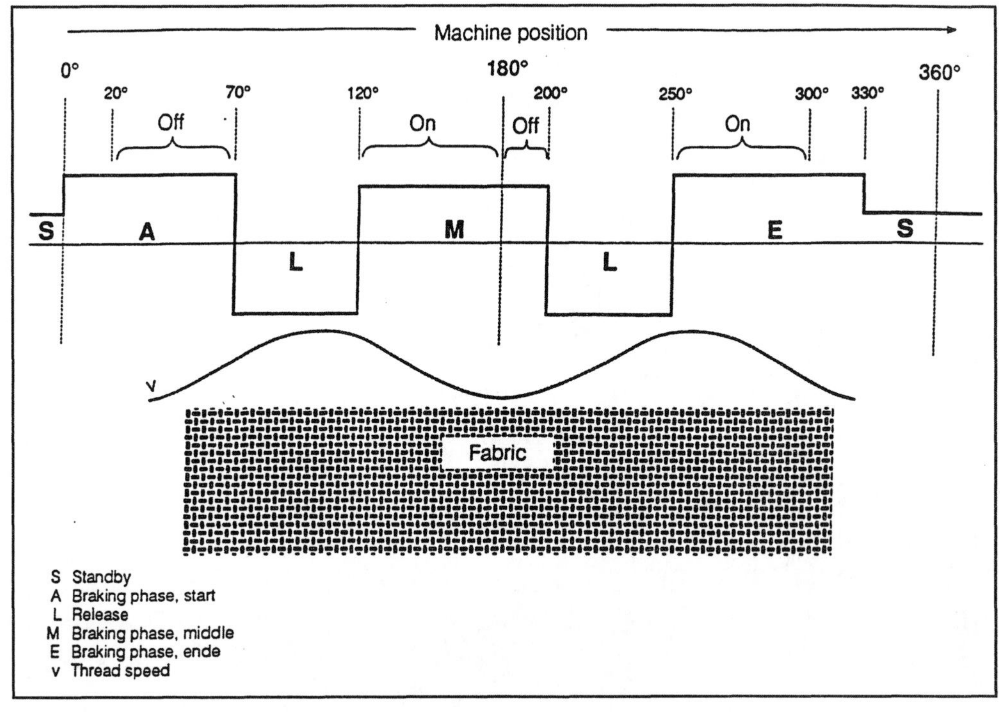

FIGURE 10.28 Braking phases for the yarn.

Braking phase S

Making angle: automatically at 330°
Breaking angle: automatically at 0°, when braking angle A switches on

Braking phase M

Making angle: from 120° to 180°
Breaking angle: from 180° to 200°

Braking phase E

Making angle: from 250° to 300°
Breaking angle: automatically at 330°

The yarn brake is automatically released after braking phases A and M.

To produce knot free fabric, the transfer from empty package to full package is monitored and the knot is removed automatically (Figure 10.29).

Pick Finding Device (Figure 10.30)

The moving weft yarn is monitored by weft monitor 2 throughout the weft insertion process. If no pick is present, the machine is stopped and then turned to the pick finding position. When the pick finding position is reached—the rapiers are outside the cloth—the actual pick finding process can be initiated automatically, reversing by one pick.

During this process, the main shaft is disengaged from the rear shaft via electromagnetic clutch 3. Electric motor 4 drives the rear shaft. The shedding mechanism and the full cross leno device are thus reversed by one pick. A sensor monitors the correct engagement of the clutch.

FIGURE 10.29 Automatic knot removal.

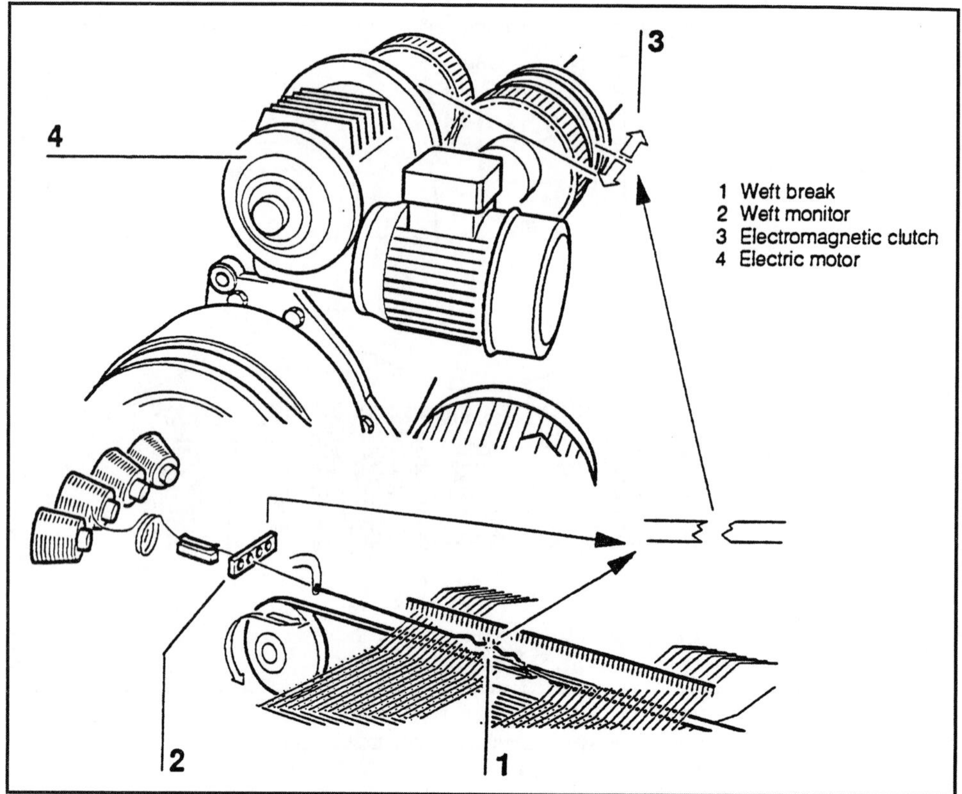

FIGURE 10.30 Pick finding device.

Whether a weft break 1 requires reversing by one, two or three picks, removing the waste yarn and then, for example, moving forward again by two picks, depends on the type of shedding mechanism. This procedure is called synchronization. Synchronization is done automatically, i.e., the correct pick is electronically preset while the weft monitor is at a standstill.

Selvages

Selvages on rapier machines can be formed by the tuck-in motion, leno or sealing unit. Figure 10.31 shows full cross leno device. The auxiliary selvage shafts and the leno needles are controlled by cams 1 with two different functions:

1) The full cross leno section 2 is used for forming the fabric selvages
2) Selvage motion 3 controls auxiliary selvages 4. The right-hand auxiliary selvage prevents the weft yarn released by the gripper from spring back into the shed.

Temples 1 keep the fabric 3 at the same width as when it is drawn into the reed (Figure 10.32). This reduces or even prevents friction between the warp ends and the reed. The temples can be equipped with a variable number of rings 2, which can have long or short needles, pointed or rounded ends and various numbers of needle rows. The temple rings are also raked progressively in order to distribute the tension of the cloth over all the rings. Selvage cutter 5 is driven by an electric motor and cuts the weft yarns. Auxiliary selvage 4 is drawn off separately.

For synthetic polymer yarns, a selvage cutting and sealing device can be fitted to fuse the weft ends. It is also equipped with a centering device.

Rapier weaving machines can be equipped with intermediate tucking units. This allows

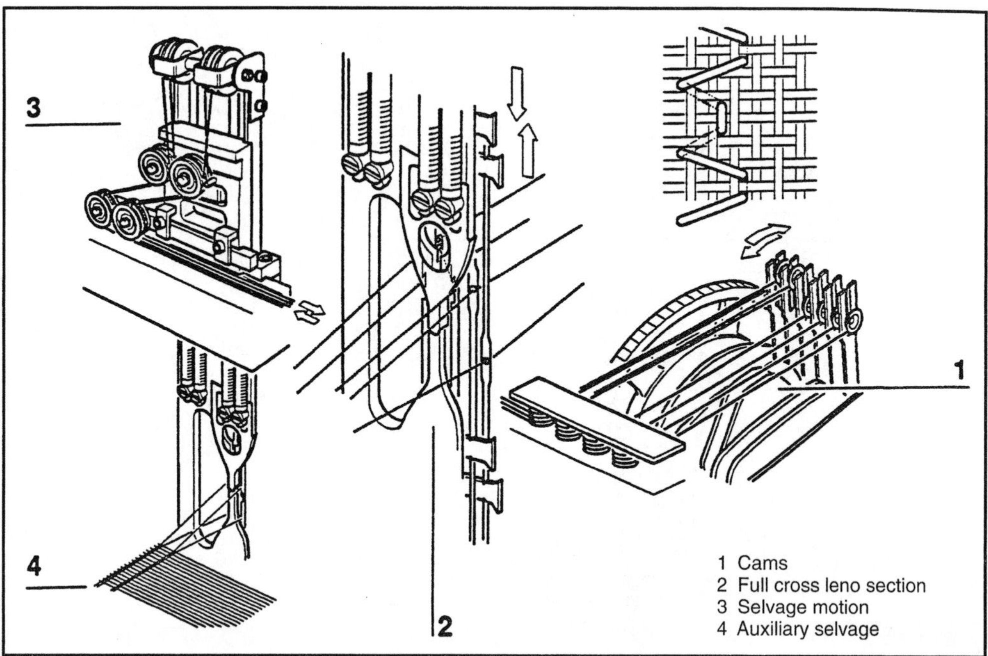

FIGURE 10.31 Full cross leno device.

several narrow panels to be woven simultaneously (Figure 10.33). Each intermediate tucking unit requires a space of 3 cm between the individual panels.

The selvage waste of rapier weaving machines can be 7 to 14 cm. Using a special device, the additional selvage on the picking side is reduced. Figure 10.34 shows the schematic diagram of a device that reduces the selvage waste to a total of 4 cm.

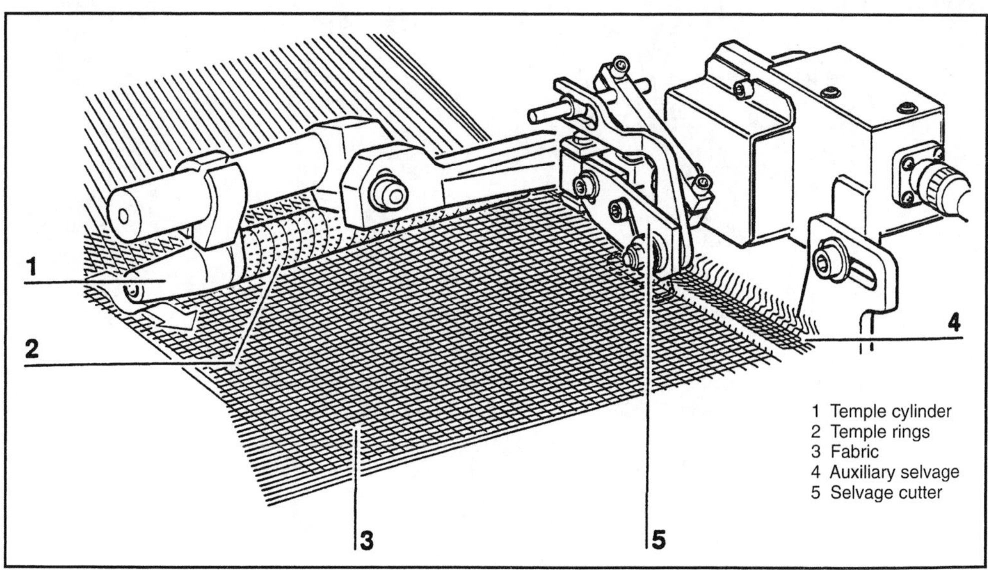

FIGURE 10.32 Temple and selvage cutter.

FIGURE 10.33 Center tucking unit on rapier machines.

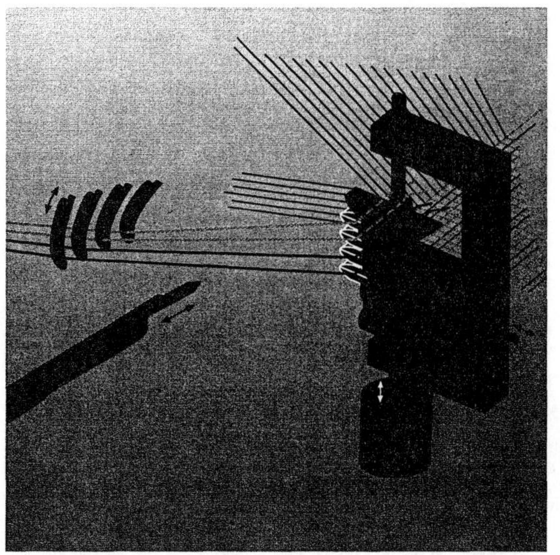

FIGURE 10.34 Schematic diagram of selvage waste saver for rapier machine.

Fabric Take-up (Figure 10.35)

Fabric take-up is electronically controlled with a highly dynamic servomotor. As a result, fine gradation of 0.1 picks/cm is possible. Weft density can be changed from 8 to 250 picks/cm. Programmable cloth fell correction and shed leveling in combination with empty pick start-up help to prevent start-up marks.

The cloth is led over spreading roller 1 to take-up roller 2. Press roller 3 prevents the cloth from slipping. Spreading roller 4, which is located immediately before the cloth wind-up, prevents creasing.

The cloth transport operates on the principle of the number of picks per unit length. Take-up is positive and continuous, by a specific amount per pick.

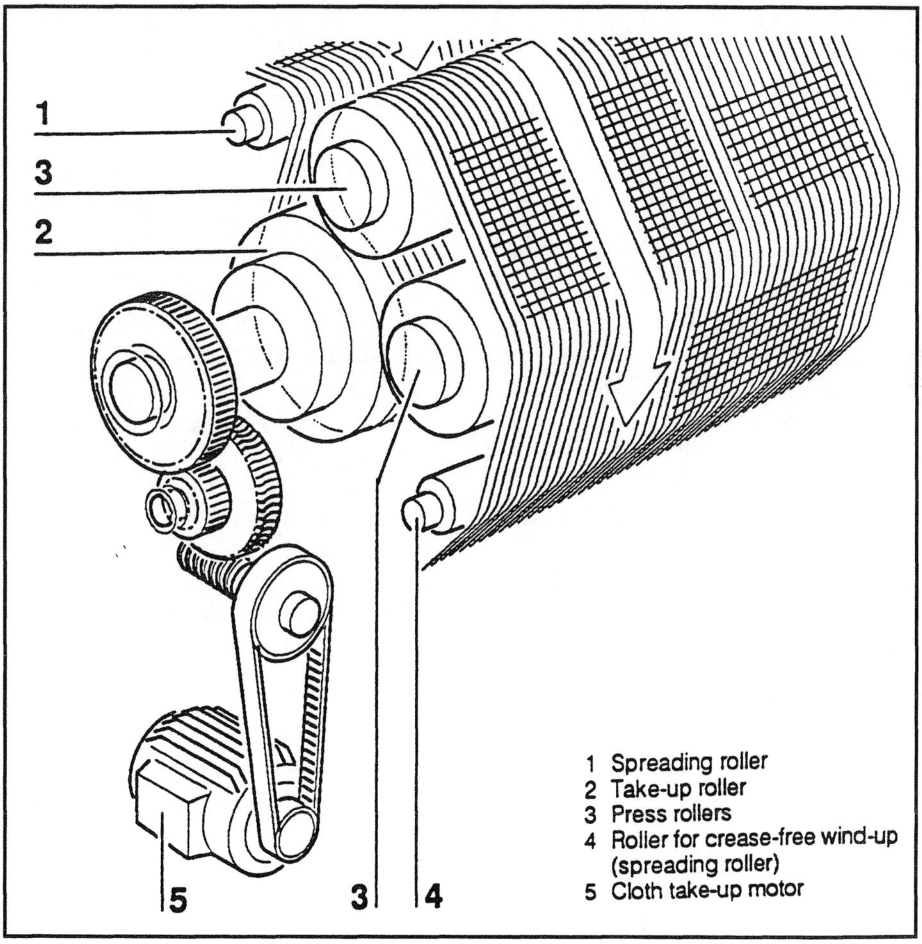

1 Spreading roller
2 Take-up roller
3 Press rollers
4 Roller for crease-free wind-up (spreading roller)
5 Cloth take-up motor

FIGURE 10.35 Schematic of fabric take up mechanism.

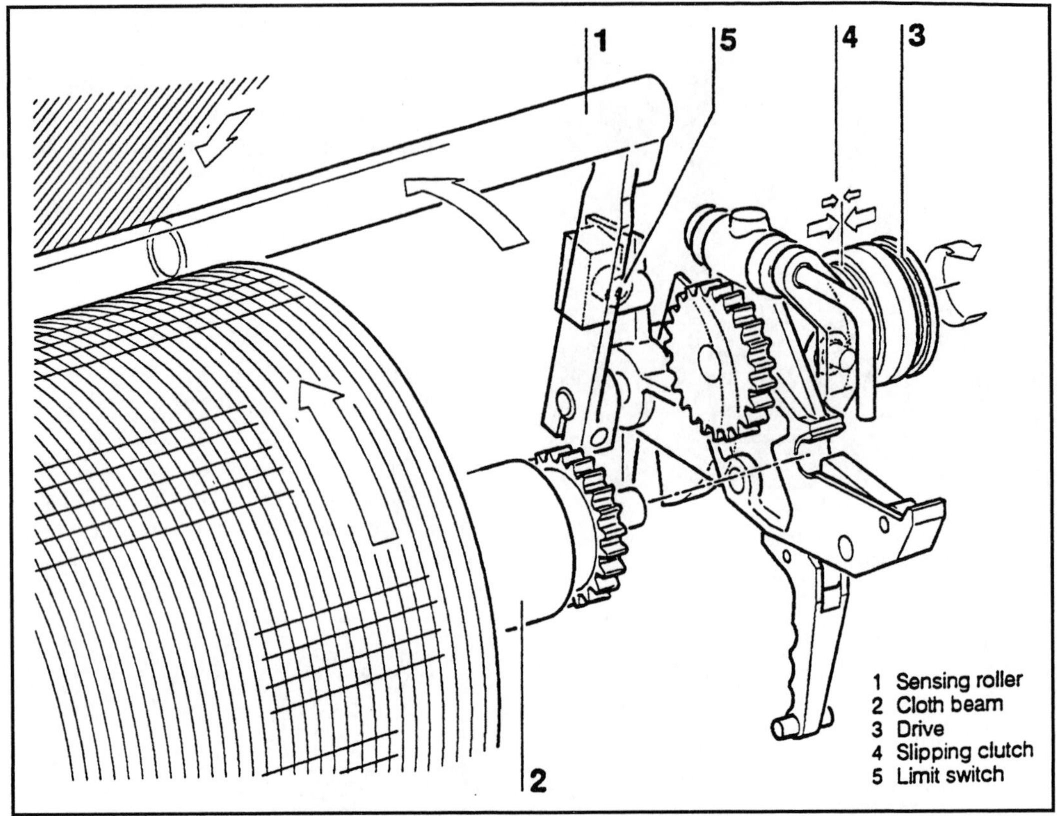

FIGURE 10.36 Schematic of fabric wind-up mechanism.

The required weft density is entered at the terminal, and cloth take-up motor 5 selects its infinitely variable speed on the basis of this input. A cloth fell correction programmed at the terminal is also performed by the cloth take-up motor.

Fabric wind-up with press roller and controlled friction coupling provides creaseless wind-up with constant, controllable fabric tension. Figure 10.36 shows the schematic of a fabric wind-up mechanism.

Drive 3 of the cloth wind-up operates via slipping clutch 4. The degree of compression of the slipping clutch is varied by the position of sensing roller 1 on cloth beam 2, so that the driving force is initially lower and increases with the diameter of the cloth beam.

The winding hardness can be set by means of a control cam. When the cloth beam reaches its maximum diameter, the sensing roller on the cloth beam actuates limit switch 5, and the weaving machine is switched off.

10.2 "WAVY TERRY" RAPIER WEAVING MACHINE

SulzerTextil developed a new dynamic pile control mechanism for rapier weaving machines in which the pile heights are freely programmable. With dynamic pile control, the pile height can be varied from one weft group to another (Figure 10.37).

Three pick and four pick groups can be programmed. The change from three to four weft terry technology together with extra picks within a pattern can be programmed freely. No mechanical conversion is necessary when changing between these two types of groups. The programmable loose pick distance assures smooth transitions at borders and with pattern changes.

Twin-rapier systems are used in face-to-face weaving, such as double-plush weaving and carpet manufacturing. In these machines, two rapiers, which are driven together, enter the shed from the same side of the weaving machine.

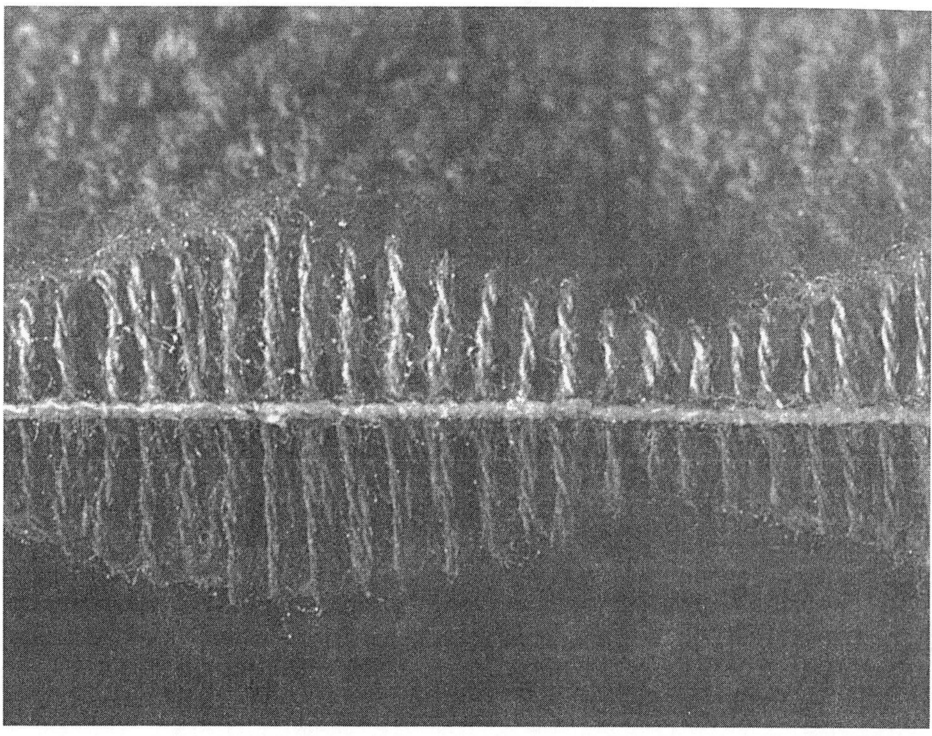

FIGURE 10.37 "Wavy" terry pattern with dynamic pile control.

10.3 ANALYSIS OF STRESSES ON WARP AND FILLING YARNS DURING RAPIER WEAVING

Gunther and Weissenberger [1] analyzed the stresses on warp and filling yarns in rapier weaving. The stress on the warp yarns is caused by the following:

- initial set up stress
- shed formation
- back rest roller movement
- reed beat-up

The dynamic force on the warp yarn is given by:

$$F = E_w \, E_{dyn} \, 10^{-2} \qquad (10.1)$$

where F: yarn force in cN/tex
E_w: warp elongation in %
E_{dyn}: dynamic elasticity modulus (E) in cN/tex

Because of prestressing, the dynamic E modulus is a little higher than the static E modulus determined by the tensile test. The warp elongation range on the weaving machine is generally below 2%.

Figure 10.38 shows the warp tension diagram during weaving of a twill fabric made of a ply yarn of count Nm 96/2 (10.4 tex × 2). Figure 10.39 shows the warp tension diagram during weaving of a plain fabric made of ply yarn of count Nm 80/2 (12.5 tex × 2). The yarns are unsized and wax treated. The highest peak loads are due to beat-up.

Figure 10.40 shows the distribution of yarn tension across the weaving width. The top curve represents the maximum peak loads and the bottom curve shows the mean tensile force of a weaving cycle.

The weft tensile force does not change much with the increase in weaving machine speed as shown in Figure 10.41. Figure 10.42 compares the peak loads of the first and last harness frames at different weaves and warp yarn counts. There is negligible difference in warp tension between different conditions. The slight difference between the first harness frame and the last one

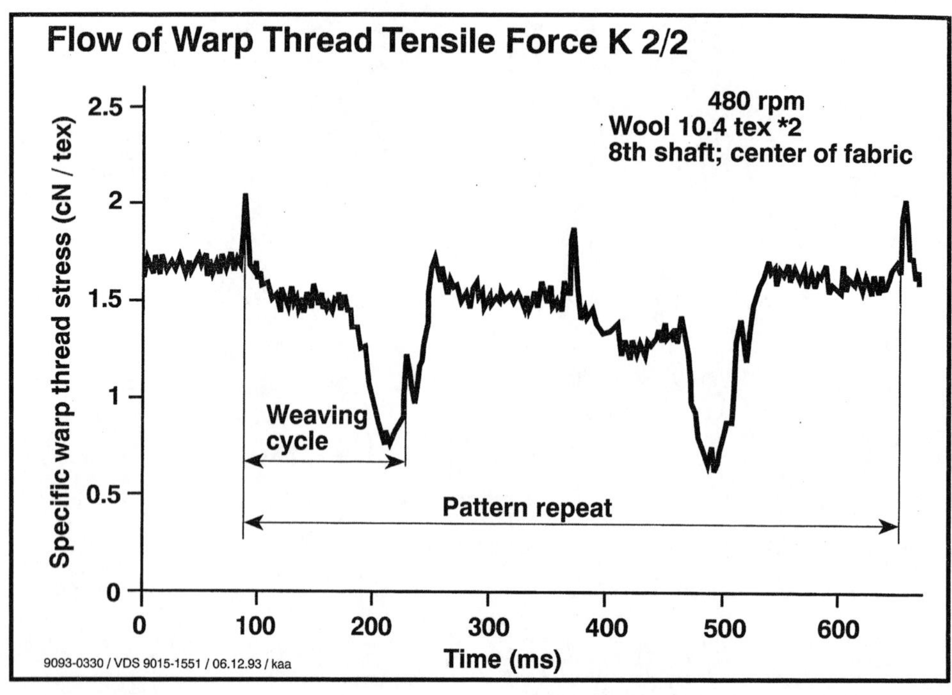

FIGURE 10.38 Warp tension trace of a Nm 96/2 woolen yarn during weaving of a twill fabric (Gunther and Weissenberger).

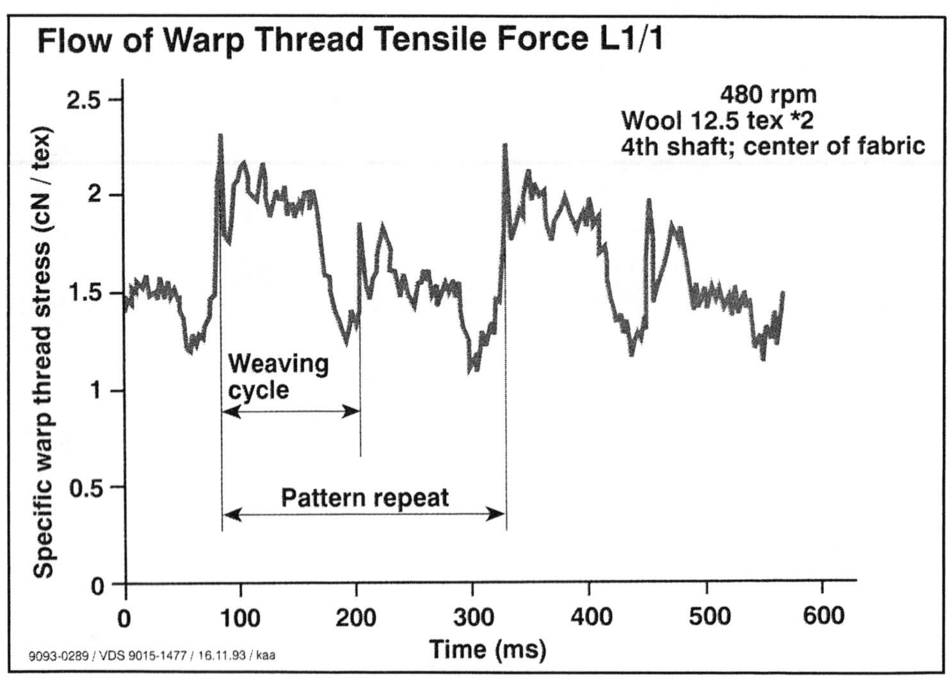

FIGURE 10.39 Warp tension trace of a Nm 80/2 woolen yarn during weaving of a plain fabric (Gunther and Weissenberger).

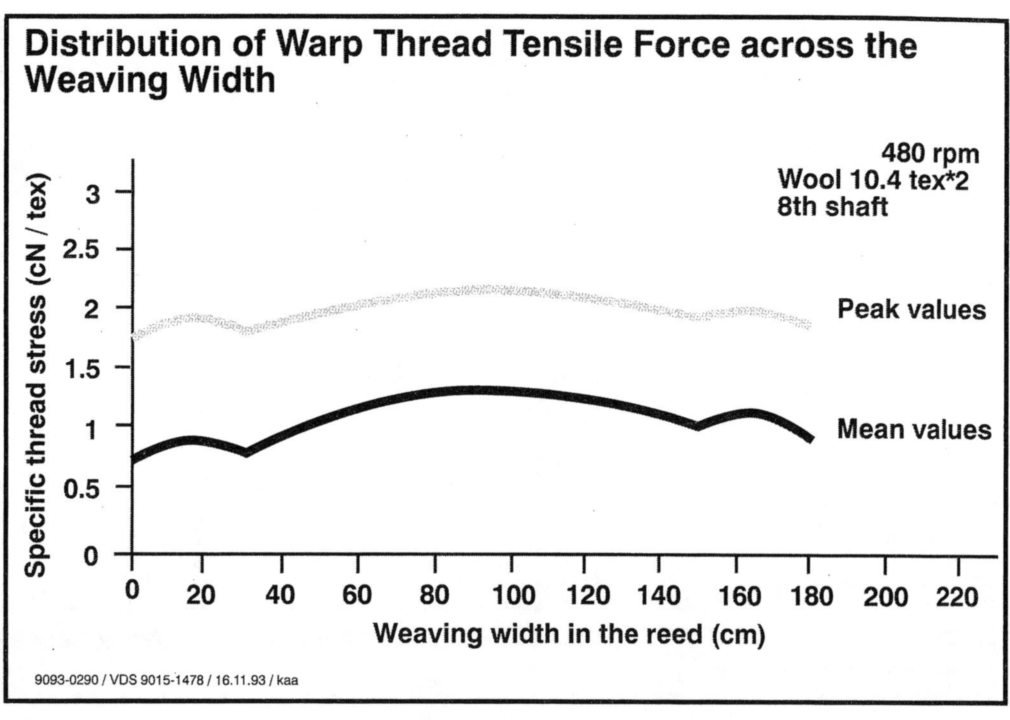

FIGURE 10.40 Yarn tension distribution across the weaving width (Gunther and Weissenberger).

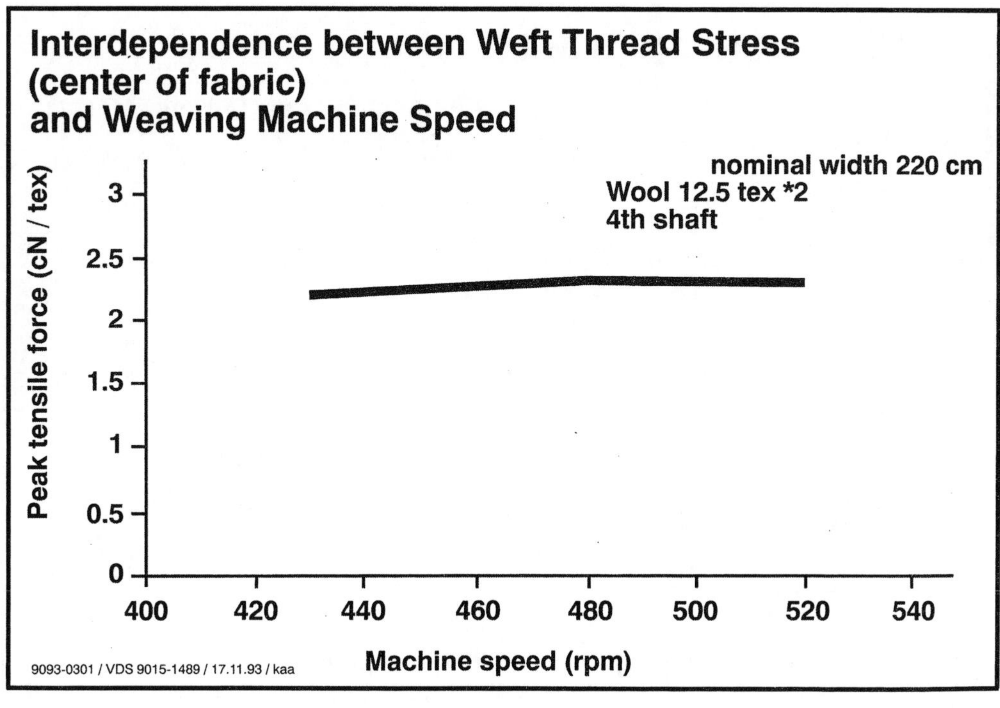

FIGURE 10.41 Effect of machine speed on filling yarn stress (Gunther and Weissenberger).

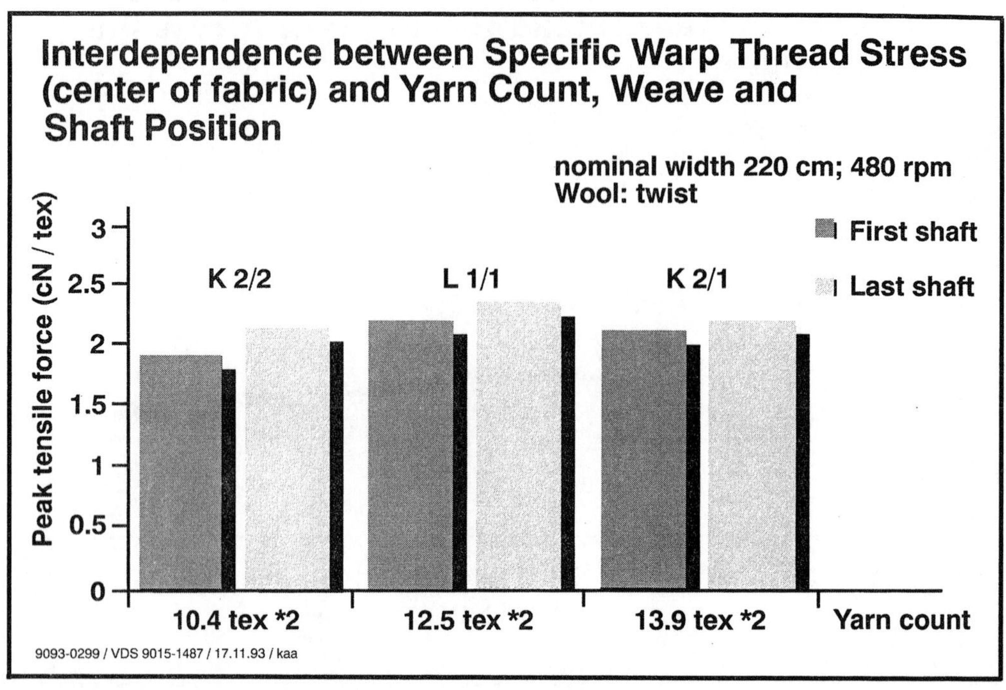

FIGURE 10.42 Effect of yarn count and harness position on warp yarn stress (Gunther and Weissenberger).

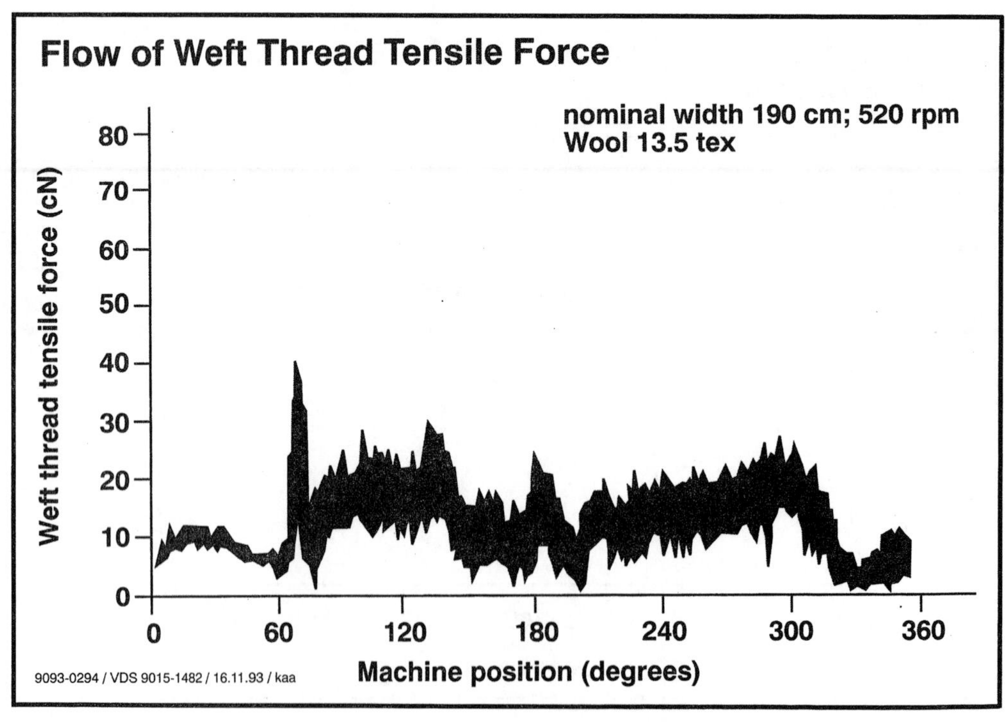

FIGURE 10.43 Filling tension diagram on a rapier machine (Gunther and Weissenberger).

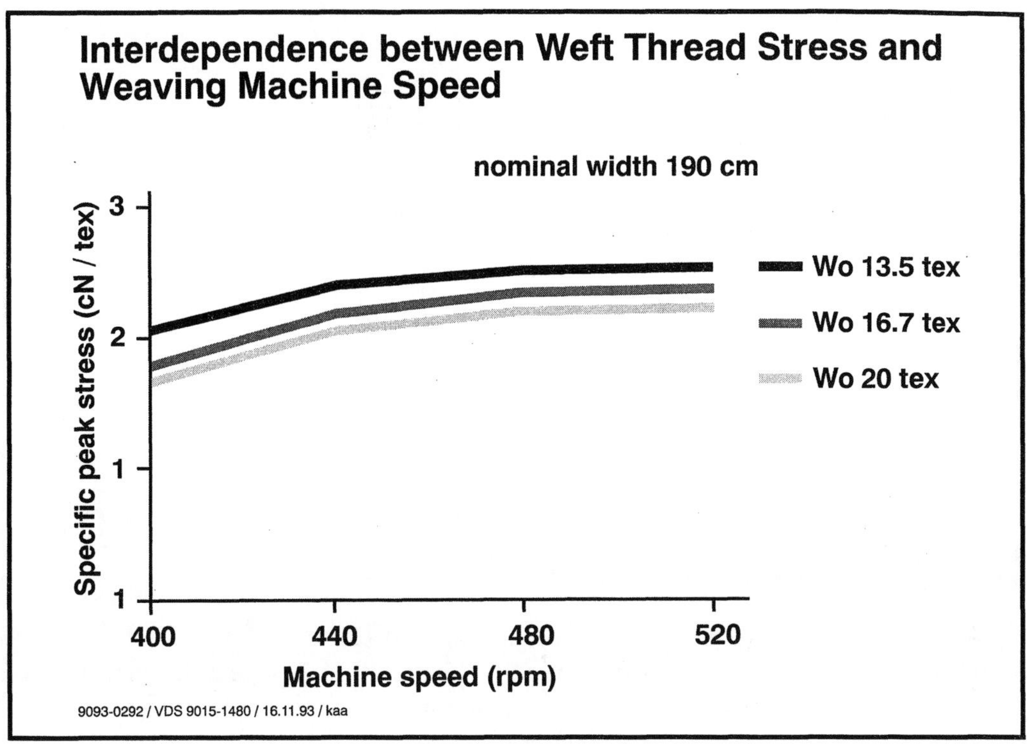

FIGURE 10.44 Effect of rapier machine speed on filling tension (Gunther and Weissenberger).

depend on the shed geometry (Figure 5.3). The fact that the differences are small is due to the high extensibility of the warp yarn.

Compared to cotton yarns, the strength of worsted woolen yarns is considerably lower, while their extensibility is markedly higher. These differences translate to the differences in weaving.

The highest values of filling yarn strain during insertion are caused by the extremely short peak forces which are caused by the maximum speed variations. The tension on the filling yarn can be approximated by [1]:

$$F = V_f \cdot \sqrt{E}\ 10^{-2} \qquad (10.2)$$

where F: filling yarn tension (cN/tex)
V_f: yarn speed (m/s)
E: elastic modulus (cN/tex)

The peak values of stress rise in proportion to the machine speed or filling insertion rate.

Figure 10.43 shows the typical filling yarn stress on a worsted woolen yarn of 13.5 tex at a machine speed of 520 rpm and 190 cm weaving width. In filling insertion by rapiers, the peak force usually takes place during the acceleration phase when the yarn is taken over. It should be noted that the stress diagrams for several filling insertion cycles are not identical and subject to change.

The filling yarn stress increases with increasing machine speed (as shown in Figure 10.44) for three different woolen yarn counts. It is also shown that the yarn tension increases with an increase in yarn fineness. Thus, the risk of yarn breakages due to overstressing during weaving of finer yarns is higher than coarser yarns of comparable quality. In other words, efficiency or speed is limited far more by the properties of the filling yarn than by those of the warp.

10.4 OPERATIONS

Filling insertion rate and machine speed are influenced by the type of raw material used, the

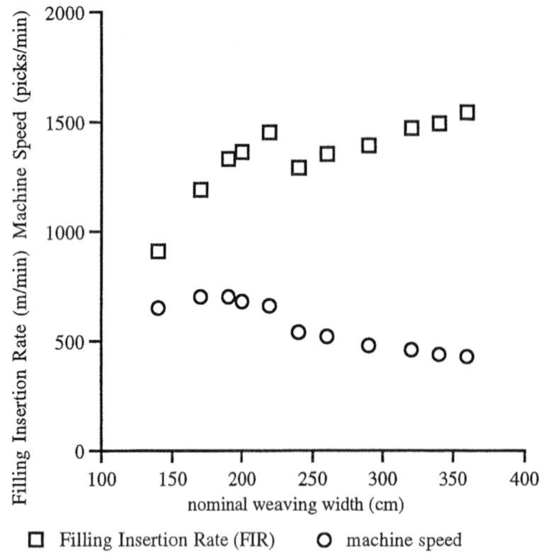

FIGURE 10.45 Effect of weaving width on filling insertion rate and machine speed for rapier machines.

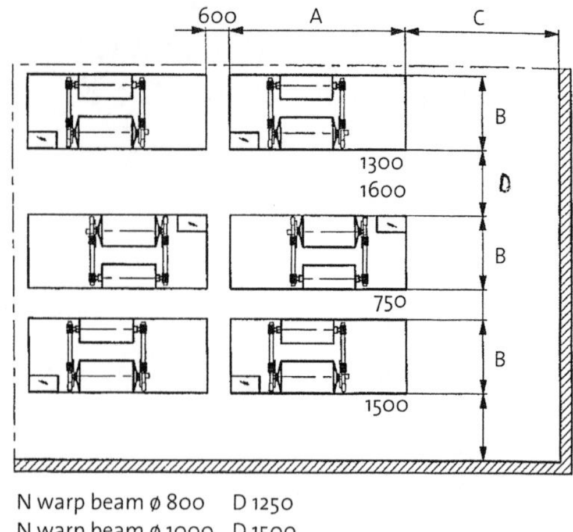

FIGURE 10.46 Layout for rapier weaving machines.

quantity of harness frames, the shed angle, yarn count and yarn quality. Figure 10.45 shows the relationship between machine width, filling insertion rate and machine speed.

Figure 10.46 shows a typical plant layout for rapier weaving machines which is designed for low noise and easy access. Figure 10.47 shows a weaving plant where rapier machines are used.

FIGURE 10.47 Rapier weaving plant.

REFERENCE

1. Gunther, K. and Weissenberger, W., "Optimized Gripper Technology for the Production of Extremely Fine and Fashionable Outerwear", 7th Weaving Colloquium, Denkendorf, October 20/21, 1993.

SUGGESTED READING

- Weissenberger, W., "Process Bridging Quality Assurance Viewed from the Fabric to the Yarn", Melliand Textile Reports, 1993, 4, pp. 278-282.
- Vonwiller, E., "Solid Weaving Technology", ATI, November 1998.

REVIEW QUESTIONS

1. How many different rapier insertion systems are there?
2. Draw the typical velocity and acceleration curves of a double flexible rapier head against the machine width.
3. Explain the driving mechanism of a flexible rapier.
4. How is the low tenacity yarn breakage frequency reduced in a rapier system?
5. What are the factors affecting the warp and filling stresses on a rapier machine? Explain.

11

Multiphase Weaving

Shuttle looms are considered to be the first generation of weaving machines. Shuttle loom development also had different stages such as hand loom, power (non-automatic) loom and automatic loom. In hand loom, every operation was performed manually. In some instances, the shedding was performed by foot operation. Power loom is a non-automatic loom on which supply in the shuttle is changed by hand. Automatic shuttle loom is the power operated machine on which the shuttles are changed automatically. All of the shuttle looms are single phase machines, i.e., each forms one shed at a time, then one filling yarn is placed in the fabric at a time (Chapter 7).

Upon reaching their capability limit for production rate, shuttle looms were replaced by "shuttleless" weaving machines, which are considered to be the second generation of weaving machines. Now, the filling insertion rate also has reached a stagnation point around 2000 m/min with the modern single phase weaving machines such as air-jet, projectile, rapier and water-jet. One reason for this stagnation is the fact that shedding, filling insertion and beat-up motions have to be done subsequently in single phase machines. This necessity to wait to insert one pick after another prevents the single phase machines to achieve significantly higher filling insertion rates than 2000 m/min. For shed formation and filling insertion, relatively large masses have to be moved fast in an oscillating pattern. The stress on the mechanisms involved and the strain on the yarn have increased continuously and in some areas it is now close to physical limits. Because of the intermittent weft insertion, the accelerations and braking of the weft yarn with every pick attained a critical value of almost 70 m/s (250 km/h) and is stressed almost to its tensile limit.

Further increases in production rates of woven fabrics (woven area/machine hour) require new technologies such as multiphase weaving. A multiphase weaving machine is one in which several phases of the working cycle take place at any instant such that several filling yarns can be inserted simultaneously. In these machines, more than one shed is formed at a time; therefore, they are also called multished weaving machines. This concept is drastically different than single phase machines in which each of the five functions takes place subsequently. Therefore, it is appropriate to consider the multiphase weaving machines as the third generation of weaving machines. Chapters 8–10 dealt with single phase shuttleless weaving machines. This chapter is about multiphase weaving machines.

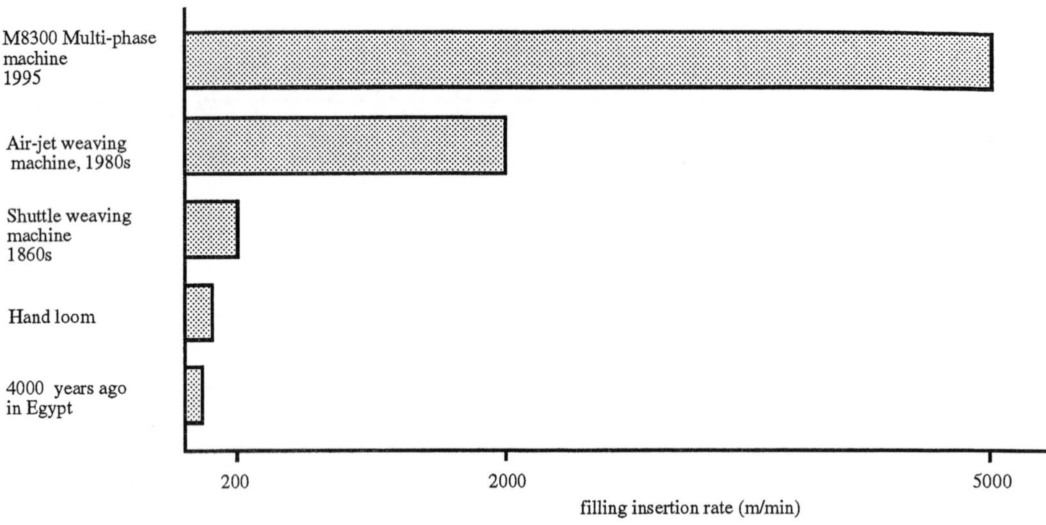

FIGURE 11.1 Major breakthroughs in filling insertion rates throughout the history.

11.1 M8300 MULTIPHASE WEAVING MACHINE

Within the last decade, SulzerTextil has developed a new multiphase weaving machine called M8300 Multi-Linear Shed Weaving Machine. M8300 is a multiphase air-jet weaving machine in which 4 picks are inserted simultaneously. It has a filling insertion rate of over 5000 m/min compared to 2000 m/min of single phase air-jet weaving machines (Figure 11.1). At ITMA'99 in Paris, SulzerTextil exhibited a show speed of 3230 ppm, which corresponds to 6088 m/min of filling insertion rate. Currently, the commercial speed is around 2800 ppm.

The highest output among single phase machines is provided by air-jet. Unlike the other single phase machines, air-jet production rates have continued to increase at a very high rate in recent years. Today, a single phase air-jet weaving machine of 190 cm width typically weaves 23 m per machine hour. The M8300 technology surpasses by far the performance of any other contemporary weaving machines including single phase air-jets. For the same width, the M8300 technology produces 69 m of fabric per hour. As the weft is inserted continuously without interruption and at an even pull off speed of around 20–25 m/s (72–90 km/h), the stress on the yarn is significantly reduced.

The modern single phase weaving machines did not completely eliminate the use of heavy reciprocating parts. The M8300 offers the potential to achieve the advantages of almost completely rotary motion and consequently high productivity.

The M8300 produces standard fabrics that amount to 65% of all fabrics produced worldwide. The M8300 technology eliminates the insertion faults and utilizes the basic requirements for uniform fabric appearance. Figure 11.2 shows the SulzerTextil M8300 weaving machine.

Functional Principle of M8300

The M8300 is based on the multilinear shed principle as shown in Figure 11.3. Several sheds, arranged in warp direction, one behind the other and in parallel, are opened across the entire weaving width. A filling yarn is inserted into each of these open sheds simultaneously; as a result, four picks are inserted at the same time. If each filling is inserted at a speed of 1250 m/min, the filling insertion rate is 5000 m/min.

Shed Formation

The warp ends pass over the rotating weaving rotor; the sheds are formed consecutively on its

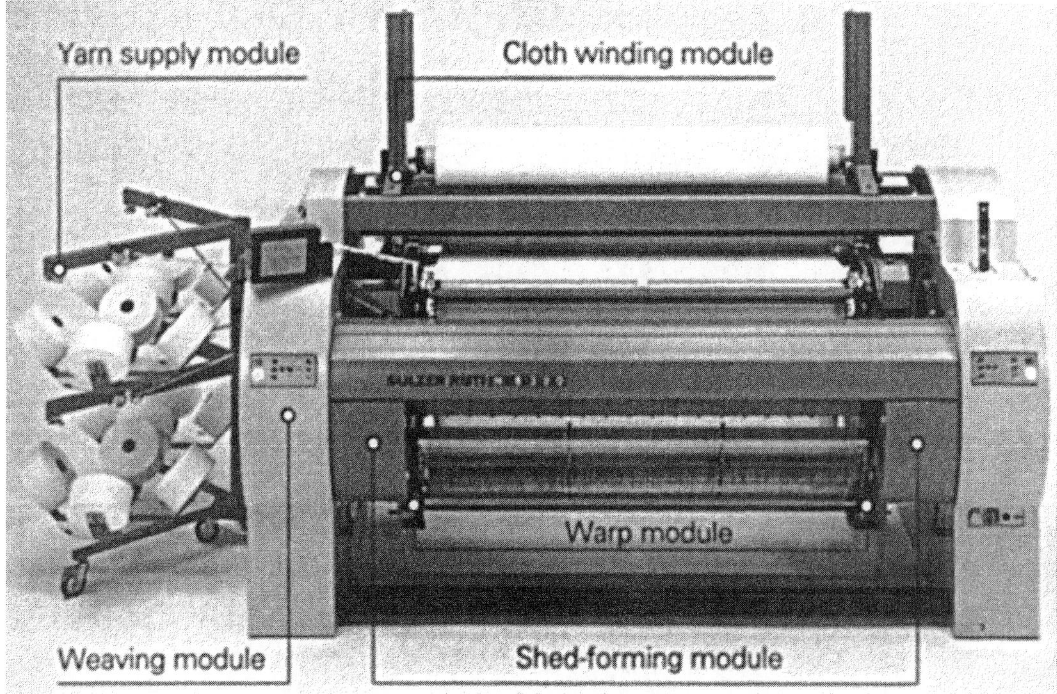

FIGURE 11.2 SulzerTextil M8300 weaving machine.

circumference by shed forming elements (Figure 11.4). The curvature and motion of the rotor causes the shed forming elements to open the sheds one after another. By minimal movements of the warp positioners (only a few millimeters) the warp ends are positioned such that they are either picked up by the shed forming elements and form the upper shed, or remain in the lower shed position (Figure 11.5). Each warp yarn is inserted individually into the eye of a warp positioner. The warp positioners are located close to the rotor and axially parallel to it. The number of warp positioners needed depends on the warp density; due to their extremely low mass and short stroke, they can move at very high speed. This is a key precondition for the performance potential of the M8300. Because the motion sequence of the warp positioners is controllable, the M8300 is capable of producing a variety of simple standard fabrics. Figure 11.6 shows the side view of the shed geometry.

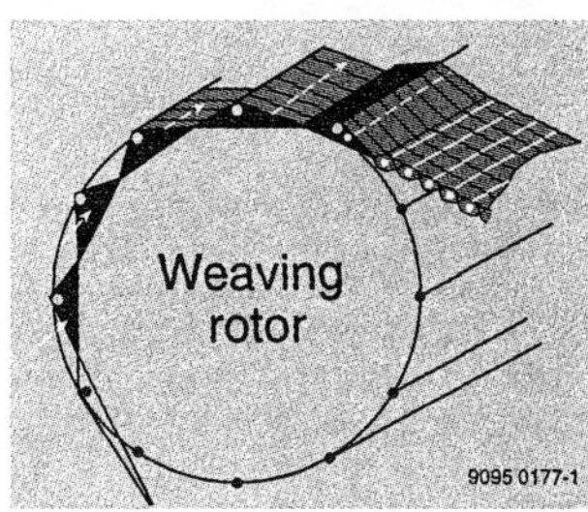

FIGURE 11.3 Multilinear shedding of M8300.

Filling Insertion

The second function of the shed forming elements on the weaving rotor is to form a filling channel (Figure 11.7). The filling is inserted into this channel by low pressure compressed air over the full width of the fabric (Figure 11.8). Relay nozzles are integrated in the shed-forming elements at regular intervals.

The four filling yarns are drawn from stationary bobbins by metering rollers, simultaneously and at a constant speed. This speed is a function of the filling insertion rate and the weaving width. It is controlled by the speed of the metering rollers. The yarn velocity is around 20 m/s at an insertion rate of 5000 m/min (Figure 11.9).

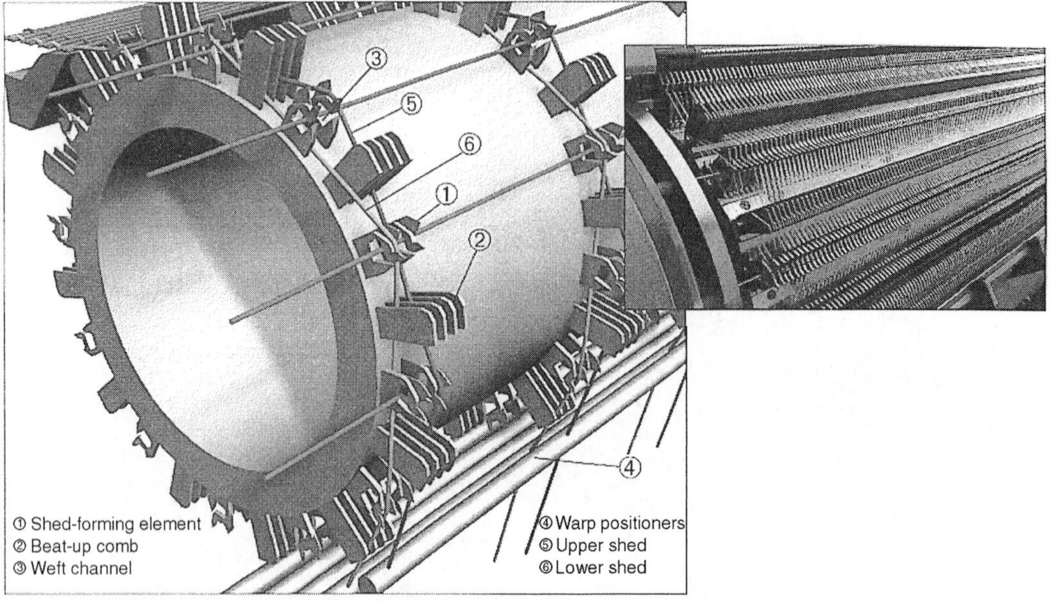

FIGURE 11.4 Shed forming elements on the M8300.

The simultaneous insertion of four fillings onto a moving rotor necessitates several technical innovations (Figure 11.10). The yarn feeder subunit is concentrically fitted to the rotor and takes on the function of yarn distribution for the constantly fed filling yarns. The filling processing unit (Figure 11.11) consists of two concentrically arranged disks that include a multichannel system. The yarns being transported with air leave the guiding channels in a synchronized sequence in order to enter the corresponding filling channel. The filling yarn controller also serves as the main nozzle. The yarn clamp and the cutter are located between filling processing and the filling channels. The clamp stops the yarn and the rotating main cutter cuts every inserted filling yarn.

During filling insertion, each yarn tip performs a screw motion as a result of longitudinal movement of the yarn and the constant rotation of the weaving rotor (Figure 11.12).

On the receiving side, arrival of the filling yarn is detected by an electronically controlled filling yarn detector. If a weft yarn is not inserted completely, the machine stops to facilitate the removal of the residual weft yarn. Additional necessary interventions to repair mispicks are signaled to the weaver via a terminal, whereby

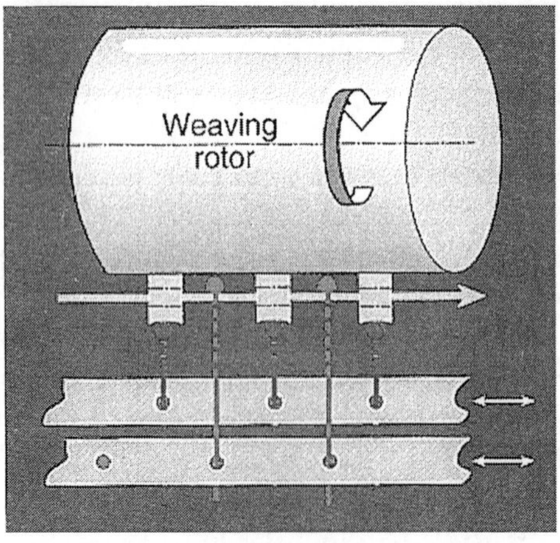

FIGURE 11.5 Warp yarn positioning.

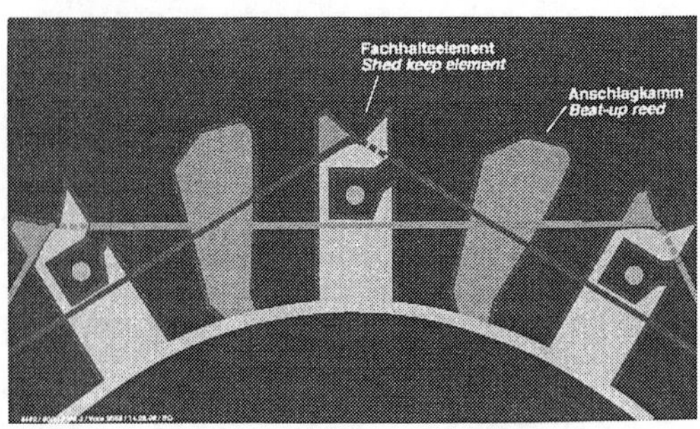

FIGURE 11.6 Shed geometry.

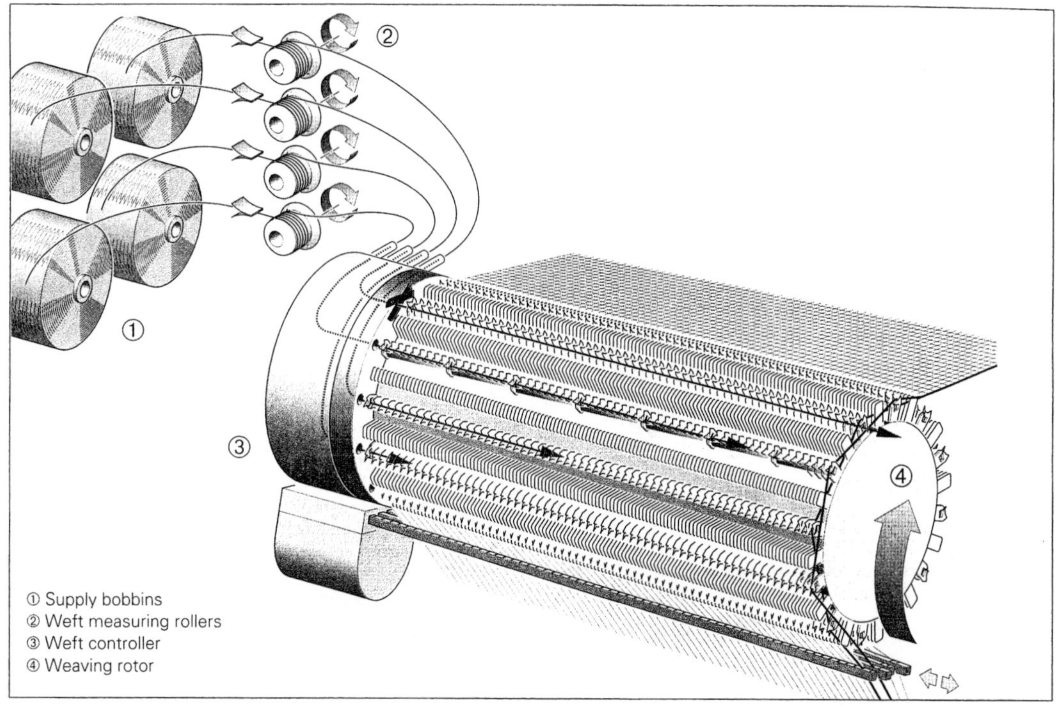

FIGURE 11.7 The filling insertion elements.

the movements necessary on the machine side are executed automatically. If yarn breakages occur in the area between the threading tube over the packages and the weft feeder, the machine repairs them automatically.

Filling Beat-up and Selvage Motion

Beat-up combs located on the weaving rotor between the rows of shed forming elements perform the function of the conventional reed (Figure 11.13). The lower shed, which rises after insertion of the filling, lifts the filling out of the filling channel over the entire weaving width. The beat-up comb behind the filling then catches up with it and beats it up. The selvages are critical for further processing of a fabric. At present, the selvages formed on the M8300 are of standard leno type. No auxiliary selvages are required, neither at the insertion nor at the receiving side.

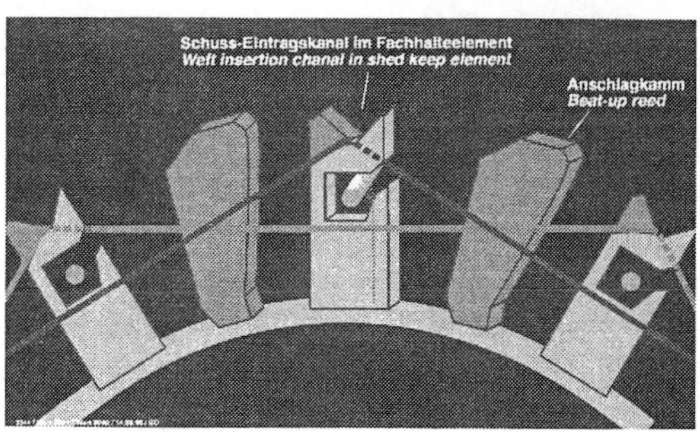

FIGURE 11.8 Side view of the rotor during filling insertion.

Filling and Warp Stress Levels

Compared to the conditions of single phase weaving, the stresses acting on warp and filling yarns are fundamentally different. The typical characteristics of the load dynamics are principally comparable to those of known weaving procedures. Differences in filling are mostly in stress level; the differences in warp are related to frictional stress, filling beat-up technique and fabric guidance. Compared to single phase weaving, the stress acting on the filling yarn is basically lower due to the lower yarn speeds (Figure 11.9).

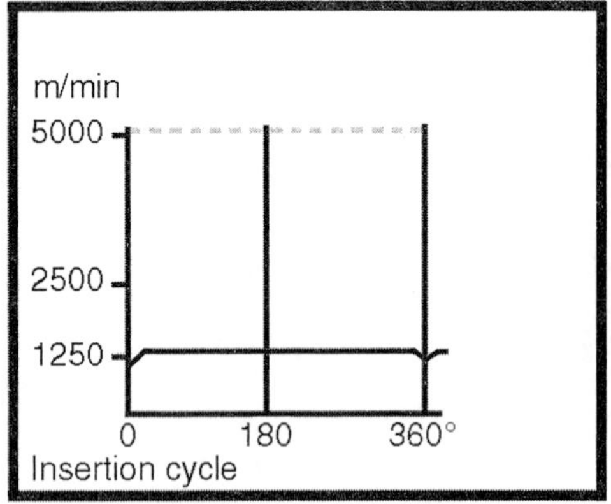

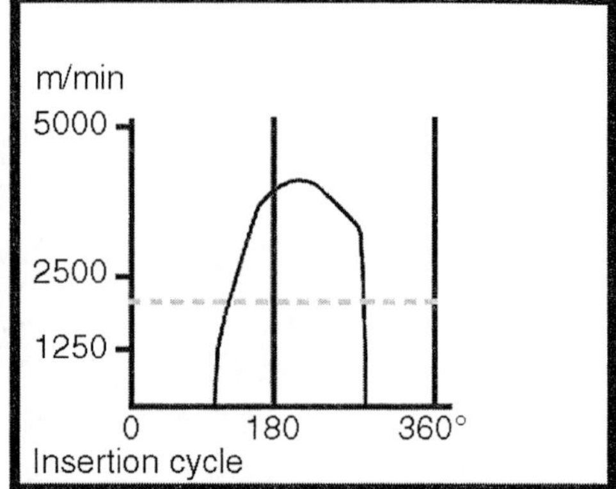

FIGURE 11.9 Comparison of filling insertion rate and filling speed of single- and multiphase weaving machines.

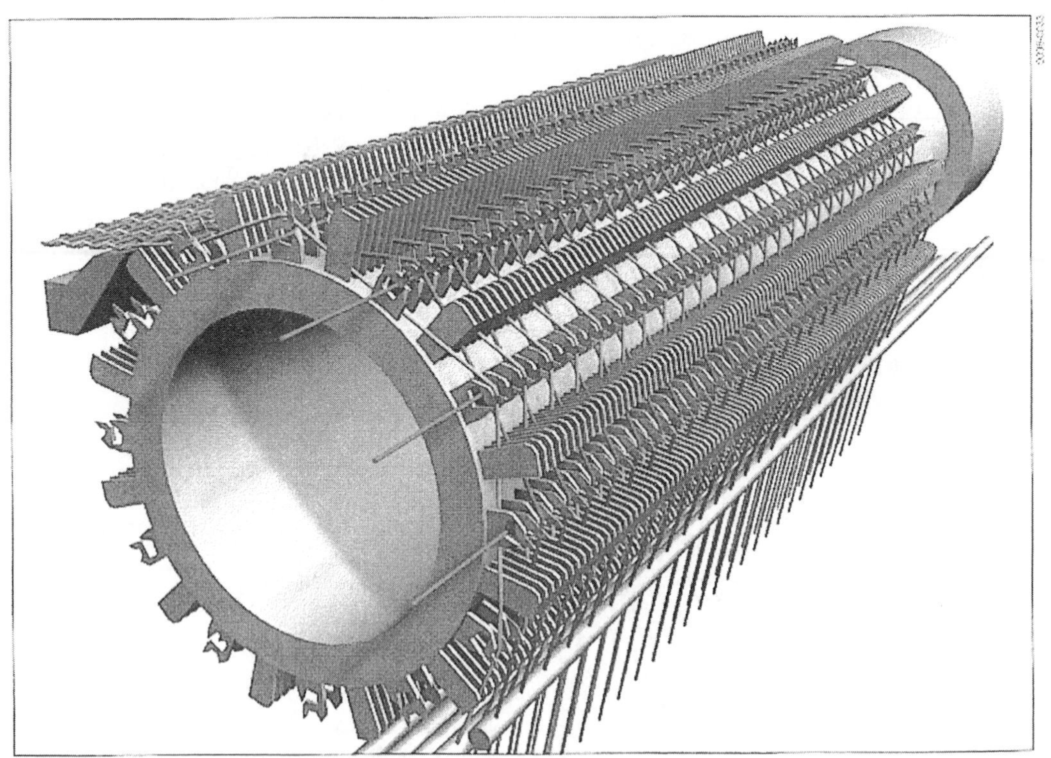

FIGURE 11.10 The fillings are inserted in staggered formation and guided to the fell by the motion of the rotor.

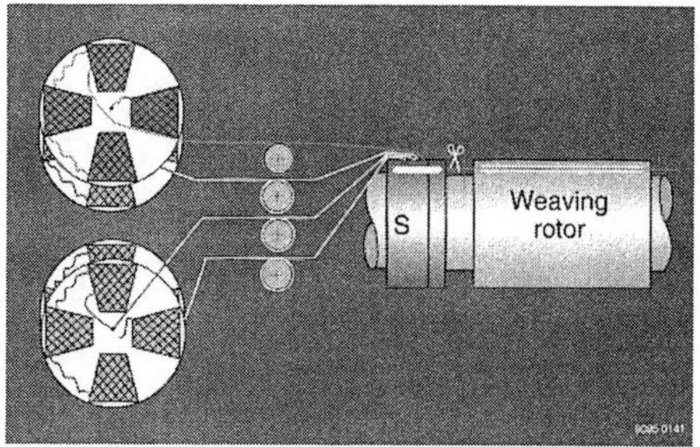

FIGURE 11.11 Filling processing unit.

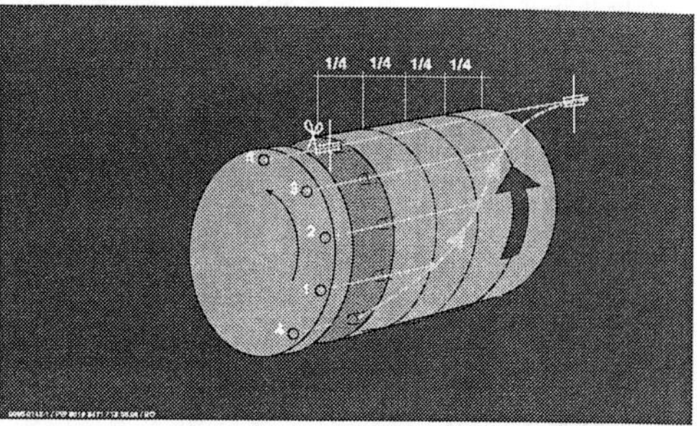

FIGURE 11.12 The path of filling yarn during insertion.

The highest filling insertion rates of 2000 m/min of the present single phase air-jet machines result in a speed of around 70 m/s. For an identical insertion rate of an M8300, speeds of only 10 m/s are required. For filling insertion rates of 5000 m/min, the yarn speed is around 20 m/s. Therefore, the maximum yarn stresses are accordingly reduced to only a fraction of the original value. Figure 11.14 shows the speed-dependent load curve of a typical yarn. According to the figure, if 40% of the output on an M8300 is run on a single phase air-jet weaving machine, the stress acting on the yarn in the air-jet would increase multifold.

Lower stress on the filling yarn potentially can reduce demands from the yarn which results in cost reductions. The frequency of yarn breaks is also reduced.

The stress on the warp yarn results from cyclic elongation caused by shed formation, filling beat-up and friction. Warp stress depends on fabric style as well. Compared to classical single phase weaving process, the shed lift and the shed dwell are reduced to less than 10%. However,

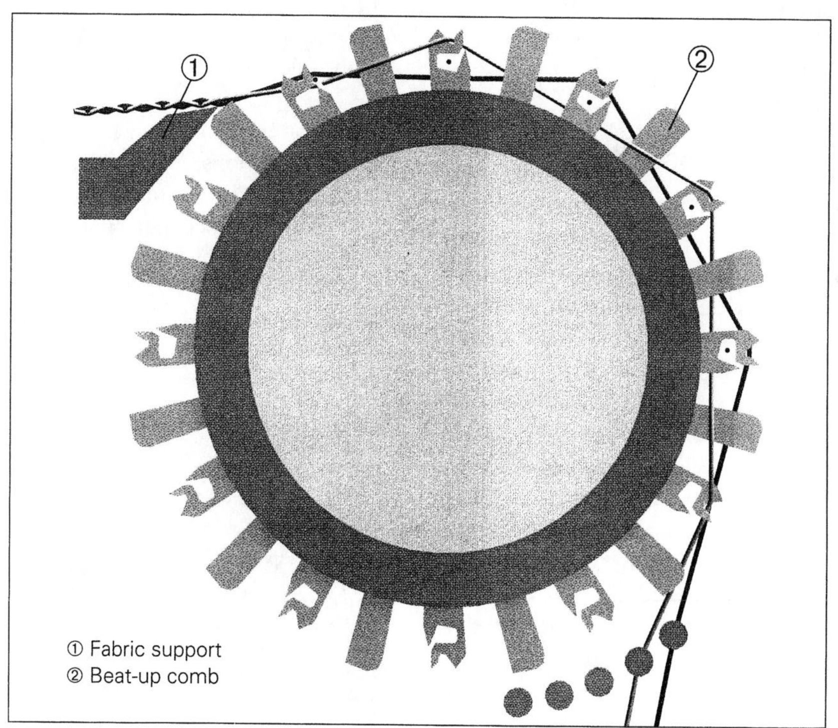

FIGURE 11.13 The filling is beaten up at the fell by the specially shaped beat-up combs.

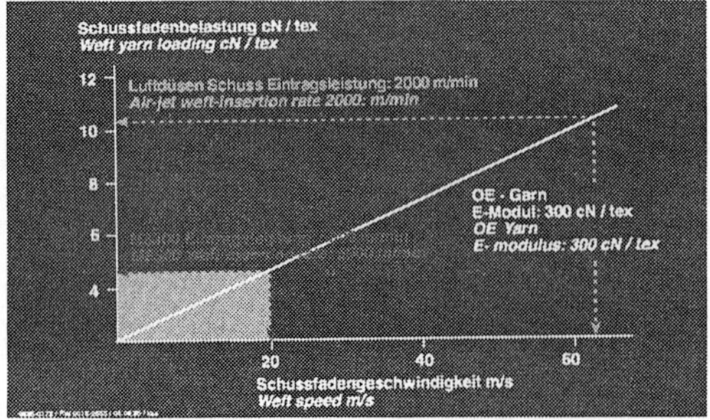

FIGURE 11.14 Comparison of stresses on filling.

the friction between yarn and metal has gained significance on account of the multished formation. The result is a reduction of the warp tensile force in the running direction of the warp, in the sequence of the sheds being formed on the circumference of the weaving rotor. Therefore, reduction of the friction coefficient becomes important in warp preparation.

Modular Machine Concept

The M8300 machine is comprised of function dependent modules (Figure 11.2):

- warp and shed forming module
- weaving module
- cloth winding module
- yarn supply and feeding module

The warp section consists of a frame which accommodates the warp beam, the rollers for guiding the warp sheet, the warp stop motion and the warp positioners. The main purpose of the division into weaving module and warp module is to allow fast warp changing (in under 45 minutes). With warp beams up to 1600 mm in diameter and winding of the fabric onto degressive batching motions, change intervals are reduced by a factor of 1:3. The weaving rotor is located in the bridge-like weaving module, which comprises a number of traversals and the machine covers. The central electronic and pneumatic control units, as well as the weft conveying module, are accommodated in the machine frame. Depending on the space available, the cloth winding module can be arranged above the machine (Figure 11.15). There are several possible cloth beam positions as shown in Figure 11.16. The filling yarn supply module comprises four large packages for each insertion unit.

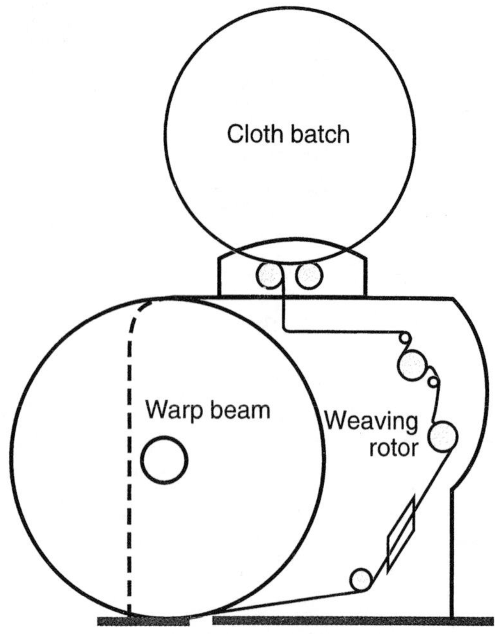

FIGURE 11.15 Flow of warp yarn in the M8300 machine.

Machine Drive and Control

Control of the M8300 is designed for support of the modular weaving system by consistent use of single motor technique. All motions are freely programmable with simple and user friendly functions. Figure 11.17 shows the actuating concept of the machine.

Instead of a mechanical guiding shaft, an electronic guiding shaft, without moving parts, is used. The guiding shaft position is transmitted to all drives via tact bus. Each drive has an electronically controlled coupling and an elec-

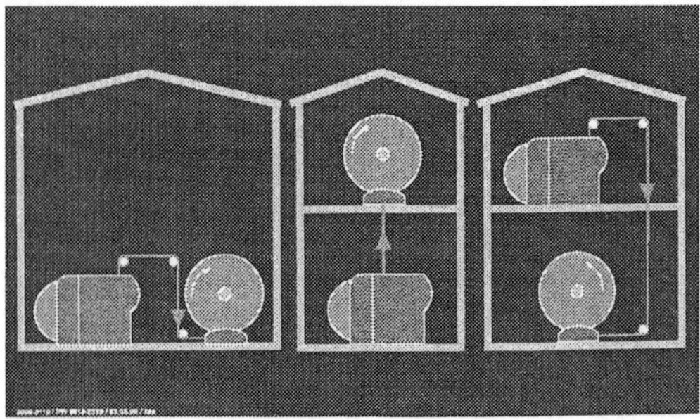

FIGURE 11.16 Possible cloth beam positions for M8300.

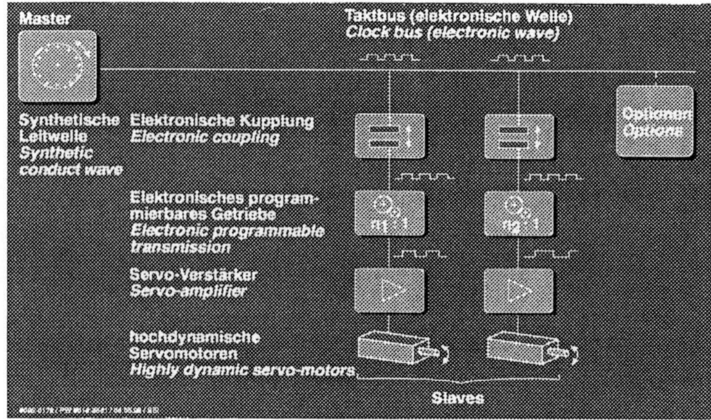

FIGURE 11.17 Actuation diagram of M8300.

tronically controlled gearbox with programmable transmission ratios. Each drive can be operated either synchronously or asynchronously in any direction of rotation. As a result, the spatial arrangement of drive shafts can be done freely. All motions can be programmed via software.

Other Operational Features of M8300

Due to small rotating masses and the weft insertion with low pressure air, the M8300 has a low consumption of energy per unit area of fabric woven. It is reported that with the M8300 machine, the energy required for the weaving of one meter of fabric is reduced by 30 to 40 %.

Similar to removal of the residual yarn, the fly and dust are removed by means of suction directly at the place of origin and routed to a central collection point for subsequent disposal. The heat resulting from friction and the main motor is guided to the return air of the air conditioning system. Table 11.1 shows some characteristic features of the M8300.

TABLE 11.1 Major characteristics of M8300.

Filling insertion rate	: up to 5,400 m/min; 2,800 picks/min
Compressed air pressure	: 3 bar
Weaving width	: 190 cm
Warp density	: up to 32 ends/cm
Filling density	: adapted to warp density
Yarn count range	: Ne 10 to 40
Selvage	: standard leno
Warp change duration	: 45 minutes
Warp beam diameter	: up to 1,600 mm
Fabric beam diameter	: up to 2,000 mm

M8300 has a large warp beam (160 cm in diameter) with approximately 18,300 meters of warp (the length depends on number of ends and yarn count). Large warp beam allows 250 hours of operation between beam replacements which is comparable to an air-jet weaving machine using a 100 cm beam.

It is reported that concerning the fabric properties, making-up of the fabric and fabric defects, there is no difference between fabric woven on M8300 and single phase machines as shown on Table 11.2.

Figure 11.18 compares M8300 and single phase machines for productivity and flexibility. The weave and style variety are the main restric-

TABLE 11.2 Comparison of fabrics made on M8300 and single phase machines.

Test		P7100	M8300
Tenacity (N)	K	320	317
	S	264	268
Elongation at break (%)	K	10.3	9.7
	S	22.2	22.3
Continued tearing strength	K	9.9	10.7
	S	6.4	8.0
Seam-slippage test (N)	K	113	118
	S	101	80
Shrinkage at washing (%)	K	−1.0	−1.5
	S	−1.2	−0.8

K : Warp
S : Weft
Shirting finishing

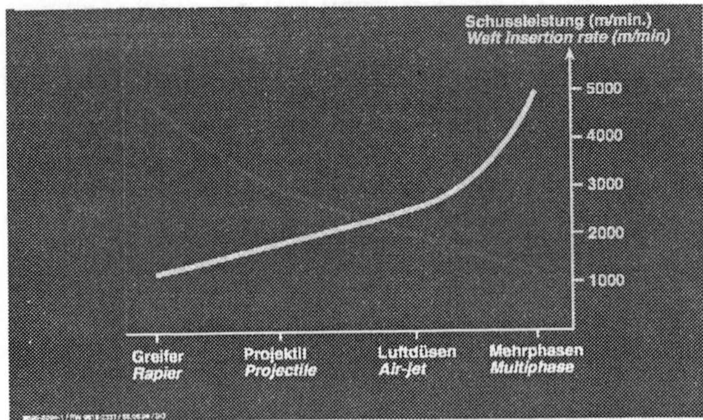

FIGURE 11.18 Productivity and flexibility of M8300 compared to single phase machines.

tions for M8300. In principle M8300 is designed for large runs due to the expenditure required to execute a style change.

Figure 11.19 shows the noise emission of M8300 and single phase machines. Compared to conventional single phase weaving machines, the noise is reduced by 10 dB(A). The noise reduction is possible due to rotating machine elements rather than reciprocating ones. Another consequence of this is the substantial reduction of the dynamic foot forces—up to 70% of the nominal value of a state-of-the art single phase air-jet weaving machine. Figure 11.20 shows the comparison of various systems for energy requirement per 100 m/min filling insertion rate.

Figure 11.21 shows a modern weaving plant with M8300 machines. In summary, the advantages of the M8300 multiphase weaving machine compared to high speed single phase weaving machines are:

- output in production of simple standard fabrics almost three times as high
- lower power consumption (approximately half)
- smaller space requirement (up to 60% depending on installation)
- lower dynamic stress on building (less expensive floor construction)
- less air conditioning needed and lower dust content in air due to air conditioning of working area
- easier to operate
- substantially lower noise level [sound pressure level reduced by 10 dB(A)]
- depending on fabric type and location, over 30% reduction in weaving costs

11.2 OTHER MULTIPHASE SYSTEMS

Prior to M8300, a series of multiphase weaving machines was developed within the last few decades (Figure 11.22). Although the first multiphase weaving patent was granted to Karl Mutter in 1926, it took 30 years to develop the first

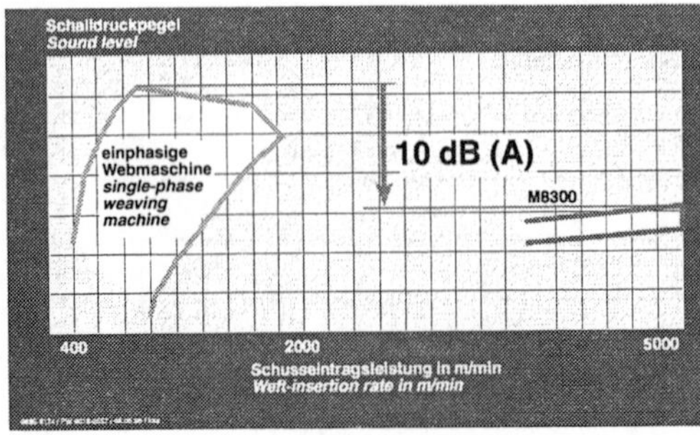

FIGURE 11.19 Noise emissions of M8300 and single phase machines.

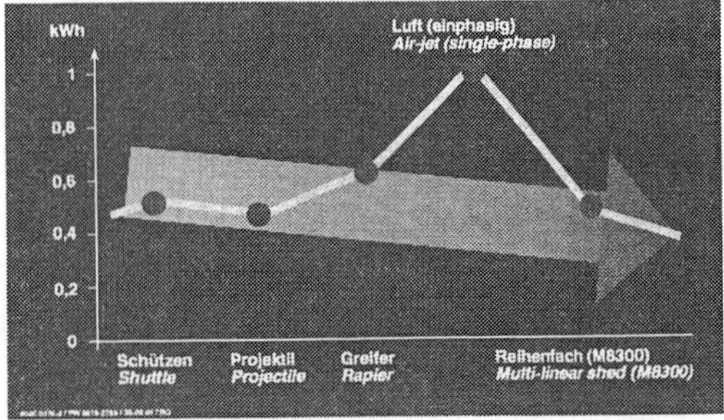

FIGURE 11.20 Energy consumption per 100 m/min filling insertion rate (Prof. Dr. Ing. Gerhard Brockel).

FIGURE 11.21 A modern weaving plant with M8300 machines.

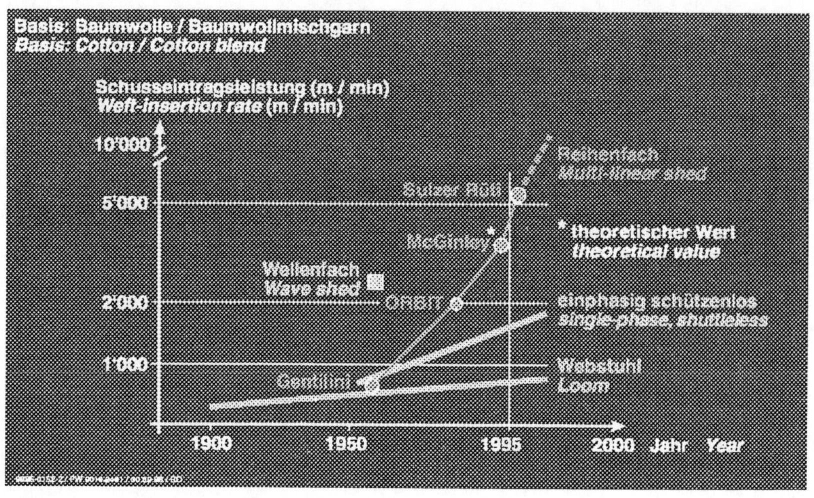

FIGURE 11.22 Development of multiphase weaving machines.

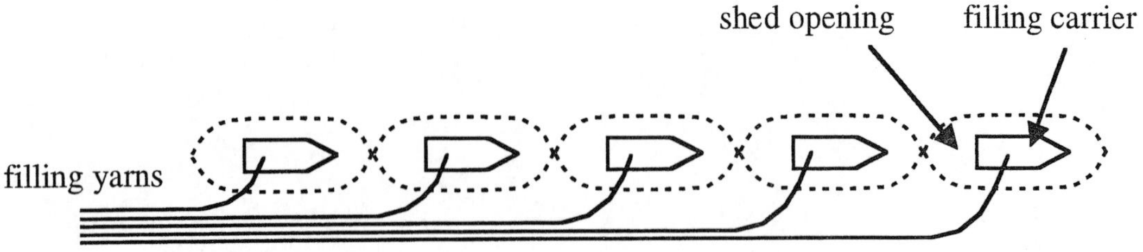

FIGURE 11.23 Schematic principle of filling direction shed wave system.

multiphase loom for industrial testing. Circular multiphase weaving machines have been in usage almost 25 years before the flat multiphase machines.

Ruti was the first loom maker to show a multiphase loom at ITMA 1971 in Paris. The Czechs, who had started work on a multiphase weaving machine in 1960, had the Kontis on display at ITMA 1975 in Milan, although they had previously shown it in Leningrad in 1972 and at ATME-I 1973 in Greenville, SC.

There were basically two principles of multiphase weaving machines developed: filling direction shed wave and warp direction shed wave.

11.2.1 Filling Direction Shed Wave Principle

In these machines, a number of sheds in the filling direction are opened subsequently for the insertion of the weft. Figure 11.23 shows the schematic of this principle. These sheds are arranged wave-like from one side to another, whereby a weft carrier slides into each shed. As a filling carrier enters one portion of the warp, a shed is formed; as the carrier leaves that area, the shed changes. This action may occur simultaneously across the width of the warp several times (Figure 11.24). As a result, at any moment, there are several shuttles (e.g., six) in the shed, each carrying a different filling yarn. This principle is also called the wave-shed principle. The machine may be rectilinear (flat) or circular.

It was claimed that the reduced speed when the filling was taken from the bobbins and the low and constant tension during the filling insertion into the shed allowed the use of standard and not particularly strong yarns. The warp, owing to the small shed opening and the type of beating up, does not require special features for the yarn quality and preparation. The filling yarn is beaten up by a rotating reed which places the ends between the discs as shown in Figure 11.25.

There has been more emphasis on filling direction shed wave principle compared to warp direction shed wave principle in the past. At ITMA 1975, four weaving machine manufacturers displayed their wave-shed multiphase weaving machines. Eight years later at ITMA 1983, there were still three manufacturers with wave shed machines.

With the exception of the circular weaving machine, none of the wave-shed weaving machines introduced to date has survived any longer than the industrial test phase. The main reasons for their failure were because mispicks

FIGURE 11.24 Simultaneous shed openings.

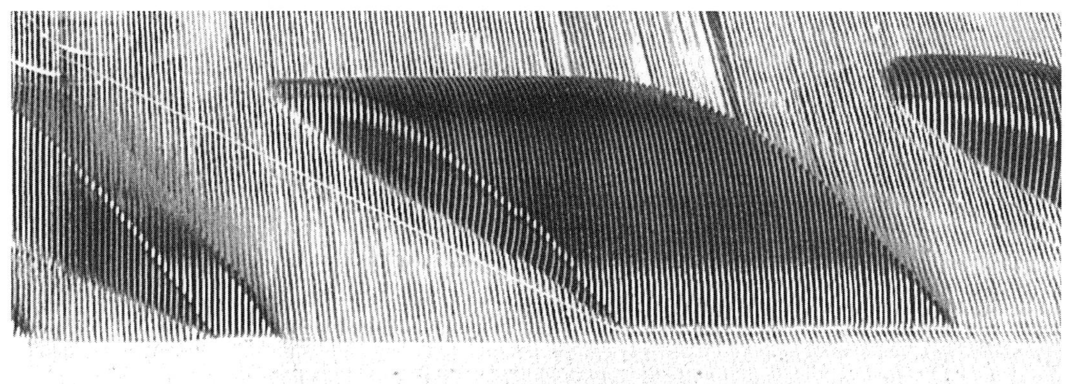

FIGURE 11.25 Rotating reed.

could not be easily repaired and the warp and filling yarns did not lay perfectly perpendicular.

Circular Weaving Machine

Although they have been in use for quite some time, circular weaving machines are not frequent in the textile industry. The main reason is the lack of flexibility in the fabric width and the narrow range of options. Only sacks and tubes are woven on circular weaving machines (Figure 11.26).

In a circular weaving machine, the warp is circular and there are continuously circulating shuttles running around the periphery in a wave or ripple shed. Circular fabrics are tubular fabrics of varying diameter without a lateral fold or edge. Tubular fabrics woven on flat weaving machines have a folded edge, as they are woven as double fabrics and are joined only at the selvages.

The weaving motion on a circular weaving machine is different than flat single phase or multiphase weaving machines. The peripheral ripple shed on circular weaving machines requires a continuous motion of the shuttle that cannot leave the shed. Shuttle drive can be mechanical or electromagnetic. There are several variations of mechanical shuttle drive as well. In an electromagnetic shuttle drive, there is no mechanical contact between the drive elements and the shuttle through the warp ends. The warp ends are, however, still located between the shuttle and the electromagnet with this type of drive. There are two types of electromagnetic shuttle drive. In the magnetic field type, the shuttle is moved by a migrating field. In the second type, the rotating electromagnet holds the shuttle firmly through the warp ends of the inner shed and carries it along as it rotates.

On circular weaving machines, a ripple shed is formed, i.e., each shuttle runs in its own shed. For this purpose, the warp is divided into segments which form sheds with small heddle frames or wires. Cams control the heddle frame or wire segments from inside via push rods. Usually only plain and twill weaves are possible.

The filling yarn is replenished through the outer shed and the warp ends. For this purpose, all heddle frame segments are moved into low shed position, leaving the shuttle accessible on the warp. An automatic shuttle change mechanism is used to remove the empty shuttle at a particular spot in the trajectory and a reserve shuttle is inserted.

The beat-up is performed by needle gears or by oscillating drop wires, rather than a reed. Needle gears reach through the warp ends, follow the circulating shuttle and press the filling yarn against the fell of the fabric. Oscillating drop wires reach through the warp ends after each shuttle pass and push the filling yarns against the fell of the fabric. They are pushed down again by the following shuttle.

11.2.2 Warp Direction Shed Wave Principle

In these machines, a number of sheds in the warp direction are opened at the same time for the insertion of the filling as shown in

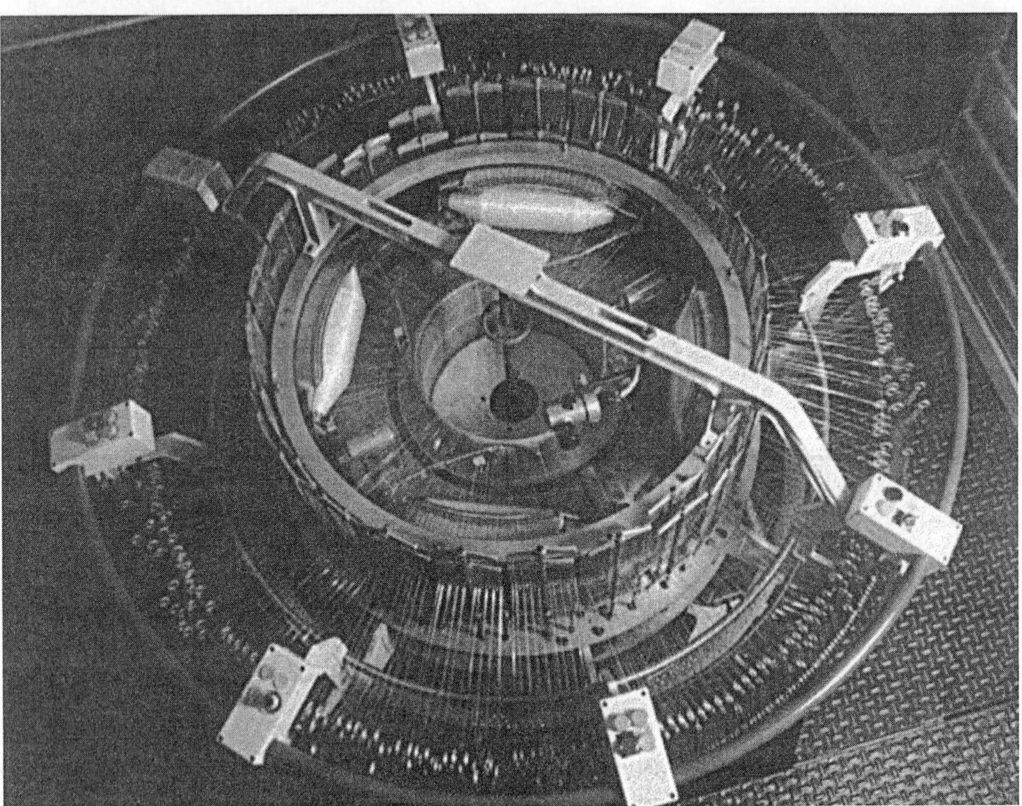

FIGURE 11.26 Circular weaving machine (photos courtesy of Saint-Freres).

Figure 11.27. These sheds are opened one after another in the warp direction; however, there are several sheds that are open at any given time. Each shed extends across the full width of the warp and moves in the warp direction. In this system, the weft yarn can be inserted in a manner similar to that of conventional weaving systems. This concept is also called

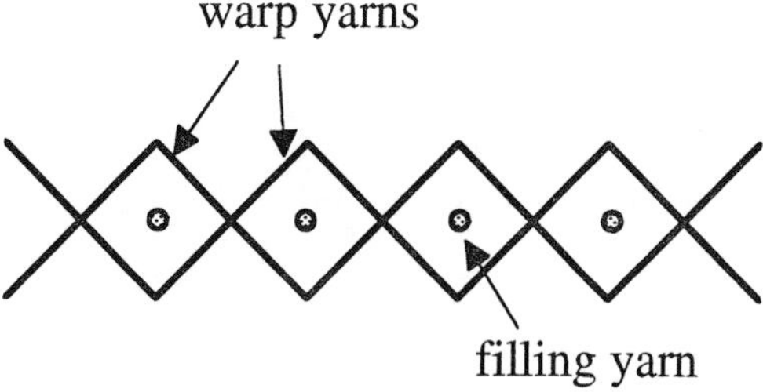

FIGURE 11.27 Schematic principle of warp direction shed wave system.

sequential-shed principle or multi-linear shed principle. The M8300 machine uses this concept.

Multi-linear shed weaving started in 1955 by Ripamonti and Gentilini which is shown in Figure 11.28. This machine had moving filling yarn feeder units with a rotating package frame and discontinuous yarn pull-off by gripper needles. At that time, 800 picks/min was thought to be feasible. Figure 11.29 shows a Sulzer patent, issued in early 1970s, which uses gripper shuttles.

In comparison with single phase air-jet weaving machines, none of these multiphase weaving machines achieved filling insertion rates that were higher, as a result, they were never commercialized. The main reasons for the failure of these machines are:

- the impossibility of repairing mispicks
- different filling yarn tensions, as a result of several filling yarn carriers being activated at the same time

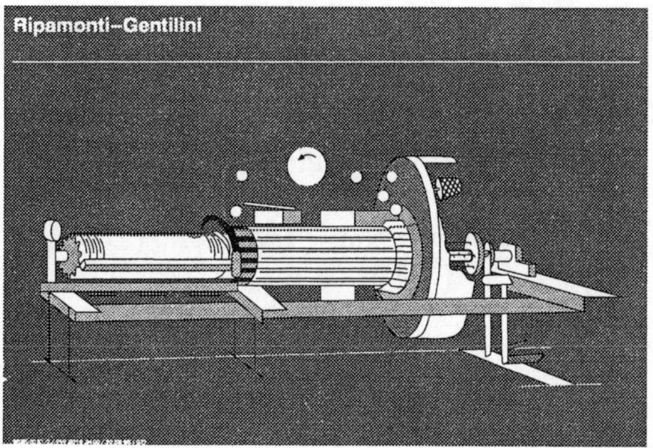

FIGURE 11.28 Schematic of Ripamonti-Gentilini machine (Prof. Dr. Ing. Gerhard Brockel).

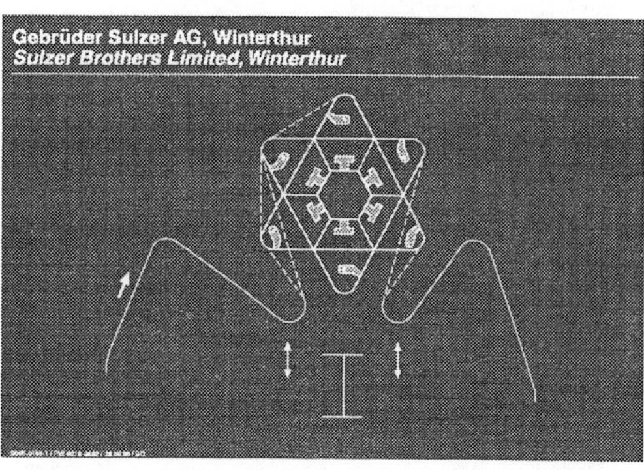

FIGURE 11.29 Filling insertion with gripper shuttles (Prof. Dr. Ing. Gerhard Brockel).

- dispensing with direct unwinding from the cross-wound package, with the disadvantage of an additional rewinding process, necessitated by the special making-up of the filling carrier
- restriction of purely mechanical filling yarn carrier systems with the disadvantage of output limitation
- impossibility to realize the required filling yarn beat-up to obtain uniform filling insertion across the entire weaving width [1–3].

Bentley produced the first commercially available multilinear shed weaving machine, the ORBIT 112 in late 1970s. Eighteen grippers were used for filling insertion. Weft insertion was mechanical and shed formation and filling feeder unit were rotational.

Thomas McGinley invented a multished weaving technology which is similar to the single phase air-jet weaving principle in many respects [4]. In the patented McGinley system, each shed formed by any conventional means is retained and moved forward to the beat-up point (cloth fell) while a filling yarn is inserted through it. Figure 11.30 shows the shed retaining system. The parts A, B, C and D are the support devices for the shed retainers. The circles above B and C are the cross sections of the closed tubes extending across the width of the machine. As the tubes B and C move toward the fabric, filling yarns are inserted through these tubes by air-jet nozzles. It is reported that insertion of the filling in nearly closed tubes reduces the amount of compressed air required for insertion by 80% of

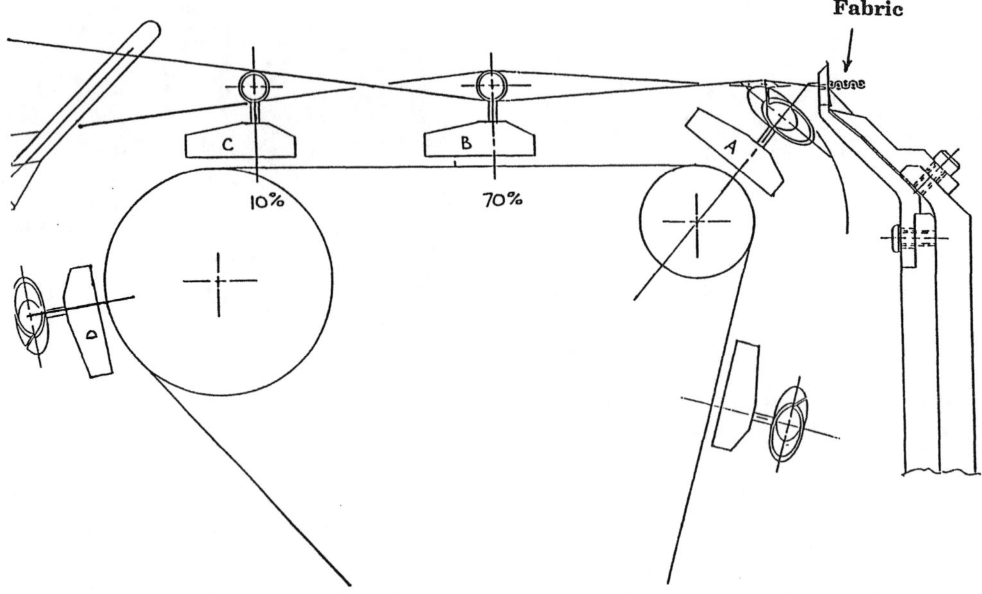

FIGURE 11.30 Shed retaining in McGinley multiphase system [4].

that used on single phase air-jet weaving machines. In Figure 11.30, the filling yarn in tube C has traveled 10% of the insertion width, and filling in tube B has traveled 70% of the insertion width. When the filling insertion is complete, the shed is released and a new type of beat-up mechanism pushes the filling into the fell. There are several sheds that exist simultaneously for inserting several filling yarns simultaneously.

The tubes through which the filling yarn is inserted consist of many small segments as shown in Figure 8.13 (Chapter 8). These segments (called gates) are individually supported by legs that extend from supports A, B, C and D as shown in Figure 11.30. These legs can rotate which in turn rotate the segments to the position shown in A and D in the figure. In this position, the segments have been rotated about 70° and they no longer form a closed tube. There is now an opening between each segment through which the warp yarns can pass. The inserted filling yarn inside the tube comes out of the tube with the help of warp yarns through a slot cut on the segments. Continuing its rotation, the tube comes out of the shed leaving the filling yarn inside the shed. Then a specially designed and patented two part beat-up mechanism beats up the filling yarn to the fabric fell.

In the McGinley system, all the functions of shedding, filling insertion and beat-up can take place simultaneously without interfering with other mechanisms. However, there are several issues to resolve so this system has not been commercialized.

REFERENCES

1. Cahill, N., "Produce Commodity Fabrics Competitively", Textile World, October 1996.
2. Perner, W. "Development, Present State and Future Prospects of Multiphase Weaving", Textil-Praxis International, Oct. 1989.
3. Offermann, P. and Reichrert, B., "Requirements and Possible Approaches for the Realization of Individual Warp End Activation in Multiphase Weaving", Textil-Praxis International, April 1993.
4. McGinley, T. F., USP 4,122,871 and 4,122,872.

SUGGESTED READING

- Steiner, A., "Breakthrough to the Next Magnitude?", Sulzer Technical Review 4/95.

- International Textile Bulletin, Fabric Forming, 3/85.
- Ormerod, A. and Sondhelm, W. S., Weaving: Technology and Operations, 1987.
- Hatch, K. L., Textile Science, West Publishing Company, 1993.
- Jungmichel, H., "Some Experiences with Multiphase Weaving", Textil-Praxis International, April 1987.
- Ellis, P., "Multiphase Weaving Goes Commercial", Textile Industries, Dec. 1977.
- Paton, D., "Multiphase Weaving Developments: Patent-based Review, Part 2", Textile Month, Dec. 1978.
- Paton, D., "Multiphase Weaving Developments: Patent-based Review, Part 2 (continued)", Textile Month, Jan. 1979.
- Oresic, F., "What Can Multiphase Weaving Offer?", Textil-Praxis International, Sept. 1983.
- Perner, H. et al, "Weft Waste in Multiphase Weaving", Textile-Praxis International, Aug. 1991.
- Perner, H. et al, "Factors Affecting the Wastage of Weft in Multiphase Weaving", Textil-Praxis International, Nov. 1991.
- "Multiphase Weaving Machine by Nuovo Pignone", Chemiefasern/Textifind (English Edition), Nov. 1982.
- Ellis, P., "ATME-I 1988 Survey 5—Projectile, Rapier and Multiphase Weaving", Textile Asia, Feb. 1982.

REVIEW QUESTIONS

1. Do you believe that single phase weaving machines have reached their physical limits? Explain why or why not.
2. Over the centuries, the basic mechanisms of weaving have not changed much in principle. Do you think that will continue to be the case in the future?
3. What were the reasons for the failure of multiphase weaving machines prior to M8300 (except the circular machine)?
4. In your opinion, what are the potential areas of concern in the McGinley concept, if any?
5. How did the increase in filling insertion rates affect the social and economical balances throughout the history? Explain why.

12

Manufacturing of Special Fabrics

Certain fabrics require special manufacturing procedures. In this chapter, the requirements and practices for manufacturing of several types of fabrics are discussed. The fabric producer has practically unlimited choices when designing and manufacturing a fabric. Table 12.1 lists the major fiber, yarn and fabric properties that can be changed in a woven structure.

12.1 DENIM

In the nineteenth century, in the Rhône Valley region of France and along the Italian Riviera, heavy cotton fabrics were produced which were known as "tissu de Nimes" and "Bleu de Genes". The modern names "denim" and "jeans" derive from these designations. The first jeans, made of heavy cotton fabric and intended as working clothes, were produced by Levi Strauss in the United States, in the 1870s.

Typical denim fabrics are woven from coarse, indigo-dyed cotton yarn. They are hard-wearing, high density fabrics with a high mass per unit area and a 3/1 twill weave. Today, more denim is produced in the world than any other type of cloth. Denim fabrics are made for a variety of applications and in a wide range of qualities and shades. They are practically standard fabrics, yet they require a considerable amount of special know-how at every stage of production.

Even today, classic denim is still dyed with indigo (produced artificially since 1897). It is a special process in which only the surface of the warp yarn is dyed; the core stays white. This is why the garment subsequently develops the typical—and desired—signs of wear.

Demand for fashion variants of classic denim will continue to grow. The most popular variants are:

- stone-washed and double stone-washed denims
- chambrays
- fancy multicolor denims
- denims with metal-effect yarns
- elastic denims
- printed denims
- jacquard-patterned denims
- denims with fancy yarns

Lightweight chambrays are used for shirts and blouses. Heavy, classic denims are made up into trousers or coats. Besides classic indigo blue, the fabric is dyed in other fashion shades and colors, the most popular being black denim. The fabrics are graded in clearly defined classes by weight, e.g. light denim 10–12 oz/sq. yd, heavy denim 14–16 oz/sq. yd.

To produce good quality denim, the conditions have to be optimal regarding the quality of all the raw materials and yarn used. For raw cotton and the carded OE (rotor) or ring spun

TABLE 12.1 Parameters that can be chosen/changed to control woven fabric structure.

Fiber	Yarn	Fabric
• coloration	• type (single, ply, spun, filament, flat, textured)	• filling number per unit length
• crimp	• spinning method (ring, open-end, air-jet, friction)	• warp number per unit length
• cross-sectional shape	• fiber content (pure or blend)	• order of interlacing
• density	• ply	• performance properties
• luster	• twist level and direction	abrasion resistance
• type (staple, filament, etc.)	• coloration	absorbency
• dye affinity	• hairiness	chemical resistance
• strength and elongation	• bulk	colorfastness
• fineness	• linear density	drape
• finish	• strength	durability
• generia	• elongation	extensibility
• grade	• abrasion resistance	flame resistance/retardance
• length and length distribution	• stiffness	mildew resistance
• moisture content	• properties due to heatset	moth resistance
• resilience		oil/grease resistance
• shrinkage		pilling resistance
• shrinkage		resilience
• solubility		shape retention
• stiffness		shrink resistance
		softness (hand)
		soil resistance
		strength
		stretch resistance
		weather resistance
		wrinkle resistance
		• weight

yarns made from it, the quality criteria are as follows:

- minimum staple length: 2.7 cm
- proportion of short fibers (less than 12 mm long): under 40%
- micronaire value: 4.0–4.5
- the Uster values for strength and elongation, the evenness CV and imperfections must conform at least to the 25% plot
- the usual count range of denim warp yarns is 50 to 90 tex and of filling yarns is 75 to 120 tex; finer yarns as fine as 25 tex in twill or plain weave are often used in denim shirts
- twist factor: 4.5 to 5.0 for warp yarns, 4.2 for filling yarns
- low yarn hairiness
- yarn strength and uniformity

In the early 1990s, most of the yarns used in denim production were OE yarns. Recently, there is a strong trend towards using more carded ring spun yarns in both warp and weft. They give the fabric a softer handle, fulfilling the requirements for "soft denims". Table 12.2 shows the acceptable values for ring spun and OE cotton yarns with a fineness of 84 tex (7 Ne) for successful denim production.

12.1.1 Warp Preparation

In denim production, warp preparation, dyeing and sizing are crucially important. Dyeing with indigo requires detailed knowledge of the physical and chemical processes involved. The various dyeing and sizing methods, and the corresponding recipes and concentrations, reaction and oxidation times not only influence the weaving process; they are also largely responsible for the appearance and quality of the denim.

Besides the classic indigo rope dyeing process, indigo sheet dyeing is also used. One variant of sheet dyeing is loop dyeing. Dyeing and sizing are either done separately, as in rope and double sheet dyeing, or the two processes are combined in a single operation, as in sheet dyeing and loop dyeing.

TABLE 12.2 *Typical properties of cotton yarns of 84 tex for denim production.*

	Ring-spun Yarn	OE Rotor Yarn
Strength (cN/tex)	17	13
Elongation (%)	8.5	8.5
Uniformity (CV%)	12	12.5
Thin places (/1000 m)	1	1
Slubs (/1000 m)	50	20
Neps (/1000 m)	50	3

The dyeing process necessitates continuous working. To produce denim economically the dyeing range must be optimally exploited because it represents a heavy investment. For this reason, it must produce 24 hours a day, seven days a week. Its output therefore dictates the optimum size of the weaving facility.

Indigo Rope Dyeing

In warp preparation for rope dyeing, 350–400 warp yarns are assembled on the ball warper to form a rope 10,000 to 15,000 meters long. Between 12 and 36 ropes are drawn through the dyeing range side by side. After dyeing they are dried on a drum drier and deposited in cans. The ends are spread out on a rebeamer or long chain beamer, and the yarn sheet is wound onto warp beams. These beams then come to the sizing machine, where they are sized, dried and assembled in accordance with the total number of ends required to make up the weaving warp beam. This process ensures optimal dyeing of the indigo, but with the disadvantages that broken ends are more frequent and yarn tensions are not always compensated. Modern indigo dyeing ranges normally operate with six dye vats. Using state-of-the-art methods, the same dyeing quality can be achieved with just three vats, with a substantial reduction in the consumption of chemicals.

Indigo Sheet Dyeing and Double Sheet Dyeing

In indigo sheet dyeing, warp beams are brought to the dyeing and sizing range instead of ropes. Dyeing, drying, sizing and after repeated drying, assembly of the warp is carried out in a single operation. In the case of double sheet dyeing, dyeing and sizing are done in two separate steps. As far as indigo dyeing is concerned neither process is always ideal, but both have the advantage that the number of broken ends is low and yarn tension can be extremely well controlled.

Indigo Loop Dyeing

In the loop dyeing process, the yarn is dyed in a single bath instead of several. The desired depth of color is attained by passing the yarn through the vat several times. Subsequently, as part of the same process, the yarn is sized. The advantages and disadvantages of loop dyeing are the same as with sheet dyeing.

Sizing Recipes

The sizing process plays a key role in further processing of the warps. The choice of sizing agents used to achieve the sizing effect is crucially important here. A relatively soft handle can be achieved using a combination of modified starch with a polyacrylate. Warps sized in this way have good running properties, helping to ensure high efficiency and an excellent final appearance. A practical example is given below:

Style data:

Warp: OE yarn, 84 tex (7 Ne), 24 ends/cm
Weft: OE yarn, 100 tex (6 Ne), 16 ends/cm
Total no. of warp ends: 3942

Size recipe for 100 liters of liquor:

- 8.0 kg modified starch
- 4.0 kg acrylate size
- 0.2 kg textile wax
- size concentration: 8.0 %
- size temperature: 85°C
- squeezing pressure: approx. 15 kN
- size pick up: 9-10%

Advantages of the denim size recipe:

Sizing room:

- easy separation of ends
- no hard size

- only slight bleeding
- no color masking
- minimal dust and fiber fly

Weaving room:

- high, constant weaving efficiency
- easy warp take-up
- minimal dust and fiber fly

12.1.2 Weaving

For finished widths of 150 to 156 cm, reed widths of 160 to 167 cm are required. Denims that are stretchable in the weft do not follow this rule. When weaving with projectile weaving machines, these fabrics are generally woven in two panels with weaving widths of 360 or 390 cm. In this way a high weft insertion rate of 1400 m/min is achieved at a machine speed of approximately 400 picks/min. For the production of heavy denims, leno selvages are advisable. Lighter denims, weighing less than 13 oz/sq. yd, can be woven with tucked selvages. The fabrics can be inspected with an electronically controlled cloth inspection machine.

The current demand on soft denims in the United States is to use lyocell yarns (e.g. Tencel®) in the weft. For structured denims, structured yarns are used for the filling. Slub and knob effects are especially popular.

Denim will continue to be in demand; but, it will come increasingly under the influence of fashion. The technologies involved in weaving of denim will also evolve. For example, the basic technology is already available for continuous dyeing of mixed fiber yarns made from cotton and polyester spun fibers in the weft, with suitable dyes. This could be the basis for producing other articles, e.g. working clothes, by the same process. A flexible weaving system, compliant with future requirements, thus represents the best possible way of keeping up with trends in denim fashions.

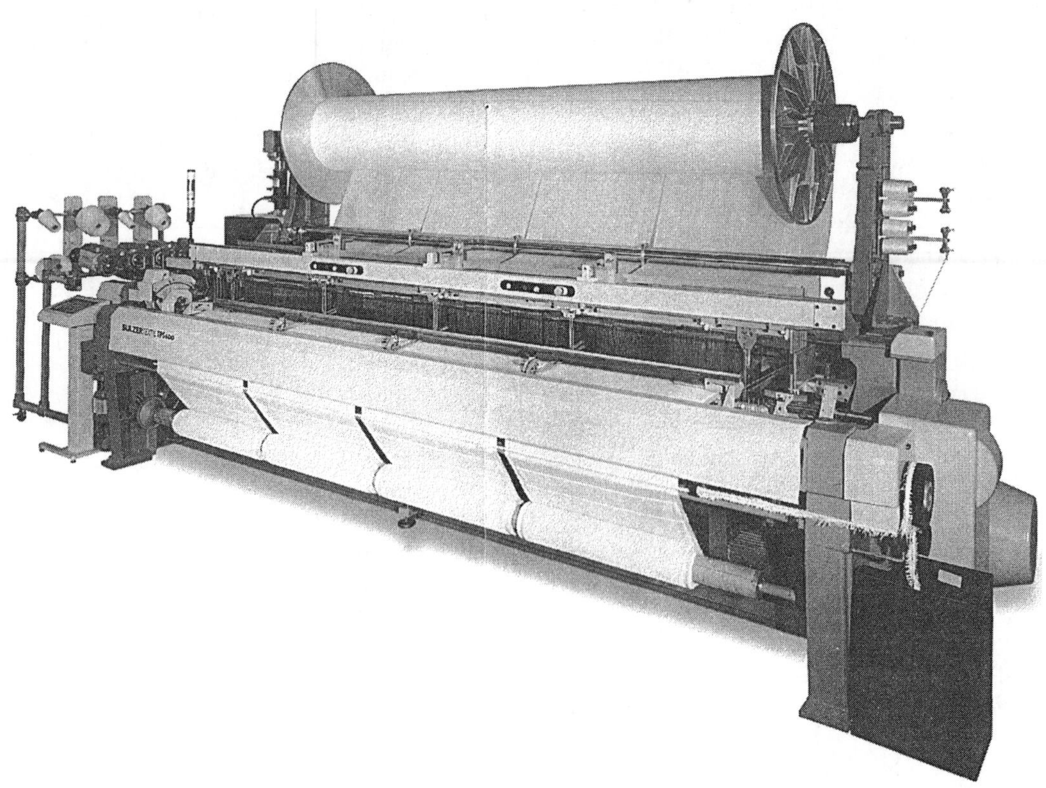

FIGURE 12.1 Rapier terry weaving machine with two warp beams.

12.2 TERRY FABRICS

Articles made of terry fabrics are used daily in many areas. They have loops at least on one side, and usually on both sides. The production of terry fabrics is a complex process and is only possible on specially equipped weaving machines. Today, terry fabrics with exclusive patterns can be produced on high speed weaving machines.

Some 2.5 billion square meters of terry are produced in the world every year. On average the weight of the fabric is between 270 g/m^2 to 600 g/m^2. The most important requirements in terry weaving are:

- low capital investment but high productivity
- impeccable fabric quality
- low spare parts consumption
- easy operation, with few mechanical settings needed
- short style changing times

There are basically two qualities of terry, according to the loop structure:

a) Classic terry, with upright loops (made of twisted yarn). These terries are usually patterned with dyed yarns
b) Fashion terry, also known as milled or fulled goods, with spiral loops (of single yarns). These are mainly piece dyed or printed fabrics.

A blend of single and twisted yarns produces additional pattern effects.

In the production of terry fabrics, two warps are processed simultaneously: the ground warp, with tightly tensioned ends and the pile warp, with lightly tensioned ends. A special weaving method enables loops to be performed with the lightly tensioned warp ends on the fabric surface. Figure 12.1 shows a rapier terry weaving machine with two warp beams.

In traditional terry weaving, by means of a special device on the weaving machine, two picks are inserted at a variable distance—the loose pick distance—from the cloth fell. The loose pick distance is varied according to the desired loop height. When the third pick is beaten up, the reed pushes the pick group, on the tightly tensioned ground warps, towards the fell and the loose pile warp ends woven into the pick group are uprighted and form loops. Depending on the weave, loops are thus formed on one or both sides of the fabric. With the basic method, known as three-pick terry (Figure 12.2), three picks form a pick group. It is possible to have pile heights up to 10 mm. Base weaves are usually 1/1, 2/1, 2/2 and 3/1. Figure 12.3 shows the X-diagram of a typical terry design.

In general, the reed has two beat-up positions which do not impose alternative movements to the warp, fabric and various components of the weaving machine (Figure 5.26). The sley has a special mechanism built in which allows differ-

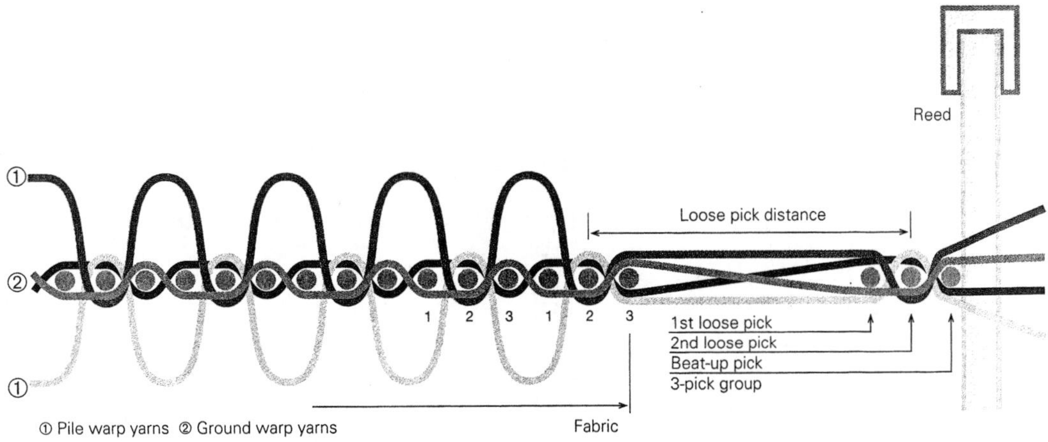

FIGURE 12.2 Structure and origin of a three-pick terry fabric.

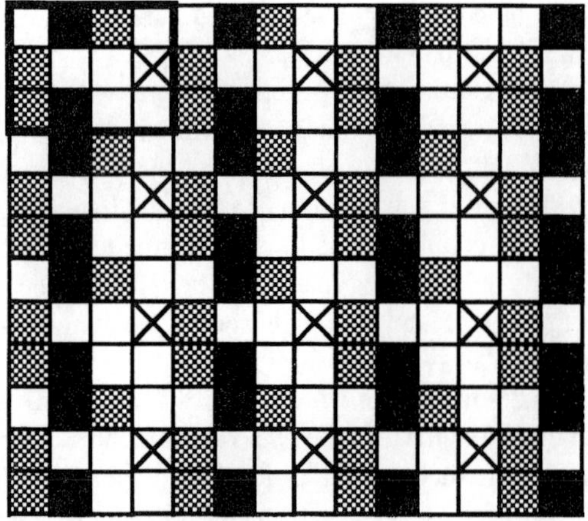

FIGURE 12.3 X diagram of a typical terry design.

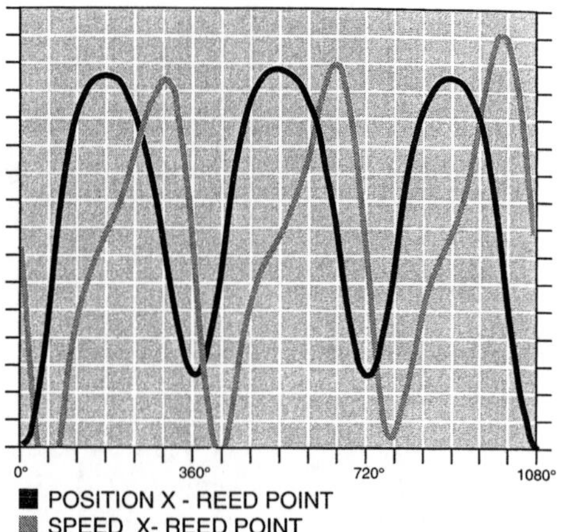

FIGURE 12.4 Position and speed of the reed of a rapier in terry weaving.

ent beat-up positions for pile formation. Figure 12.4 shows the position and speed of the reed of a rapier machine to produce terry cloth.

Tucked in selvage or leno selvage are commonly used. The warp is evenly let-off by a system of constant tension control from full to empty beam, this is controlled by a highly sensitive electronic device. Figure 12.5 shows a typical warp tension diagram in terry cloth weaving.

Air-jet machines are also successfully used in terry manufacturing. The individual components of an air-jet terry weaving machine are shown in Figure 12.6.

The individual inverter enables the speed to be regulated while the machine is running, e.g., when weaving borders. The tensions of the ground and pile warps are detected by force sensors and electronically regulated. In this way warp tension is kept uniform from full to the empty warp beam. To prevent starting marks or pulling back of the pile loops the pile warp tension can be reduced during machine standstill. This can be programmed at the operator touch screen terminal. An automatic increase in tension can be programmed for weaving borders.

The way the back rest roller system is controlled depends on the weave. During insertion of the loose picks and during border or plain weaving the warp tension between open and closed shed is compensated for by negative con-

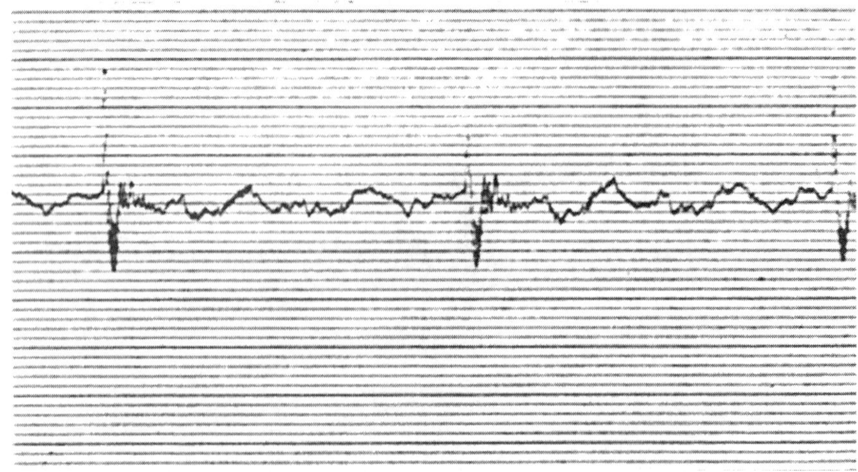

FIGURE 12.5 Warp tension diagram in terry weaving.

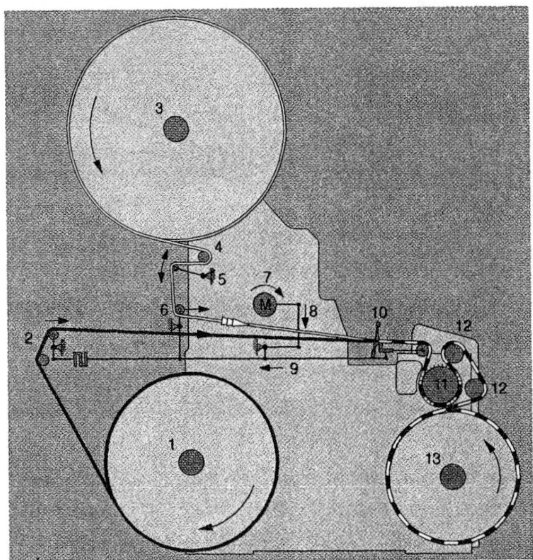

1	Ground warp beam	6	Pile guide roller
2	Warp tensioner	7	Servo motor
3	Pile warp beam	8+9	Lever system
4	Guide roller for terry motion	10	Fabric table
5	Tension compensating roll	11	Take-up roller
		12	Nip rollers
		13	Cloth beam

FIGURE 12.6 Components of an air-jet terry weaving machine.

trol. A warp tensioner with torsion bar is used for the ground warp, and a special tension compensating roll for the pile warp. Low mass design is a distinctive feature shared by both systems.

During full beat-up the warp tensioner is moved forward positively via a lever system, thus avoiding an increase in warp tension. Tension in the pile warp is compensated for by positively controlled motion of the pile guide roller.

Multipile Heights

The dynamic pile height control allows the pile height to be altered while the machine is running. The desired pile height is programmed at the operator terminal and activated by corresponding control signals from the dobby or jacquard machine. The loose pick distance from one pick group to another can be adjusted from 0.1 to 18 mm in 0.1 mm steps. No gauges or tools are needed. If necessary, the loose pick distance between individual picks can be adjusted in 0.1 mm steps up to a maximum of 3 mm to prevent loops being flattened. It is possible to switch between three-, four- and five-pick terry while the machine is in operation. Automatic switching is freely programmable. These functions are performed by a quick response digitally controlled servo motor. The weaving table is driven on both sides by a rigid, solid construction. This results in a uniform pile height over the entire fabric width. The pile height setting is easily reproducible, as the data can be transferred to other machines by memory card. A mechanical pile height control is optionally available. The electronically controlled fabric take-up allows eight different filling densities to be set per terry cloth. Separate warp stop motions are provided for ground and pile warp. A warp stop motion with section indicator is available. It facilitates location of broken ends, particularly on wide weaving machines.

For complex patterns the weaving machine has to be equipped with a jacquard machine. For less demanding patterns a dobby is sufficient, and very simple, non-patterned fabrics can be woven with a cam motion.

Terry fabrics are often very complex, with different colored warp ends in combination with loop patterns. They are subject to changing fashions, the market is constantly demanding new qualities and designs. The rapid development of electronics, with microprocessor controls and highly dynamic stepping motors in combination with modern mechanisms, has enabled fabric designers to produce completely new patterns. Figure 12.7 shows the special terry sley gear with dynamic pile control. Via a servo motor, the beat-up position for each pick and, thus, the

FIGURE 12.7 The terry sley gear with dynamic drive.

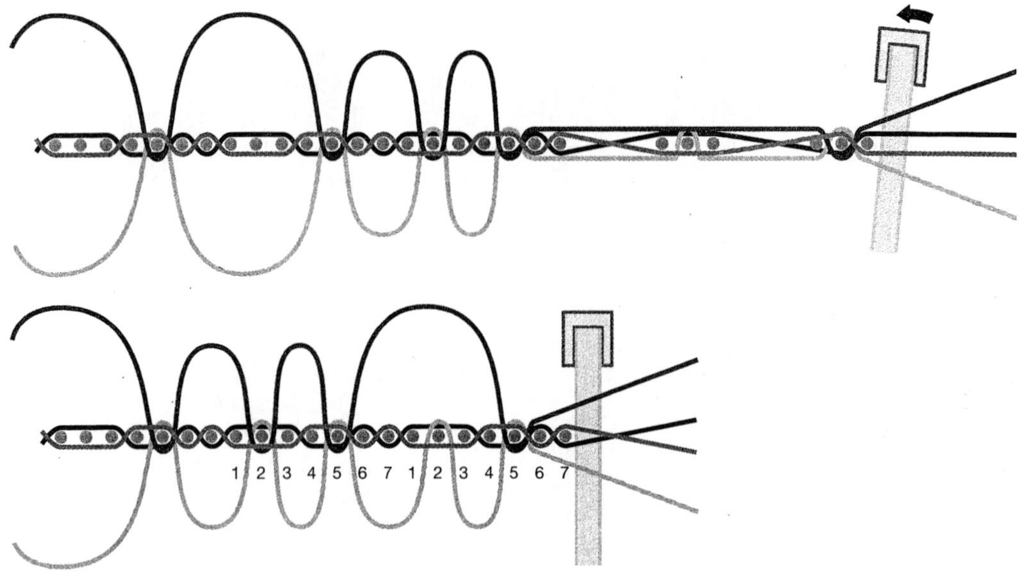

FIGURE 12.8 Special seven filling terry design with two-pick groups and full beat-up.

type of terry and the pile height can be freely programmed from one pick group to another. In this way, 200 different loose pick distances, and hence the same number of pile heights, can be programmed in any order desired. For example, three- and four-pick terry, and even fancy types of terry can be combined in the same length of fabric. This gives the fabric designer a broad range of patterning options, and the weaving engineer a technology for improving the fabric structure, because the transition from one pattern element to the next can be woven with greater precision.

With these elements, a new patterning method, called sculptured terry, has been developed. At each full beat-up, two pile loops of different heights are formed in the filling direction. The secret of this method of pattern formation lies in the fact that two loose pick groups formed at distances corresponding to the pile heights are beaten up to the cloth fell together.

For two short loops the pile yarns are woven into both loose pick groups and for one large loop into the second loose pick group only. The greatest difficulty was to develop a basic weave which resulted in neat loops without excessive friction between warp and filling at full beat-up. The solution was found in a special seven pick weave combined with full beat up at the sixth and seventh pick (Figure 12.8). In this way, a second pile height is also formed in filling direction, making sculptured patterning possible by the difference in pile height in warp and filling direction (Figure 12.9).

A precondition for this kind of pattern formation is freely programmable sley travel on a rapier weaving machine. Microprocessor control allows the loose pick distance to be programmed easily and individually for each pick.

FIGURE 12.9 Terry fabric with variable pile heights.

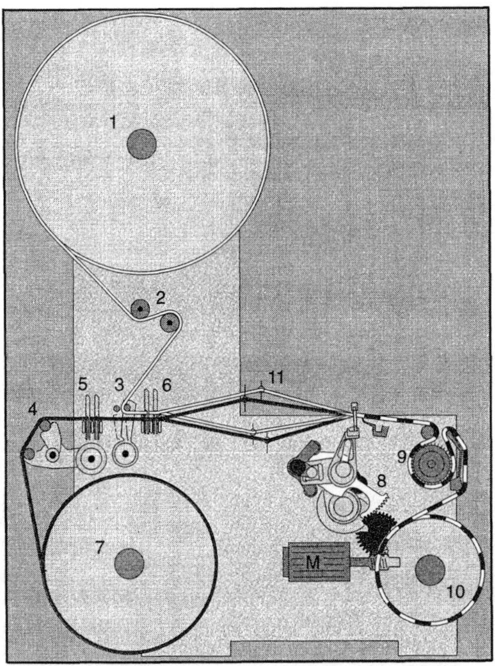

1 Pile warp beam
2 Guide roller
3 Pile tensioning beam
4 Ground warp tensioning
5 Ground warp stop motion
6 Pile warp stop motion
7 Ground warp beam
8 Sley reduction drive
9 Cloth take-up
10 Cloth beam
11 Shed formation by jacquard machine, dobby or tappet motion

FIGURE 12.10 The components of the rapier terry weaving machine with pile formation by shortening of sley travel.

Adaptations can be carried out at any time, for instance when a pattern is woven for the first time. The terry version of the rapier weaving machine (Figures 12.10 and 12.11) can be equipped with a control system for a maximum of eight different filling colors or yarns, and a jacquard machine; thus giving fabric designers practically unlimited scope in the design of terry fabrics.

Woven Pleated Fabrics

Wash proof pleated fabrics usually need to have more than 50% synthetic fibers such that the pleats do not fall out during wearing or washing. Pure cotton and wool fabrics also can be made pleated by applying synthetic resin finishes. It is now also possible to make permanent pleats during weaving without synthetic fibers or finishing [1].

To make pleated fabrics by the weaving process, an electronically controlled sley drive is used. In this mechanism, the beating up point is changed by small but precise steps from the normal beating up point. Figure 12.12 shows the formation of woven pleats. Only some of the warp yarns are used for weaving a pleat, which are stored in a separate warp beam. A greater filling density is used in the pleat area which compensates for the small proportion of warp yarns. The other warp yarns lie underneath the fabric when the pleat is woven [Figure 12.12(a)]. When the pleated area is woven, the fabric is not taken up and the filling density is achieved by shifting the beat-up point of the sley [Figure 12.12(b)]. For example, with a filling density of 40 picks/cm, the beat-up position has to be shifted only 0.25 mm filling by filling. Once the pleated portion is woven, which can be up to 20 mm long, the sley is returned to the normal beat up position (full beating) within two milliseconds. All the warp yarns are used again. The warp yarns wound on the separate warp beam are let off and the pleat falls into line [Figure 12.12(c)].

12.3 TECHNICAL TEXTILES

Technical fabrics feature a great variety of applications. In today's technology, the significance and use of industrial fabrics are increasing in agriculture and civil engineering, protection and safety, automotive industry and transportation, storage and packaging, medical and ecological sectors, sports and recreation, etc. (Table 12.3). Special fibers and yarn types with specific properties, ultrastrong synthetics, polyester and polyamide, polyolefin, polypropylene, polyethylene, glass fiber yarns and new heavy duty fibers like aramids and carbon fibers, have been taking the places of non-textile or conventional textile materials [2]. Innovative and successful product development, new processes and technologies broaden the existing markets and open a constantly growing field of applications for industrial textiles. For example, there are more than 25 textile applications in automotive engineering ranging from filter fabrics to tire cord. The same applies to other industrial markets. In

FIGURE 12.11 A rapier weaving machine equipped with a jacquard machine and a control system for eight filling colors.

civil engineering, there are more than 80 application areas of geotextiles [3]. There is probably no other industrial product that is used as widely as textiles. This shows the great diversity and expansion of this market and its importance for the textile and weaving industries. At present, industrial textiles constitute 9% of the worldwide fabric production.

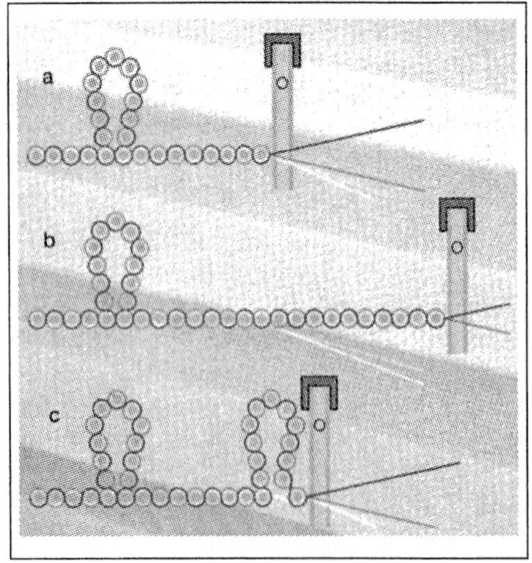

FIGURE 12.12 Schematic of weaving pleats [1].

TABLE 12.3 Major application areas of technical textiles [2].

• advertising	• leather
• agriculture	• mechanical engineering
• automotive	• medical
• aviation	• mining
• building	• oil industry
• ceramic	• packaging
• chemical	• paper
• computer	• pharmaceutical
• electrical	• plastics
• environmental protection	• printing
	• recycling
• fishing	• rubber
• food	• space
• furniture	• textile
• home textile	• transportation
• horticulture	• wire
• landscaping	• wood processing

Technical textiles replace not only conventional textile raw materials but also traditional building materials and metals. Due to their low weight, high tenacity and their indifference to attacks of corrosion, they keep conquering new fields of applications. Today, with the exception of a small portion of natural fibers, the industrial textile markets are dominated by synthetic fibers. The exceptional demands made on industrial fabrics are best met by yarns consisting of man-made fibers with their different specific properties.

Many weaving mills complement their production program with industrial fabrics to have some flexibility in the market. However, weaving of some industrial fabrics usually requires heavier duty weaving machines due to extraordinarily high mechanical and dynamic stresses and wide widths of fabrics. For example in the area of warp, enormous tensile forces are encountered. Due to the very high fabric densities, great beat-up forces of the sley result.

The production of industrial textiles requires modern, high performance weaving machines. In many areas, standard weaving machines cannot fulfill these special requirements. The range of heavy and wide fabrics requiring specific weaving machines is constantly increasing. Therefore, the production of these fabrics requires auxiliary attachments to customary weaving machines. High density, extremely heavy fabrics and weaving widths in excess of 540 cm require more extensive equipment or even new machine concepts.

Many of today's industrial fabrics can be woven on projectile, rapier and air-jet weaving machines. Projectile weaving machines are suitable for economic and efficient production of medium to heavy industrial fabrics. The rapier and air-jet weaving machines with special equipment are adapted to weave industrial textiles. For example, the rapier weaving machines are used for weaving airbag fabrics and for processing delicate yarns, while weft insertion with air-jet is widely used to produce faultless glass fiber fabrics. Table 12.4 gives examples of fabric types woven on each type of weaving machine for optimum performance.

When processing high tenacity or high modulus yarns, enormous tensile forces act on the weaving machine, that, over the entire working width, may amount to 10 kN/m. For this reason, certain component parts and subunits have to be designed to be more robust. For example, the machine frame has to be reinforced in the center and at the ends to achieve the required stiffness. The coupling of the main drive must be able to withstand the high stresses that are caused by the transmission of such high forces. The warp tensioner needs to be strengthened with the required number of supports (Figure 12.13). Usually two or three deflection rollers are used to take up the high tensile forces generated by the reed beat-up (Figure 12.14).

Shifting of the cloth beam is achieved by gearwheels with latches. The high tensile forces are taken care of by an additional safety latch on the hand wheel and a broadened enveloping wormwheel (Figure 12.15). The weft density can

TABLE 12.4 Examples of fabrics woven on different weaving machines.

Weaving Machine Type	Typical Fabrics Woven
Projectile	Base fabrics Glass fiber fabrics for coating, reinforcing and insulation
Air-jet	Precision engineering fabrics as coated substrates for electronic circuits
Rapier	Glass fiber fabrics for coating, reinforcing and insulation

FIGURE 12.13 Warp tensioner with extra supports.

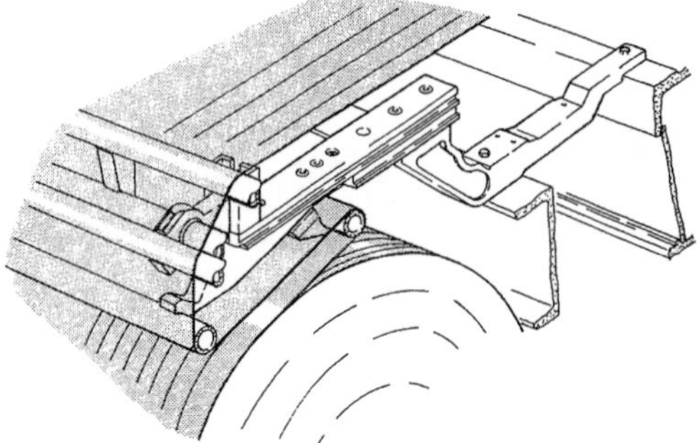

FIGURE 12.14 Use of three deflection rollers.

be altered by means of case-hardened, exchangeable change wheels. The entire cloth take-up gear is automatically oil lubricated.

Floating take-up roller enhances the fabric quality of heavy fabrics by ensuring reliable drawing-off of the fabric. With the self-regulating contact pressure of the two pressing rollers, slipping back of the fabric is prevented, even at high warp tensile forces. When producing heavy fabrics, the length of the recoiling fell can be shortened. Damage to the fabric surface is avoided with the floating take-up roller. Another advantage of this subunit is that it allows winding of fabric on big batches while leveling tension differences between batch winder and cloth take-up. Besides, fabric beams can be changed while the machine continues to operate.

FIGURE 12.15 Broadened enveloping wormwheel.

Projectile Weaving Machines for Weaving Technical Textiles

Special projectile weaving machines for industrial fabric weaving are built in two versions. The type P7, a mechanically controlled machine, is designed for weaving light to medium heavy fabrics. The microprocessor controlled weaving machine P7M is designed for manufacturing extremely heavy fabrics in widths of up to 12 m. These machines are ideal for weaving cotton felts, agrotextiles, geotextiles, conveyor belts, sailcloth, tire cord, cinema and theater screens, filter fabrics, tarpaulin, paper machine clothing and wire fabrics.

For weft insertion there are two projectiles with different weights. They can also be fitted with clamps that have larger clamping surfaces and higher clamping forces. Usually, the supporting of sley is enhanced for heavy fabrics (Figure 12.16). In designing the reinforced main shaft and the square sley drive couplings, the heavy forces should be considered.

The special equipment of P7M mainly includes the warp let-off, cloth take up, sley with sley drive and the transmission of the driving forces. To control the strong creep speed motion and to regulate the different subunits, the P7M is fitted with microprocessor control.

The warp let-off motion and the cloth take-up can be actuated independently of each other

FIGURE 12.16 Extra support of sley for heavy fabric weaving.

to permit accurate adaptation to the fabric specifications. The two subunits are designed so that they are able to resist high warp tensile forces and enable highly accurate weft densities. The warp tensioner and the floating take up roller, that are to absorb the heavy forces, can be supported in any place over the entire weaving width. Warp tensile forces of 20 kN/m—relative to a weaving machine of 360 cm weaving width—have been reached.

With certain monofilament fabrics, the weft density is designed so that the round diameters of the picks are distorted. To attain the necessary beat-up forces, sleys fixed with wedges, and rigid reeds with wide collars must be used. The quantity of sley actuators is increased, and the drive shafts with couplings are given larger diameters. Since heavy and dense fabrics cannot be woven with normal temples, specific clamping temples acting on the entire width are employed. These temples also guarantee that all picks rest straight in the fabric.

In addition to the standard weaving width of 540 cm, SulzerTextil developed projectile weaving machines with 8.46 m (Figure 12.17) and 12 m working width. On these machines, medium weight cotton fabrics are woven, e.g., movie screens. Production of other types of industrial fabrics such as geotextiles and agricultural textiles is also possible.

Extremely heavy multilayer forming fabrics for papermaking require warp tensile forces of 30 to 50 kN/m and can be woven only on very specific heavy duty weaving machines. Until a few years ago, specially built massive shuttle looms were the only machines available to make these types of fabrics for papermaking [4]. Recently, projectile and rapier weft insertion systems are being used in weaving machines for papermaking fabrics. Projectile machines with a speed of 130 ppm and reed width of 8.46 m are available today to produce paper machine clothing. Some of these machines are also equipped with a quick style change unit which allows style change within four hours.

12.3.1 Textiles in Transportation

Presently there are more than 25 different textile applications in the automotive industry, ranging from tire cord to filter cloth and interior trim. The situation is similar in the aircraft industry and in shipbuilding [2].

Tire Cords

More than 800,000,000 tires are produced each year worldwide. Tire cord fabrics are a significant sector of technical textiles. Tire cord is produced successfully from viscose, polyester and aramids on weaving machines, preferably machines 190 cm (75 inch) wide and equipped with weft mixer. Tire cord is a fabric, that, practically, consists of warp cords only. The number of warps is 6 to 13 ends per centimeter. The warp yarns are normally held in their parallel position by only one filling yarn per centimeter. The warp comprises high tenacity, twisted filament yarns made of polyester, polyamide, aramid or rayon. The counts of the yarns range from 940 × 2 to 1840 × 3 dtex. The filling material is comprised, in most cases, of cotton or core yarn of counts 20 to 30 dtex. Tires for road vehicles, construction machines and airplanes must withstand high mechanical and dynamic forces and feature high levels of elasticity and shock resistance. The tire of a medium size

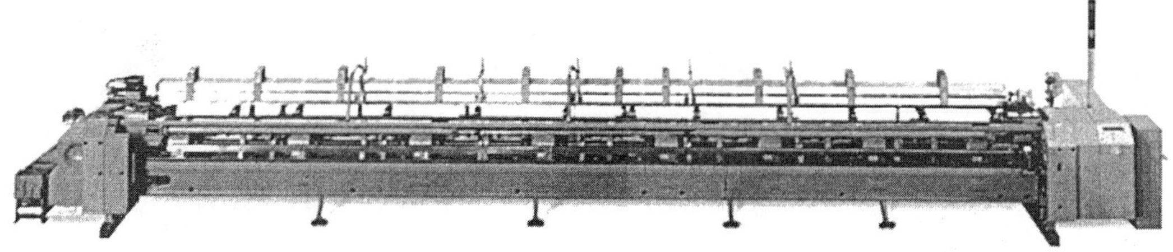

FIGURE 12.17 Projectile weaving machine with a weaving width of 846 cm.

TABLE 12.5 Typical tire cord fabric constructions [5].

Warp Material	Yarn Count Warp (dtex)	Ends / cm	Weft Material	Yarn Count Weft (tex)	Picks / cm
Rayon	1840 × 2	8.8	Cotton	20	1.0
Rayon	1840 × 2	9.9	Cotton	20	0.8
Rayon	1840 × 3	9.1	Olboplast	22	0.8
Polyamid	940 × 2	10.0	Cotton	30	0.6
Polyamid	940 × 2	9.3	Cotton	30	0.8
Polyamid	1400 × 2	4.9	Cotton	30	1.4
Polyamid	1400 × 2	6.3	Cotton	30	0.8
Polyamid	1400 × 2	11.0	Cotton	30	0.6
Polyamid	1400 × 3	6.3	Cotton	30	0.8
Polyamid	1880 × 2	6.1	Cotton	30	0.8
Polyamid	1880 × 2	9.8	Cotton	30	0.8
Polyester	1100 × 2	11.6	Cotton	30	0.6
Polyester	1100 × 2	13.9	Cotton	30	0.6
Polyester	1670 × 2	11.3	Cotton	30	0.6

automobile contains up to 0.5 kg of fabric. Table 12.5 gives examples of tire cord fabric constructions.

The major fiber suppliers for tire cord fabrics are Acordis, DuPont, ICI and Rhône Poulenc. The structure of a radial tire consists of fabric layers that are arranged radially and diagonally to the tire axis which improves comfort and safety (Figure 12.18).

There are several distinguished steps in tire cord fabric manufacturing compared to conventional fabrics, as shown in Figure 12.19. Carcass plies made of textile materials are woven cord fabrics made of cord warps and light filling yarns. A cord is produced by plying several twisted filament yarns together.

In order to fulfill the high physical demands, special yarns must be used. Table 12.6 lists some of the most widely used yarn materials and yarn counts in tire cord fabrics. For bicycle tires, usually multifilaments with around 200 turns per meter are used, e.g., PA 940 dtex Z 200.

To have the properties of these yarns maintained, or even improved, they are twisted in a specific process. By means of twisting, essential properties of the warp material in the carcass of the tire can be influenced. The desired strength is reached by combining the necessary quantity

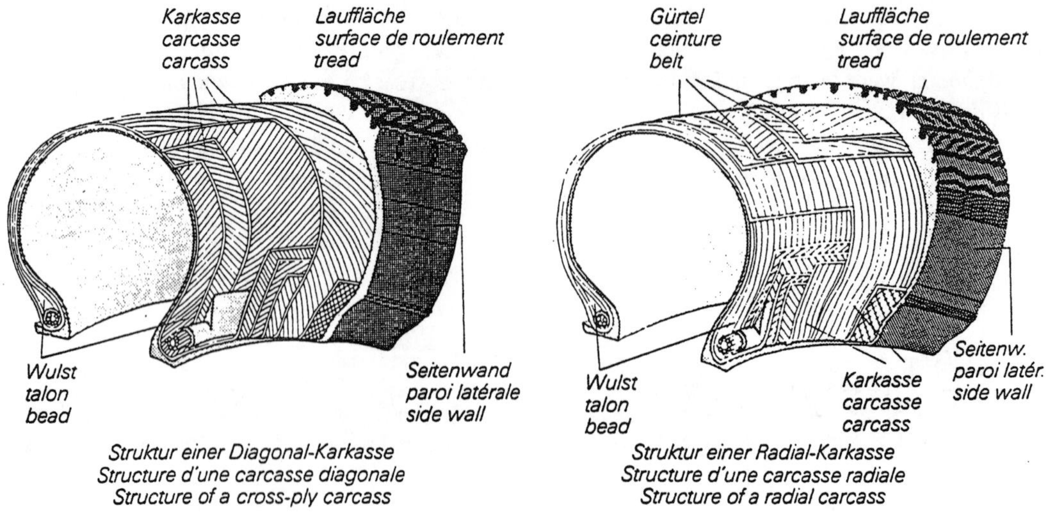

FIGURE 12.18 Schematic structures of major tire designs [5].

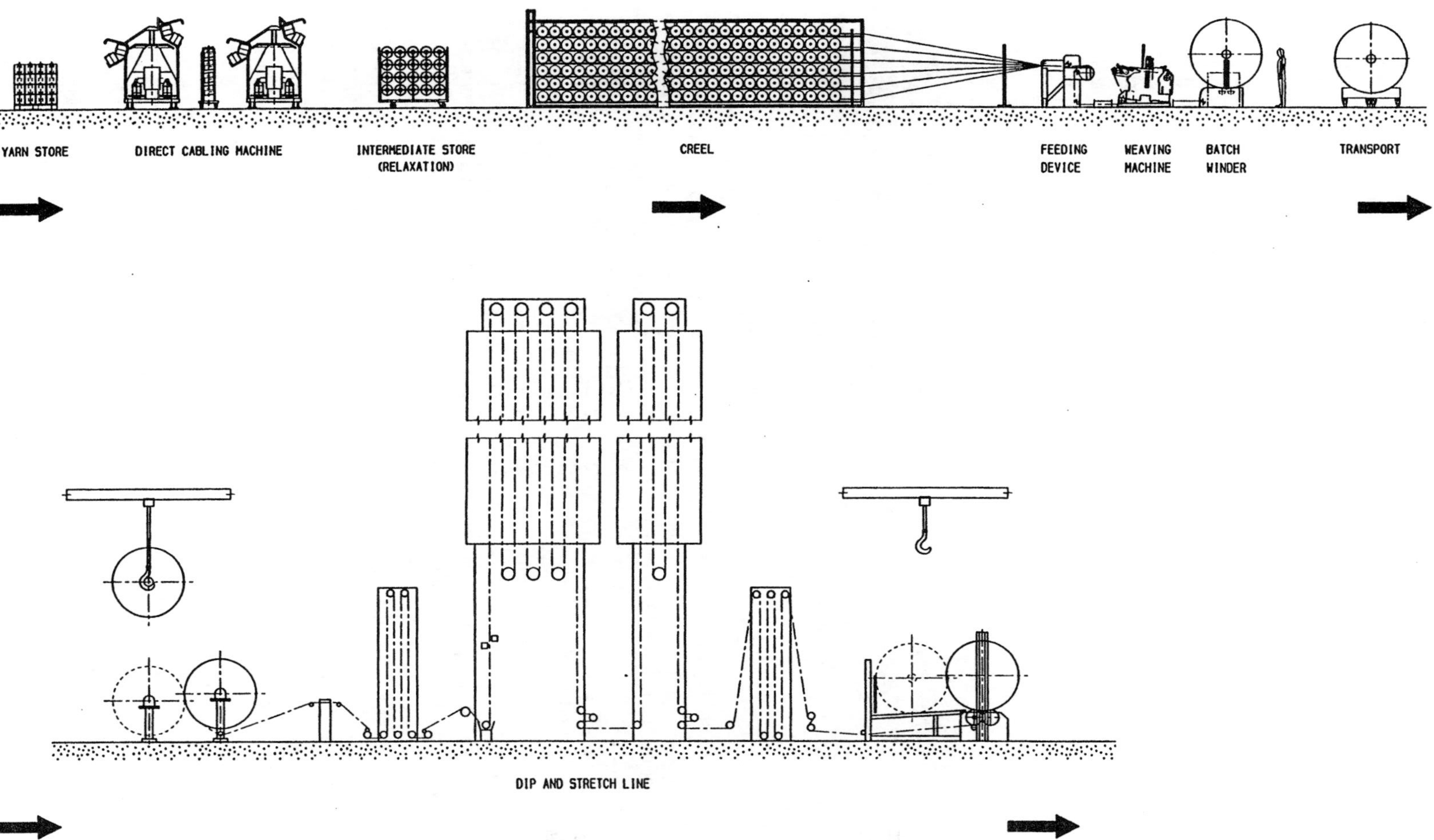

FIGURE 12.19 Major steps in tire cord manufacturing [5].

TABLE 12.6 Some of the yarn materials and counts used in tire cord manufacturing [5].

Material	Reference	Yarn Count (dtex)	Twist (T/m)
Rayon	CV	1220 × 2	590
Rayon	CV	1840 × 2	472
Rayon	CV	2440 × 2	410
Polyamide	PA	940 × 2	472
Polyamide	PA	1400 × 2	394
Polyamide	PA	1880 × 2	335
Polyamide	PA	2100 × 2	315
Polyester	PES	1100 × 2	472
Polyester	PES	1100 × 3	354
Polyester	PES	1440 × 2	433
Polyester	PES	1440 × 3	316
Polyester	PES	1670 × 2	410
Polyester	PES	2200 × 2	360
p-Aramid	AR	1680 × 2	330
p-Aramid	AR	2520 × 2	280

of individual warp yarns. It is essential that symptoms of fatigue as well as the necessary elasticity are improved.

The filament yarn has very little twisting, i.e., the individual filaments are arranged almost in parallel inside the yarn so that the latter can absorb the entire tensile force. Only two filament yarns are twisted with each other. This process is called cabling. There are various processes for producing cabled yarns.

a) Ring twisting. The classical procedure for producing cabled yarns is the two stage twisting process on the ring twisting frame. In the first stage, the ply twisting, untwisted filament yarn is given a twist in the Z-direction. In the second stage, the cable twisting, two such twisted filament yarns are again twisted with each other on a ring twisting frame; however, twisting is carried out in the other direction (S twist). The twist of the individual yarn is thus nullified, and the filaments are again aligned in parallel inside the yarn.
b) Direct cabling. In this method, the cabled yarns are produced in one step. In every spindle position, there are two packages. The inner yarn is combined with the outside yarn in a specially arranged mechanism. This system has several advantages over the ring twisting process; since it is a one step process, only half of the machinery is needed which reduces floor space, energy and manpower; large packages of up to 12 kg can be produced; yarn tensions can be controlled better which results in more uniform twist; and waste can be reduced by producing packages of preset yarn length.

For twist relaxation and to avoid uneven shrinkage of the warp material, the yarn needs to be stored for a period of two to four days. If this is not done, problems will occur when weaving from creel or when producing warp beams because the unrelaxed yarn tends to snarl. Intermediate storage can be done either on pallets, special storing trolleys or on the mobile creel modules.

Type of weaving machine is a critical part of tire cord manufacturing that influences the quality; usually air-jet and projectile weaving machines are used. Tire cord weaving requires special mechanical and electronic equipment. A typical machine can produce 300 m of tire cord fabric in one hour (Figure 12.20). At the beginning and end of the fabric, there are special sections called headers. Headers are woven with increased number of picks per unit length. Transition from header to ground fabric can be done without stopping the machine. Fabric rolls can weigh up to 4 tons. Usually a tire cord fabric piece is woven with a length of 1000–2000 m.

Warp Preparation and Weaving—Due to coarse warp yarns, dense warp settings of tire cord fabrics as well as high production of the weaving machines, the consumption of warp material is enormous. The conventional way of weaving from a warp beam with the available standard dimensions of flange diameters is not economical. The warp beam capacity, depending on yarn count and flange diameter, is in the range of 1000–3000 m. This means within a very short time (4–10 hours), the warp is woven off and new warp beams have to be brought to the machine. For an economical further processing, large lengths are of advantage. In tire cord weaving plants, the following two working methods are applied.

a) Weaving from Assembly Beams

Weaving of tire cord and conveyor belts from assembly beams is done successfully. According

FIGURE 12.20 Tire cord production on a projectile machine.

to the required total number of ends, four to six assembly beams are placed directly behind the weaving machine (Figure 12.21). On each assembly beam, 400 to 600 ends can be warped on. The warp yarn is fed to the weaving machine via the feeding device.

b) Weaving from the Creel

Drawing the warp yarns directly from the creel is the most widely used system today. Weaving off the creel increases efficiency, improves handling and fabric length and therefore running time.

The weight of an individual package can change from 2.5 to 6 kg, which, for heavy tire cord, corresponds to a yarn length of approximately 7,000 to 16,000 meters. Changing from 2.5 to 6 kg packages reduces weaving costs by more than one-third. That is why there is a trend toward package weights of 8 to 12 kg. The yarn is drawn from the package tangentially, i.e., no additional twist is introduced to the yarn. A feeder unit draws the warp yarn sheet off the package and introduces it into the weaving machine with the correct tension that corresponds to the weft density.

Based upon the special characteristics of tire cord fabrics and the up and down stream processes, certain measures must be taken to comply with the requirements of the weaving system:

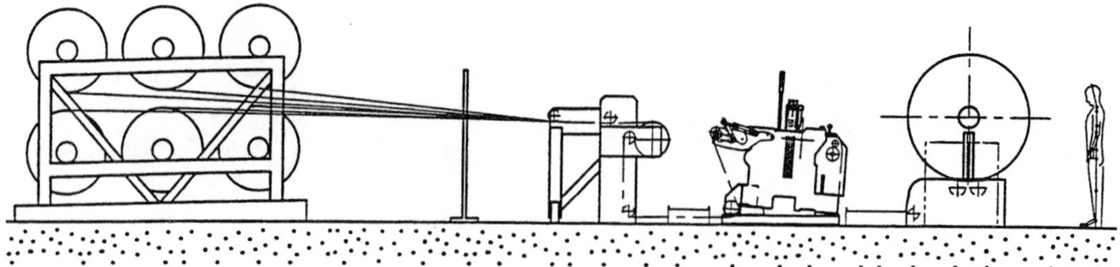

FIGURE 12.21 Weaving tire cord fabrics from beams [5].

- processing coarse warp yarns economically
- even distribution of the warp yarns over the whole weaving width
- stable tucked-in selvages, i.e., evenly rounded tucked-in tails
- monitoring of the tucked-in selvage
- careful guidance of this unstable fabric from the weaving machine to the batching motion by means of guide and deflection elements
- winding longer piece lengths on the batching device with perfect selvages, which means the selvage edges and warp yarns, due to the increasing weight, are not deformed, thus creating a cauliflower effect
- simple operation of the weaving machine, e.g., by automatic header weaving
- friendly operational software allowing easy programming and data transfer to the weaving machine terminal for the respective tire cord article
- totally controlled filling insertion, which means the filling pick, from filling insertion to reed beat-up is guided and the yarn held under a constant even tension
- reliable proven weaving system resulting in high productivity
- inserting different filling yarns reliably at a high machine performance
- flexibility of the weaving machine allowing article changes, even by large deviations in economic fabric specification

The warp tension in the weaving machine is achieved by means of back rest and spring tension. Shed formation is via a cam (tappet) motion, with four harnesses for the base weave and two harnesses for the selvage. Filling density is regulated by electronically controlled take-up motion. Weaving of the header and piece length is automatically achieved according to the preset program.

The fabric is wound onto batches with a maximum winding diameter of 2 m. The density of the fabric batch must remain constant from inside to outside. For this, degressively working batch winders are used. With the degressive winding technology with surface-driven winder or winder with axial drive, the surface pressure remains constant from beginning to the end of a fabric batch. It can be individually adjusted to any type of fabric. This helps to avoid the so-called cauliflower effect. Fluctuations of widths and densities, particularly in the selvage zones, are thus eliminated. This has a beneficial effect on the further processing in the dipping plant and also on the coating of the fabric with rubber. The large running length of a giant batch ensures an absolutely uniform fabric stretching across the entire fabric length in the dipping plant, depending on the count of the warp yarn of 6,000 to 12,000 m. The large piece lengths also eliminate latex tailings at the necessary interfaces.

Weaving the Header (Tabby)—Since tire cord fabrics are subject to after treatment, such as impregnating and hot-drawing, large piece lengths are required. At the beginning and end of each fabric, a header for transition and test samples must be woven. This header is produced with a different filling yarn and at increased density (Figure 12.22). Depending on the manufacturer, the header is approximately 5 to 60 cm in length. In this section, the filling density is increased to about five picks per centimeter. The header is useful to join tire cord pieces by sewing them together on the dipping and hot stretching line and also as reference pieces and pieces for quality control. Therefore, at least three headers

FIGURE 12.22 Header [5].

and control pieces are woven per tire cord piece. In today's machines, weaving of header and test samples has been automated using electronically controlled fabric take-up and a programmable microprocessor. With the microprocessor, the filling density of the main fabric, filling density of the header, fabric length, and filling yarn selection for main fabric and header can be freely programmed. In addition, the positions of temple (high or low) and tucking needle can be automatically altered. Tucked-in selvage offers several advantages for after treatment.

Tire Cord Weaving Machines—Projectile weaving machines are very suitable for tire cord weaving. The optimum characteristics of a such a machine are:

- filament execution
- weaving machine nominal width 190 cm
- two color electronic filling selector
- positive cam motion
- projectile type D12

Coarse yarns, for example, cotton 34 tex × 3 or even glass yarn 300 tex are inserted during the header weaving process. The D12 projectile can handle coarse yarns.

The breast beam with revolving deflection roller and PVC coated take-up roller provides a slippage free take-up without displacing picks.

With standard tuck-in units, a tuck-in tail of 15 mm is obtained. The tire cord unit that is specially designed for tire cord weaving achieves a tuck-in tail of up to 40 mm. The tuck-in tail is controlled during the complete tuck-in procedure. The tuck-in needle is inserted into the shed; any displacement of the warp ends during this operation is practically impossible, hence, an exact positioning of the tucked-in tails is achieved. The selvage monitoring device stops the weaving machine in case a tucked-in tail has not been tucked in properly. Defective selvages are avoided.

With the use of a double reed, warp yarns can be drawn-in singly (one end per dent) ensuring an absolute even dispersion of the warp yarns over the full width of the fabric. For tire cord fabrics with a lesser warp density, a normal reed can be used.

With the automatic header weaving unit, it is possible, via a programming device, to define the piece length, choice of filling yarns, length and number of headers to be woven. Changes are not done manually, thereby increasing the efficiency of the weaving machine and the quality of the fabric roll.

Automatic header weaving includes the following:

- Free programmable filling densities in the range of 0.4 to 12 picks/cm (1 to 30 picks/inch); selection of filling yarn for the base and header weaves
- Programming done by means of a simple PC or laptop for data transfer
- In the terminal on the weaving machine, two different programs with 200 sets can be stored, and on request recalled.
- When weaving header or base, the temples are moved automatically in or out (Figure 12.23)
- Automatic adjustment of the tucking needles when weaving base or header

The tire cord fabric can be woven within very tight tolerances due to the combination of the automatic header and fabric take-up.

Dipping and Stretching—The grey fabric is dipped in a specially formulated adhesive solution (Resorzin, Formaldehyde and Latex = RFL) to facilitate the adhesion of the fabric to the rubber compound in tire manufacturing. At the same time, depending on the yarn types used, the fabric is stretched under very accurate condi-

FIGURE 12.23 Controllable temple [5].

tions of time, temperature and tension to modify the physical properties of the cord yarn.

During the final dipping and stretching process, the fabric is automatically controlled for filling straightness and bagginess with the aid of optically or mechanically controlled fabric guides. Depending on the utilized raw material, the dipping and stretching process is done in different combinations. Next to stretching the fabric lengthwise, the fabric width is also controlled.

A dipping and heat stretching line generally consists of the following components:

- let-off station for batches with up to 2 m diameter
- sewing station to connect two batches
- fabric accumulators to guarantee a continuous operation of the line
- dip station for dipping the fabric into the RFL solvent
- squeeze rollers for better impregnation of the fabric
- dewebber or vacuum station to remove surplus RFL solvent and determine the dip pick-up
- drying and heating zones
- pulling and stretching units to transport the fabric and for controlled stretching of the fabric
- winding or take-up unit for fabric rolls with a diameter of 2 m

The resulting fabric roll is ready for further processing at the calender line where it is treated with a rubber mixture.

Chafer Fabrics—Chafer fabrics are only a small percentage of textiles which are used for reinforcement of tires. The purpose of chafer fabric is to reinforce the bead of the tire and cover its circumference. Damages due to friction between the bead and the rim during operation as well as when mounting the tire onto the rim are avoided.

High dimensional stability, flexibility as well as good adhesion to rubber are prerequisites for chafer fabrics. The fabric itself is often held in a square setting with a relatively light construction. The main material utilized is 940 dtex nylon 6,6. Depending on the tire use, e.g. trucks or cars, the constructions of the chafer fabrics vary significantly. The warp and filling densities are in a range from 4 to 11 picks/cm. Multifilament yarns and air textured yarns are used. The same fabric is also produced from monofilaments; 460 dtex in warp and filling.

Table 12.7 shows the variety of possible fabric constructions of chafer fabrics. The twist of the multifilaments is usually 200 turns per meter. For inserting the twist into the multifilaments, both two-for-one twisting and ring twisting are suitable. The fabric widths of the dipped and stretched fabrics are in the range of 145 to 150 cm.

Conveyor Belt Fabrics

Conveyor belts serve as transportation roads for loose bulk material, such as pebble stones, sand, ores, etc. They are also used in distribution, sorting and filling plants for small and large unit loads on railway, mail, supermarkets and airports. Unlike transporting bulk goods, the conveyor belts may not be cleated, i.e., the belt must be flat. Depending on the tasks, the demands are versatile and rather high [2].

Traditionally, these fabrics have been woven on projectile weaving machines with tucked-in selvages, without warp and filling waste, or with leno selvages, in single or multiple widths. The

TABLE 12.7 Examples of chafer fabric constructions.

Yarn Count Warp (dtex)	Yarn Count Weft (dtex)	Raw Material	Warp Ends per cm	Picks per cm	Remark
940	940	Nylon	6.3	6.3	
940	940	Nylon	8.3	8.3	Car
1400	1400	Nylon	7.1	7.1	Lorry
1200	1200	Rayon	8.3	8.3	Car
1840	1840	Rayon	7.1	7.1	Lorry
460	460	Nylon	9.8	9.8	Monofil

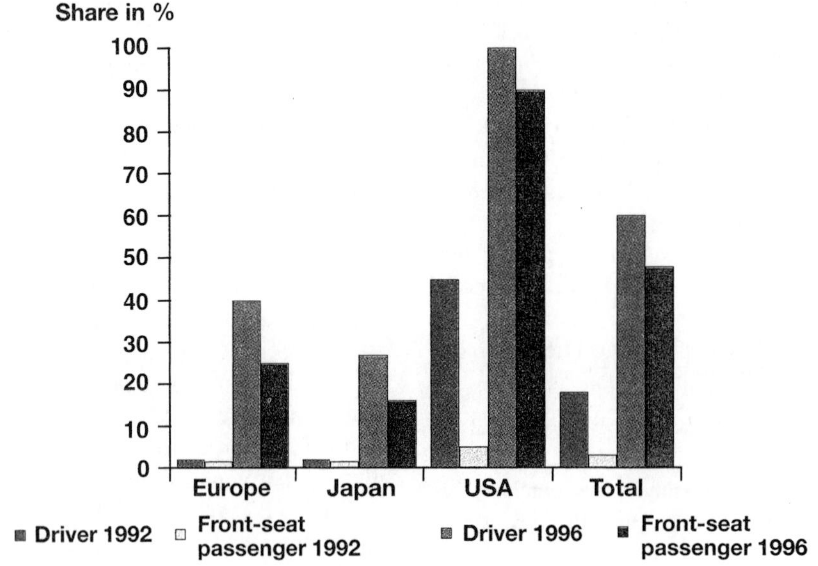

FIGURE 12.24 New cars with air bag.

common fabric is made with polyester in the warp and polyamide in the filling. Other materials are also used for specific applications.

Air Bag Fabrics

In contrast to safety belts which provide only limited protection under unfavorable conditions, the air bag system protects head and body altogether. Today, the air bag is regarded as a well tried, passive retaining system which has proven its reliability and effectiveness in many accidents [2]. As a result of the success of the drive side air bag, other systems protecting the front seat passenger (passenger bag) or the passengers in the back seat have been developed. Other air bags are also available to protect the head and knees. Research work is going on to develop air bags to protect the passengers from side impacts. The demand for air bag fabrics is constantly rising as can be seen in Figure 12.24.

Due to existing international standards and the necessary certification procedures that have to be completed to make air bags, the manufacturing of them requires special considerations. Table 12.8 lists the typical quality specifications for air bag fabrics. Major requirements of air bag fabrics are (Figure 12.25):

- low fabric thickness
- high strength in warp and weft direction
- high tear propagation resistance
- high anti-slip properties of the seams
- resistance against aging
- defined dimensional stability
- defined air permeability (10 lt/dm^2/min at 500 Pascal)
- product liability guaranteed for 15 years

TABLE 12.8 Typical quality specifications for air bag fabrics.

			Target
Weight (g/m^2)	240		240
Weave	linen		
Air permeability (L/100 cm^2/min)	3.6		<10
Grey width (cm)	177		
Reed width (cm)	186		
	Warp	Weft	
Density (1/cm)	23	22	
Tensile strength (N/5 cm)	3495	3025	2500
Elongation (%)	41.4	25.1	35/25
Tear propagation resistance (N)	171	170	130

Yarn: Rhonel Tech polyamide 6.6 dtex 470 f68 Z60 tangled or dtex 470 f68 0-0 tangled/sized finished fabric

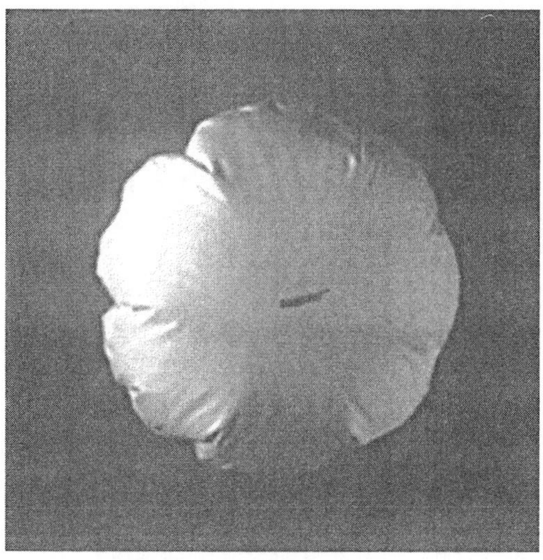

FIGURE 12.25 Air bag.

Air bag fabrics are made of polyamide 6,6 multifilament yarns with counts from 235 to 940 tex. Air bag fabrics are generally dense which is a challenging task to weave. Today, these fabrics are woven successfully with warp and filling stoppage rates below 1.0 stoppage per 10^5 picks. The following configuration is recommended when weaving air bag fabrics on rapier weaving machines:

- full width temple
- sley race instead of supporting teeth
- controlled bolt clamp
- selvage rollers (with 24 yarns, right and left)
- knots stop motion
- weft yarn feeder
- controlled weft yarn brake
- take-up roller with rubber coating
- selvage sealing device, programmable via microprocessor

Air bag fabric styles are also woven in double width on projectile weaving machines with the following recommended configuration:

- robust design
- cam motion
- filament equipment
- electronically controlled warp let-off
- warp tensioner
- projectiles D12 with large clamping surfaces
- sley with oblique position of the reed of 4°
- reed with 7 mm dents, bright polished and stainless
- full-width temple across entire fabric width
- selvage sealing device

With the rapier weaving machines, the warp yarns do not need to be twisted or sized, but only intermingled. This applies to coarse fibril types f68 and f72 as well as for the fine fibril yarns f136 and f144. This is a major advantage in terms of economy and ecology.

Tensile strength, elongation, tear propagation resistance and weight requirements of air bag fabrics are critical. The main concern of fabric tests, however, is uniform air permeability across the entire width of the fabric. Figure 12.26 shows the air permeability of grey and finished fabrics woven on rapier and projectile weaving machines.

1. Fabric woven on rapier weaving machine

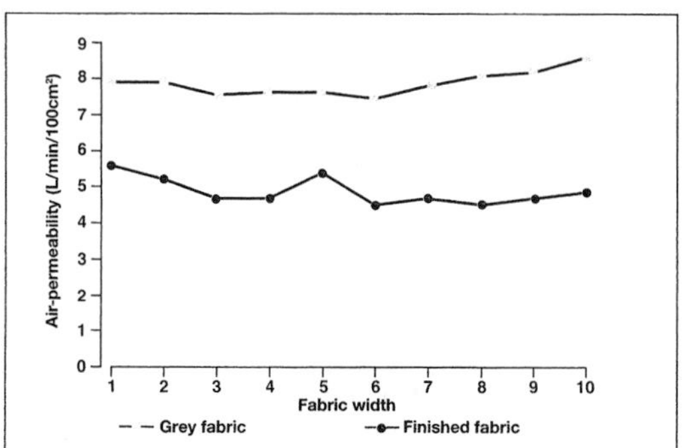

2. Fabric woven on projectile weaving machine

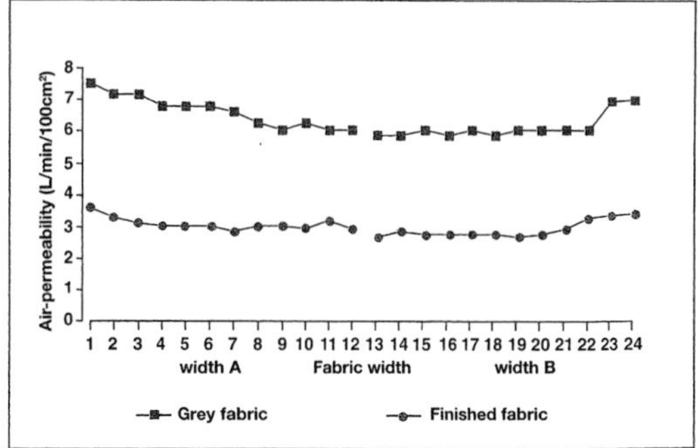

FIGURE 12.26 Air permeability of grey and finished fabrics woven on rapier and projectile weaving machines.

FIGURE 12.27 12 m wide weaving machines for forming and dryer fabric weaving (courtesy of Jager).

12.3.2 Paper Machine Clothing

Forming fabrics that are used on paper machines run at up to 2000 m/min (120 km/h) speeds. This necessitates a strong and stable high density fabric made of monofilaments [4]. Due to heavy monofilaments used in fabrics, paper machine clothing is woven on very massive weaving machines. Weaving of these fabrics requires extremely high beat-up forces (up to 10 tonnes) on heavy duty weaving machines [6]. These machines are designed such that the forces developed during weaving are well balanced. "Zero defects" is literally a must in these fabrics; geometrically, the weft and warp have to be accurately positioned in the fabric in all three dimensions. Satisfying these criteria involves a substantial effort in mechanical and electronic design. For instance, the drive for the shed forming unit is separate from that for the weaving machine. Until recently, shuttle looms were the main machines to make forming fabrics of up to 12 m width. Figure 12.27 shows a shuttle loom for weaving of forming and dryer fabrics. Figure 12.28 shows a 28 m wide shuttle weaving machine for circular press fabric manufacturing. Today, projectile machines of 8.46 m width, weighing around 50 metric tons, are being used to make forming fabrics. These new machines run at 130 picks per minute (compared to 30–40 ppm of shuttle loom) which is equivalent to 1000 m/min filling insertion rate (Figure 12.17).

The elements of the weaving machines used to make paper machine clothing must be reinforced. In particular, sley drive has to be heavy duty for the required beat-up force. The warp tension is around 1500 kg/m which requires special let-off and take-up. Paper machine fabrics made of monofilament yarns demand high quality and evenness; warp let-off and fabric take-up must be highly accurate and uniform. Projectile weaving machines are equipped with the floating fabric take-up roller for non-slip fabric take-up, a sley with a key groove for special reeds, special cams for reduced shed stroke, specially arranged temples and projectiles with increased clamping power. Fine forming fabrics made of monofilaments are very smooth and are prone to slip

FIGURE 12.28 28 m wide weaving machine for press fabric manufacturing (courtesy of Jager).

during fabric take-up, particularly at a highly dense setting. The floating take-up roller is coated with a special rubber compound for fine, dense and heavy fabrics. Special gearing is used to precisely control reversing of the fabric when a weft break must be repaired.

With fabrics made of monofilaments, selvages must be carefully handled. With coarser monofilaments with diameters of up to 0.4 mm, a firm selvage is formed by means of a selvage sealing device. Fine monofilament fabrics with densities of up to 60 picks/cm can be fitted with tucked-in selvages. Sectional warp beams are usually used for monofilament yarns as shown in Figure 12.29.

12.3.3 Geotextiles

Geotextiles are industrial textiles employed in civil and hydraulic engineering. Development and application of geotextiles have increased drastically within the last 20 years [2].

FIGURE 12.29 Sectional warp beam for monofilament fabrics.

The materials used in geotextiles are polypropylene, polyethylene, polyester and polyamide. Geotextile fabrics are used in various applications such as membranes for lining sewage treatment plants, swimming pools, drinking water reservoirs and refuse dumps. For woven geotextiles, mostly polypropylene yarns of 550 to 3300 dtex are used. Sometimes monofilament yarns are also used.

The main functions of geotextiles in civil engineering applications are separation, reinforcement, filtration and drainage. As an example, Figure 12.30 shows the separation function of geotextiles. Woven geotextiles are ideal for the separation of two solid but different materials such as sand and crushed stone. Soil is good in compression but weak in tensile strength. Therefore, woven geotextiles are ideal to reinforce the soil since they have very good tensile strength. Nonwoven geotextiles in general are not used for reinforcement of soil.

Geotextiles must meet several mechanical, hydraulic, degradation and durability properties. The important mechanical properties of geotextiles include surface related dimensions, thickness, tensile strength and puncture resistance. Important hydraulic properties include permeability, permittivity (flow across the plane of fabric) and transmissivity (flow in the plane of fabric). Geotextiles should resist UV radiation, chemical substances and biological influences such as mildew and fungi.

Geotextiles are preferably woven in the largest possible fabric widths. For serial production, projectile weaving machines with a nominal weaving width of up to 540 cm and a maximum weft insertion rate of 1200 m/min are used. Wider projectile weaving machines are manufactured as special weaving machines with a maximum nominal width of 12 m (Figure 12.31). The effective weft insertion rate and speed depend on yarn count, yarn quality, style and the real working width.

The warp and weft yarns should have a protective twist Z60 to ensure problem-free processing. When untwisted weft yarn is used, a weft sealing device is required to prevent loss of fibrils. Untwisted yarns lead to improved covering and affect the permeability.

12.3.4 Safety and Protective Fabrics

New materials are replacing conventional fabrics in the field of working, protective and safety clothing, in rescue and ambulance service applications and the leisure and sports sectors. In addition to fabrics woven of high strength polyester and polyamide yarns, more cloths are now

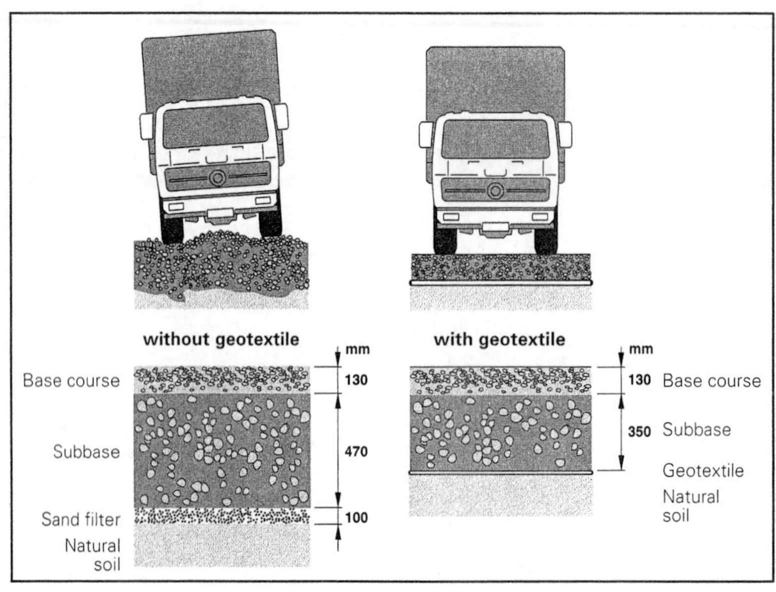

FIGURE 12.30 Separation function of geotextiles.

FIGURE 12.31 Geotextiles woven on projectile weaving machines with up to a width of 12 m.

being made from aramid yarns (primarily for protection against heat), in hard and soft ballistics and in sports.

It has always been the function of clothing to protect bodies from external influences. Protective garments for professional and leisure purposes combine active and passive safety aspects and provide high-level protection for the person who wears them. By employing the latest development yarns and suitable design, fabrics can be produced that meet the demands of the most varied protective functions. In many instances, fabrics of this kind are able to serve one or more protective functions ranging from protection from cold, heat, dirt, water and aggressive chemicals, down to injuries caused by tools, weapons and sudden falls (Figure 12.32).

For obtaining sufficient protective properties, such clothes often were very thick, thus severely hampering the person who had to wear them in doing his/her job. By using new fibers, modern protection textiles could be made less heavy and more function targeted. Fashionable cut and coloring made them not readily recognized as protection means.

The fibers that are used in safety and protective textiles can be grouped as follows:

a. Ultrastrong fibers of low temperature stability: high density polyethylene (HDPE), polypropylene (PP), polyvinyl alcohol (PVA)
b. Ultrastrong fibers of high temperature stability: para-aramids, polyetheretherketone (PEEK)
c. Hardly combustible organic fibers of high temperature stability: m-aramids, phenolic resin fibers
d. Inorganic fibers of high temperature stability: silicone dioxide (SiO_2) fibers, carbon fibers

Today all the manufacturers of man-made fibers of importance offer fibers and filament yarns featuring specific properties. Also, high tenacity polyamide, polyester and cellulose fibers are employed either alone or in blends with other fibers. Microfibers are used in various application areas including sports and leisure wear. Special fabrics made of aramid fibers are effectively used in the areas of safety and protection. Protective clothing made of aramids has proven very reliable in heavy duty professional use.

Aramids belong to the group of aromatic polyamides. They are known under brand names such as Kevlar® and Nomex® from DuPont, Twaron® from Acordis, Spectra® from Al-

FIGURE 12.32 Fabrics with specific properties protect from most severe injuries.

liedSignal, among others. These materials feature very specific properties. Kevlar®, Twaron® and Spectra® are used wherever high tensile strength is required, while Nomex® is known for its exceptional heat resistance. Compared with other man-made fibers, the tensile strength of aramids exceeds that of other high tenacity synthetic and glass fibers. Another interesting safety material is the melamine resins like Basofil®. For extreme conditions, heavy fire protection suits are made of Basofil®, which features outstanding thermal insulation properties and the advantage that it does not melt. Normally, Basofil® suits are used by fire fighting personnel, welders, laboratory people and industrial workers that are exposed to high temperatures. Fabrics made of yarns consisting of fiber blends, such as Basofil®/cotton or Basofil®/wool are used for professional groups that are only occasionally exposed to unforeseen flames and heat.

Density is another important criterion for the suitability of materials for certain applications. Aramids have shown to be superior to glass and steel. The fibers feature an extremely high degree of impact strength and toughness and do not melt; they oxidize to ashes at approximately 400–430°C. Table 12.9 shows some of the application areas of high performance aramid fibers. Figure 12.33 shows several fabrics made of high performance fibers and their application areas.

TABLE 12.9 Examples of application areas of high performance aramid fibers.

Ballistic	Bulletproof vests, helmets, motorcyclists' helmets, hard armorings, shields
Personal protection	Safety gloves, protective clothes, flameproof motorcyclists' dresses, bulletproof sheetings for automobiles
Building sector	Retaining walls, concrete reinforcement
Reinforced synthetics	Boat building, canoes, high powered racing boats, sporting goods, ski, golf clubs, tennis racquets
Aviation	Seats, rudder units, body parts
Automotive industry	Radial tires, chassis frames for automobiles, clutch and brake linings, gaskets, plastic reinforcements
General industry	Conveyor belts, driving belts, flameproof tarpaulins, stadium roofs, oil barriers, ventilating shafts for mines (air conduits), boat building fabrics (inflatable boats)

FIGURE 12.33 High performance fabrics and their application areas.

Timbermen working with chain saws are extremely prone to accidents. The most common accidents are injuries of the legs. The conventional protection of the lower extremities consisted of 28 layers of polyamide yarn knittings that were fixed to the trouser legs as cushion-shaped bags. Even though this sort of protection fulfilled its purpose perfectly, it was bulky, stiff and heavy such that the timbermen found it uncomfortable enough to refrain from using it.

Today, chain saw protective clothing consists of a four layer fabric made of Kevlar® aramid yarns (Figure 12.34). It is lightweight, soft, plain yet strong. The tensile strength of the Kevlar® aramid is five times higher than steel fibers of identical weight and count. Due to its low mass per unit area, the manufacturers of protective clothing are able to produce entire trouser legs of four layer fabrics of Kevlar® aramid yarns, reaching from the thighs to the ankles. This results in a fundamental increase of protection and wearing comfort.

Fabrics made of high tenacity fibers are used to produce mittens and boots for timbermen, featuring outstanding protection from cuts. Stainless steel fibers, core twisted with man-made yarns, are particularly used for weaving cut protective fabrics.

There are different kinds of fabrics and suits against different levels of fire exposure, depending on the purpose. Regular firefighter suits are different than fire entry suits. Figure 12.35 shows a fire entry suit which can be used to enter right to the source of the fire without great risk, to rescue people or to close an emergency valve in a burning drilling platform.

The fiber used in the suit shown in Figure 12.35 is a Basofil®/E fiber by BASF. The fiber characteristics are:

Type of fiber	: melamine resin staple fiber, hardly combustible, heat resisting
Fiber diameter	: 8 to 20 μm
Specific density	: 1.4 g/cm^3
Permanent heat resistance	: 200°C
	No melting point
	No formation of drops under influence of flames

Fire protective suits are made of highly varied textile structures. The schematic of the structure

FIGURE 12.34 Timbermen wear versatile protective clothing to avoid injuries through cuts.

FIGURE 12.35 Fire entry suit made of high temperature fibers.

of the fabric of Figure 12.35 is shown in Figure 12.36. The coating of the outer fabric reflects the heat radiation. The intermediate film blocks off steam. This prevents water, evaporating on the hot surface, from penetrating the textile system and causing scalding when the fireman is cooled by spraying him with water. A fleece layer offers additional isolation. The inner fabric layer serves as lining.

There is a tremendous amount of research and development work behind every safety and protective textile fabric. Before a protective suit reaches the testing phase, it has to pass many of the following production and control stages: fiber production, spinning, weaving, finishing, fleece manufacturing, coating, making-up, film making, production of protective panes, sewing yarn production and quality control. Protective

a: Fabric made of Basofil/aramide, aluminium coated, approx. 580 g/m²
b: Steam tight, heat-resisting protective film
c: Needlefelt 100%, Basofil, approx. 400 g/m²
d: Lining made of Basofil/aramide, approx. 250 g/m²

FIGURE 12.36 Construction of the heavy fire protective suit of Figure 12.35.

clothing of this type may well be compared to a chain which is only as strong as its weakest link. For this reason, the design and construction of a protective suit is tried and optimized in large scale and expensive test runs.

The importance of smooth interaction of all links of the production chain is demonstrated in tests as shown in Figure 12.37. In this case, the melting of an unsuitable eye-protecting pane would have led to heavy burning. The test sequence of protective clothing must contain a theoretical component as well as a practical one. Accordingly, theoretical tests are conducted in the laboratory on the textile surface and on the dummy, while the practical trials are carried out as field tests with a living person in the protective suit.

Fabrics made of reflecting yarns are a novelty. Strips of these fabrics, sewn onto protective garments, contribute decisively to protect the person who wears them so that he/she is seen at a great distance during dusk and darkness. Clothing with these reflecting strips is mainly used for the fire brigade and police but also for sportswear, e.g., for bicycle and motorcycle sports as well as for pedestrians (Figure 12.38).

Weaving of Aramid and Other High Performance Fibers

Due to the specific properties of aramids, aramid warps must be prepared with great care. The following preconditions must be met for the successful weaving of these fibers:

- uniform yarn tension over the entire width of the warp beam
- uniform diameter over the entire width of the warping beam
- perfect transitions of sections (in most cases warping is done on sectional warping machines)
- mat chromium plated or ceramic yarn deflections (the rolling yarn brake is best suited).

Air-tangled, twisted and spun yarns in the count range of 215–3300 dtex are woven on projectile and rapier weaving machines for the most varied applications. Some modifications must be made to the weaving machines so as to enable problem-free weaving of the mostly untwisted yarns in warp and weft. For best re-

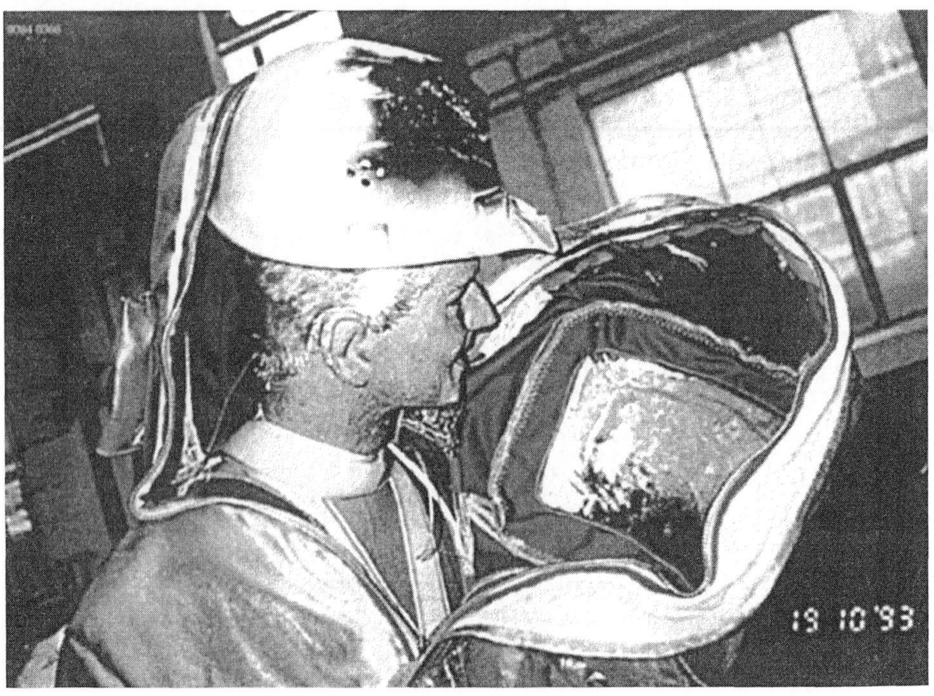

FIGURE 12.37 Testing of a protective suit.

FIGURE 12.38 Reflective fabrics for safety (courtesy of Schoeller Textile Limited).

sults in weaving aramids, the following are recommended for projectile weaving machines:

- warp beams with tube diameter of 215 mm as a minimum
- whip roller or warp tensioner either mat chromium plated or plush coated
- floating take-up roller for special styles
- take-up roller with rubber band for light and heavy styles
- cutter with increased pressure
- projectile clamps D12 with clamping surface of 2.2 × 4 mm as a minimum
- weft feeders with separated yarn layers

The following components of the rapier weaving machines are specially adapted for weaving friction-sensitive p-aramid yarns:

- low friction surfaces of the guide rollers
- special sley without supporting teeth for reduced yarn friction
- yarn-friendly rubber rollers in the cloth take-up system

12.3.5 Agrotextiles

Agrotextiles are increasingly being used in farming and horticulture. Agrotextiles are either woven fabrics or nonwovens, mainly made of synthetic materials that are indifferent to rotting. Fabrics are preferably used where tensile strength and dimensional stability are required, e.g., protecting the plantations from extreme, natural conditions. The results are products of improved quality, increased yields, fewer losses and decreased damages. This permits substantially reduced usage of weed killers and pesticides.

Agrotextiles are used as soil cover, sack cloths, hail, insect, rain, sun and wind protection. Normally, they are made of narrow polypropylene ribbons or fibrillated yarns. On the other hand, monofilaments mainly consisting of high density polyethylene (HDPE) and polyester can also be used for manufacturing agrotextiles.

Fabrics for Soil Covering

Fabrics for soil covering are, above all, used in greenhouses. They prevent the formation of weeds, while the woven-in marking yarns help to keep constant distances between plants (Figure 12.39). Open air application stops the growth of weeds, prevents erosion and parching of the ground. Also, the application of herbicides can be dramatically reduced.

Hail Protection Fabrics

Hail protection fabrics are made of ultraviolet (UV) stabilized polyethylene monofilaments

FIGURE 12.39 Fabrics covering the ground, reduce parching the soil and prevent weed growth.

with typical diameters of 0.25–0.30 mm, with a mesh width of 10 × 4 mm. The fabric features antislip properties due to the employed leno weaving technique as well as lengthwise and crosswise stability (Figure 12.40). As a rule, these fabrics have a service life of 15 to 20 years; however, for economical reasons it is recommended that they be changed after 10 years.

Tests have shown that dark nets have the longest service life, but they also absorb the largest portion of sunlight, namely 7%. With the hail protection system, the principal costs are the structure and its erection.

Insect Repellent Fabrics

Vegetables, such as cucumbers, tomatoes and aubergines as well as some ornamental plants are often attacked by pests like whitefly or tobacco or scale insect. These are able to infect the plants with up to 25 different types of viruses and they leave so called honey dew on the leaves that later gives rise to the formation of a black fungus.

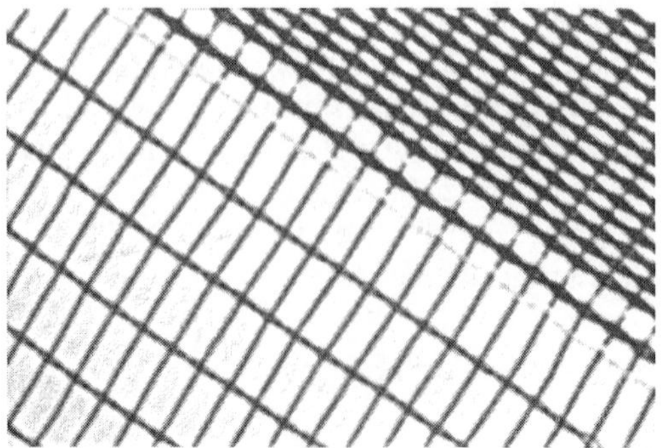

FIGURE 12.40 Structure of hail protection fabric.

This constitutes an opportunity to employ insect repellent fabrics making insecticides superfluous. Openings for doors and windows of greenhouses are covered with these fabrics. Sometimes, the fabrics are stretched across the open-air plantations so that the pests can no longer get to the plants (Figure 12.41). Another advantage of such application is that the climate is not disturbed in any way. The best experience was made with linen-weave fabrics of UV stabilized polyethylene monofilament yarns with a diameter of 0.25 mm. Warp density is 24 ends/cm and weft density is 11 picks/cm.

Rain Protection Fabrics

Rain protection fabrics feature a grid structure and protect flowers and berries from damage through rainfall.

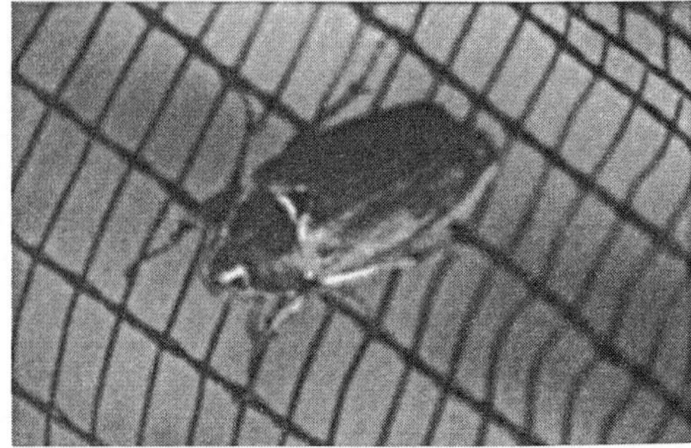

FIGURE 12.41 Insect repellent fabrics.

Sunshade

Sunshade fabrics serve to develop a microclimate for plantations of flowers, ornamental plants and fruit. The insulation enhances the conditions of growth and the soil is not parched. The result is savings in irrigation costs.

Protection from Wind

Wind protection fabrics are produced in the form of fine grid work. In most cases, they are made of polyethylene monofilament yarns (Figure 12.42).

Manufacturing of Agrotextiles

Agrotextiles are mainly woven on projectile weaving machines with weaving widths of 430, 460 or 540 cm. The projectile weaving machines offer the following advantages for this purpose:

- large weaving widths
- usage of warp beams with large diameters reducing the number of stoppages due to warp changes
- universal applicability
- the possibility to operate with high and low shed or only with high shed leno
- quick switching to other weft yarns
- low energy consumption

12.3.6 Sail Cloth

Old sail cloths were genuine cotton fabric. In winter they used to freeze and in summer thunderstorm, the sails used to become wet, heavy and gigantic rags. In calm weather, the sailors had to hoist up water to wet the sails. This was the only solution to have the early sailing ships move just fast enough to allow the ship to be maneuvered.

Today's sail cloths consist almost exclusively of polyester or polyamide multifilament yarns that are often reinforced with monofilaments or aramid yarns. They are made in different weight categories, i.e., approximately 120 to 500 g/m^2. These textiles are normally dense fabrics that make special demands on the weaving machine. Today, the expectations from a sail cloth include storm and tear resistance, light weight and optimal cover for exploiting the slightest breeze (Figure 12.43). Sail cloth should be lightweight, e.g., the area of the spinnaker fabric of the racing yacht Black Magic, winner of America's cup 1995, was 650 m^2 and it weighed only 19.5 kg.

FIGURE 12.42 Sensitive plantations must be protected from strong wind.

FIGURE 12.43 Sail cloth.

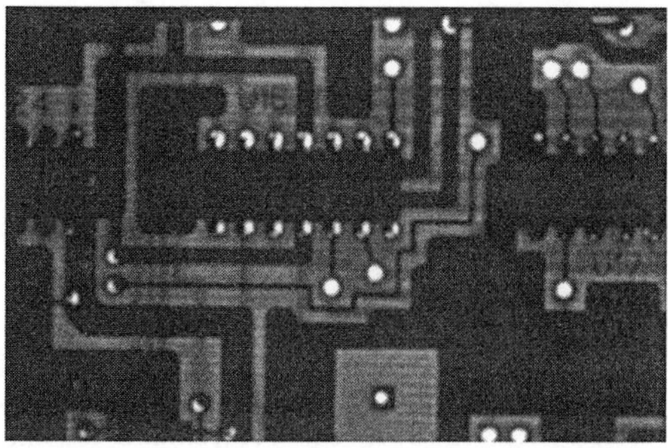

FIGURE 12.44 Fabrics made of glass fiber yarn are used to reinforce printed circuit boards.

To weave sail cloth, several portions of a regular weaving machine should be reinforced including drive, drive shafts, warp let-off, warp tensioner, cloth take-up and sley. To be able to successfully produce dense sail cloths of satisfactory quality, the following equipment is recommended:

- robust structure
- electronically controlled warp let-off
- sley with oblique position of 4°
- reed with 7 mm dents, bright polished
- temple with pressure from above
- selvage sealing device
- full width temple

12.3.7 Other Technical Textiles

Circuit Boards (Fiber Glass Weaving)

Printed circuit boards as used in electronics industry require high mechanical strength and dimensional stability; glass fabric reinforced composites are ideal for this type of application. Two to seven layers of carrier fabric are coated with synthetic resin. It is an application that makes extreme demands on fabric quality, because practically no fibrillation is tolerated (Figure 12.44).

Figure 12.45 shows the stress-strain diagrams of major textile fibers. Glass fibers are extremely strong and stretch only minimally when subjected to strain.

Low knot and loop strength of glass fibers causes the fibers to break which presents a challenge for weaving these yarns. This is why broken ends are repaired by gluing. Glass fiber yarns can be subjected to extreme stresses; but they consist of a large number of highly sensitive individual filaments, which for electronics applications, for example, are no more than 5–9 micrometers in diameter (0.6–1.7 dtex). Therefore, preventing yarn damage at high filling insertion rates is essential for glass fabric weaving. All weft glass fibers used in fabrics for circuit boards, from 2.8 to 136 tex, can be woven; the range for other applications extends up to 272 tex.

Air-jet machines have proven to be ideal for weaving glass fiber fabrics. However, industrial fabrics made of glass fibers make special demands on the weaving machine. The masses of all moving parts of the back rest roller system, including the extremely light whip roller, need to be minimized. Together with the reduced shed height, weaving is thus possible with the lowest possible warp tension.

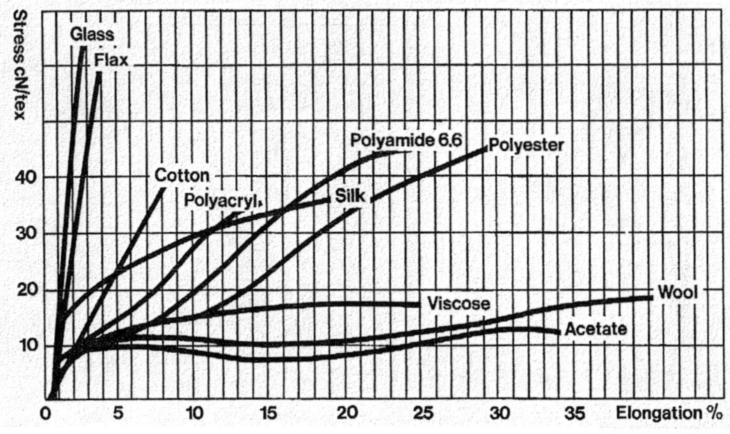

FIGURE 12.45 Stress-strain behavior of major textile fibers.

A vulcanized feed roller with smooth chrome-plated pressing rollers helps with cloth take-up without displacement of the weft. Not only is the number of starting marks minimized, but also creasing is prevented. For economical further processing, batching motions with central drive are normally used. Due to the low strain, fabric contraction is minimal and temples are therefore unnecessary.

An inside treadle motion or an electronically controlled negative dobby motion can be used as shedding. In contrast to the weaving machines for spun fiber and filament yarns, the dobby motion is mounted on a superstructure. The accumulation of glass fibers on it, which would result in heavier wear of the mechanical components, is thus avoided.

Machines for weaving glass fiber yarns are equipped with special electronically controlled weft yarn feeders (Figure 12.46). Balloon breakers reduce the take-up tension, enabling higher weft insertion rates to be achieved at lower air pressures. The integrated withdrawal device reduces tension peaks at the end of weft insertion.

The active weft clamp holds the left end of the weft, while on the right the stretching nozzle keeps the yarn tight until the warp is taken up.

The blades of the weft scissors are diamond coated, which greatly increases their service life. The selvage trimmers are of the chopper type, consisting of a knife that strikes a small plastic anvil—because good fabric quality includes clean cut selvages (Figure 12.47). The electric selvage trimmers can be programmed via the terminal. They require minimal maintenance. Constant weft arrival times are achieved by the automatic pick control glass system, which regulates the pressure of the main nozzle automatically.

Since the final windings of a weft package usually have a large number of filament defects, an optical sensor is used to stop the machine before the package runs out, so that the weaver can change the package. To avoid additional fibrillation, the drop wires are often removed. The weft is monitored with a profile sensor. Standard leno devices are used for selvage formation. As the selvages thus generated are flat-

FIGURE 12.46 Special weft yarn feeders ensure low tension weft insertion and thus high fabric quality.

FIGURE 12.47 Clean cutting and long life are typical characteristics of chopper-type cutters.

ter, less thickness is added when winding on. False selvages on the left and right can be generated or dispensed.

Construction Textiles

The use of textiles in buildings has long been accepted. Besides the rapidly expanding market for lightweight load bearing surfaces for everyday applications and hangar type structures, they are increasingly used in the field of larger scale roof structures.

Coated fabrics made of high strength polyester and polyamide yarns, aramids and glass fibers have developed into reliable, versatile and extremely economical construction materials [2].

Medical Textiles

In addition to the familiar applications of industrial fabrics in the medical sector ranging from object textiles for hospitals and doctors' practices to plasters and dressing gauze, new areas of application have been developed. Of special interest in this context are membranes, for example, blood filters and carbon fiber fabrics used in human medicine and for composite materials.

Fabrics woven from polyvinyl alcohol (PVA) are used in medical technology to treat internal and external wounds. They dissolve in hot water at 80°C with almost no residue. In making these fabrics, the warp yarns are twisted but they are not sized. There is too much accumulation of dust and abnormal running performance during weaving. Projectile weaving machines have been proven to be successful in weaving PVA fabrics [7].

Sports and Recreation Fabrics

Each type of sport has its own functional garment. Today's sporting clothing must be able to fulfill several protective functions, e.g. safety from injuries through grazes, abrasion resistance, resistance to cutting, permeability to vapor and impermeability to water. Up to date high tech fabrics meet these requirements optimally.

Other leisure time activities, like snowboarding or roller skating, attribute great importance to functional dresses. These sports can give rise to unexpected falls with extended sliding that creates frictional heat. This frictional heat can lead to burning of the skin. That is why this kind of clothing must feature perfect thermal insulation. The outer layer of such high tech fabric consists of an aramide fiber. This fiber is highly indifferent to chemical attacks, abrasion and heat of up to 450°C. Besides, yarns of high tenacity polyamide 6,6 and elastic yarns are processed into this fabric, lending it the required elasticity, resistance to tearing and wear resistance. For winter sports, wind and waterproof fabrics are important.

12.4 CARPET WEAVING

Woven carpets are produced on specially designed shuttle and rapier machines. Figure 12.48 shows a single shuttle carpet weaving machine.

Figure 12.49 shows a carpet weaving machine with double rapier. Figure 12.50 shows the schematic of the top view and cross section of a carpet machine. Two carpets are woven together and then they are separated by cutting the piles between them. It is also possible to weave single rapier weave structures where top and bottom rapiers operate alternately, providing single filling insertion.

In a typical double carpet machine, the filling yarns are drawn from cones. It is possible to have up to eight cones on a frame. Filling densities can

FIGURE 12.48 Single shuttle carpet weaving machine (courtesy of Hemaks).

FIGURE 12.49 Double rapier carpet weaving machine (courtesy of Michel Van De Wiele).

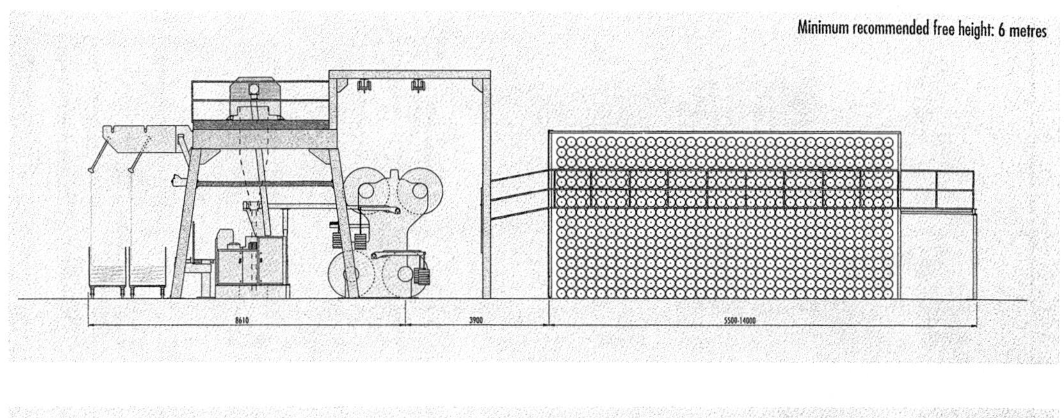

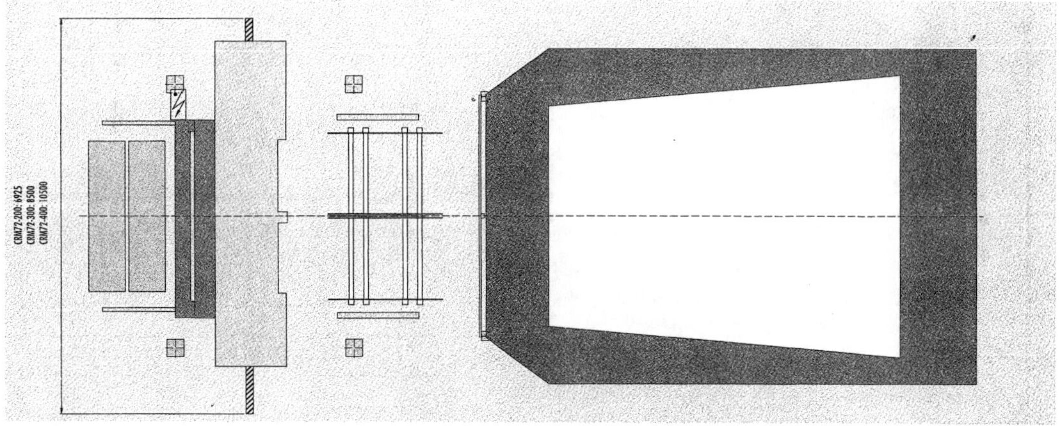

FIGURE 12.50 Schematics of a carpet machine (courtesy of Michel Van De Wiele).

be steplessly varied, typical values being 40 and 200 picks/dm. Typical warp densities used are between 25–50 ends/dm; 60,000–1,000,000 tufts per meter square is possible. The typical weight is 1000–5000 g/m^2. Double pile height can be between 9 to 30 mm. Up to 8 section warp beams can be accommodated in modern machines. Single-tier and double-tier models are available.

Figure 12.51 shows the schematic of a parallelogram reed motion mechanism that assists with the production of pile surfaces. Figure 12.52 shows some examples of carpet weave structures. Carpets can be produced in cut and loop piles as well. Figure 12.53 shows various carpet designs with combinations of cut and loop piles. Figure 12.54 shows the relations between insertion speed, pile density and production rate of a typical rapier carpet weaving machine. Today double carpet weaving machines have a speed of up to 130 ppm at 4 m width. Machine speed can be changed depending on the weaving conditions. Mechanical and electronic jacquard systems provide a high degree of design variability. Electronic jacquard systems can provide up to 5120 hooks. Recent high capacity jacquard carpet weaving machines have 4480 and 6720 solenoids. Weaving ten color carpets is possible. The productivity is around 50 m^2/hour.

The major pile construction techniques for woven carpets are [8]:

- 2 picks/2 revolutions, weave-through, with and without mixed contours
- 2 picks/2 revolutions, weave-through, dead pile partly floating
- 2 picks/2 revolutions, semi-weave through
- 2 picks/2 revolutions, not woven through
- 1 pick/2 revolutions, semi-weave-through, single gripper rod
- 1 pick/1 revolution, symmetrically woven
- 1 pick/1 revolution, symmetrical, combined with 2 picks/2 revolutions
- 3 picks/3 revolutions, weave-through
- 2 picks/3 revolutions, weave-through

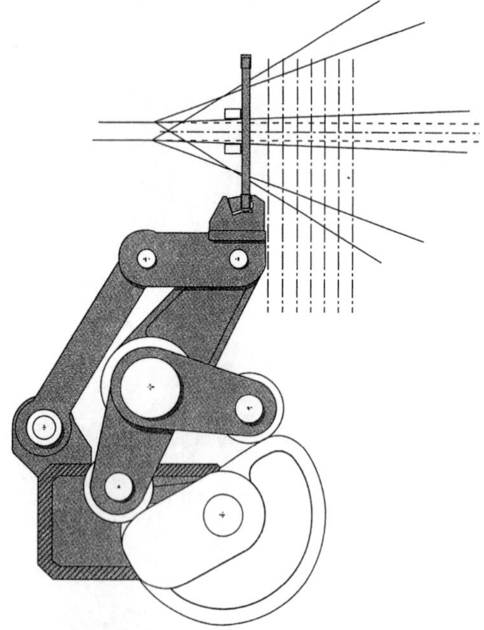

FIGURE 12.51 Schematic of parallelogram reed motion mechanism (courtesy of Michel Van De Wiele).

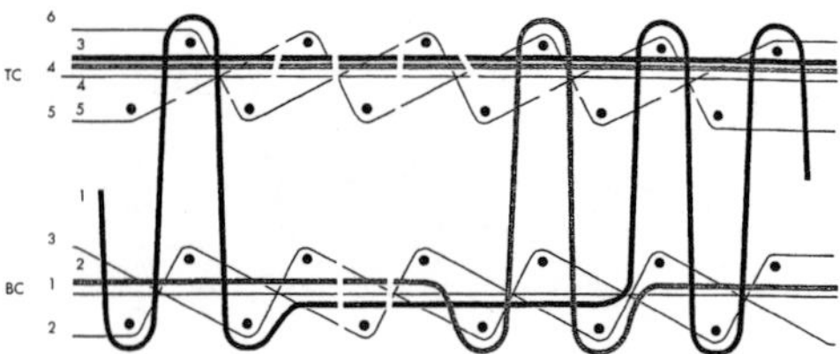

1/2 V weave structure without mixed colours

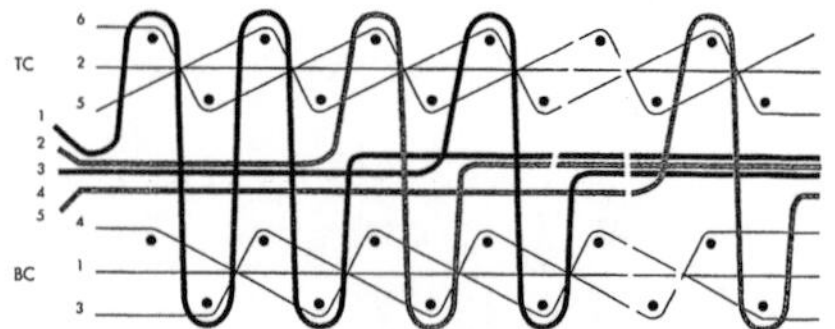

1/2 V weave structure with dead pile floating in the centre

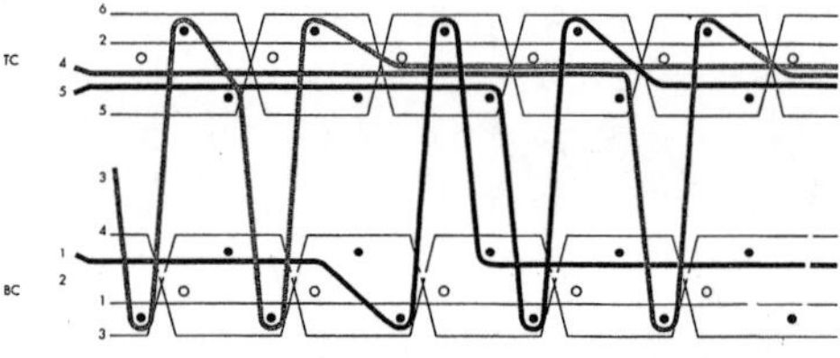

• WEFT INSERTION
○ NO WEFT INSERTION

1/3 V → 1/2 V weave structure without mixed colours

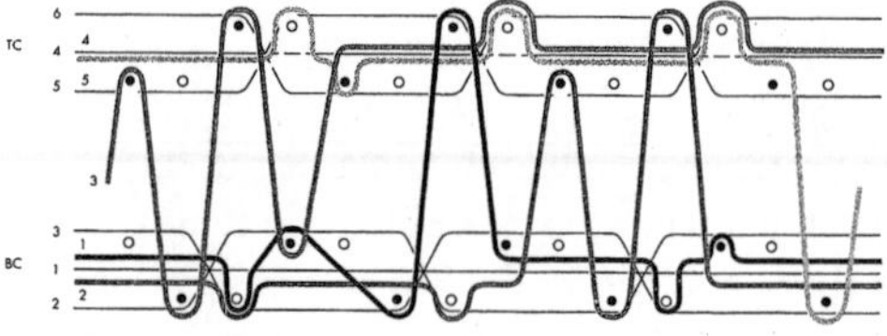

• RAPIER INSERTION
○ NO RAPIER INSERTION

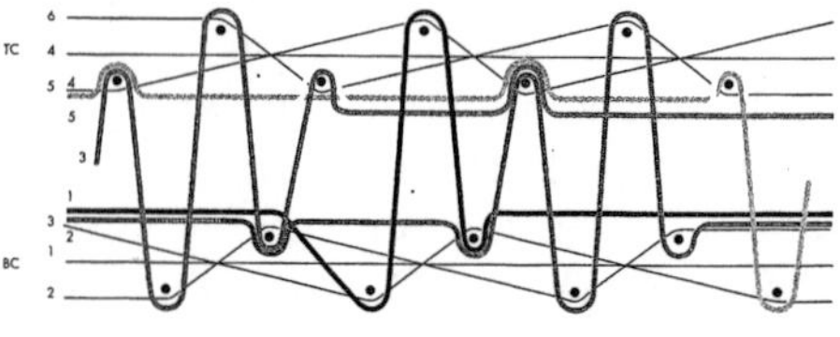

1/2 V → 1/1 V weave structure by alternating rapier interruption

FIGURE 12.52 Examples of carpet weave structures (courtesy of Michel Van De Wiele).

Three revolution constructions are considered to be classical constructions. Two revolution constructions are used for tight woven carpets due to their wear resistance and aesthetic properties. Today 3 rapier weaving machines are used to make 3 weft carpets by the 2 revolution method.

12.5 VELVET WEAVING

Velvet fabrics are used in a variety of applications such as jewelry boxes, film sealing, bathroom carpets, curtains, upholstery, automotive interiors, prayer mats and wall rugs. Long pile velvet is used to imitate fur and technical fabrics.

Velvet fabrics have a fluffy surface due to cut loops. Two layers of fabrics are woven together with a yarn binding them. When the binder yarn is cut, the layers are separated resulting in a fluffy surface on one side of each fabric. Figure 12.55 shows the schematic of velvet fabric weaving and formation. A bent reed and lancets are used to allow weaving with minimal shed dimensions to reduce the tension peaks on the pile warp ends.

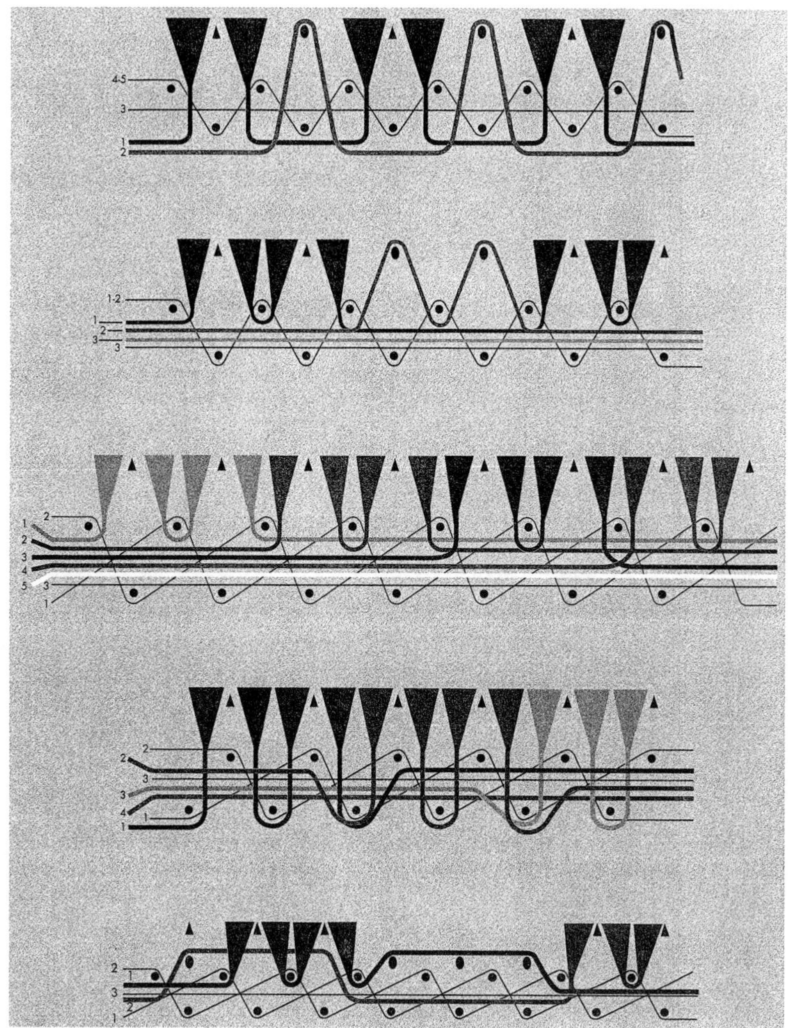

FIGURE 12.53 Examples of carpet designs with cut and loop piles (courtesy of Michel Van De Wiele).

Figure 12.56 shows machines for plain and dobby design velvet weaving and Figure 12.57 shows a jacquard design velvet machine.

One method to form loop pile is to use lancets. False picks are inserted above the lancets to form the pile; then, the false picks are automatically removed during weaving (Figure 12.58).

By controlling the pile and base warp yarns, various weave structures can be obtained. For example, sculptured loop pile velvet can be produced with flat woven filling and warp effects as shown in Figure 12.59. Shadow loop pile velvet is obtained by giving a direction to the loop pile. Color effects can be achieved using a filling selector. Any type of natural and man-made yarns are used including cotton, polypropylene, acrylic, polyamide, etc.

12.6 NARROW FABRIC WEAVING

Ribbons, tapes and webbings are all considered woven narrow fabrics if they contain woven selvages and are less than 30 cm. They are woven on special narrow fabric weaving machines,

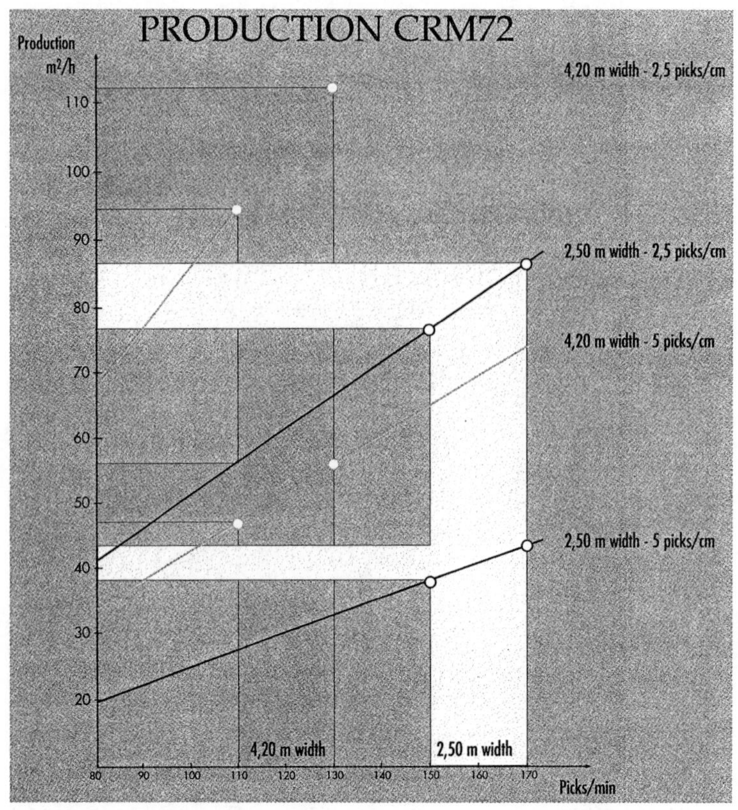

FIGURE 12.54 Relationships between insertion speed, pile density and production rate of a typical rapier carpet weaving machine (courtesy of Michel Van De Wiele).

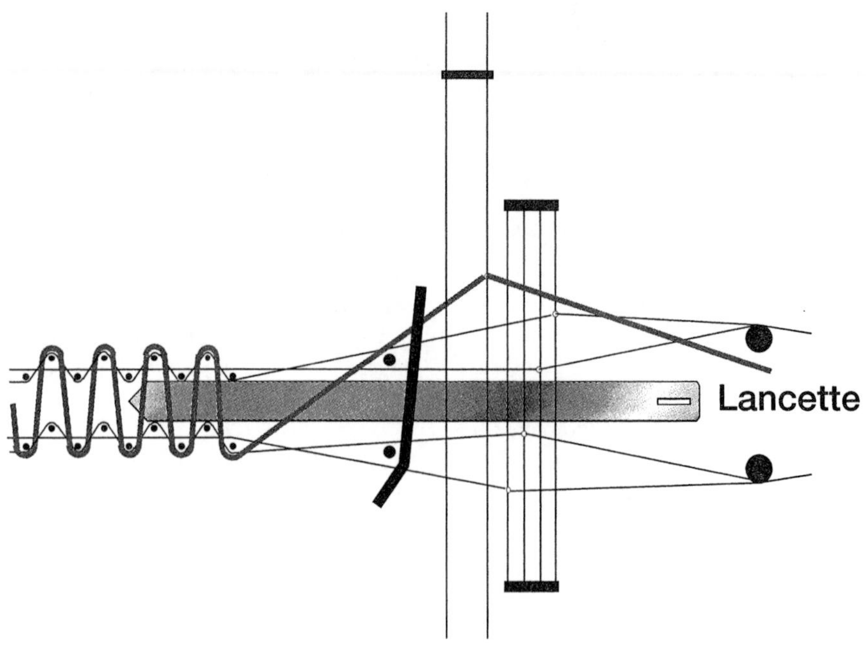

FIGURE 12.55 Schematic of velvet weaving (courtesy of Michel Van De Wiele).

FIGURE 12.56 Machines for plain and dobby velvet weaving (courtesy of Michel Van De Wiele).

using the basic principles of warp and filling interlacing. Several sets of warp yarns may be beamed to make several narrow fabrics, side by side, on the same weaving machine. Some tapes and ribbons are prepared by cutting full width fabrics into strips, and sealing the edges. Thermoplastic fiber ribbons can be made this way. Elastic webbing or tape is made by using bare or wrapped rubber warp yarn. Heavy webbings are often made with stuffer warps embedded between the webbing face and back. In essence, such webbings are double or tubular fabrics with reinforcing warp yarns lying between the two layers, and intersecting both sufficiently to bind the composite structure together, forming an integral unit.

FIGURE 12.57 Jacquard velvet machine (courtesy of Michel Van De Wiele).

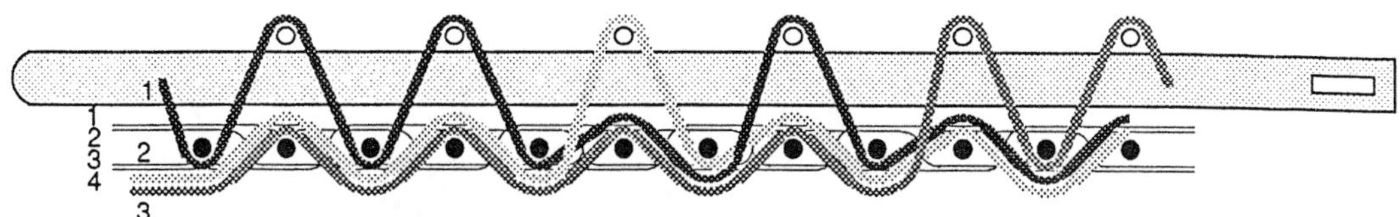

FIGURE 12.58 Forming loop pile with false picks (courtesy of Michel Van De Wiele).

Although some shuttle looms are still used, narrow woven fabrics are generally manufactured on needle weaving machines. Because of the importance of a good selvage, shuttleless methods are generally not used. Figure 12.60 shows a needle weaving machine for weaving light to medium heavy belts. Needle weaving machines usually can have up to 20 harnesses. In narrow fabric weaving, a single warp beam or multiple warp beams can be used. The warp yarn can also be taken directly from the creel for simple designs with small number of ends. Figure 12.61 shows a warping machine for small beams.

In a narrow fabric needle weaving machine, a hairpin loop of filling is inserted into the shed at very high speed by the needle. Then the filling yarn is locked-in at the opposite selvage by a knitting latch needle. Due to the loop of the filling yarn, the fabric always has double picks. The selvage on the filling insertion side is a woven selvage and the selvage on the latch needle side is a knit selvage. However, it is possible to have a regular selvage on both sides (Figure 12.62). Narrow fabric needle weaving machines can have up to 4000 picks per minute. Electronically controlled jacquard shedding is possible.

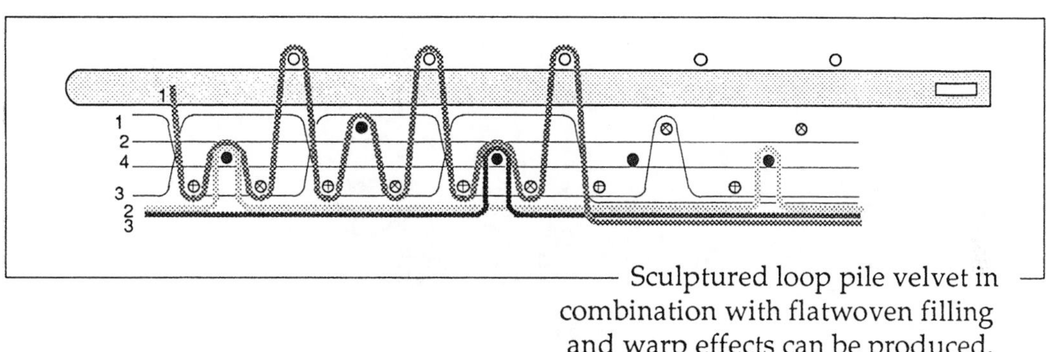

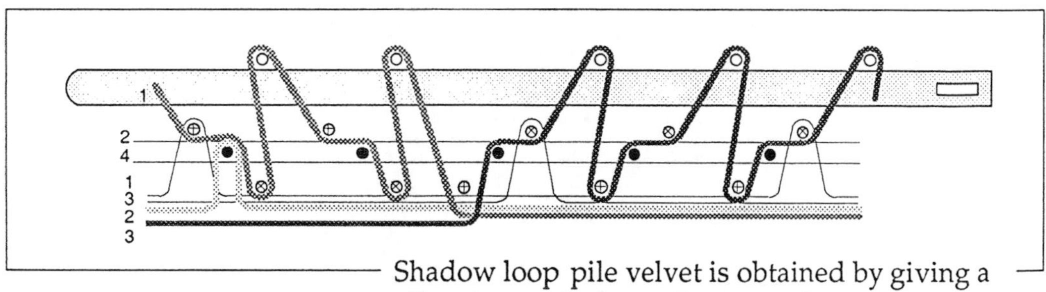

FIGURE 12.59 Various forms of loop piles (courtesy of Michel Van De Wiele).

FIGURE 12.60 Needle weaving machine (courtesy of Jakob Muller AG).

FIGURE 12.61 Warping machine for needle weaving machine (courtesy of Jakob Muller AG).

Fabrics that are more than 30 cm but less than 1 m wide are woven on narrow rapier or air-jet weaving machines. These machines are small scale replicas of regular shuttleless weaving machines. Figure 12.63 shows a narrow rapier weaving machine that is used for labels, woven images, ties, narrow fabrics for technical use, etc. The machine width is 115 cm.

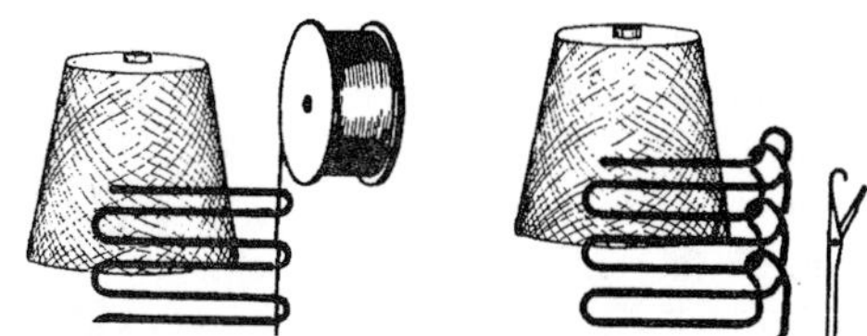

FIGURE 12.62 Selvage types (courtesy of Jakob Muller).

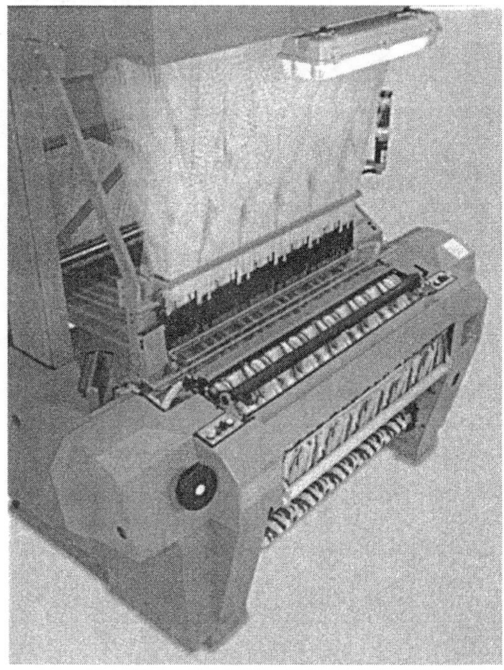

FIGURE 12.63 Rapier narrow weaving machine (Jakob Muller AG).

REFERENCES

1. Marfurt, P., "Rapier Weaving Machine for Pleated Fabrics", Melliand International (2), 1998.
2. Adanur, S., Wellington Sears Handbook of Industrial Textiles, Technomic Publishing Co., Inc., 1995.
3. Koerner, R., Designing with Geosynthetics, Prentice Hall, 1994.
4. Adanur, S., Paper Machine Clothing, Technomic Publishing Company, Inc., 1997.
5. Nick, U., Tire Cord Woven with Sulzer Ruti Projectile Weaving Machines P7100, August 1997.
6. Baumann, H., "10-Tonne Beating-up Force", Technical Review Sulzer, 3/1997.
7. Jamison, H., "Water Successfully Solves Disposal Problem", Sulzer Technical Review, 3/97.
8. Gossl, R., "CTM 640 Double Carpet Weaving Machine—Technical Parameters and Potential Uses", Melliand International (1), 1996.

SUGGESTED READING

- Smith, B. et al., "Industrials: Opportunities to Consider", Textile World, February 1998.
- Zuecker, L., "Carpet Weaving and Jacquard Machines—A Trend Report on ITMA '95", Melliand International (1), 1996.
- Vogel, R., "Servo Controlled Reed Beat-up: New Patterning Options", ITB International Textile Bulletin, 1/99.
- Smith, W. C. and McCurry, J. W., "Industrial Textiles Thrive Through Technology", Textile World, February 1999.
- Goldberg, S., "Textiles Race for Safety", ATI, February 1999.
- Smith, W. C., "High-Performance Fibers Protect, Improve Lives", Textile World, October 1998.

- Leifeld, M. et al., "Development of Intermingled Combination Yarns for Elastic Fabrics", Melliand International (2), 1998.
- "Flexible Air-Jet Weaving Machines for Terry Fabrics", Melliand International (2), 1998.
- Havich, M. M., "It's All in the Jeans", ATI, April 1998.
- McCurry, J. W., "Upholstery Mills Chase Soft Fabric", Textile World, 1998.
- LaForce, R. A., "Developing New Fabrics From Microfibers", ATI, November 1991.
- Fulmer, T. D., "Medical Applications of Textiles", ATI, May 1991.
- Bowen, D. A., "How Microdenier Fiber Will Change Your Processes", Textile World, October 1991.
- Smith, W., "Techtextil Spotlights Product Innovation", Textile World, November 1997.
- Smith, W., "Textiles in Automotives: Market of Significance", Textile World, June 1996.
- Page, D., "High-Tech Fabrics Aim Even Higher", Textile Horizons, February 1997.
- Moulin, G., "Optimized Pile Weaving Technologies for Carpets and Velvet", Textil Praxis International, March 1993.
- McCurry, J. W., "Towel Mills Modernize to Compete", Textile World, May 1999.

REVIEW QUESTIONS

1. Explain indigo dyeing processes for denims.
2. Why the size material is not removed from denim fabrics.
3. How are the multiheight loops formed on terry designs? What kind of mechanisms are used for this purpose?
4. What are the distinguishing characteristics of tire cord fabrics? Explain the reasons of these characteristics.
5. What kinds of weaving machines are the most suitable for tire cord weaving? Why?
6. What is the most important property of air bag fabrics? Why?
7. Find out what the circular weaving is for press fabrics? How is it done?
8. What are the five major functions that geotextiles play? Give application examples of these functions.
9. What are the common yarns and designs used in geotextiles?
10. What type of special fibers are used for safety and protective fabrics?
11. What are the special requirements for weaving aramid yarns?
12. What is the most important property of a sail cloth? How is it achieved?
13. Explain the challenges in fiberglass weaving and how they are addressed.
14. Compare the structures of woven carpets and tufted carpets.
15. What is the limitation of a narrow needle loom in fabric manufacturing?

13

Fabric Structure, Properties and Testing

After covering the design and manufacturing of woven fabrics in the previous chapters, it would be proper to analyze the fundamentals of woven structure in this chapter. The most important fabric properties are explained. Testing principles, methods and equipment for fabric testing are summarized. The purpose of this chapter is to relate the fabric properties to the manufacturing process and end use performance.

Fabrics are flexible yet strong. Flexibility is one of the most important characteristics of woven fabrics. The fabric flexibility is mostly due to flexible fibers and yarns in the fabric. Due to their polymeric nature and fine diameters, fibers are quite flexible. A staple yarn is a lot more flexible than a monofilament yarn of the same count. A multifilament or staple yarn may consist of several hundred or thousand fibers in its cross section. Although the fibers are twisted together in a staple yarn, there is still room for the fibers to move relative to each other (called fiber migration) under different types of loading, including bending, which results in a flexible structure of yarn. Increasing twist increases the stiffness of the yarn and therefore of fabric. Restricting fiber movement or slippage in the yarn, which is the case in sizing of warp yarns, increases the stiffness of the yarn. Fabric weave structure also affects flexibility to a certain extent.

The major factors that contribute to fabric strength are fiber inherent strength and yarn strength. Weave design also affects fabric strength, especially fabric modulus.

13.1 WOVEN FABRIC STRUCTURE

Woven fabric technology is deeply rooted in geometry. A fabric consists of millions of fibers assembled together in a particular geometry. The properties of a fabric depend on material properties, fiber and yarn structure and properties and fabric structure and geometry as shown in Table 12.1.

Peirce [1] specified eleven structural parameters that represent fabric construction as shown in Figure 13.1:

- L : length of yarn between yarn intersections
- P : projected length of yarn between the intersections
- C : yarn crimp
- H : distance between the center of yarn and fabric plane
- α : angle between horizontal direction and the yarn axis
- D : sum of the warp and filling yarn diameters

The first five parameters are for both warp and filling. These eleven interdependent parameters can be used to impart the major properties to the fabric. Although fabrics are considered to

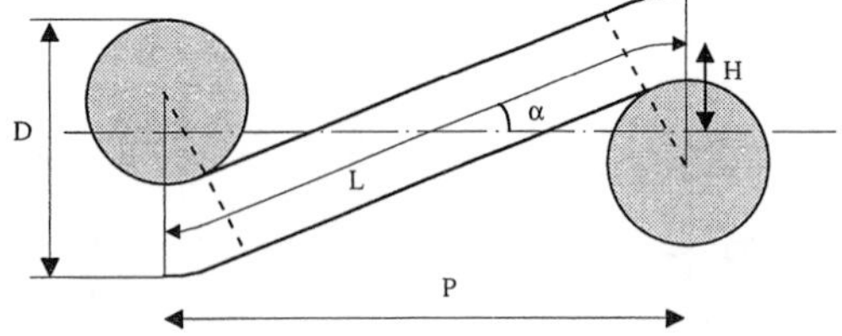

FIGURE 13.1 Geometry of the unit cell for a plain weave.

be two-dimensional for practical purposes, they are indeed three-dimensional.

Warp count is the number of warp yarns per unit width of fabric and filling count is the number of filling yarns per unit length of the fabric. Today, using optical/electronic technology, fabric yarns per unit length can be measured rapidly with extreme accuracy and repeatability.

13.2 WOVEN FABRIC PROPERTIES

Fabric Weight and Thickness

Fabric weight can be expressed in two ways: direct and indirect system. In the direct system, the fabric weight per area is given, e.g., g/m^2 or oz/yd^2. In the indirect system, which is used less in practice, usually the running length per weight is given. However, in this case the fabric width should also be specified. Fabric weight is affected by the following:

- fiber density
- yarn size
- contractions
- fabric construction
- weave pattern
- tensions during weaving
- finishing

Fabric thickness is important since it affects permeability and insulation characteristics of fabric. The standard test method for measuring thickness of textile materials is given in ASTM D1777 [2]. The thickness measurement is done at a specified pressure of the thickness gauge. The gauge pressure and area are usually reported along with the measured thickness.

Fabric Cover Factor and Density

One of the most important functions of fabrics is the covering function. There are two types of cover functions of fabrics: optical cover and geometrical cover.

The reflection and scattering of the incident light by the fabric surface is called optical cover function. Optical cover characteristics of fabric depend on the fiber material and fabric surface. Dyeing and finishing can change the optical cover properties. Geometrical cover is the area of fabric covered by fibers and yarns and is characterized by fabric cover factor.

Fabric cover factor (CF) is defined as the ratio of projected fabric surface area covered by yarns to the total fabric surface area and given by the following equation (Figure 13.2):

$$CF\ (\%) = (C_w + C_f - C_w \cdot C_f) \times 100 \qquad (13.1)$$

where C_w is the warp cover factor and C_f is the filling cover factor.

$$C_w = n_w \cdot d_w$$

$$C_f = n_f \cdot d_f$$

where n_w : warp count
n_f : filling count
d_w : diameter of warp yarn
d_f : diameter of filling yarn

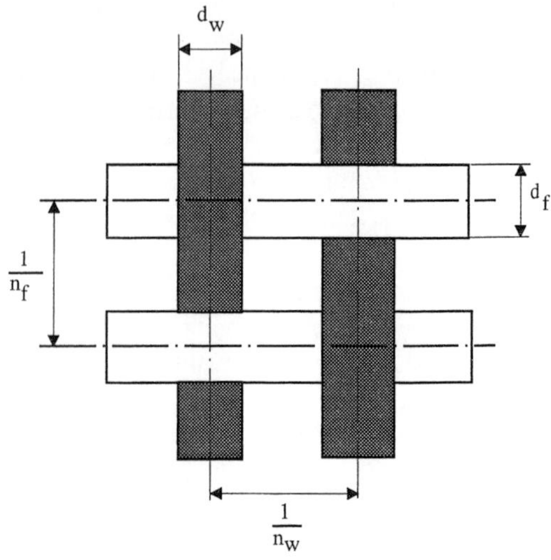

FIGURE 13.2 Cover factor diagram of a plain weave.

The maximum cover factor is 1 in which the yarns touch each other. This situation is called the jamming state. Theoretically, the cover factor can be larger than 1 in which the yarns pile up on each other giving multilayers of yarns. The property of a fabric that is affected most by the cover factor is the permeability. Liquid and gas (air) permeability depend on the cover factor to a great extent.

The density of a fabric is the weight per unit volume of the fabric. The unit volume is calculated by multiplying the unit area with fabric thickness. Theoretically, the fabric density can be very close to fiber density.

Crimp

Due to the necessity of interlacing in a woven fabric structure, at least one of the two yarn sets must have crimp. In most of the cases, both warp and filling have crimp. However, theoretically it is possible to have straight filling and crimped warp and vice versa as shown in Figure 13.3. The amount of crimp in each yarn can be largely controlled by controlling the yarn tensions during weaving. Crimp ratio between warp and filling can also be changed to a certain extent for some yarns such as monofilaments with heatsetting or finishing processes. Tensioning a yarn causes its crimp to be reduced. This will result in tension increase in the other yarn set. Another method to change crimp ratio is to relax and shrink the fabric with water or heat. Changing the ratio of crimp between the warp and filling yarn is called crimp transfer or crimp interchange.

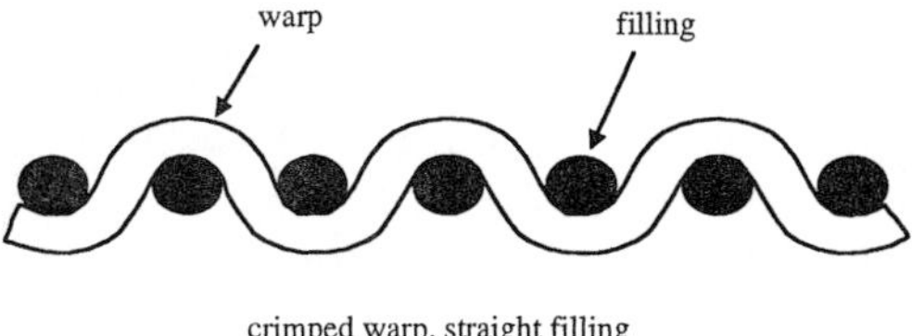

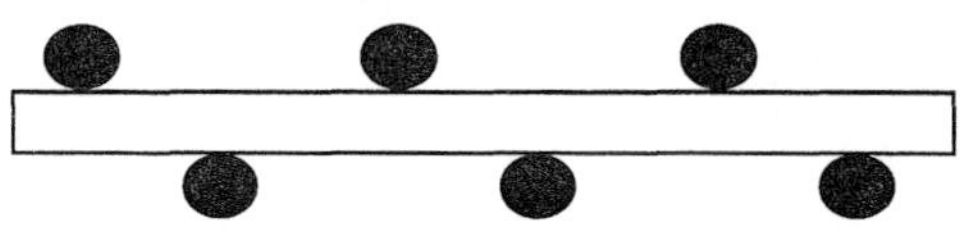

FIGURE 13.3 Possible yarn crimp variations in fabrics.

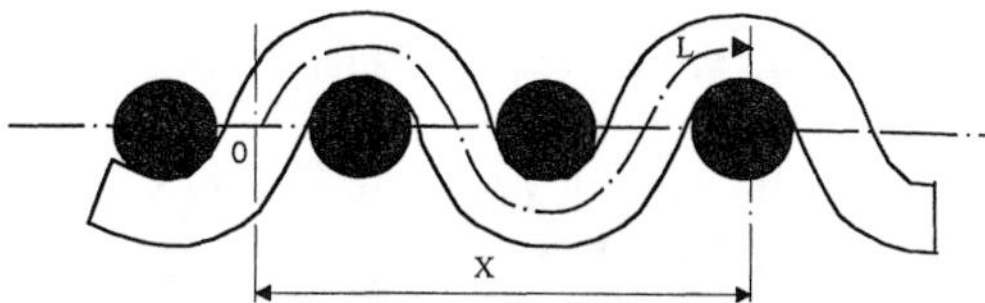

FIGURE 13.4 Crimped structure of a plain fabric.

Yarn crimp is also affected by the pattern of yarn interlacing in a fabric; high frequency of interlacing increases yarn crimp. For example, the plain weave has the highest frequency of interlacing and therefore the highest yarn crimp level in both warp and filling yarn. Satin (or sateen) weaves have the lowest frequency of interlacing and hence lowest degree of yarn crimp. Increasing yarn crimp in a particular direction decreases the fabric modulus and increases the elongation in that direction. This is because the tensile load is initially used to decrimp the yarn which is relatively easier than extending the yarn.

It should be noted that the fibers in a staple yarn also have some crimp. Moreover, in addition to the crimp due to interlacing, the yarn itself can also be wavy which could be considered another type of crimp. Crimp affects the weight, thickness, cover, flexibility and hand of fabric.

The American Society for Testing Materials (ASTM) makes two types of definitions related to crimp: percent crimp and percent take-up as shown in Figure 13.4 [2].

$$\%\ \text{Crimp (crimp factor)} = [(L - X)/X] \times 100 \quad (13.2)$$

$$\%\ \text{Take-up (contraction)} = [(L - X)/L] \times 100 \quad (13.3)$$

In practice, another related term, yield, is used quite often during manufacturing. Yield is defined as the ratio of woven fabric length to the warp length. Yield affects the fabric modulus. The yield, along with fabric design, material type, weave, thickness, warp diameter and tension, determines how much a fabric will stretch in a heatsetting operation. This will affect the fabric modulus and the fabric final sizing to achieve the proper length. This, in turn, determines fabric stability.

The uniformity of the yield is important because of the interaction among fabric properties. Changes in yield can cause differences in fabric thickness, air permeability, filling count, fabric width and appearance, as well as in fabric modulus.

Yield can be measured with a simple method: make a mark on the warp yarns and on the fabric at the same time; after, say, 10 cm of warp has come off the warp beam then make a mark again on the fabric. Measure the amount of fabric which was made in 10 cm of warp and divide this by 10 which gives the yield in the fabric.

Tensile Strength

Tensile strength is the most important property of a fabric. In almost every fabric development and manufacturing, tensile properties are reported. Modulus, breaking strength and elongation at break are widely used for quality control.

There are different types of fabric tensile tests that are used depending on the fabric and purpose: strip tensile test, grab tensile test and wide width tensile test. In the strip tensile test, a narrow strip of fabric sample is used (ASTM D5035 Breaking Force and Elongation of Textile Fabrics). The jaws of the tensile testing machine, which are wider than the fabric sample, clamp the sample on both ends and a tensile load is applied until fabric breaks. In the grab tensile test, the jaws are narrower than the fabric width to reduce the effect of Poisson's ratio (ASTM D5034 Breaking Force and Elongation of Textile Fabrics). Grab tensile test is more widely used for heavy industrial fabrics such as geotextiles. Wide width tensile tests are also used mostly for industrial textiles (e.g. ASTM D4595 for geotextiles). Narrow fabrics such as webbings, ribbons, etc., are usually tested at full width. Fabric modulus is measured using ASTM Test Method D 885 [2]. Specifications of textile machines for tensile testing are described in ASTM D 76. The terminology of tensile properties of textiles is given in ASTM D4848. Other ASTM test methods related to tensile testing include:

- ASTM D 1775 Standard Test Method for Tension and Elongation of Wide Elastic Fabrics (Constant Rate-of-Load Type Tensile Testing Machine)
- ASTM D 4964 Tension/Elongation of Wide/Narrow Elastic Fabrics by Constant Rate of Elongation Type Tensile Testing Machine

When fabric is extended in one direction (uniaxial load), first, crimp in that direction decreases. Fabric is relatively easy to extend during crimp decrease. After that, the yarn material starts bearing the load which would reduce the extension rate of the fabric. While crimp is decreasing in one direction (load direction), it increases in the opposite direction. Crimp interchange continues until a force equilibrium is attained. In biaxial loading, force is applied in two directions. In this case, the crimp interchange depends on the magnitude of the forces.

Warp and filling yarns exert forces onto each other at the crossover points. Since yarns are compressible, these forces cause the yarns to deform and take an elliptic shape in the fabric structure rather than a near round shape. The height and width of the yarn's elliptic shape depend on the twist level. The ratio of height/width is called the aspect ratio. Fabric cover factor is affected by the width and the crimp is affected by the height.

Tear Resistance

In a relatively dense fabric, individual yarns oppose to the tearing load one by one, that is why propagation of tear is relatively easy. If the number of yarns per unit length is low, then the yarns are allowed to displace themselves and form groups to resist the tear in groups rather than individually. This increases tear resistance of the fabric, that is why loosely woven fabrics have better tear resistance than dense or coated fabrics. For example, a gauze fabric is difficult to tear because of low number of yarns. Fabric weave also has an effect on tear resistance. A 2×2 basket weave has higher tear resistance than a plain weave since two yarns act together against tear.

The tests that are used to measure tear resistance of fabrics are tongue (ASTM D 2261), and Elmendorf test (ASTM D 1424). The first method uses a tensile testing machine to measure the tearing force. Elmendorf uses a pendulum and measures the tear energy. The trapezoid tear test, which uses another method, is not used anymore.

Fabric Bow and Skew (Figure 13.5)

The condition in which the filling yarns lie in the fabric in the shape of an arc is called fabric bow. Bow is expressed as:

$$\% \text{ Bow} = (KL/MN) \times 100 \qquad (13.4)$$

Fabric bow can be symmetrical or non-symmetrical.

The condition in which the filling yarns in a fabric do not lie perpendicular to the warp yarns is called skew. Skew is expressed as:

$$\% \text{ Skew} = (PQ/QR) \times 100 \qquad (13.5)$$

Fabrics with bow or skew are not acceptable in today's quality conscious business environment. If not corrected, bowed or skewed materials become seconds or reruns. Bow and skew detection and correction machines are used for this purpose. Figure 13.6 shows a single width terry towel bow and skew control machine. This particular machine detects and corrects bow and skew error by monitoring the cut bands oriented in the filling direction of the single width web. Using an optical detection system, the exposed portion of the cut band's filling yarn is electronically monitored. Correction signals are sent to the machine where straightening of bow and skew distortions occurs automatically [3].

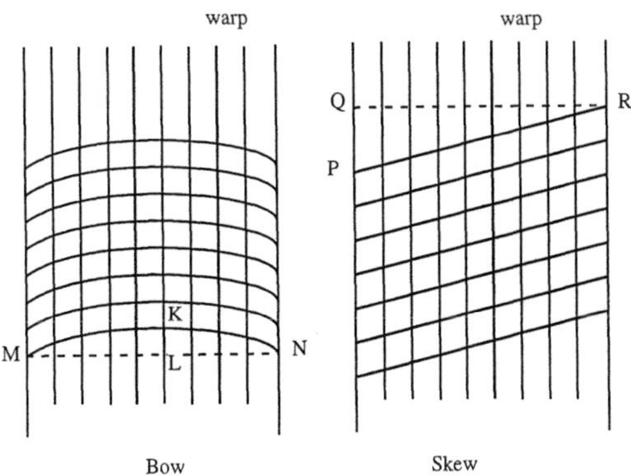

FIGURE 13.5 Fabric bow and skew.

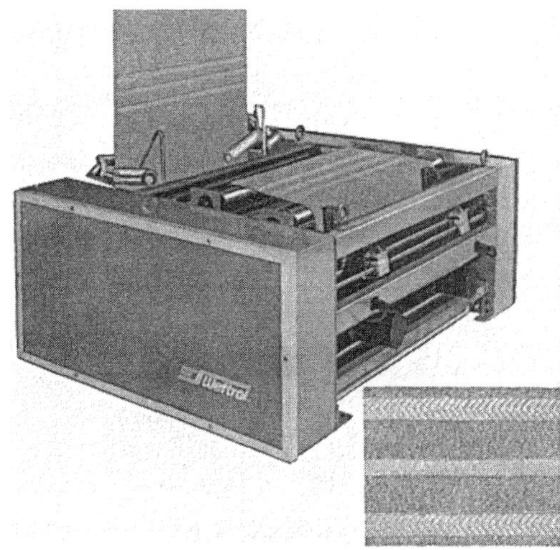

FIGURE 13.6 Terry cloth bow and skew control machine (courtesy of Mount Hope).

A bow and skew optical sensing system consists of a sensing unit through which the fabric is threaded. The sensing unit consists of light sources that project light beams through the fabric and optical receivers that read the filling angle at each sensing point. The light sources and receivers are mounted at a fixed angle to the fabric to optimize the signals through dense plain and twill fabrics such as denims. Light intensity can be automatically adjusted based on the density of the fabric for proper signal generation. The receiver heads on the opposite side of the fabric have lenses to focus the image of the fabric on a semiconductor silicon chip. The chip has an array of radially oriented photosensitive lines that generate electrical signals proportional to the light received through the fabric. These signals are amplified and sent to the angle computer card for calculation of the filling angle. Figure 13.7 shows a high intensity optical sensing system for denim and other dense fabrics.

Air Permeability

Air flow through a fabric is very complex due to the complicated structure of the fabric. The air flow through a fabric at a pressure difference

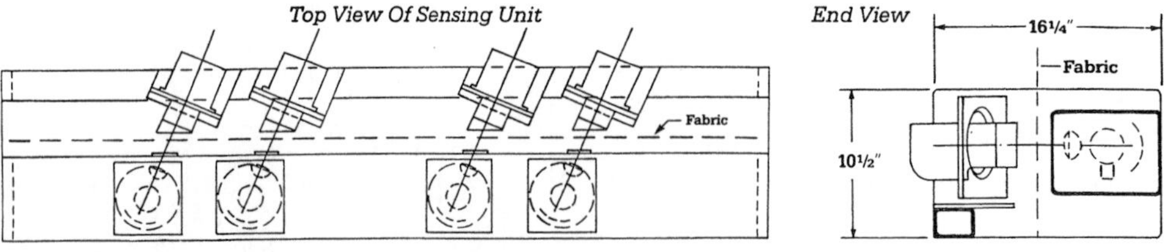

FIGURE 13.7 High intensity optical bow and skew sensing system for denims and other dense fabrics (courtesy of Mount Hope).

between the two surfaces is directly proportional to the open area of the fabric and square root of the differential pressure. High differential pressure further complicates the air flow since the viscoelastic fibers may deform under pressure.

Fabric air permeability is a measure of air flow through the fabric at the standard pressure drop. The air permeability is measured as the air flow in cubic feet of air per square foot of fabric per minute in the standard system and in cubic meter of air per square meter of fabric per minute in the metric system.

Air permeability is of considerable value in predicting insulation characteristics of a fabric. Increasing twist in the yarn increases the air permeability of the fabric. Air permeability test of textile fabrics is described in ASTM D737.

Void Volume

Void volume is the amount of space in a volume of fabric that is not occupied by solid material (Figure 13.8). Both fabric thickness and internal structure affect absolute void volume.

Void volume can be calculated as follows:

Total volume
= length x width × thickness

Calculated void
= total volume – (weight of sample/density of material)

% void volume
= (calculated void/total volume) × 100 (13.6)

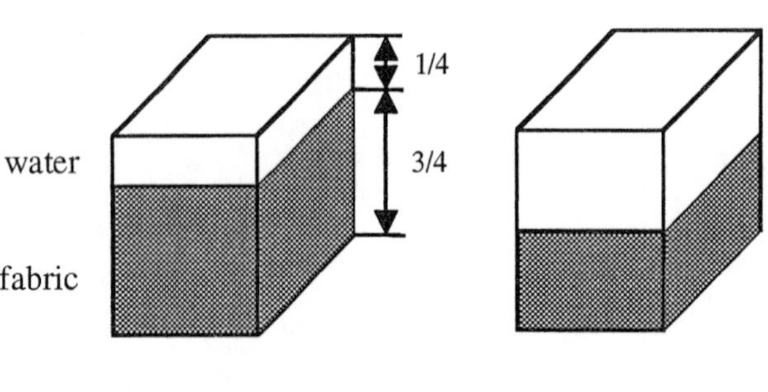

FIGURE 13.8 Void volume.

Abrasion Resistance

Both the fiber material and fabric geometry affect the abrasion resistance of a fabric. Some polymers are intrinsically better abrasion resistant than others. The twist level, yarn crimp and weave design affect the abrasion resistance of the fabric. The amount of fiber and yarn surface that is in contact with the abradant is important. Increasing surface contact increases the abrasion resistance of the fabric. Low twist yarns may present greater surface to the abradant; however, too little twist may leave loose fibers protruding from the yarn body which may be snagged or broken during abrasion. High twist reduces the abrasion resistance of the yarn. With today's technology, it is possible to arrange abrasion resistant fibers on the sheath while having fibers with high tensile strength in the core.

Abrasion resistance of fabrics is measured in terms of visual appearance, number of cycles to open a hole in the fabric and residual strength of the fabric. There are several tests for abrasion resistance:

- Inflated Diaphragm Test (ASTM D3886)
- Flexing and Abrasion Method (ASTM D3885)
- Oscillatory Cylinder Method (ASTM D4157)
- Rotary Platform Double Head Method (ASTM D3884)
- Uniform Abrasion Method (ASTM D4158)
- The Accelerotor (AATCC 93)
- Martindale Abrasion Tester (ASTM D4966)
- Special Webbing Abrader

Burst and Impact Resistance

Some applications require resistance of fabrics against pressure forces which are perpendicular to the fabric plane. Filter fabrics, geotextiles, parachutes, transportation bags, air and tension structure fabrics must often withstand considerable bursting pressure. Bursting loading is similar to biaxial tensile loading in which fiber and yarn moduli play an important role. For better burst resistance, fabrics are designed to have equal properties in warp and filling directions. There are two types of tests to measure burst resistance of fabrics: the diaphragm test (ASTM D3786 Mullen burst test) and the ball burst test (ASTM D3787).

Some fabrics are designed to withstand impact loading. Ballistic protective fabrics, airbags and seat belts are examples of these types of fabrics. The key to high impact resistance of a fabric is good energy absorption in a short time. The energy absorbing capability of a fabric is indicated by the area under the load-elongation curve. There are different test methods to measure impact resistance of fabrics including free falling weights, dropping pendulums and shooting devices [4].

Flexibility and Stiffness

Strength and flexibility are the two properties that make textiles unique. Fabric flexibility is affected by the flexibility properties of the constituent fibers and by the yarn and fabric structure.

To measure the stiffness of fabrics, two methods are used: either the fabric is bent under its own weight or an external load is applied to the fabric. In the cantilever test (ASTM D1388), a strip of fabric is bent under its own weight. Bending length is one-half of the resulting overhanging length. The stiffness (flexural rigidity) is obtained by multiplying the cube of the bending length by the fabric weight per unit area. The Heart Loop Test method is described in ASTM D1388 which requires no commercial tester. Another fabric stiffness test, the Circular Bend test, is done according to ASTM D 4032.

Drape and Hand

Drape and hand are extremely important for apparel fabrics. Drape can be defined as the ability of a fabric to bend under its own weight to form folds. Hand or handle is a subjective property that can be related to the comfort perception of the fabric.

An analysis of fabric hand has been described by the ASTM as being composed of eight components: compressibility, flexibility, extensibility, density, resilience, surface contour, surface friction and thermal properties. The measurement of these properties does not give one an evaluation of hand. Sueo Kawabata of Japan approached the task of providing a single value for hand by starting with the development of instruments that would be capable of evaluating the desired fabric properties under low load conditions. He believed that this would more closely relate to the human concept of hand. His instruments were designed to measure the hand related properties: tensile and shear behavior, bending behavior, compressive behavior, and surface roughness and friction. These properties are similar to those listed by the ASTM with the exception of thermal characteristics. Kawabata developed an equation that gives a weighing to each of the measured properties and called the resultant summation Total Hand Value. The weighing factors were developed through extensive human subjective evaluations of a range of fabric types and the ranking of characteristics. The weighing factors are believed to be appropriate for the population within which the data were taken but there is some question as to the application of the same weighing factors in a different culture. The instruments of the Kawabata Evaluation System (KES) can be used for determining the listed fabric properties and are useful in providing relative data for fabric comparisons.

Compared to other types of fabrics such as knit, braided and nonwoven, woven fabrics wrinkle and retain crease more. This is a good property when crease is wanted, e.g., ironing. However, for the most of the time besides ironing, wrinkle resistant fabrics are desired. Fabrics with high extensible fibers that have good elastic recovery usually have good wrinkle resistance. Fibers with high secondary creep have low wrinkle resistance. Fabric wrinkle resistance is also affected by temperature and relative humidity. Densely woven fabrics have less wrinkle resistance due to low freedom for fiber movement.

Flame Resistance

Flame resistance can be obtained in two ways:

a) by using inherently flame resistant fibers such as Nomex® aramid
b) by treating (coating) the fiber or fabric with flame resistant chemicals

The disadvantage of using fiber/fabric coating is the decrease in flame resistance with repeated washings. However, this method may be less expensive than using inherently flame resistant fibers.

There are numerous standard test methods that deal with fire and flammability. A compilation of over 100 ASTM standards is given in the book Fire Test Standards published by the ASTM.

One of the beauties of textile technology is that it allows mixing of materials in almost every stage of fabric production. In staple yarn manufacturing, different fibers can be intimately blended together to improve certain properties of yarns. For example a cotton, nylon and carbon fiber blend yarn can have comfort properties due to cotton, good abrasion resistance due to nylon and flame resistance due to carbon fiber. Another example is Nomex® III by DuPont, which is made of 95% Nomex® aramid for flame resistance and 5% Kevlar® aramid for strength. Morever, during yarn manufacturing, different single yarns can be plied together to alter properties. In fabric manufacturing, different warp and filling yarns can be chosen. Moreover, filling mixing of several filling yarns is possible with today's technology.

13.3 WOVEN FABRIC IDENTIFICATION

Table 13.1 lists the more commonly produced fabric types by name. Woven fabric structures can be identified with the naked eye or with microscope. The following guidelines are generally applicable in identification of various characteristics of woven fabrics.

Determination of Warp Direction

If one set of yarns have ply in the fabric, it is usually the warp yarns. Warp yarn needs to be stronger than the filling yarn due to heavy forces acting upon them. In general, the warp density (ends/unit length) is more than the filling density (fillings/unit length). In the fabric, warp yarns are usually straighter than filling yarns since filling yarns may have more tendency for bow and skewness. The selvage of the fabric runs parallel to the warp direction. In greige fabrics, the warps may still have the size material on them which makes the yarns stiffer. Prominent stripes or marks are usually in the warp direction. Reed marks also run in the warp direction. If the crimp levels are different, it is a high

TABLE 13.1 Common fabric type names.

Fabric	Description
Batiste	: Fine, transparent, smooth hand (mercerized cotton)
Bedford cord	: Warpwise cords; uncut
Bengoline	: Coarse warp, fine fill, pronounced warp ribs
Broad cloth	: Finer warp; warp density (number) is twice of filling density
Chambray	: Colored warp, white fill, fine to medium yarns
Corduroy	: Uncut is like bedford; cut is unique
Crepe	: Rough hand
Crinnoline	: Stiff, open
Denim	: Colored warp, white or colored filling, twill weave, stiff
Drill	: Like denim but not blue and medium weight
Duck	: Fabric made of heavy, plied yarns
Flannel	: Brushed, soft hand
Gabardine	: Twill, finer yarns, armed services dress uniform
Gingham	: Plaid or checks
Huck	: Flat toweling; floats checkboard motif
Lawn	: Fine, transparent, unmercerized
Muslin sheeting	: Same as percale, but coarser, more open
Osnaburg	: Cheap printcloth, coarser yarns; lower number of warps and filling
Oxford	: Fine warp, coarse filling, 2/2 filling rib
Percale sheeting	: Balanced warp and weft density (numbers), balanced yarn counts (density); more yarns than muslin
Plaid	: Color pattern both ways
Poplin	: Finer warp than weft; slightly pronounced ribs in weft direction
Print	: Like percale but with print
Sateen	: A satin weave where filling yarns dominate on top
Sport denim	: Lighter weight, less stiff

probability that there is more crimp in warp yarns.

Determination of Face and Back of Woven Fabrics

In general, the fabric design is more visible on the face. For example, in a twill fabric, twill lines are more prominent on the face. Ribs are more visible on the face in a ribbed fabric. Satins are smoother on the face than the back. Slub yarn fabrics are more distinct on the face. The face of the napped fabrics is fuzzier and softer. The face is usually finer and more lustrous on double fabrics. The face would have less reed marks than the back. In finished fabrics, the face has better finish quality. In printed fabrics, the prints on the face are more clear and the colors predominate.

Determination of the Order of Interlacing (Weave)

Order of interlacing can be determined with the naked eye for coarse fabrics or using a magnifying glass or a microscope for fine fabrics. It is important that an undistorted sample that is larger than the repeat unit (by estimation) is examined from the main body of the fabric for this purpose. Starting at a randomly selected point on the lower left side of the fabric, the interlacing pattern of the warp and filling yarns is determined until a repeat is found in both directions. Warp yarns are numbered from left to right and filling yarns are counted from bottom to top. The selvage design is determined in a similar way. However, it is usually drastically different than the rest of the fabric.

Determination of the Presence of Size and Finish

Sometimes observation by the naked eye is enough to detect the size or finish on the fabric. The next step would be to determine the hand properties of fabrics such as stiffness, smoothness, etc. If necessary, the sample can be observed under a microscope.

Standard Test Methods

Tables 13.2 and 13.3 list the ASTM and AATCC (American Association for Textile Chemists and Colorists) standard fabric test methods. Company test methods are not included in these tables.

TABLE 13.2 Standard test methods used for fabric testing (copyright ASTM).

Standard	Title
ASTM D434	Resistance to Slippage of Yarns in Woven Fabrics Using a Standard Seam
ASTM D737	Air Permeability of Textile Fabrics
ASTM D885	Testing of Tire Cords, Tire Cord Fabrics, and Industrial Filament Yarns Made from Man-Made Organic-Base Fibers (D885 M is in metric units)
ASTM D1117	Testing Nonwoven Fabrics
ASTM D1230	Test Method for Flammability of Clothing Textiles
ASTM D1388	Stiffness of Fabrics
ASTM D1424	Tear Resistance of Woven Fabrics by Falling Pendulum (Elmendorf Apparatus)
ASTM D1683	Failure in Sewn Seams of Woven Fabrics
ASTM D1777	Thickness of Textile Materials
ASTM D1908	Needle-Related Damage Due to Sewing in Woven Fabrics
ASTM D2261	Tearing Strength of Woven Fabrics by the Tongue (Single Rip) Method (CRE Tensile Testing Machine)
ASTM D2262	Tearing Strength of Woven Fabrics by the Tongue (Single Rip) Method (CRT Tensile Testing Machine)
ASTM D2475	Standard Specification for Wool Felt
ASTM D2594	Stretch Properties of Knitted Fabrics Having Low Power
ASTM D2646	Backing Fabrics
ASTM D3107	Stretch Properties of Fabrics Woven from Stretch Yarns
ASTM D3511	Pilling Resistance and Other Related Surface Changes of Textile Fabrics; Brush Pilling Tester Method
ASTM D3512	Pilling Resistance and Other Related Surface Changes of Textile Fabrics; Random Tumble Pilling Tester Method
ASTM D3514	Resistance of Apparel Fabrics to Pilling (Elastomeric Pad Method)
ASTM D3597	Woven Upholstery Fabrics—Plain, Tufted or Flucked
ASTM D3773	Length of Woven Fabrics
ASTM D3774	Width of Woven Fabrics
ASTM D3775	Fabric Count of Woven Fabric
ASTM D3776	Mass per Unit Area (Weight) of Woven Fabric
ASTM D3786	Hydraulic Bursting Strength of Knitted Goods and Nonwoven Fabrics—Diaphragm Bursting Strength Tester Method
ASTM D3787	Bursting Strength of Knitted Goods—Constant-Rate-of-Traverse (CRT) Ball Burst Test
ASTM D3884	Abrasion Resistance of Textile Fabrics (Rotary Platform, Double—Head Method)
ASTM D3885	Abrasion Resistance of Textile Fabrics (Flexing and Abrasion Method)
ASTM D3886	Abrasion Resistance of Textile Fabrics (Inflated Diaphragm Method)
ASTM D3887	Standard Specification for Knitted Fabrics
ASTM D3936	Delamination Strength of Secondary Backing of Pile Floor Coverings
ASTM D3940	Bursting Strength (Load) and Elongation of Sewn Seams of Knit or Woven Stretch Textile Fabrics
ASTM D4033	Resistance to Yarn Slippage at the Sewn Seam in Upholstery Fabrics (Dynamic Fatigue Method)
ASTM D4034	Resistance to Yarn Slippage at the Sewn Seam in Woven Upholstery Fabrics
ASTM D4032	Stiffness of Fabric by the Circular Bend Procedure
ASTM D4157	Abrasion Resistance of Textile Fabrics (Oscillatory Cylinder Method)
ASTM D4158	Abrasion Resistance of Textile Fabrics (Uniform Abrasion Method)
ASTM D4848	Standard Terminology of Force, Deformation and Related Properties of Textiles
ASTM D4852	Coated and Laminated Fabrics for Architectural Use
ASTM D4966	Abrasion Resistance of Textile Fabrics (Martindale Abrasion Tester Method)
ASTM D4970	Pilling Resistance and other Related Surface Changes of Textile Fabrics (Martindale Pressure Tester Method)

TABLE 13.3 Standard AATCC test methods used for fabric testing [5] (copyright AATCC).

Test Title	Method Number
Abrasion Resistance of Fabrics: Accelerotor Method	93-1989
Absorbency of Bleached Textiles	79-1992
Aging of Sulfur-Dyed Textiles: Accelerated	26-1989
Alkali in Bleach Baths Containing Hydrogen Peroxide	98-1989
Alkali in Wet Processed Textiles: Total	144-1992
Analysis of Textiles: Finishes, Identification of	94-1992
Antibacterial Activity Assessment of Textile Materials: Parallel Streak Method	147-1993
Antibacterial Finishes on Textile Materials, Assessment of	100-1993
Antifungal Activity, Assessment on Textile Materials: Mildew and Rot Resistance of Textile Materials	30-1993
Antimicrobial Activity Assessment of Carpets	174-1993
Appearance of Apparel and Other Textile End Products after Repeated Home Laundering	143-1992
Appearance of Fabrics after Repeated Home Launderings	124-1992
Appearance of Flocked Fabrics after Repeated Home Laundering and/or Coin-Op Dry-Cleaning	142-1989
Ash Content of Bleached Cellulosic Textiles	78-1989
Bacterial Alpha-Amylase Enzymes Used in Desizing, Assay of	103-1989
Barre': Visual Assessment and Grading	178-1993
Bond Strength of Bonded and Laminated Fabrics	136-1989
CMC: Calculation of Small Color Differences for Acceptability	173-1992
Carpets: Cleaning of; Hot Water (Steam) Extraction Method	171-1989
Carpet Soiling:	
Accelerated Soiling Method	123-1989
Service Soiling Method	122-1989
Visual Rating Method	121-1989
Chelating Agents: Active Ingredient Content of Polyaminopolycarboxylic Acids and Their Salts; Copper PAN Method	168-1992
Chelating Agents: Disperse Dye Shade Change Caused by Metals; Control of	161-1992
Chelation Agents: Chelation Value of Aminopolycarboxylic Acids and Their Salts; Calcium Oxalate Method	149-1992
Chlorine, Retained, Tensile Loss: Multiple Sample Method	114-1989
Chlorine, Retained, Tensile Loss: Single Sample Method	92-1989
Chromatic Transference Scale	EP3
Color Change Due to Flat Abrasion (Frosting):	
Emery Method	120-1989
Screen Wire Method	119-1989
Color Measurement of Textiles: Instrumental	153-1985
Color Measurement of the Blue Wool Lightfastness Standards: Instrumental	145-1985
Colorfastness to:	
Acids and Alkalis	6-1989
Bleaching with Chlorine	3-1989
Bleaching with Peroxide	101-1989
Burnt Gas Fumes	23-1989
Carbonizing	11-1989
Crocking: Carpets-AATCC Crockmeter Method	165-1993
Crocking: AATCC Crockmeter Method	8-1989
Crocking: Rotary Vertical Crockmeter Method	116-1989
Degumming	7-1989
Drycleaning	132-1993
Dye Transfer in Storage: Fabric-to-Fabric	163-1992
Fulling	2-1989
Heat: Dry (Excluding Pressing)	117-1989
Heat: Hot Pressing	133-1989
Laundering, Home and Commercial: Accelerated	61-1993
Light	16-1993
Light at Elevated Temperatures and Humidity: Water Cooled Xenon Lamp Apparatus	177-1993
Light: Detection of Photochromism	139-1989
Non-Chlorine Bleach in Home Laundering	172-1990

TABLE 13.3 (continued).

Test Title	Method Number
Oxides of Nitrogen in the Atmosphere under High Humidities	164-1992
Ozone in the Atmosphere under Low Humidities	109-1992
Ozone in the Atmosphere under High Humidities	129-1990
Perspiration	15-1989
Pleating, Steam	131-1990
Solvent Spotting: Perchloroethylene	157-1990
Stoving	9-1989
Water	107-1991
Water: Chlorinated Pool	162-1991
Water: Sea	106-1991
Water Spotting	104-1989
Water and Light: Alternate Exposure	125-1991
Water (High Humidity) and Light: Alternate Exposure	126-1991
Compatibility of Basic Dyes for Acrylic Fibers	141-1989
Creases; in Fabrics, Retention of, after Repeated Home Laundering	88C-1992
Dimensional Changes in Automatic Home Laundering of Woven or Knit Fabrics	135-1992
Dimensional Changes in Automatic Home Laundering of Garments	150-1992
Dimensional Changes in Commercial Laundering of Woven and Knitted Fabrics Except Wool	96-1993
Dimensional Changes on Drycleaning in Perchloroethylene: Machine Method	158-1990
Dimensional Changes of Woven or Knitted Textiles: Relaxation, Consolidation and Felting	99-1993
Dimensional Restoration of Knitted and Woven Fabrics after Laundering	160-1992
Disperse and Vat Dye Migration: Evaluation of	140-1992
Dispersibility of Disperse Dyes: Filter Test	146-1989
Dispersion Stability of Disperse Dyes at High Temperature	166-1993
Drycleaning: Durability of Applied Designs and Finishes	86-1989
Dusting Propensity of Powder Dyes: Evaluation of	170-1989
Electrical Resistivity of Fabrics	76-1989
Electrostatic Clinging of Fabrics: Fabric-to-Metal Test	115-1989
Electrostatic Propensity of Carpets	134-1991
Extractable Content of Greige and/or Prepared Textiles	97-1989
Fabric Hand: Subjective Evaluation of	EP5
Fabrics; Appearance of, after Repeated Home Laundering	124-1992
Finishes in Textiles: Identification	94-1992
Fluidity of Dispersions of Cellulose from Bleached Cotton Cloth	82-1989
Foaming Propensity of Disperse Dyes	167-1993
Formaldehyde Release from Fabric, Determination of: Sealed Jar Method	112-1993
Frosting (Color Change Due to Flat Abrasion)	
Emery Method	120-1989
Screen Method	119-1989
Gray Scale for Color Change	EP1
Gray Scale for Staining	EP2
Hydrogen Peroxide: by Potassium Titration: Determination of	102-1992
Insect Pest Deterrents on Textiles	28-1989
Insects, Resistance of Textiles to	24-1989
Light Blocking Effect of Curtain Materials	148-1989
Migration: Disperse and Vat Dye: Evaluation of	140-1992
Mildew and Rot Resistance of Textiles: Fungicides	30-1993
Oil Repellency: Hydrocarbon Resistance Test	118-1992
Oils, Wool: Oxidation in Storage	62-1989
pH of the Water-Extract from Bleached Textiles	81-1989
Photochromism, Detection of	139-1989
Retention of Creases in Fabrics after Repeated Home Laundering	88C-1992
Rug Back Staining on Vinyl Tile	137-1989
Seams; in Fabrics; Smoothness of, after Repeated Home Laundering	88B-1992
Shampooing: Washing of Textile Floor Coverings	138-1992
Soil Redeposition, Resistance to: Launder-Ometer Method	151-1990
Soil Redeposition, Resistance to: Terg-O-Tometer Method	152-1990
Soil Release: Oily Stain Release Method	130-1990

TABLE 13.3 (continued).

Test Title	Method Number
Speckiness of Liquid Colorant Dispersions: Evaluation of	176-1993
Stain Resistance: Pile Floor Coverings	175-1993
Standard Depth Scales for Depth Determination	EP4
Thermal Fixation Properties of Disperse Dyes	154-1991
Transfer of Acid and Premetallized Acid Dyes on Nylon	159-1989
Transfer of Basic Dyes on Acrylics	156-1991
Transfer of Disperse Dyes on Polyester	155-1991
Water Repellency: Spray Test	22-1989
Water Repellency: Tumble Jar Dynamic Absorption Test	70-1989
Water Resistance: Hydrostatic Pressure Test	127-1989
Water Resistance: Impact Penetration Test	42-1989
Water Resistance: Rain Test	35-1989
Weather Resistance:	
General Information	111-1990
Sunshine Arc Lamp Exposure with Wetting	111A-1990
Sunshine Arc Lamp Exposure without Wetting	111C-1990
Exposure to Natural Light and Weather	111B-1990
Exposure to Natural Light and Weather through Glass	111D-1990
Weather Resistance of Textiles: Xenon Lamp Exposure	169-1990
Wetting Agents, Evaluation of	17-1989
Wetting Agents: Evaluation of Rewetting Agents	27-1989
Wetting Agents for Mercerization	43-1989
Whiteness of Textiles	110-1989
Wrinkle Recovery of Fabrics: Appearance Method	128-1989
Wrinkle Recovery of Fabrics: Recovery Angle Method	66-1990

REFERENCES

1. Peirce, F. T., "The Geometry of Cloth Structure", JTI, Vol. 28, T45, 1937.
2. American Society for Testing Materials, Annual Book of ASTM Standards (Volumes 7.01 and 7.02), published yearly by the ASTM, Philadelphia, PA.
3. "Application of Weftrol™ for Single Width Terry Towel Bow and Skew Control", Mount Hope, 1996.
4. Adanur, S., Wellington Sears Handbook of Industrial Textiles, Technomic Publishing Co., Inc., 1995.
5. American Association of Textile Chemists and Colorists, Technical Manual of the American Association of Textile Chemists and Colorists, published yearly by the AATCC, Research Triangle Park, NC.

SUGGESTED READING

- Ganssauge, D. et al, "How Do Fabric Attributes Influence the Handle Characteristics of a Fabric?", Melliand International (2), 1998.

REVIEW QUESTIONS

1. Derive Equation (13.1).
2. What fabric properties are affected by the crimp? Explain.
3. Which fabric design would have the highest modulus? Why? Assume that all the other fabric properties are the same.
4. How can you increase the tensile strength of a fabric?
5. What is rip stop fabric? Explain.
6. What equipment is used to correct fabric bow and skew?
7. How is air permeability of fabrics measured? What is the significance of air permeability?
8. What are the factors affecting abrasion resistance?
9. For what kind of application of fabrics are burst and impact resistance important? Why?
10. How can you measure the hand of a fabric?

14

Weaving Plant Operations

Rising quality requirements placed on fabric, efficiency, cost pressure, quick response and new market demands as well as safety, technological and ecological requirements, are criteria demanding increasing attention in weaving plant operations.

The requirements of increased performance and improvement of productivity and quality are associated with a decrease of machine stoppages caused by yarn breaks, fabric defects and the share of seconds. Fabric quality can be improved by minimizing yarn stress, reduction of downtimes and error rate. Process cost can be minimized by reducing downtimes, simplifying operation, improving flexibility and expanding the application area.

Today's competitive market conditions require flexible weaving mills regarding selection of styles, lot sizes and delivery times. As a result, just-in-time supply has become a necessity. Today, it is possible for one person to carry out a style or pattern change in less than 30 minutes.

The factors that affect the profitability of a weaving plant can be grouped under three categories: (a) Process, (b) Location and (c) User. The major process performance parameters are:

- investment
- filling insertion rate (m/min)
- system related factors
  —operating costs (time/operation)
  —filling waste (cm/pick)
- stop frequency
- energy consumption
- space requirement

The factors that affect the weaving manufacturing cost most are machine hours, building space, labor hours and energy (kWh). Weaving downtime depends on four basic factors [1]:

- machine stop rate
- maintenance reliability
- changeover time
- service delay

Machine stop rate is the most important factor among the four. Efficiency of high-throughput weaving depends on stops. High-throughput weaving requires very low stop rates. Yarn quality and good warp and filling preparation are required for this. For example, in denim manufacturing, the yarn break frequency in warping lies between 0.1 and 0.3 per million meters. Due to speed, air-jet weaving machines demand a lower warp stop rate than shuttle, rapier and projectile machines. This low yarn break frequency is a decisive criterion for ensuring that the yarn runs break-free in the subsequent stages.

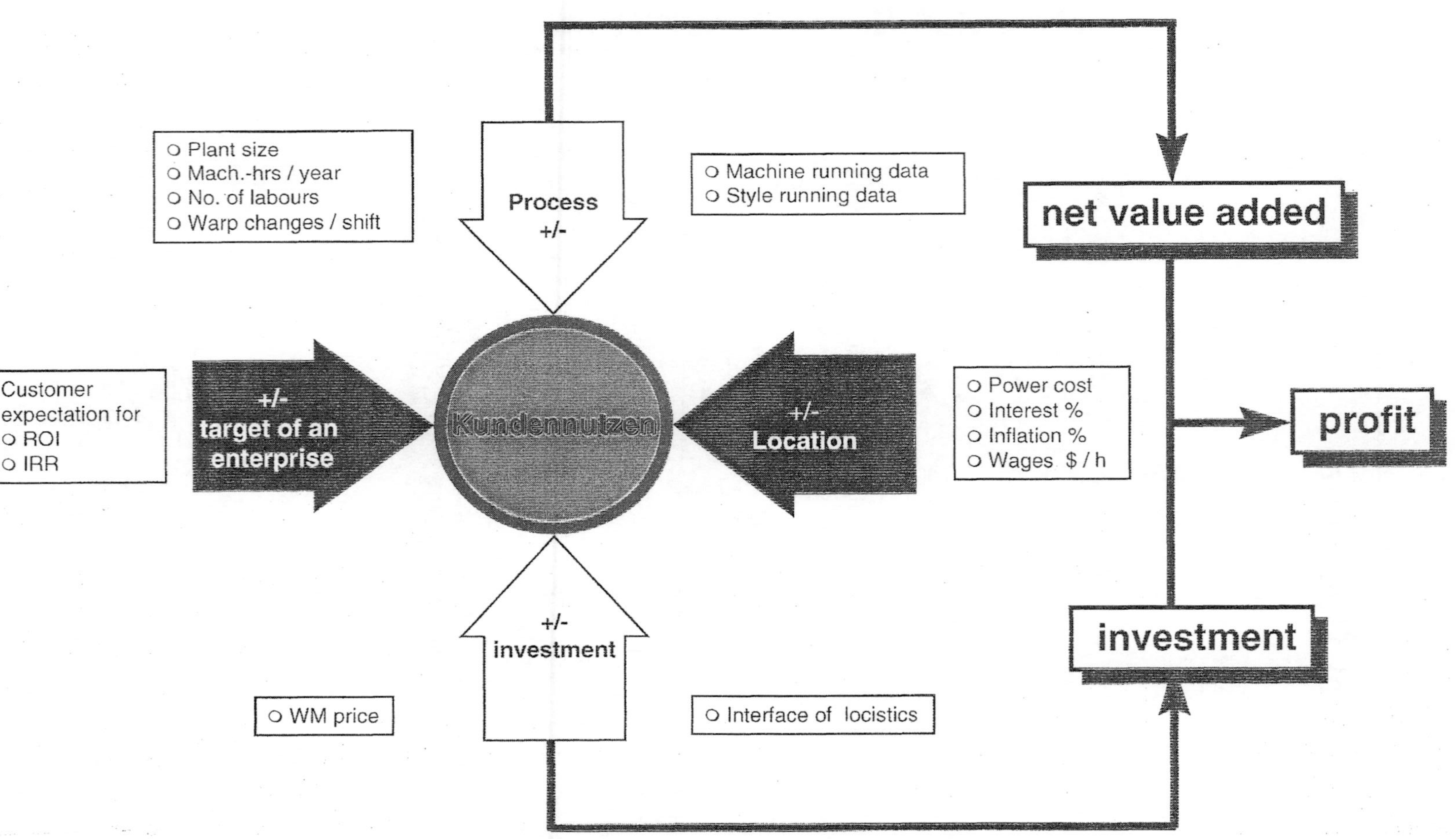

FIGURE 14.1 Factors affecting the customer's benefits (Kundennutzen: customer benefit).

The yarn uniformity takes highest priority and is even more important than yarn strength, because the filling insertion system places considerable demands on the yarn, and also on the quality of the yarn package. The yarn break frequency of 0.4 to 0.7 or less breaks per hour in the warp and less than 0.3 breaks per hour in the filling during weaving is extremely small for today's high filling insertion rates. Regular machine maintenance is necessary to minimize downtime and keep the throughput high. One way to reduce changeover time is to use large warp beams. This allows the machine to run longer between beam replacements.

Location valuing parameters are:

- tax on profit (%)
- subsidy (%)
- duty (%)
- capital interest (%)
- wages (per hour)
- social costs (%)
- power price (per kW)
- building costs (per m^2)
- governmental regulations

User utilization parameters are:

- style program
  —stops/10^5 picks
  —warp running time
  —number of warp/order
- diversity of articles
- volume of production (m/year)
- depreciation period (years)
- machine hours/years (hours)
- return on investment (ROI) expectancy (%/year)

Figure 14.1 shows the factors that contribute to the benefit of the customer. Figure 14.2 shows the elements of a typical weaving machine environment. The supplies include the warp yarn, the filling yarn, power, air (for air-jet machines), climate, service and maintenance. The output of the machine is the fabric. Dust and waste are also produced during the process. Figure 14.3 shows the weaving mill environment which may consist of hundreds of machines. The purpose of the weaving room manager is to optimize the weaving process to maximize the profit. In the following sections, the technological factors related to optimizing the weaving process are considered.

14.1 IMPROVING PRODUCTIVITY

Reduced Setting Time, More Production Time

Shorter running times per style, wider variety in the product mix and the demand for higher productivity in the weaving process have necessitated a situation in which the setting up times are gaining increasing significance.

With the high productivity of today's weaving machines, the weaving process suffers from the long stoppage times caused by changes of style and warp on the weaving machine. Optimization of the warp and style change process results in higher productivity and savings in production costs. Such measures involve capital expenditure and they must therefore be economically viable.

To reduce downtimes and the frequency of changes, there are various options and possible systems which include extended warp running time, optimized style change (quick style change) and optimized warp change (quick warp change).

It is hardly possible to generalize which system will produce the best results. The solution varies from one weaving plant to another, depending on the machines in the plant and the range of styles woven, and must be individually checked for every application.

Extended Warp Running Time

Today, on projectile and air-jet weaving machines, warp beam diameters of 1100 mm are available with standard arrangement in the weaving machine. Warp beams of 1600 mm maximum diameter can be placed outside the weaving machine (Figure 14.4), overhead (Figure 14.5) or in a cradle underneath the weaving machine. The beneficial influence of the larger warp beam on the length of time the warp runs and the frequency of warp changes is evident from Figures 14.6 and 14.7.

The warp capacity is increased while the efficiency of the installation is enhanced owing to the reduction in down time. This concept brings benefits especially with short warp running

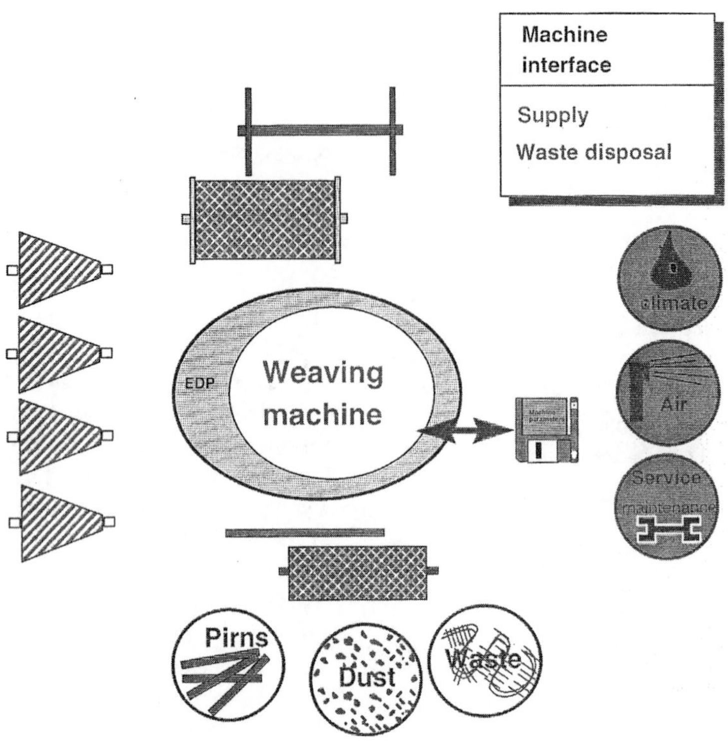

FIGURE 14.2 Elements of machine interface.

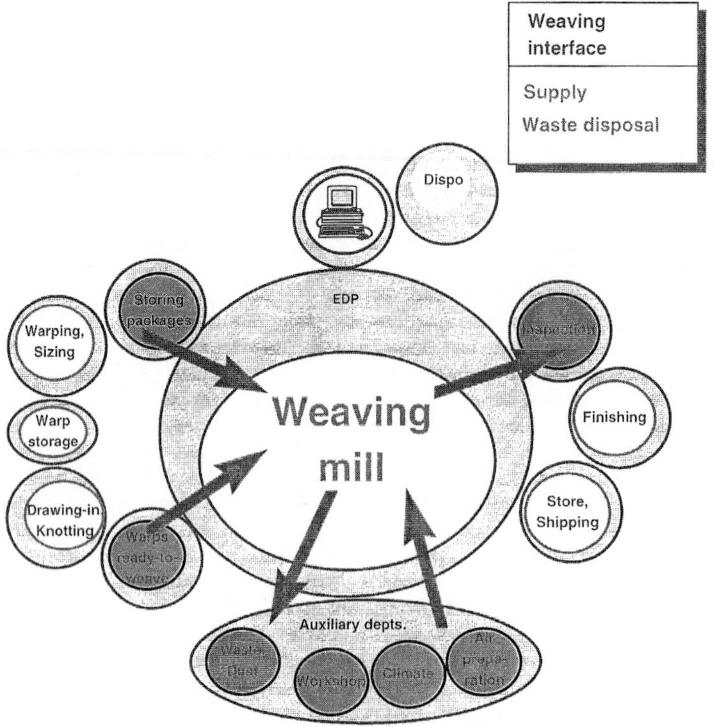

FIGURE 14.3 Weaving mill environment.

FIGURE 14.4 A 1600 mm diameter warp beam mounted outside the projectile weaving machine.

times. It should be noted that transport equipment and alleys should be designed to take into account the size and weight of the warp beams. There are, however, some drawbacks to large warp beam diameters. These are shown in Table 14.1.

Table 14.2 shows an exemplary comparison of a conventional denim plant with regular weaving machines and weaving machines with large beams. The warp length and warp running time are doubled with the new system. The number of warp changes is reduced by half. As a result, fewer personnel is needed with higher efficiency and lower weaving costs.

Due to the reduction of machine downtimes as a result of the increased warp length, plant efficiency is increased, as well. The reduction of the number of warp changes allows a reduction in manpower and personnel costs. These types of large beams are especially beneficial for denim beams.

Quick Warp Change (QWC) and Quick Style Change (QSC)

Weaving mills today are forced to adapt to rapidly changing market conditions with respect

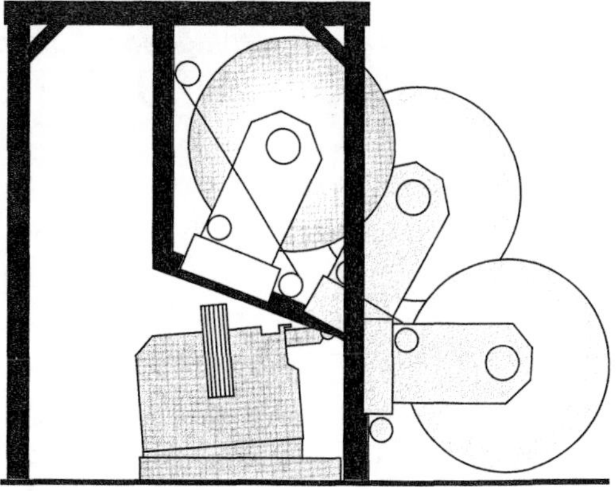

FIGURE 14.5 Mounting the large warp beam overhead.

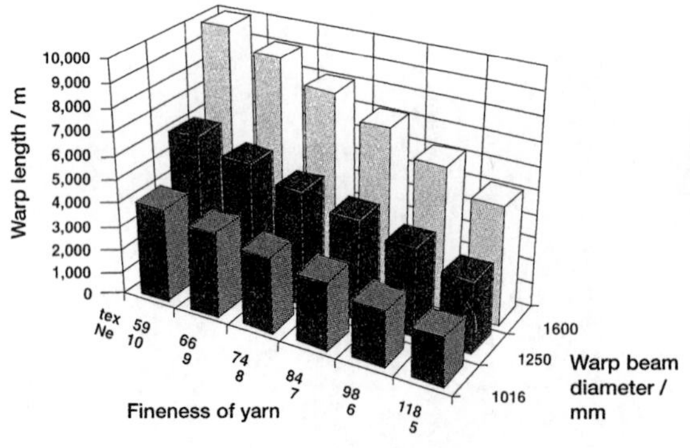

FIGURE 14.6 Effect of yarn count and warp beam diameter on warp length.

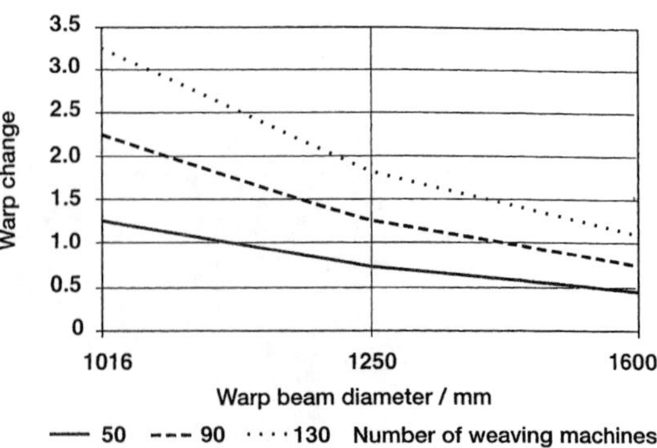

FIGURE 14.7 Effect of warp beam diameter on frequency of warp changes per 8 hours shift.

to style choice, batch size and delivery deadlines. Just-in-time delivery has become a necessity. Optimization of the resetting procedure of the weaving machine, particularly warp and style changes, is critical to reduce downtime and improve productivity.

There has been no clear distinction between warp change and style change. When a warp of different pattern but with the same number of ends and the same yarn count is tied on the weaving machine, the style woven is in fact a different one, but in weaving terms this is "just another warp change". Some weavers already talk about a style change in case only the number of picks gets altered. Table 14.3 shows the differences between warp change and style change. Figure 14.8 shows the flow diagrams of quick warp and style changes.

Calculations have shown that quick warp change offers a possible time savings of 30–60 minutes over conventional warp change. It is evident from Figure 14.9 that in a given weaving plant with a given number of warp changes per shift, the gain in production hours is also considerable.

"Quick style change" is an integrated, overall concept. In order to reduce machine downtimes to an absolute minimum, the resetting work must be organized in the preparation department where the harness package for the new style is completely prepared, assembled and adjusted.

This separation of the procedures "style preparation" and "machine preparation" enables the just-in-time style change in less than 30 minutes. Preconditions for such a quick style change are automatic functional units and operational aids

TABLE 14.1 Advantages and disadvantages of large warp beam diameters.

Advantages	Disadvantages
• Increase warp capacity by 53% at 1250 mm and 150% at 1600 mm warp beam diameter	• More space required depending on the material handling system
• Number of warp changes reduced by 33% and 66% respectively	• Handling equipment suitably designed for large warp beam diameter
• Higher weaving machine capacity utilization	• Restricted warp monitoring by the weaver
• Lower labor requirement for warp changing	• More complicating machine operating depending on the system
• Less warp handling	
• Equal warp dyeing over longer warp lengths	
• Less second quality	
• Less warp waste	

TABLE 14.2 Exemplary comparison of weaving machines with conventional and large beams.

	Regular Beam	Large Beam
Warp beam diameter	1016 mm	1500 mm
Warp length	2013 m	4280 m
Warp running time	149 hours	317 hours
No of warp changes/day	11.6	5.45
Ratio warp beam/batch winder	1:1	1:2
Personnel required/ 4 shifts:		
Total	60	44
Warp change	24	12
Installation efficiency	90.9%	92.2%
Investments	100%	119%
Weaving costs	100%	91%

on the weaving machine. Figure 14.10 shows the system flow chart of the "quick style change". Figure 14.11 shows the style changing on a rapier weaving machine where a transport truck is used for supplying warp beam, warp stop motion, harness frames and reed to the weaving machine.

Advantages of Quick Style Change:

- The set-up work that used to be performed in the weaving machine while the machine was idle can now be dissociated from the machine. It can be organized and accomplished more quickly in a quiet dust-free room as an independent process.
- Inter-departmental planning and standardized handling ancillaries enable material handling and warp preparation to be rationalized.

TABLE 14.3 Comparison of warp change and style change.

Style/Warp Change from Organisational Aspect	QSC	QWC
Warp change		■
Change in reed width	■	
Change in number of shafts	■	
Drawing-in change	■	
Change in number of warp ends	■	

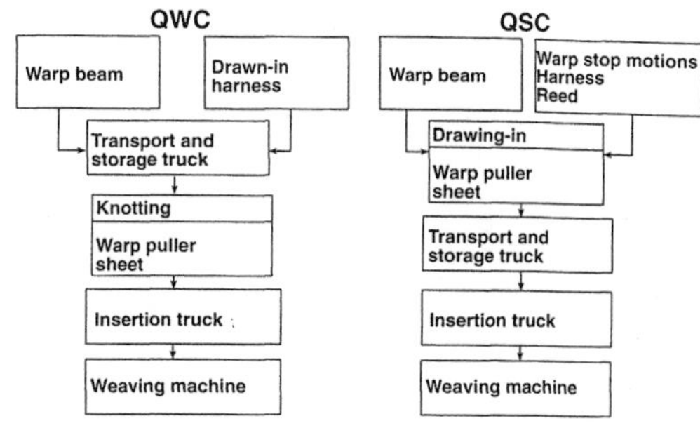

FIGURE 14.8 Flow diagrams of quick warp and style changes.

- Operations planning and logistics can be rationalized by dissociating the preparatory processes for warp and style from work on the actual weaving machine.
- The plant gains extra machine hours and consequent opportunities for quicker order completion, increased turnover and better management of overhead.

Employing QWC and QSC in a plant having around 150 weaving machines and a style change to warp change ratio of 1:3, the same production can be achieved with 150 weaving machines instead of 154 weaving machines. This is accompanied by the work force being simultaneously reduced by ten. If the size of the plant is kept at 154 weaving machines, the additional production amounts to 200,000 meters per year.

14.2 PLANT LAYOUT

A weaving mill, planned to the last detail, with sufficiently dimensioned transport alleys,

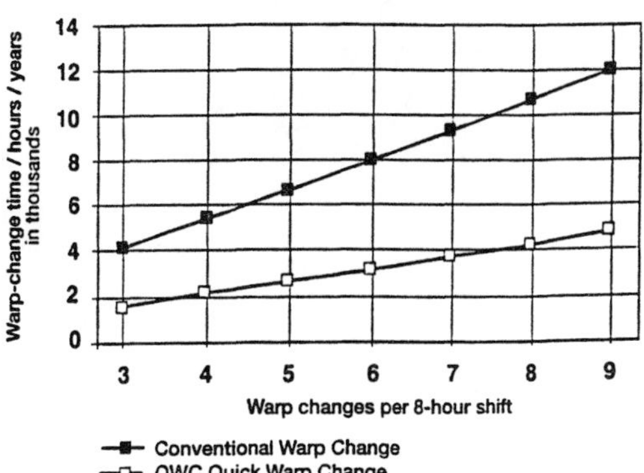

FIGURE 14.9 Warp change time requirements per year.

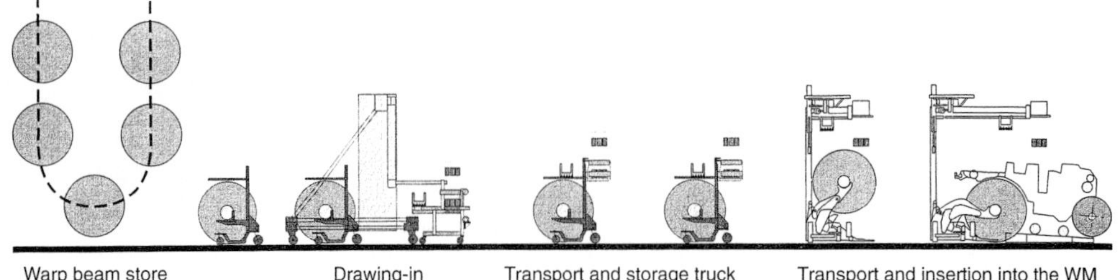

FIGURE 14.10 Process flow chart of the quick style change.

correct lighting, optimum air conditioning, sound insulation and overhead cleaners, etc., is the prerequisite to economical fabric production.

With the preliminary project studies, the requirements for the erection of a new weaving plant are determined. The project study is generally based on a certain style, e.g. for denim. Evolving from a well-balanced production unit comprised of weaving preparation and weave room, the following parameters are indicated:

- quantities to be produced
- required yarn quantities
- specifications of weaving preparation machinery and its utilization rate
- required media
- manpower requirement
- investment in machinery
- plant layout drawing

These fundamental data allow carrying out feasibility studies which are used as bases for decision making. A layout plan shows the positions of the weaving machines as well as the alleys for supply and waste removal transports. The dimensions of the main and weaver's alleys and of the warp gangways are chosen to suit the transportation devices to ensure problem free traffic flow at optimum utilization of the available space.

FIGURE 14.11 Quick style changing on a double-width projectile weaving machine.

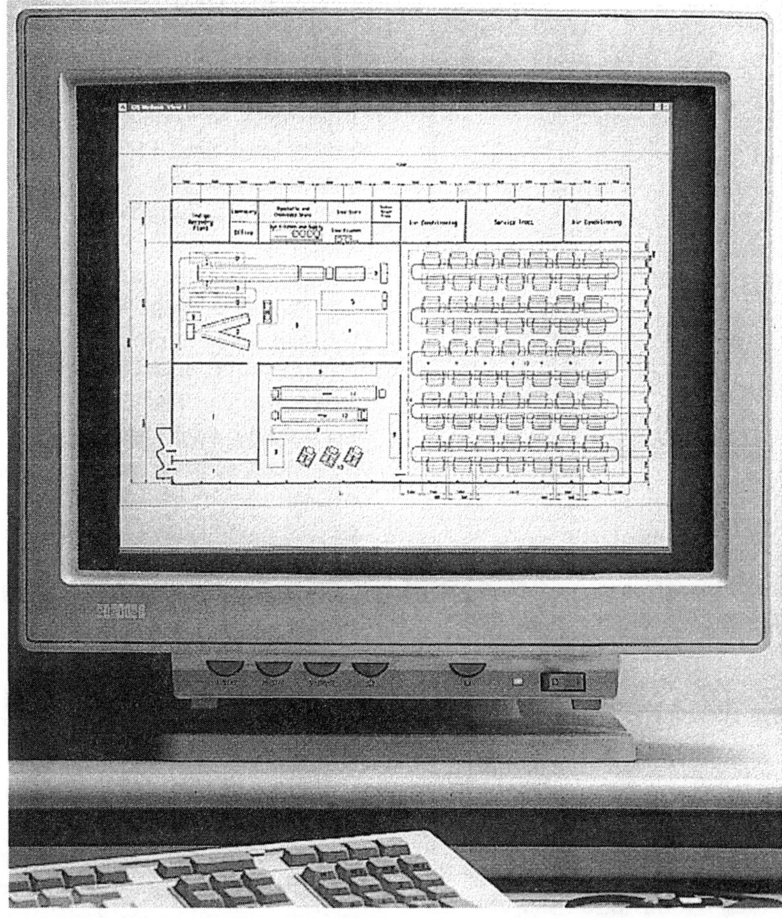

FIGURE 14.12 The layout and installation drawings developed using a CAD system.

The layout drawing is the basis for the installation drawing which contains all the relevant information needed for the planning work and for the final arrangement. The position of each individual weaving machine in the weave room is shown in correct scale in a system of coordinates as well as the connections for electrical power and compressed air (Figure 14.12).

For installation of machines in existing buildings, vibration measurements should be conducted to determine the suitability of the buildings for machine installation. Noise emission should also be considered.

14.3 AIR CONDITIONING IN TEXTILE PLANTS

In the past, humidification was the only major concern in textile plants. Today, a modern textile air conditioning system must provide adequate control in temperature, humidity, pressure, suction and cleanliness which is called a total air control (TAC) system [2].

The purpose of the air conditioning plant is to produce the temperatures and humidity needed for optimal textile production and maintain them regardless of indoor and outdoor effects, at the same time providing the best possible conditions for the operators as well. Since the lint and dust are produced in relatively high concentrations during textile processing, the air conditioning plant must provide adequately clean air.

The type and configuration of the air conditioning system are determined by outdoor and indoor climatic conditions, site and buildings, the machine plant and the production itself. It

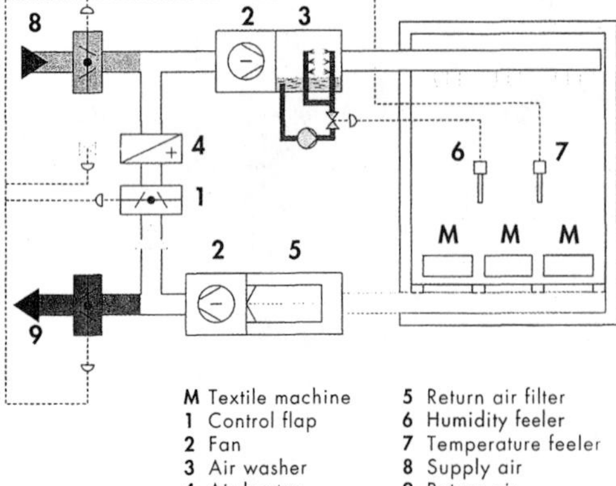

FIGURE 14.13 Components of a weave room air conditioning system.

should also be noted that modern high speed machines consume more energy, which means that air conditioning plants must efficiently remove increasing amounts of heat.

Figure 14.13 shows the components of a typical weave room air conditioning system. Conventional room air conditioning plants are humidifying or partial air conditioning plants with year round room air humidity and adaptive air temperature in summer. Full air conditioning plants keep both the air humidity as well as the room temperature uniform year-round. For this purpose, they are usually equipped with a refrigeration unit. The treated air is generally introduced into the production facilities via ceiling ducts. The return air permeated with fiber fly and fine dust is suctioned off by an extract air fan through floor gratings and ducts, filtered and fed to the outside air or if necessary recycled as recirculated air. It thus leaves the room with the temperature and humidity required for the production area.

In so-called supersaturation plants, finely atomized water is also introduced into the room. As this water evaporates, the air is cooled, so that the amount of supply air fed into the room and thus the energy consumption of the air conditioning plant can be reduced. Supersaturation plants can range from simple ceiling units to systems which treat the supply and return air in central units in the same way as with conventional room air conditioning, post humidifying the supply air with pressurized water jets before it enters the room.

In all cases, however, the dust concentration in the room is inevitably higher than with conventional room air conditioning on account of the reduced air change rate. Moreover, in low production rooms, the danger of machine corrosion as well as precipitation of mineral components of the humidifying water cannot be ruled out.

Supply air treatment systems have no extract fan, so that a slight overpressure builds up in the room. The return air is discharged to the atmosphere by this overpressure via automatic louvers fitted in the outside walls, or else recirculated by the supply air fan.

Systems of this kind are especially economical and cost effective. They are thus ideal for air conditioning smaller production rooms and for retrofitting in existing buildings. Uneven air flows occurring in the room when operating with outside and recirculated air are, however, a concern.

14.4 COMPUTERS AND AUTOMATION

Integrated artificial intelligence is incorporated in most of the modern weaving machines. Modern electronics with microprocessors, integrated in the machine, monitor, control, regulate and optimize all major weaving machine functions. The central microprocessor monitors and controls the weaving process. The touch screen terminal is used as the interface between the operator and the machine (Figure 14.14). The

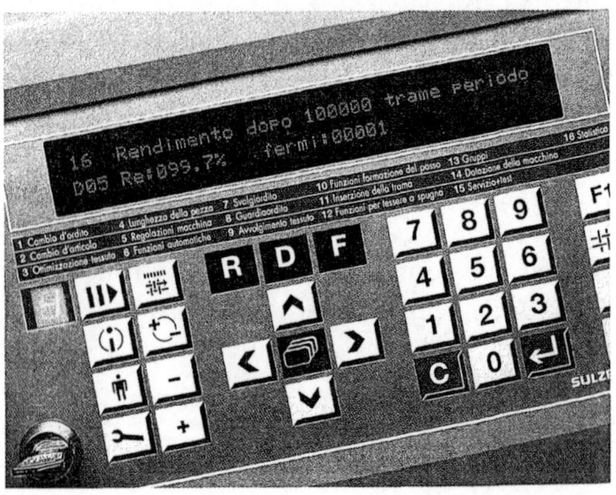

FIGURE 14.14 The microprocessor terminal with its menu operated display to form the operator-machine interface.

terminal acts as the interface between personnel and the machine. Corresponding to the varying information requirements, the microprocessor control provides specific terminal functions for operators, supervisors and managers. Setting data may be transferred from one machine to another with a memory card. If no setting data is available for a style which had not been woven previously, the microprocessor can calculate standard parameters from the entered fabric specification and transfer them automatically to the electronically controlled function units. Two-way communication with production management systems or production data collection systems are possible with suitable interfaces.

Today's modern weaving machines offer high operational reliability with centralized microprocessor control. These operations were included in the previous chapters for each particular machine. Some examples of automation and electronic control are:

- Electronically controlled warp let off provides constant warp tension.
- Electronically controlled cloth take-up maintains the accurate cloth weight.
- Electronically controlled color selection allows picking of multifilling colors.
- Electronically controlled yarn brake reduces filling breaks and waste and provides constant filling tension.
- Menu guided operation at the terminal assists the Quick Style Change.
- Interfaces for two-way communication with modern systems for plant management and production data collection.

Programming and Archiving Systems

Weaving machine manufacturers developed user friendly programming and archiving systems for the weave room (Figure 14.15). With these systems, the pattern data and the machine settings can be programmed in the office with the aid of a PC from where they are transferred onto the machine. These systems offer the following advantages:

- shortening the resetting and cycling times
- increasing efficiency
- improving reproducibility

FIGURE 14.15 Programming and archiving system (PAS).

Electronic Spare Parts Catalog

Today, weaving machine manufacturers provide spare part catalogs in electronic formats such as CD-ROM. A CD-ROM weighing 20 g and featuring a diameter of 12 cm can store information of a catalog comprising of some 13,000 pages of paper, weighing 50 kilograms. From an ecological point of view, this new development offers a big advantage. Figure 14.16 shows a sample window of electronic spare parts catalog.

Electronic Direct Ordering System (EDOS)

The just-in-time philosophy is becoming increasingly evident and manufacturers are reducing their spare part reserves to absolute minimum levels. With today's modern, capital intensive and highly productive weaving machines, the cost of lost production is much higher than that incurred for spares. Furthermore, the requirement—which depends on the employed raw weaving materials and external influences—can be very irregular. The consequences are shorter deadlines with greater responsibility, which can only be overcome by tightening up the flow of information and material between the customer and the supplier.

Electronic Direct Ordering System makes it easier for customers to order parts and supplementary parts. The electronic transfer of data prevents transmission errors and offers the assur-

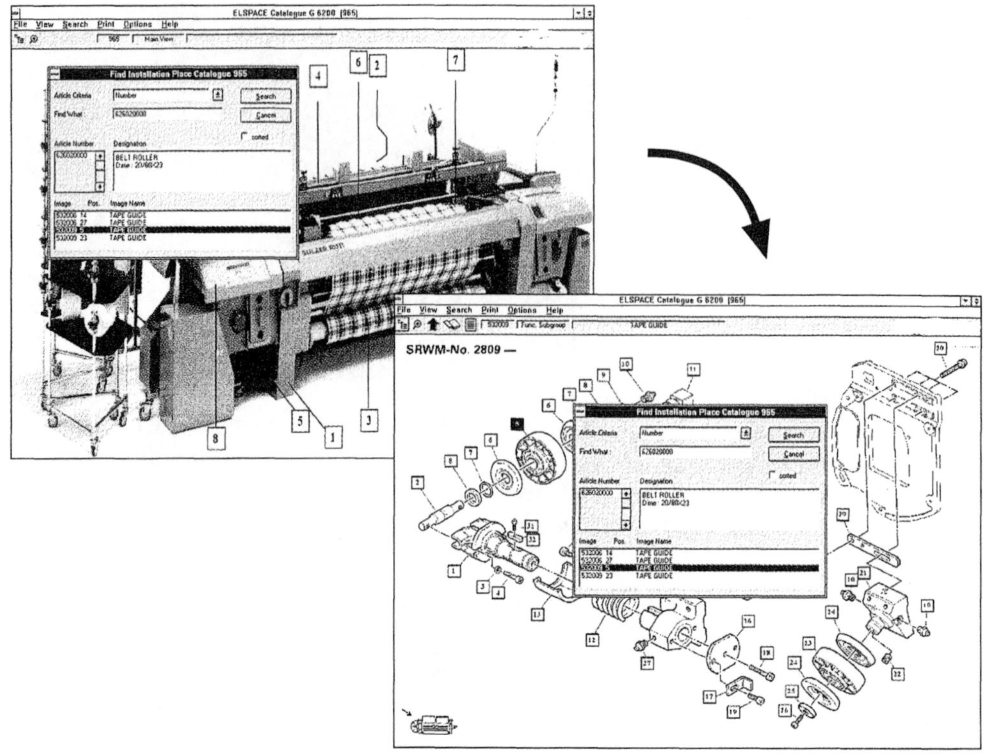

FIGURE 14.16 A sample page of a CD-ROM parts catalog.

ance that the customer will quickly receive the correct parts (Figure 14.17). The system accelerates and channels the flow of information and goods of an order on its way from the customer to the supplier and back again. For the electronic ordering of spare parts, the customer needs a commercial type PC, a modem, a printer and the necessary software (Figure 14.18).

With the EDOS systems, the customer can receive complete, up-to-date and accurate information quickly. The high transmission quality is also beneficial. Ultimately, the customer receives his urgently required parts quicker: the order cycle is reduced. The electronic processing of the order data reduce the possibility of typographical and transmission errors. The customer does not have to enter and check the majority of the data with each other. The simplified bookkeeping and order monitoring leads to a reduction in costs.

Figure 14.19 shows the operation of EDOS. The customer selects the parts that he wishes to order from the standard set of data; he sees all the necessary data and complements them with the required amount and mode of delivery. The data are then transmitted by modem to the subsidiary and passed onto the head office. After a short time and along the same route, he receives the confirmation of order, with price and information concerning the readiness for dispatch. He knows already that he will receive the right parts and when he can expect them.

Computer Aided Design (CAD) and Computer Aided Manufacturing (CAM)

Today's CAD/CAM technology allows transmission of fabric design from computer screen to the machine. CAD/CAM saves product development and manufacturing time, provides quicker response with good accuracy and quality. Pattern changes and modifications are done easier and quicker. Figure 14.20 shows the designing of jacquard fabric on the computer screen.

FIGURE 14.17 Electronic direct ordering system.

There are various configurations and setups of CAD/CAM systems. As an example, in one configuration for woven terry fabrics, the artwork produced by a stylist is scanned into a workstation. Then, the design parameters are entered for a specific fabric style. Parameters for each fabric style are stored in a database. The scanned-in artwork is edited and a weave is selected from a database. Colors are displayed on the computer screen. The pattern file developed is digitized and stored in a diskette for subsequent downloading into the jacquard machine controller on the weaving machine. Through the workstation-printer interface, a simulation of the fabric can be printed out. Thus, the terry towel can be seen in virtual reality. Figure 14.21 shows the computer simulation of a terry towel fabric.

In a more scientific approach, CAD of fiber, yarn and fabric simulations can be done based on mathematical and physical principles. Thin and thick places and other irregularities can be represented in the yarn depending on the simulated manufacturing model. Using the 3-D image of the yarn, the resulting woven fabric can be simulated (Figure 3.1). Visualizing the physical characteristics of yarn on the screen before the yarn is manufactured provides significant advantages in reducing R&D and prototype manufacturing time of fabrics. The manufacturer can determine the suitability of a yarn in a fabric a lot faster with less cost. Even the washdown effects on fabrics can be simulated with computers.

With the growing demands made on fabric quality and efficiency of the weaving machines, their functionality rests increasingly on electronics and information systems along with mechanics. Today, fabric design is developed on the

FIGURE 14.18 Equipment needed for electronic ordering of spare parts.

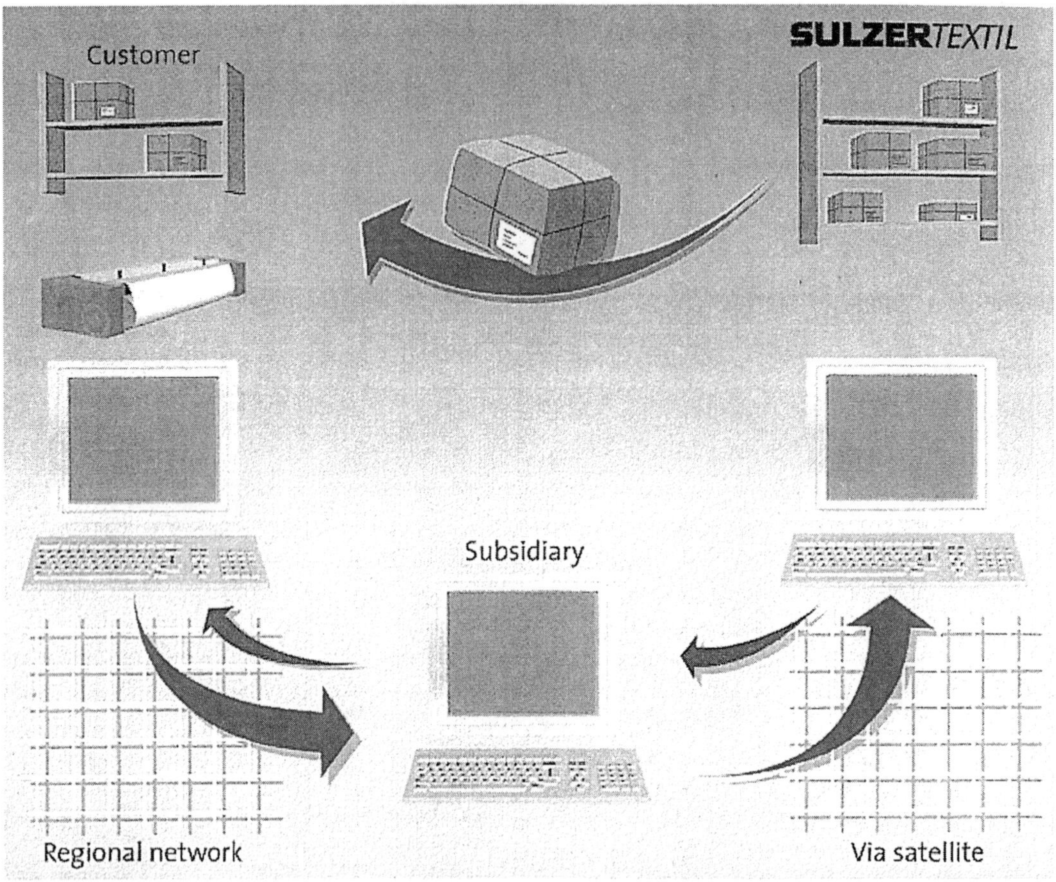

FIGURE 14.19 Operation of EDOS.

FIGURE 14.20 CAD of a jacquard design (courtesy of Staubli).

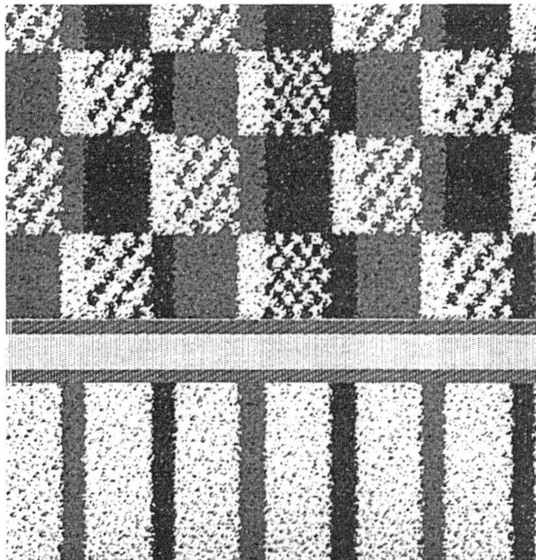

FIGURE 14.21 Computer simulation of a terry towel fabric [3].

computer and transferred to the weaving machine in a memory card. Figure 14.22 shows the functional principle of a programming and archiving system that can be linked with a CAD/CAM system.

The building blocks of such a CAD/CAM system are as follows:

Weaving Data Archive

- managing and archiving all weaving data records
- creating new weaving data
- interface for importing new pattern data from various CAD programs

Weaving Data Record

Diagrammatic representation of harness frame and color control as well as special functions. Numeric representation of filling density and pile height. Developing and modifying pattern data.

- pattern for card punching (also electronic)
- color pattern
- special functions
- electronically controlled cloth take-up
- dynamic pile control

Parameters

Programming of the electronically controlled functional units on the weaving machine. Developing or modifying the electronic machine settings.

- warp let-off functions
- data for the electronically controlled filling yarn brake
- controlled fabric fell correction

Data Transfer

Transferring the weaving data to the memory card (MC) or back. With the aid of the data carrier, the weaving data are transferred to a memory card. After the memory card has been loaded with one or more sets of weaving data, the memory card is passed on to the weave room staff who transfer the data to the weaving machines by means of the data carrier.

After fine optimizing of the weaving data on the machine terminal, the data can be archived in the PAS via the data carrier, thus ensuring the reproducibility of the fabric quality.

Computer Based Training [4]

Today's fast and powerful computers enable multimedia-based training programs. These programs allow improvements of the quality standard, motivation of the staff members as well as reduction of training costs.

Multimedia based training offers the possibility of passing on knowledge within the industrial sector with large quantities of visual material. The computer based training (CBT) system teaches operation and maintenance of the weaving machine by means of full-screen video, animation, audio, pictures, diagrams and text. It enables job-targeted learning and the material taught can be immediately rested on the machine. It is also possible to repeat such exercises as often as necessary and each student is given the same explanation, irrespective of the environment. Comprehensive training courses can thus be considerably shortened. Function, operation and maintenance of the weaving machines are presented in an effective way, since the students have normally no opportunity to watch these on the machine. The entire material is stored on CD-ROM disks and can be adaptedindividually and customer-specifically on training PCs. The interactive CBT can be used at any time and anywhere, independent of the instructor, be it on the machine (Figure 14.23), in

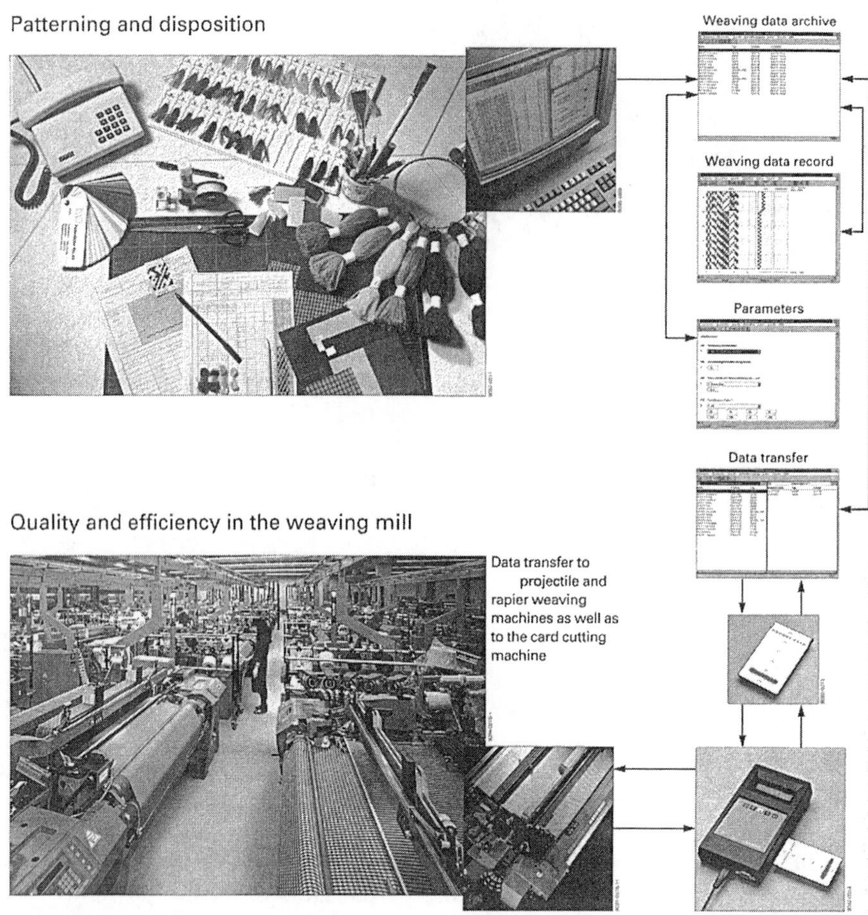

FIGURE 14.22 CAD/CAM integrated with the programming and archiving system.

FIGURE 14.23 The interactive Computer Based Training provides learning with a laptop PC at the machine on which the shown work can be directly carried out.

the office or at home. Service and sales engineers can take CBT with them in their hand luggage.

The modular structure of the CBT system affords learning in appropriate steps. The students are able to call the individual learning steps and to repeat them as often as they wish, without specific PC experience. Concerning presentation, they can choose between video film (Figure 14.24), animation (Figure 14.25) or still photography, each one accompanied by spoken text. The latter can be displayed and printed.

A multitude of test and qualification possibilities enable detailed control of the learning success. The lecturer who is responsible for

FIGURE 14.24 Video sequences show assembly and setting procedures on the weaving machine with explanatory texts from the operating instructions.

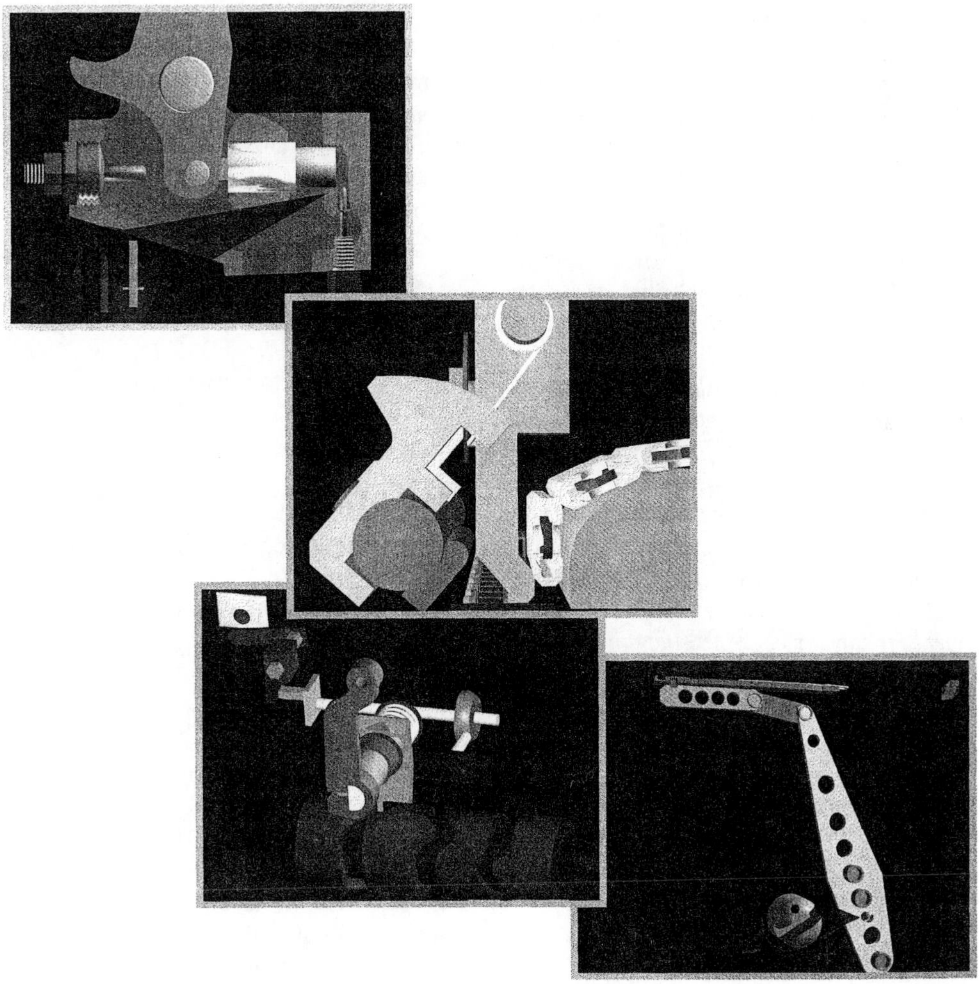

FIGURE 14.25 Animations present the functions as well as setting and assembly steps in a clear manner. Shown from varying perspectives, the processes are rendered readily understandable.

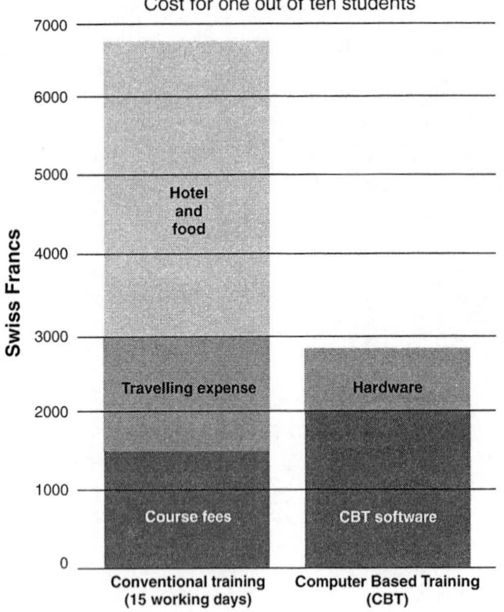

FIGURE 14.26 Cost comparison between conventional training and training at the PC.

the training obtains an up-to-date and objective survey on the learning success of all participants of the course, who, in turn, have the possibility to watch the actions carried on the machine through laptop PC.

Even though the preparation of CBT programs involves considerable expenditure, the overall costs for the entire learning package are far below those for training the customer personnel in the conventional way (Figure 14.26).

For CBT training, a PC configuration is required on which CD-ROMs can be run. Several PCs can be networked with each other. A training manager supervises the learning progress of the trainees. An advantage of CBT is that it is always available when necessary and when there is time for learning. CBT constitutes a tool that enables internationally active enterprises to train their coworkers and customer employees, on a global basis, with always the same standard. This guarantees that each production location delivers the same product quality.

14.5 TECHNOLOGY SUPPORT CENTERS

Some weaving machine builders are offering technical support to their customers through feasibility studies, weaving trials, etc., in research weaving facilities. Service and customer weaving trials are conducted in these facilities (Figure 14.27).

With the weaving trials, fiber and yarn manufacturers can determine whether or not, and at which weaving machine speeds, the desired styles can be optimally produced on the weaving machines. In these research centers, new fabrics can be developed and weaving problems can be solved.

The filling insertion trials reveal the expected frequency of filling stoppages as well as the suitability of the yarn for the corresponding type of weaving machine. The yarns and fabrics are inspected for possible defects as well.

Machine manufacturers also help customers to meet the new demands with the existing machinery. They provide services and studies for retrofitting, conversion, customer oriented solutions for special equipment, special fabrics and accessory and auxiliary components. The following changes of machine specifications are demanded most frequently:

- conversion from single pick to weft mixer
- increasing the number of harnesses
- conversion from dobby to jacquard machine
- equipment for heavy fabrics
- increasing the warp beam diameter

14.6 STANDARDS AND REGULATIONS

14.6.1 ISO 9000 Quality Standards

In recent years ISO 9000 standards have gained considerable attention in the textile industry. ISO is important for international trade as well as domestic business. An ISO 9000 certificate acknowledges that the certified company runs a comprehensive quality assurance system and that they fulfill the requirements of the international ISO standards. The ISO 9000 series of standards sets precise preconditions concerning the hierarchy of a quality assurance system and demands proof of the system's capability to develop and maintain quality.

ISO stands for the International Organization for Standardization (in French) headquartered in Geneva, Switzerland. Its purpose is to develop and promote common standards worldwide. ISO

FIGURE 14.27 Weave room with various machines for weaving trials for customers.

has hundreds of technical committees and thousands of subcommittees and working groups to prepare standards applicable worldwide.

ISO 9000 series is a set of international standards for quality assurance and management. It is a written set of good practice procedures which, when followed, will result in consistency for production, design, or provision of services. There are five standards in the set: ISO 9000, ISO 9001, ISO 9002, ISO 9003 and ISO 9004. These standards are generic and adaptable to quality systems already in place.

It should be emphasized that ISO 9000 is not a set of "product standards". ISO 9000 applies to all industries and is not product specific. It does not address the technological issues; i.e. it does not specify or teach any existing or new technology to be used in order to produce or improve the quality of a particular product. ISO 9000 complements the product standards and refers to quality system elements that should be implemented to have consistent and therefore high quality products and services.

ISO 9000 provides a uniform approach for registering quality systems and is acceptable worldwide. The standards evolved from the 1979 British Standards, BS 5750. Numerous standardization bodies have adopted the contents of these standards. In the US, ISO 9000 series of standards was adopted jointly by the ANSI (American National Standards Institute) and ASQC (American Society for Quality Control) as Q90 series of standards. As a result, ISO 9000-9004 series corresponds to Q 90-94 in the US. The European equivalent of these standards is EN 29000 series.

ISO 9000 standards are designed to be used for establishing and maintaining quality management and systems for company use and to satisfy outside contracts. ISO 9000 requires extensive documentation by mandating management to record all changes and decisions. The unwritten rule concerning the application of ISO 9000 is that if all personnel were suddenly replaced, the new people, properly trained, could use the documentation to continue making the product or providing the services as before.

ISO 9000 standards were developed by the ISO Technical Committee 176 (TC 176). The committee was formed in 1980 and issued the first ISO 9000–9004 series of standards in 1987. The standards are reviewed and updated if necessary, every five years.

ISO 9000, the first standard in the series, is the road map for understanding the rest of the series and provides key quality definitions. This helps an organization to choose which contractual standard (9001, 9002 or 9003) is best suited for their business. Each element of ISO 9000 series is suitable for different types of business:

ISO 9000: —Quality management and quality assurance standards
—Guidelines for selection and use

ISO 9001: —Quality Systems—Model for quality assurance in design/development, production, installation and servicing

ISO 9002: —Quality Systems—Model for quality assurance in production and installation

ISO 9003: —Quality Systems—Model for quality assurance in final inspection and testing

ISO 9004: —Quality management and quality system elements
—Guidelines for interpretation standards

ISO 9001 is a quality system model for quality assurance in design, development, production, installation and servicing. It is like a contract between supplier and customer that demonstrates the supplier's capabilities in the above areas. This is the highest ISO qualification an organization can have. It addresses all of the operations that affect the quality, or fitness for use of a good or service. The elements of ISO 9001 are shown in Table 14.4 along with 9002 and 9003.

TABLE 14.4 The elements of ISO 9001–9003 systems.

	ISO 9001	ISO 9002	ISO 9003
1. Management Responsibility	x	x	x
2. Quality System	x	x	x
3. Contract Review	x	x	
4. Design Control	x		
5. Document Control	x	x	x
6. Process Control	x	x	
7. Purchasing	x	x	
8. Purchaser Supplied Product	x	x	
9. Product Identification Traceability	x	x	x
10. Inspection and Testing	x	x	x
11. Inspection, Measuring and Test Equipment	x	x	x
12. Inspection and Test Status	x	x	x
13. Control of Nonconforming Product	x	x	x
14. Corrective Action	x	x	
15. Handling, Storage, Packaging and Delivery	x	x	x
16. Quality Records	x	x	x
17. Internal Quality Audits	x	x	
18. Training	x	x	x
19. After-sales Servicing	x		
20. Statistical Techniques	x	x	x

ISO 9002 is a quality system model for quality assurance in production and installation. It differs from ISO 9001 in that it does not cover Design Control (i.e. R&D) and Servicing after sales. ISO 9002 assures that a company has the capability to supply acceptable products and services to its customers. It also certifies that the company is able to document and control all of the processes and procedures from manufacture to installation.

ISO 9003 is a quality system model for quality assurance in final inspection and testing. It is the least strict of all the series that mainly deals with detecting, documenting, controlling nonconforming material in the final inspection and test of a product or service. In addition to design control and after-sales servicing, ISO 9003 does not cover contract review, corrective action, internal quality audit, purchasing, purchaser supplier product and process control.

ISO 9004 describes guidelines for quality management and quality system elements. It explains a basic set of elements by which quality management systems can be developed and implemented.

The elements included in ISO 9004 are:

- quality in marketing, production, specification, and design
- product safety and liability
- control in production
- product verification
- personnel
- poor quality costs

14.6.2 ISO 14000 Environmental Standards

ISO 14000 is an international standard that is developed for Environmental Management System (EMS). What ISO 9000 is to quality management is what ISO 14000 to environmental management. Therefore, there are some similarities between ISO 9000 and 14000 in concept. In fact, some of the elements of both systems, such as document control and statistical methods, are the same. ISO 9000 and 14000 can exist simultaneously to form one uniform management system for quality and the environment. ISO 14000 has the potential to apply to every manufacturing company. It is supposed to help any company in any country to meet the goal of "sustainable development" and environmental friendliness [5]. As in the case of ISO 9000, ISO 14000 is a series of standards. The family of ISO 14000 standards are listed in Table 14.5.

The members of the ISO approved ISO 14001 and 14004 as official international standards in the summer of 1996. ISO 14001 is the registration standard for an Environmental Management System (EMS) and ISO 14004 is the guideline standard for more guidance on implementation. These two new standards have also been approved by the European Union as a European standard. The official approval of the complete

TABLE 14.5 ISO 14000 series of standards.

ISO 14001	Environmental Management Systems—Specification with guidance for use
ISO 14004	Environmental Management Systems—General guidelines on principles, systems and supporting techniques
ISO 14010	Guidelines for Environmental Auditing—General principles on environmental auditing
ISO 14011/1	Guidelines for Environmental Auditing—Audit Procedures. Auditing of Environmental Management Systems
ISO 14011/2	Guidelines for Environmental Auditing—Compliance audits
ISO 14011/3	Guidelines for Environmental Auditing—Statement audits
ISO 14012	Guidelines for Environmental Auditing—Qualification criteria for environmental auditors
ISO 14013	Guidelines for Environmental Auditing—Management of environmental audit programs
ISO 14014	Guidelines for Initial Reviews
ISO 14015	Guidelines for Environmental Site Assessments
ISO 14020	Principles of Environmental Labeling
ISO 14021	Environmental Labeling—Terms and definitions of self-declaration environmental claims
ISO 14022	Symbols of Environmental Labeling
ISO 14023	Testing and Verification of Environmental Labeling
ISO 14024	Environmental Labeling—Guiding principles, practices and certification procedures for multiple criteria
ISO 14030	Environmental Performance Evaluation
ISO 14040	Principles and Guidelines of Life Cycle Assessment
ISO 14041	Life Cycle Assessment—Life cycle inventory analysis
ISO 14042	Life Cycle Assessment—Impact Assessment
ISO 14043	Life Cycle Assessment—Evaluation and interpretation
ISO 14050	Terms and Definitions

ISO 14000 family of standards is still in progress. However, many countries already adopted the complete series. ISO 14000 was adopted by the U.S. as a national standard in early 1996. ISO Technical Committee TC 207 is responsible for the development and revision of the ISO 14000 standards.

The following steps are useful for the implementation process of ISO 14000.

- Senior management must support the implementation of the system.
- Choose the external environmental regulations to which your company must comply.
- Make an internal audit of the environmental system.
- Identify the environmental regulations and issues.
- Develop and write a strategic environmental plan.
- Start implementing the strategic plan.
- Continually monitor the progress.

ISO 14000 can be used as a powerful marketing tool since environmental friendliness is popular with individuals, companies, and government agencies. However, it should be noted that ISO 14000 does not guarantee an environmentally friendly company as ISO 9000 does not guarantee quality. Moreover, ISO 14000 does not automatically provide compliance with the local, state or federal regulations.

Associations for Standards, Regulations and Specifications

International Standards Organization (ISO)
Rue de Varembe 1
CH-1211 Geneva 20
Switzerland
Phone: 41 22 749 0111
Fax: 41 22 733 3430
http://www.iso.ch

American National Standards Institute (ANSI)
11 West 42nd Str., 13th Floor
New York, NY 10036
USA
Phone: (212) 642-4900
Fax: (212) 302-1286
http://www.ansi.org

American Society for Quality Control (ASQC)
P.O. Box 3005
Milwaukee, WI 53201
USA
Phone: 800-248-1946
http://www.asqc.org

American Society of Testing and Materials (ASTM)
100 Barr Harbor Drive
West Conshohocken, PA 19428-2959
USA
Phone: (610) 832-9585
Fax: (610) 832-9555
http://www.astm.org

American Association of Textile Chemists and Colorists (AATCC)
P.O. Box 12215
Research Triangle Park, NC 27709
USA

Environmental Protection Agency (EPA)
National Headquarters
401 M-Street, SW
Washington, DC 20460
USA
Phone: (202) 260-5917
Fax: (202) 260-3923
http://www.epa.gov

Industrial Fabrics Association International (IFAI)
345 Cedar St., Suite 800
St. Paul, MN 55101-1088
USA
Phone: (612) 222-2508
Fax: (612) 222-8215

National Institute of Standards and Technology (NIST)
US Department of Commerce
Administration Building, Room A629
Gaithersburg, MD 20899
USA
Phone: (301) 975-5923
http://www.nist.gov

National Standards Association
1200 Quince Orchard Boulevard
Gaithersburg, MD 20878
USA

Occupational Safety and Health Administration (OSHA)
401 M Street SW

Washington, DC 20460
USA
Phone: (202) 219-4667
http://www.osha.gov

Registrar Accreditation Board (RAB)
611 East Wisconsin Avenue
P.O. Box 3005
Milwaukee, WI 53201-3005
USA
Phone: (414) 272-8575
Fax: (414) 765-8661

REFERENCES

1. Cahill, N., "Produce Commodity Fabrics Competitively", Textile World, October 1996.
2. "How to Choose Air Conditioning to Maximize Profitability", Textile Maintenance and Engineering, Textile World, May 1998, page 67.
3. Melling, K. G., "From Screen to Machine: Profitability by Design", Textile World, May 1998.
4. Heer, A. and Scapin, A., An Up-front Involvement, Sulzer Ruti Technical Review, 1/97.
5. Clements, R. B., Complete Guide to ISO 14000, Prentice Hall, 1996.

SUGGESTED READING

- "CAD Takes Over", Textile World, August 1997.
- Cahill, N., "Plan Manufacturing Strategy with Common Company Vision", Textile World, April 1993.
- Cahill, N., "Produce Commodity Fabrics Competitively", Textile World, October 1996.
- Cahill, N., "Strategy for Designing World Class Textile Plants", ATI, October 1990.
- Dockery, A., "CAD Speeds Quick Response", ATI, March 1991.
- Dockery, A., "Electronic Purchasing", ATI, February 1991.
- Fiorino, S., "ISO Registration Changing Global Business", ATI, May 1993.
- Fulmer, T. D., "World Class Textile Manufacturing", ATI, August 1990.
- Isaacs, III, M., "Weaving Strides Into 21st Century", Textile World, February 1997.
- Jackson, S. L., "ISO 14000: What You Need to Know", ATI, March 1997.
- Jayaraman, S., "Making Information a Competitive Edge", ATI, March 1997.
- Little, T. J., "Quick Response 90", ATI, June 1990.
- Lomax Jr., J. F., "Information Technology and Workplace Privacy", ATI, February 1997.
- Magloth, A., "Virtual 3-D Design of Close-Fitting Clothing", ITB International Textile Bulletin, 6/98.
- Melling, K. G., "CAD/CAM: Screen to Machine", Textile World, February 1998.
- Melling, K. G., "Print Patterns: Creativity, Precision and Skill", Textile World, November 1998.
- Mohamed, M. H., "The Word in Weaving at ITMA: Automation", ATI, December 1991.
- Perkins, W. S., "Making Textiles A Little Greener", ATI, April 1998.
- Sawhney, A. P. S., "How to Improve Air-jet Weaving Efficiency", Textile World, May 1990.
- Schwendimann, R. and Plaschy, M., "Reducing Costs at Weaving and Warp Tying", Textile Horizons, May 1990.
- Seabrook, B. and Seabrook, C., "Cutting Plant Operating Costs", ATI, May 1993.
- Sprinkle, S. D. and McCoy, D. W., "Corporate CIM Data", Textile World, April 1991.
- West, J., "CIM—Changing the Image and Output of Manufacturing", ATI, October 1990.

REVIEW QUESTIONS

1. What is the difference between quick warp change and quick style change?
2. Why is air conditioning important in textile plants? Explain.
3. What is EDOS? Explain how it works.
4. What are the latest CAD/CAM techniques used in weaving?
5. Is there any relationship between the quality of a fabric produced in a plant and its being ISO 9000 certified?

15

Future of Weaving

In today's fast moving technological world, it is hardly possible to predict the future of any industry even six months down the road. However, if one gets some encouragement from the relatively stagnant principles of weaving over the history, an attempt can be made to make some guesses with reasonable accuracy for the near future which is the purpose of this short chapter.

Weaving is an increasingly demanding process. Economical manufacturing of woven fabrics is becoming more and more important. As a result, high performance and flexible electronic systems are being used in weaving. Multiprocessors are used to control, monitor and communicate functions. Electronics increased the processing speed, flexibility and reliability of the weaving machines over the years. Control systems are equipped with production statistics, efficiency calculation and various other counters. The data can be retrieved from the machine with an interface. Fabric parameters, patterns, colors and control functions can be entered at the communication panel available on the machines. The pattern can also be entered with a computer, serial interface or a memory card.

Weaving will continue to be the dominating way of fabric formation in the world. This is due to the advantages of woven fabrics compared to the other fabric types. The principles of interlacing yarns to make a woven fabric have not been changed since the beginning of this industry. However, there has been drastic changes in the equipment used in weaving especially within the last half century.

The speed of weaving machines has increased dramatically within the last two decades. As a result, the productivity per machine has been increased considerably. For example, between 1990 and 1998, the area of fabric produced per weaving machine hour was more than doubled, i.e., 14 m^2 per machine hour in 1990 versus 32 m^2 per machine hour in 1998 [1].

Weaving machine speeds continue to increase. Of course, there is no need to say that the existing some 2.5 million shuttle looms in the world will be replaced by shuttleless looms. It is expected that the machine speeds and filling insertion rates of single phase projectile, air-jet and flexible rapiers will also increase. However, it is reasonable to state that there may be not much room for speed improvement in single phase machines due to physical limitations. Therefore, the major improvements are to be expected in the multiphase weaving area. M8300, in particular, offers great potential to increase the machine speed and filling insertion rate from current levels of around 2850 ppm and 5400 mpm, respectively. Increasing reed width will help to increase filling insertion rate.

Nevertheless, there are research programs to further improve the operation of single phase weaving machines. For example, pneumatic beat-up is being studied to replace the traditional reed in air-jet weaving. In pneumatic beat-up, compressed air is to be used to push the filling yarn into the cloth fell. However, the quality of the beat-up remains to be seen in this new concept. As another example, some shedding mechanisms can be placed directly under the machine.

Higher speed weaving machines consume warp beams faster. Therefore, to reduce the frequency of warp-outs, very large warp beams are becoming popular. This necessitates larger slasher heads, improved material handling and beam storage [1].

Color selection capabilities of weaving machines is also expected to increase. Currently, air-jets, projectiles and flexible and rigid rapiers offer six or even more colors.

Automation through electronics and computers will increase. With the exception of the insertion systems, machine builders are making parts of different type weaving machines interchangeable. For example, air-jet machine and rapier machine parts are made interchangeable to reduce the spare parts inventory in the mill. This trend is expected to continue.

Weaving productivity and flexibility have been increased considerably as a result of systems such as quick style change, off-loom take ups, inverter drives, filling feeders, electronic let-off and take-up, automatic filling repair and new monitoring systems. Innovations to further increase productivity and flexibility are expected to continue which may include automatic warp stop repair, automatic fabric doffing, automatic filling supply system, etc. [1].

With the ever expanding application areas of high performance industrial fabrics, the weaving machines will be further modified to meet the market demands. Table 15.1 shows the increase in industrial market segments in the 1990s. As a result of this, the machines to make industrial textiles will be wider and stronger. As an example, the rapier and projectile principles are now being used in weaving machines to produce paper machine clothing which requires up to 30 m wide weaving machines. Until a few years ago, shuttle was the only insertion method used in those machines.

It is expected that in the future on a global basis, rapier weaving machines will have the largest market share of single phase weaving machines followed by air-jet, water-jet and projectile [3].

Further discoveries are expected in new polymers and fibers that are even higher performance than the current ones. The strength of textile fibers will be increased further, resulting in "hyperstrong" fibers. Parallel to this, new yarn structures may be developed. Therefore, these new materials will open new markets for textiles. This may require new challenges from the weaving machines.

Environmental awareness in public and regulations by the governments will open up new growth areas in "environmentally friendly" materials. Currently, research is being done in this area to develop "biodegradable" fibers. In this respect, sizing is not a value added process and therefore, any reduction or complete elimination of sizing would be a dream come true for woven fabric manufacturers.

Fabric versatility will improve further. One area that has great potential for improvement is the manufacture of weaving machines for complex three-dimensional (3D) shapes. Traditional woven fabrics have been planar structures, i.e., the dimension in the thickness direction can be neglected compared to the dimensions in the X-Y plane. However, with the industrial textiles penetrating into almost every industry in the world almost daily, 3D woven fabric structures are gaining more importance. Interestingly enough, to the best of the author's knowledge, traditional weaving machine manufacturers have not shown enough interest in building weaving machines that can weave 3D structures. Development of 3D weaving machines so far has been confined to universities and research centers such as NASA. As a result, there has been no commercially available 3D weaving machine—yet. However, this is expected to change eventually with increasing market share of 3D fabrics.

New fabric development will be a key factor for survival and success in the 21st century. It is expected that new products will account for 30% of the profits and 37% of the sales growth

TABLE 15.1 Fabric usage for some industrial market segments (estimated; units in millions of square yards, except where indicated) [2].

Industrial Market Segment	1991	1992	1993	1994	1995
Airbag	4	7	12	20	32
Architectural fabrics	3	3	3.25	3.5	3.7
Automotive	285	290	300	292	347
Awning and canopies	19	19.5	19.5	20	22
Banners and flags	9	10	11	13	15
Casual furniture	49	50	50	50	52
Geosynthetics	360	380	400	435	466
Marine products	19	18	18.5	19.2	17
Medical (in billions of sq. yards)	2.1	2.3	2.4	2.6	2.8
Safety and protective	225	250	275	300	322
Single-ply roofing	115	110	110	115	117
Tarpaulin	105	110	110	112	113
Tent and tent rental	10	10	10	10.5	11

over the next several years [4]. However, introduction of new fabrics is not an easy task. On average, only one in 20 new products has been successful for practical purposes.

It is possible that the weaving machine suppliers may have to become a system supplier. They may have to get involved in the logistics of the weaving machine environment, such as information flow and material flow. Some of this is already happening with quick style change and similar processes.

Efforts to reduce vibration, noise level and energy requirements of weaving machines will continue. A fabric produced on a weaving machine is very light and flexible compared to the mass and strength of the weaving machine. Maybe, there is no need for such an over-kill machine to make a light, flexible structure such as a woven fabric once we understand the textile materials, weaving process and the interaction between the two better.

REFERENCES

1. "Taking Stock of Weaving Innovation", ATI, February 1999.
2. Strzetelski, A., "Here Comes the Future", Industrial Fabric Products Review, January 1995.
3. "Sulzer Scales New Heights", ATI Special Report, February 1999.
4. Smith, B., private communication

SUGGESTED READING

- "What are the Main Priorities for Weaving in the Future?", ITB International Textile Bulletin, 3/98.
- "World Textiles: The Future", Textile Horizons, March 1990.
- Delporte, C. H., "Textiles on the Edge of 2000", ATI, May 1999.
- Mohamed, M. H., "The Future of Fabric Formation Technology, Weaving: Up Front at ITMA", ATI, November 1998.
- Seidl, R., "Current Trends in Weaving Machine Construction", Melliand International (1), 1998.

QUESTION

In your opinion, what will be the highest weaving machine speed and the highest filling insertion rate in year 2007? Which insertion system(s) will achieve them?

Appendix 1

Characteristics of Major Weaving Machines Produced in the World

(Source: Textile Horizons)

Rapier Looms At ITMA 1999 and Their Performance.

Manufacturer	Model	Style	Speed PPM	Reed Width (Cloth Width cm)	WIR m/min	Shedding
Dornier *(Germany)*	PTV 4 1516	Shirting	570 x2	190 (175.5) Double pick insertion	2,000	Stäubli Dobby 2861
	HTV S 4 /S20	Technical Fabric	500	220(198)	990	Stäubli Dobby 2861
	HTVS 81518	Worsted	520	220(183.5)	954	Stäubli Dobby 2861
	HTVS121,1	Upholstery	500	180(155)	775	Stäubli JacquardLX320
Somet	Super Excel HTP	Apparel	430	360(350)	1,505	Stäubli Dobby 2622
	Super Excel HTP	Upholstery	600	230(210)	1,260	Bonas DSJ 2688
	Super Excel HTP	Apparel	700	190(175)	1,225	Saubli Rotary 2861
	Thema Super Excel	Technical Fab.	350	340(330)	1,150	Stäubli Rotary Dobby
	Super Excel HTP	Cool Wool	550	210(176)	968	Stäubli Rotary 2670
	Thema Super Excel	Shirting	550	190(172.2)	947	Fimtextiles E.Jacquard
	Thema Super Excel	Upholstery	600	190(149.6)	898	Stäubli JacquardLX 320
Picanol *(Belgium)*	Gamma-4-R-340	Fancy Voile	440	340(312)	1,372	—
	Gamma-8RA90	Shirting	700	190(188)	1,316	Positive Dobby
	Gamma-FF-4-R-190	Ladieswear	600	190(176)	1,056	—
	Gamma-8-J-190	Furnishing	615	190(150)	922	Stäubli Jacquard
	Gamma-4-R-220	Mens outerwear	530	220(192)	1,017	Positive Dobby
	Gamma-4-R-190	Awning Fabric	630	190(128)	806	—
Sulzer *(Switzerland)*	G6300-13200-N 8 SP20	Ladieswear	700	200(190)	1,330	Stäubli Dobby
	G6300-B340 N 8 Sp20	Bedspread	430	340(297)	1,277	Stäubli Dobby
	G6300-B190 N 8 SP20	Shirting	700	190(172)	1,206	Stäubli Dobby
	G6300 S320 N 8 J	Upholstery	360	320(306)	1,100	Jacquard
	G6250 W1 90 N 8 SP1 6	Menswear	600	190(175)	1,050	Stäubli Dobby
	G6250 B190 N 8 SP16	Blouses	600	190(167)	1,000	Stäubli Dobby
	G6300 W220 N8 Sp20	Menswear	540	220(179)	966	Stäubli Jacquard
	G6200 W220 N6 SP28	Menswear	540	220(179)	966	Elitex Jacquard
	G6200 B190 N 8 J	Brocade	600	190(156)	936	Stäubli Jacquard
	G6200 B190 N 6 J	Hometextiles	600	190(155.3)	932	Schleicher Jacquard
	G6200 B190 F 8 J	Terry Towel	500	190(3x61.5)	922	Grosse Jacquard

(continued)

Rapier Looms at ITMA 1999 and Their Performance (continued).

Manufacturer	Model	Style	Speed PPM	Reed Width (Cloth Width cm)	WIR m/min	Shedding
	G6300 S170 N 8 J	Scarves	600	170(150)	900	Jacquard
	G6200 S140 N 8 J	Labels	630	140(132.6)	835	Jacquard
Vamatex *(Italy)*	9000 PLUS	Blanket	325	300(260)	845	Stäubli E. Jacquard
	9000 PLUS	Conveyor Belt	480	192(162)	777	Fimtextile R. Dobby
Vamatex Negative Rapiers *(Italy)*	LEONARDO	Cotton Fabric	430	360(164x2)	1,410	Fimtextile R. Dobby 3010
	LEONARDO	Table Linen	450	320(307)	1,381	Stäubli Jacquard LX3200
	LEONARDO	Curtains	400	360(311)	1,368	Fimtextile RD 3010
	LEONARDO	Lining	830	170(150)	1,245	Fimtextile RD 3010
	SP 1151	Terry	345	360(334)	1,152	Grosse Jacquard
	LEONARDO	Shirting	660	190(170)	1,122	Fimtextile R. Dobby
	LEONARDO	Denim	640	190(175)	1,120	Stäubli R. Dobby 2861
	LEONARDO	Woollen Fab.	560	220(198)	1,111	Stäubli R. Dobby 2867
	LEONARDO	Clothing	650	190(159)	1,033	Bobbio Jacquard
	LEONARDO FTS	Ind. Fabric	560	190(183)	1,024	Stäubli R. Dobby 2861
	P1001SUPER	Womenswear	550	190(180)	990	Stäubli R. Dobby 2668
	LEONARDO	Furnishing	650	190(148)	962	Gross Jacquard
	LEONARDO	Furnishing	650	190(147)	955	Bonas Jacquard
	LEONARDO FTS	Ties	560	190(144)	806	Bonas Jacquard
Panter Negative Rapiers *(Italy)*	E 4X H 1900	Sportswear	820	190(160)	1,307	Stäubli Dobby
	E 4X H 3400	Furnishing	420	340(303)	1,272	Stäubli E. Jacquard
	E 4X H 1900	Furnishing	650	190(157)	1,020	Bonas Jacquard
	E 4X H 2100	Curtains	462	210(157)	725	Stäubli E. Dobby
	E 4X H 1900	Shirting	650	190(172)	1,118	Stäubli Dobby
Jacob Muller *(Switzerland)*	Mugrip 3 M13J3 111150	Labels	730	115	840	Jacqu Mughr 3 M13J3
	111150	Labels	700	(115)	805	Jacquard
Sapa Textil *(Spain)*	LEADER	Furnishing	625	1190(175)	1,096	Stäubli Dobby 2670
	LEADER		675	190(177)	1,194	Stäubli Dobby 2670
Panter *(Italy)*	E 4X H 3400	Curtains	320	340(340)	1,088	Stäubli E. Dobby
ICBT *(France)*	Proton	Voile	270	360(356)	961	Stäubli Dobby 2668
	Proton	Check	600	190(155)	930	—
Vaupel *(Germany)*	EWM 90 180	Labels	800	189 (2x84.4)	1,350	Jacquard LX16001JC5
	EWM 80 - 130	Labels	654	130(125)	817	Jacquard IBJ/S250
	EWM - 100 E	Labels	600	(100)	600	—
CTM *(China)*	SINIL		504	200(165)	841	Stäubli Dobby 2668

Air-Jet Looms at ITMA 1999 and Their Performance.

Manufacturer	Model	Style	Speed PPM	Reed Width (Cloth Width cm)	WIR m/min	Shedding
Sulzer *(Switzerland)*	M8300 Multiphase	Cretonne	3,230	190(188.5)	6,088	(Demonstration) Cam
	M8300 Multiphase	Workwear	2,430	190(170)	4,118	Cam
	L5300 B250-F4-SPTE	Terry Towel	600	250(4x58)	1,390	Positive Stäubli 2861
	T4300-B260 F 6 J	Terry Towel	540	260 Z50	1,350	Stäubli Jacquard

Air-Jet Looms at ITMA 1999 and Their Performance (continued).

Manufacturer	Model	Style	Speed PPM	Reed Width (Cloth Width cm)	WIR m/min	Shedding
Tsudokoma	ZAX-1190-2C4S	Filament	1,800	190(179)	3,222	—
(Japan)	ZAX-390-2C-C4	Sheeting	800	390(372)	2,950	Positive Cam
	ZAX-340-6C-D16	Curtain	700	340(310)	2,170	E. Positive Dobby
	ZAX-340-8C-J	Towel	700	340(304)	2,128	Electronic Jacquard
	ZAX-230-6C-D16	Spandex Wvg.	900	230(212)	1,908	E. Negative Dobby
	ZAX-190-4C-C6	Dyed fabric	1,000	190(170)	1,7001	Positive Cam
	ZAX-190-2C-C8	Denim	1,000	190(164)	1,640	Positive Cam
	ZAX-210-6CD16	Worsted	850	210(190)	1,615	—
Picanol	OMNI-4-P-380	Bedsheeting	720	380(378)	2,721	Electronic Dobby
(Belgium)	OMNI-F-2E-190	Lining Fabric	1,600	190(149)	2,384	Crank Motion
	OMNI-F-2-P-340	Voil	750	340(312)	2,340	Electronic Dobby
	OMNI-4-J-250	Mattress	950	250(220)	2,090	Stäubli E. Jacquard
	OMNI-6-J-340	Upholstery	600	340(302)	1,812	Bonas E. Jacquard
	OMNI-6-R-190	Upholstery	900	190(150)	1,350	Electronic Dobby
	OMNI-4-J-190	Fancy Lining	—	190(155)	—	Electronic Jacquard
	DELTA-X-F-2-E	Ladies Outer Wear	900	190(175)	1,575	Crank Symmetric
	DELTA-X-4-R-190	Shirting	750	190(170)	1,275	Positive Dobby
Toyoda	JA2S-390DE-MT-T610	Dobby Fabric	600	390(2xl 75)	2,100	Electronic Dobby
(Japan)	JA2S-190TN-MF-T610	Stretch Fabric	1,200	190(165)	1,980	Cam Shedding
	JA2SA90TP-EF-T610	Denim	1,110	190(165)	1,831	Positive Cam
	JA6FA 90DE-EF-T610	Value Added	900	190(170)	1,530	Negative Electronic
	JA4FA90DE-MF-T6119	Stretch Fabric	850	190(175)	1,487	N-Electronic Dobby
	JA4SF-1 90TE-EF-T610	Different Weaves	800	190(170)	1,360	Electronic Dobby
Dornier	LWV 8 1 J	Sheeting	600	430(420)	2,520	Stäubli Jacquard LX
(Germany)	LWV 2 1 E 4	Workwear	600	380(334)	2,000	Positive Cam
	L~ 2 1 E 4	Twill Fabric	1,00	190(175)	1,750	Stäubli Positive Cam
	DWL 8	Labels	1,530	168	1,530	Stäubli Jacquard CX880
	LTNF 8 1 J	Terry Towel	620	260(245)	1,519	Gross Jacquard
	DWT 8	Labels	1,100	126	1,380	Stäubli CX 880
	ATV 41 S 20	Upholstery	700	190(188)	1,316	Stäubli CX 880
Somet	Clipper	Voile	750	340(315)	2,363	Tappet Fimtextile 2024
(Italy)	Clipper	Voile	650	340(315)	2,048	Tappet Fimtextile 2024
	Clipper	Denim	1,000	190(175.5)	1,755	Tappet Fimtextile 2024
	Clipper	Upholstery	580	340(296)	1,718	Stäubli Jacquard CX880
	Clipper	Linning	1,000	190(156.4)	1,564	Stäubli Jacquard CX880
	Clipper	Apparel	900	190(169)	1,521	Stäubli Rotary 2871
	Clipper	Sportswear	900	190(163)	1,467	Stäubli Rotary 2871
Trustfin	Techno 440 PT Servo		400	440(400)	1,672	Dobby
(Czech)	Techno 240 PT		650	240(230)	1,495	Crank
	190 CT		650	190(182)	1,831	Cam Shedding
	Techno 150 GTS		700	150(120)	840	Crank
Gunne	190 TC	Tyre Cord	700	190(173)	1,211	—
(Germany)	260 AIR-F	Towel	620	260(243)	1,506	Stäubli Rotary Dobby 2670
Muller AG	MWET MBJL111 150	Labels	1,400	(115)	1,610	Jacquard
(Switzerland)	MWET M13,11-111150	Labels	1,000	(115)	1,150	Jacquard SPE
Vaupel	EWM 90-180 E	Labels	800	2x84.4	1,350	Jacquard LX1600JC5
(Germany)						

Water-Jet Looms at ITMA 1999 and Their Performance.

Manufacturer	Model	Style	Speed PPM	Reed Width (Cloth width cm)	WIR m/min	Shedding
Tsudlakoma	ZW-405CA90AC-4S	Taffeta	1,600	190(170)	2,700	—
(Japan)	ZW405-210-2C D 16	Pattern	1,000	210(196)	1,960	Electronic N. Dobby
Toyoda	LW1F-19-CS-EF-602	Heavy Quality	1,500	190(173)	2,595	Crank
(Japan)	LW2F-2110TP-1V117-603	Heavy Weught	900	210(190)	1,710	Positive Cam
	ILW-4F-211 OFE-EF-601	Decorative Fabric	700	210(197.6)	1,383	N. Electronic Dobby

Projective Looms at ITMA 1999 and Their Performance.

Manufacturer	Model	Style	Speed PPM	Reed Width (Cloth width cm)	WIR m/min	Shedding
Sulzer	P7300-B390-N2 EP-R-DI	Stretch Denim	360	390(388)	1,400	Tappet (Demonstration)
	P7250B360NMSASPD12	Curtain	365	360(349)	1,273	Stäubli Dobby
	P7150B360N2EPRROD1	Geotextile	300	430(418.5)	1,255	Tappet
	P7150B360N2EPRROD1	Leisure	370	360(2x168.5)	1,250	Tappet
	P7250B360N4MSASPD12	Fitter Fabric	300	360(2xl 80)	1,080	Stäubli Dobby

Appendix 2

Troubleshooting in Warping and Sizing

(Source: Phil-Chem, Inc. and Southern Phenix Textiles)

Creel Section Checklist (Section Beams):

- Section beams should be kept in good condition
- Distance between heads should be checked
- Heads should be polished and checked for warping
- Barrels should be checked for burrs
- Density of prepared section beam should be checked periodically
- Align section beams
- Do not "jam" any beam bearings
- Check for proper horizontal and vertical centers
- Check distance from creel to size box on each size

Check List for Brakes on the Slasher:

- Ropes, weights or bearings must be maintained
- All bearings should be the same type
- Air pressure should be the same if using a disc system
- Creel stretch should be checked using a portable stretch transducer and extension cord of mechanical clocks

Running Out Old Set on the Slasher:

- Watch section beams closely
- Use all good yarn on beams
- Do not let any beams run out
- Mark yarn when set is out

Winding the Slasher Out:

- Pump size from boxes
- Wash size from rolls
- Raise rolls
- Turn off steam
- Pull over and doff warp

While Set Is Off:

- Clean slasher, especially size boxes
- Clean filters
- Check rolls
- Perform minor maintenance

Laying in a New Set:

- Check each beam for the correct:
  —number of ends
  —yarn number
  —blend
  —ticket vs. beam info

Counting—Striking:

- Yarn straight
- Do not roll yarn
- Pick light and heavy dents
- Count selvage ends

Dry Lease:

- Do not start slasher without lease

Storage Kettles:

- Coils—no size build-up, etc.
- Temperature—slightly lower than cooking
- Agitation—normally around 45 rpm

Restarting Slasher:

- Check all settings such as:
  —box temperature
  —roll pressure
  —can temperatures
  —moisture setting
  —stretch settings

Patrolling the Slasher:

- Size level in boxes
- Size is foaming or not
- Laps—boxes and/or beams
- Drawbacks
- Solids in boxes

Leasing:

- Rods should be spaced properly, level and parallel
- The rods should contain no burrs, etc.
- Splits should be even to ensure proper dry yarn splitting
- Do not skip
- Prevents stuck ends
- Prevents crossed ends

Excessive Ends Out of Lease Caused by:

- Too much stretch
- Too much size in yarn
- Lumps in size
- Defective squeeze roll
- Misalignment
- Worn bearings
- Grooved or rough roll
- Too little stretch
- Worn delivery roll covering
- Weak yarn
- Rolled ends

Comb Checklist:

- Should have the amount of dents
- Dents should be straight and free of burrs
- Dents should have proper height and spacing
- Yarn should be adjusted so that it does not drag the bottom of the comb

Breakouts in Comb Caused by:

- Hard size
- Wild yarn
- Rolled or crossed ends
- Gouts or lint sized to yarn
- Waste falling from hood
- Lap in size box too long

Excessive Shedding Caused by:

- Size not properly cooked
- Size temperature too low
- Squeeze roll pressure incorrect
- Defective squeeze rolls not properly dressing yarn
- Improper moisture regain setting
- Excessive stretch in lease section

Delivery Roll/Front End Checklist:

- Delivery roll should be covered properly and clean
- Should be free of burrs, scratches, etc.
- Run correct tension between delivery roll assembly and the loom beam
- Pack roll should have correct pressure and free of cuts and burrs

High and Low Selvages on Loom Beams:

- Warped heads on loom beams
- Journal worn on loom beams
- Arbor bearings worn on slasher
- Expansion comb improperly set
- Press roll not traveling properly

Slasher Oil Caused by:

- Size boxes not properly cleaned

- Wrong packing in stuffing gland
- Over greased squeeze roll bearings
- Dripping from hood or duct
- Oil leaking from motor on cooking or storage kettle
- Over oiling drive chain

Doffing:

- Place tape as instructed
- Do not roll selvage
- Store properly

Size Preparation:

- Must have correct water to start
- Correct type of size
- Correct temperature
- Correct cooking time

Laps on Section Beams and Rolls:

- Section beam weight vs. ends per beam
- Weak yarn
- Drawbacks or crossed ends from warper
- Rough drum at warper cutting yarn
- Lost ends at warper
- Incorrect slasher stretch
- Rough or grooved rolls and lease rods
- Misalignment of slasher
- Worn bearings at slasher
- Uneven tension at warper
- Rough or scarred section beam heads

Cutting Laps:

- Knife flat on section beam
- Cut high first pass
- Cut lower next, but not too low

Breaking Out Drawbacks:

- Tie back to same end
- Tie back to adjacent end is not good

Creep Speed:

- Necessary "evil"
- Do not creep for breaks
- Let helper watch

Creep speed causes the following:

- reduced yarn strength
- reduced yarn elongation
- reduced yarn abrasion resistance
- increased hairiness
- significantly higher warp stops

Drying Section Checklist:

- Run correct steam pressure in cylinders
- Use just enough tension between cylinder sections and delivery roll assembly to prevent yarn from looping back on split rods
- Check traps and pigtails for each section
- Temperature controllers and recorders should be maintained
- Cylinders should be inspected for cuts, dents, buildup or bad coating
- Cylinder bearings should be checked
- Chain drives should be lubricated and tight
- Siphon tubes should be checked

Hoods/Exhaust System:

- Should be clean to prevent material from falling onto the sized warp yarn
- Should have a small negative pressure so that the vapor will be exhausted
- Should be dry so that moisture does not drip on the dried yarn causing a dye resist spot

Rolled Ends Are Caused by:

- Ends out of lease
- Crossed ends in comb
- High or low selvages on section beams
- Section beams not aligned correctly

Stuck Ends Are Caused by:

- Not leasing
- Ends out of lease
- Loose ends on section beams
- Beam alignment in creel
- Excessive moisture in warp
- Size box temperature too low
- Low or slack selvages on section beams
- Ridged section beams
- Ends not counted in correctly
- Ends broken and not put in right dent

Ends Running in and out Are Caused by:

- Laps on section beams
- Very high stretch
- Mechanical problems
- Leasing procedures
- Hard size

Hard Size Is Caused by:

- Size boiling or excessive agitation in boxes
- Size skimming in boxes
- Slasher standing too long without jogging
- Low temperature in size box
- Grooved or cracked squeeze roll
- Lumps in size due to improper agitation and cooking
- Size box not properly cleaned

Finished Volume High—Solids Are Low:

- Water meter error
- Water meter leak
- Defective seal
- Excessive condensate

Finished Volume Low—Solids Are High:

- Possible water meter error
- Check by measuring water

Finished Volume Low—Solids Correct:

- Drain valve under kettle leaking
- Defective three way valve
- Pumping raw size to storage

Finished Volume Slightly Low—Solids High:

- Size let out of mix

Solids Correct at Cook—Low at Storage:

- Coil in storage may be leaking
- Empty the kettle
- Cover coil with water
- Turn on steam
- Look for bubbles

Solids Correct at Cook and Storage—Low at Boxes:

- Possible condensate problem for traps

Soft Warps:

- Size level too low
- Size temperature incorrect
- Squeeze roll pressure too high
- Excessive foam in size box
- Sizing ingredients not weighed correctly
- Size not properly cooked

Wet Warp:

- Moisture control not operating properly
- Condensation dripping from hood
- Defective squeeze roll

Appendix 3

Typical Sizing Examples of Common Fabric Styles

(Source: Robinson, G. D., "An Overview of Spun Yarn Size Formulation Selection", ATMA Slashing Short Course, Auburn University, Auburn, AL, Sept. 1993)

FIBER: 100% COTTON **WEAVE: LIGHT TO MEDIUM WEAVE**
YARN: 20/1 TO 30/1

Size Formula	Shuttle	Rapier	Projectile	Air-jet
% PVA	25–60	50–100	50–100	80–100
% Starch	75–40	50–0	50–0	20–0
% CMC				
% Wax	7–10	7–10	7–10	7–10
% Urea	5–0	5–0	5–0	6–5
% Acrylic				
% Polyester				
% Other				
% Add-On:	14–12	14–13	14–13	12–10

FIBER: 100% COTTON **WEAVE: MEDIUM TO TIGHT**
YARN: 35/1 TO 50/1

Size Formula	Shuttle	Rapier	Projectile	Air-jet
% PVA	80	80–100	80–100	65–100
% Starch	20	20–0	20–0	35–0
% CMC				
% Wax	7–10	7–10	7–10	7–10
% Urea	0–6	6–0	6–0	6–0
% Acrylic		0–6	0–6	0–6
% Polyester				
% Other				
% Add-On:	11–12	14–11	14–11	13–10

FIBER: 65/35 POLY/COTTON **WEAVE: LIGHT TO MEDIUM**
YARN: 20/1 TO 30/1

Size Formula	Shuttle	Rapier	Projectile	Air-jet
% PVA	50–100	75–100	75–100	75–100
% Starch	50–0	25–0	25–0	25–0
% CMC				
% Wax	8	8–10	8–10	8–10
% Urea		5–8	5–8	
% Acrylic				
% Polyester				
% Other	Consider Antistat			
% Add-On:	11–9	12–10	12–10	11–10

FIBER: 65/35 POLY/COTTON **WEAVE: MEDIUM TO TIGHT**
YARN: 35/1 TO 50/1

Size Formula	Shuttle	Rapier	Projectile	Air-jet
% PVA	75–100	75–100	75–100	100
% Starch	25–0	25–0	25–0	0
% CMC	Consider Using CMC Instead of Starch			
% Wax	8–10	8–10	8–10	8–10
% Urea		5–8	5–8	5–8
% Acrylic		0–6	0–6	0–6
% Polyester				
% Other				
% Add-On:	15–14	13–11	13–11	10–11

FIBER: 40/60 POLY/COTTON **WEAVE: MEDIUM TO TIGHT**
YARN: 35/1 TO 50/1

Size Formula	Shuttle	Rapier	Projectile	Air-jet
% PVA	50–75	75–100	75–100	75–100
% Starch	50–25	25–0	25–0	25–0
% CMC				
% Wax	8	8–10	8–10	8–10
% Urea	8–5	8–5	8–5	8–5
% Acrylic		0–6	0–6	0–6
% Polyester				
% Other	Consider Antistat			
% Add-On:	12–10	14–12	14–12	12–10

FIBER: 100% POLYESTER OR NOMEX® ARAMID **WEAVE: MEDIUM**
YARN: 20/1 TO 40/1

Size Formula	Shuttle	Rapier	Projectile	Air-jet
% PVA	100	100	100	100
% Starch				
% CMC				
% Wax	6–8	6–8	6–8	6–8
% Urea				
% Acrylic				
% Polyester	0–5	0–5	0–5	0–5
% Other	Consider Antistat ———			
% Add-On:	6–9	10–12	10–12	9–11

FIBER: 100% VISCOSE RAYON **WEAVE: LIGHT TO MEDIUM AND MEDIUM TO TIGHT**
YARN: 20/1 TO 30/1 AND 35/1 TO 50/1

Size Formula	Shuttle	Rapier	Projectile	Air-jet
% PVA		30–100	30–100	60–100
% Starch		70–0	70–0	40–0
% CMC				
% Wax		7–10	7–10	7–10
% Urea				
% Acrylic		0–6	0–6	0–6
% Polyester				
% Other				
% Add-On:		8–4	8–4	7–5

FIBER: 100% HWM RAYON **WEAVE: LIGHT TO MEDIUM AND MEDIUM TO TIGHT**
YARN: 20/1 TO 30/1 OR 35/1 TO 50/1

Size Formula	Shuttle	Rapier	Projectile	Air-jet
% PVA		30–100	30–100	60–100
% Starch		70–0	70–0	40–0
% CMC				
% Wax		7–10	7–10	7–10
% Urea				
% Acrylic		0–6	0–6	0–6
% Polyester				
% Other				
% Add-On:		9–5	9–5	8–6

FIBER: 65/35 POLY/COTTON **WEAVE: MUSLIN SHEETING**
YARN: 20/1 TO 30/1

Size Formula	Shuttle	Rapier	Projectile	Air-jet
% PVA	75–100	75–100	75–100	75–100
% Starch	25–0	25–0	25–0	25–0
% CMC				
% Wax	8–10	8–10	8–10	8–10
% Urea				
% Acrylic				
% Polyester				
% Other	Consider Antistat ———			
% Add-On:	10–9	12–10	12–10	11–10

FIBER: POLY/COTTON **WEAVE: PERCALE SHEETING**
YARN: 35/1 TO 45/1

Size Formula	Shuttle	Rapier	Projectile	Air-jet
% PVA	75–100	75–100	75–100	75–100
% Starch	25–0	25–0	25–0	25–0
% CMC				
% Wax	8–10	8–10	8–10	8–10
% Urea				
% Acrylic				
% Polyester				
% Other				
% Add-On:	15–13	16–14	16–14	15–12

FIBER: 100% COTTON **WEAVE: DENIM**
YARN: 10/1 TO 20/1

Size Formula	Shuttle	Rapier	Projectile	Air-jet
% PVA		0–15	0–15	
% Starch		100–85	100–85	
% CMC				
% Wax		7	7	
% Urea				
% Acrylic		15–0	15–0	
% Polyester				
% Other				
% Add-On:		13–11	13–11	

FIBER: 100% COTTON OR POLY/COTTON **WEAVE: LIGHT TOWEL, GROUND WARP**
YARN: 10/1 TO 20/1

Size Formula	Shuttle	Rapier	Projectile	Air-jet
% PVA	50–100	50–100	50–100	Proprietary
% Starch	50–0	50–0	50–0	
% CMC				
% Wax				
% Urea	5–8	5–8	5–8	
% Acrylic	6	6	6	
% Polyester	0–5	0–5	0–5	
% Other				
% Add-On:	12–11	14–12	14–12	

FIBER: 100% COTTON **WEAVE: LIGHT TOWEL, PILE WARP**
YARN: 8/1 TO 15/1

Size Formula	Shuttle	Rapier	Projectile	Air-jet
% PVA			0–100	Proprietary
% Starch	100	100	100–0	
% CMC				
% Wax	7–10	7–10		
% Urea	0–5	0–5		
% Acrylic	Various Binders Used, Usually Acrylic at 5%			
% Polyester				
% Other				
% Add-On:	2–3	3–5	5–1.5	

FIBER: 100% COTTON OR POLY/COTTON **WEAVE: HEAVY TOWEL, GROUND WARP**
YARN: 10/1 TO 20/1

Size Formula	Shuttle	Rapier	Projectile	Air-jet
% PVA	50–100	50–100	50–100	Proprietary
% Starch	50–0	50–0	50–0	
% CMC				
% Wax	5–8	5–8	5–8	
% Urea	5	5	5	
% Acrylic	0–5	0–5	0–5	
% Polyester	0–5	0–5	0–5	
% Other				
% Add-On:	14–12	16–14	16–14	

FIBER: 100% COTTON **WEAVE: HEAVY TOWEL, PILE WARP**
YARN: 8/1 TO 15/1

Size Formula	Shuttle	Rapier	Projectile	Air-jet
% PVA	0–100	0–100	0–100	Proprietary
% Starch	100–0	100–0	100–0	
% CMC				
% Wax	5–10	5–10	5–10	
% Urea	5	5	5	
% Acrylic	0–5	0–5	0–5	
% Polyester	0–5	0–5	0–5	
% Other				
% Add-On:	3–2	5–1.5	5–1.5	

Appendix 4

Examples of Temples Used in the Textile Industry

(courtesy of Broll)

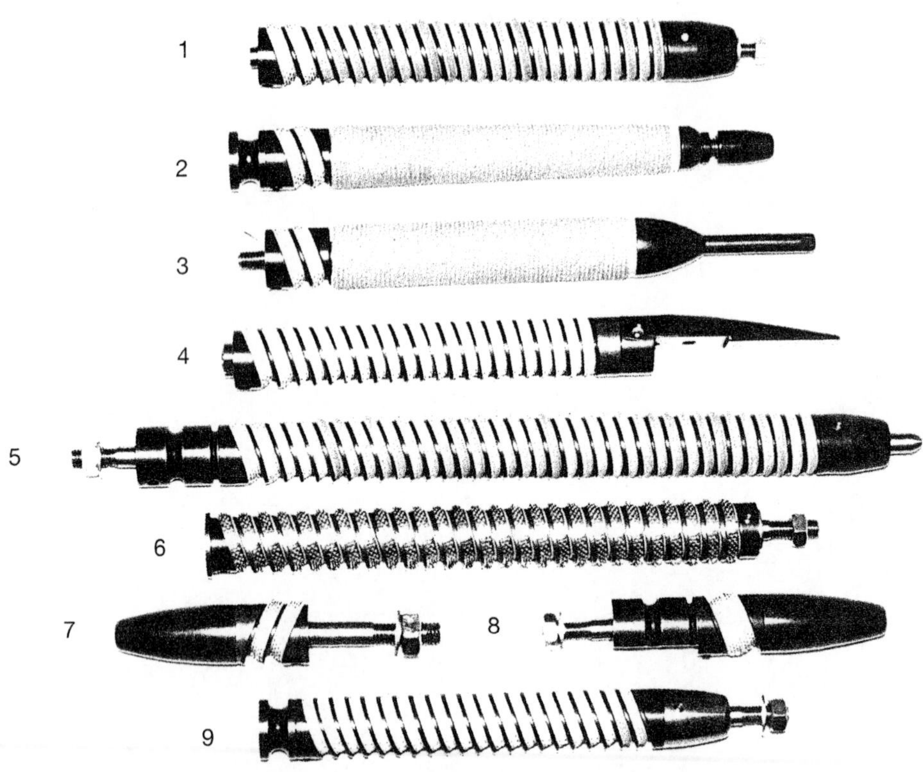

1. Breithalterzylinder mit 2 Nadelrädchen und 23 Kunstgummirädchen, differenziert, parallel oder Torsion
Temple with 2 selvage burrs and 23 rubber rings, progressive, parallel or torsion
2. + 3. Breithalterzylinder mit 2 Kantenrädchen und gerillter Kunstgummiwalze, konisch oder zylindrisch
Temple with 2 selvage burrs and threaded rubber barrel, conical or cylindrical
4. Breithalterzylinder mit 2 Kantenrädchen und 20 Nadelrädchen, differenziert, parallel oder Torsion
Temple with 2 selvage burrs and 20 burrs, progressive, parallel or torsion
5. Breithalterzylinder mit 5 Kantenrädchen und 32 Kunstgummirädchen, differenziert, parallel oder Torsion
Temple with 5 selvage burrs and 32 rubber rings, progressive, parallel or torsion
6. Breithalterzylinder mit 27 Nadelrädchen, differenziert, parallel oder Torsion, speziell für Frottier
Temple with 27 burrs, progressive, parallel or torsion, special for terry-cloth
7. + 8. Stummelbreithalter mit 2 Kantenrädchen, normal oder conterschon
Short temple with 2 selvage burrs, regular or conterschon
9. Breithalterzylinder mit 19 Nadelrädchen und 2 Kunstgummirädchen, differenziert, parallel oder Torsion
Temple with 19 burrs and 2 rubber rings, progressive, parallel or torsion

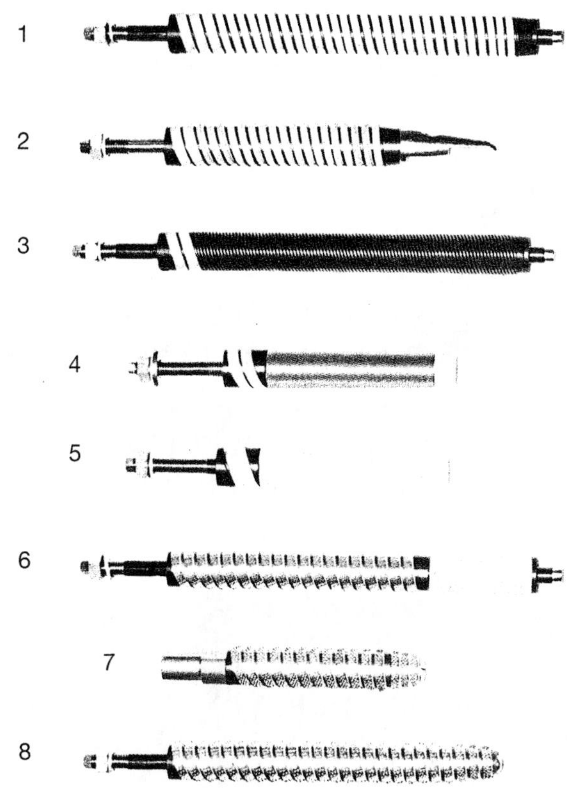

1. Breithalterzylinder mit feinen Nadelrädchen, differenzial, parallel oder Torsion
   Temple with fine pinned burrs, progressive, parallel or torsion
2. Breithalterzylinder mit 2 Kantenrädchen und Kunstgummirädchen
   Temple with 2 selvage burrs and threaded rubber barrel
3. Breithalterzylinder mit 2 Kantenrädchen und 9-gängig-gerillter Stahlwalze; z.B. für Jeans
   Temple with 2 selvage burrs and steel barrel threaded in 9 directions; e.g. for Denim
4. Breithalterzylinder mit 2 Kantenrädchen und fein behauener Stahlwalze mit Innenbefestigung
   Temple with 2 selvage burrs and fine punched steel barrel with inside-mounting
5. Breithalterzylinder mit 2 Kantenrädchen conterschon und gerillter Kunstgummiwalze mit Innenbefestigung
   Temple with 2 selvage burrs conterschon/bulldog and threaded rubber barrel with inside-mounting
6. Breithalterzylinder mit 19 Rädchen und gerillter Kunstgummiwalze mit Flachgewinde
   Temple with 19 burrs and threaded rubber barrel with flat-thread
7. Breithalterzylinder mit 15 Nadelrädchen, in der Spitze konisch zulaufend, speziell für Frottier
   Temple with 15 burrs, narrowing at the end, special for terry-cloth
8. Breithalterzylinder mit 25 Nadelrädchen, in der Spitze konisch, schräggestellt zulaufend, speziell für Frottier
   Temple with 25 burrs, inclined narrowing at the end, particularly suited for terry-cloth

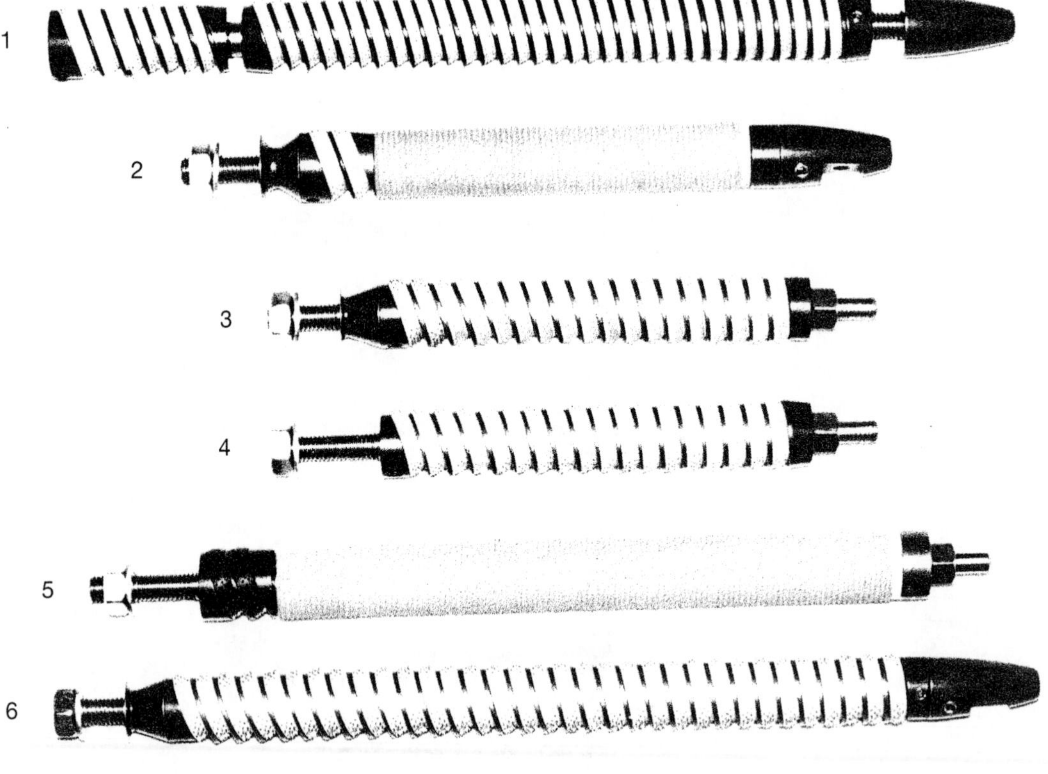

1. besonders langer Spezialbreithalterzylinder mit feinen Nadeln für Einleger
   Extra long temple with fine pinned burrs, special setup for tuck-in
2. Breithalterzylinder mit 2 Kantenrädchen und gerillter Kunstgummiwalze
   Temple with 2 selvage burrs and threaded rubber barrel
3. Breithalterzylinder mit 2 Kantenrädchen und Kunstgummirädchen, differenziert, parallel oder Torsion
   Temple with 2 selvage burrs and rubber rings, progressive, parallel or torsion
4. Breithalterzylinder mit 18 Nadelrädchen, differenziert, parallel oder Torsion
   Temple with 18 burrs, progressive, parallel or torsion
5. Breithalterzylinder mit 2 Kantenrädchen aus Kohlefaser und gerillter Kunstgummiwalze, speziell für Luftdüsenwebmaschinen
   Special temple for air-jet looms with 2 carbon fiber selvage burrs and threaded rubber barrel
6. Breithalterzylinder mit 33 Nadelrädchen mit Kugelspitzen differenziert, parallel oder Torsion, speziell für Cordgewebe
   Temple with 33 burrs with ball pins progressive, parallel or torsion particularly suited for cordfabrics

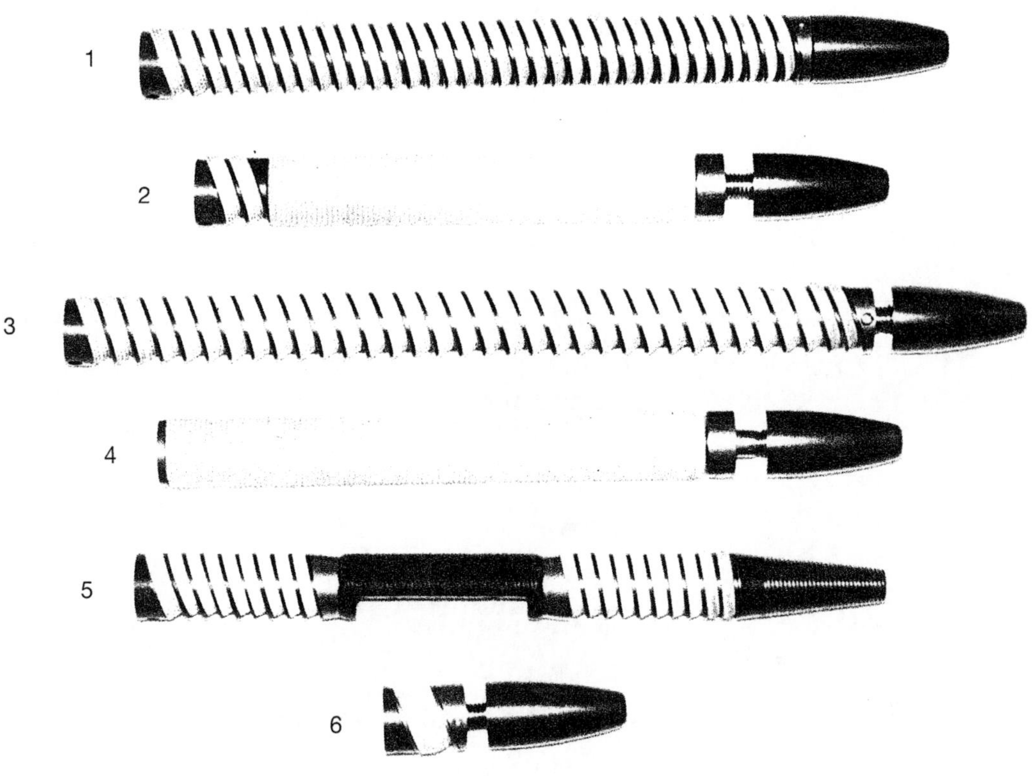

1. Breithalterzylinder mit 2 Kantenrädchen und 34 Kunstgummirädchen, differenziert, parallel oder Torsion
   Temple with 2 selvage burrs and 34 rubber rings, progressive, parallel or torsion
2. Breithalterzylinder mit 2 Kantenrädchen und gerillter Kunstgummiwalze, konisch oder zylindrisch
   Temple with 2 selvage burrs and threaded rubber barrel, conical or cylindrical
3. Breithalterzylinder mit 32 Doppelrädchen und 2 Kunstgummirädchen, parallel oder Torsion, für besonders heikle Artikel
   Temple with 32 double-burrs and 2 rubber rings, parallel or torsion, for delicate fabrics
4. Breithalterzylinder mit durchgehender gerillter Kunstgummiwalze
   Temple with rubber barrel
5. Breithalterzylinder mit Nadeleinsätzen und gerilltem Zwischenstück, differenziert oder parallel
   Temple with needle-inserts and threaded axle and nose piece, progressive or parallel inclined
6. Stummelbreithalter mit 2 Nadelrädchen conterschon
   Short temple with 2 pinned burrs conterschon

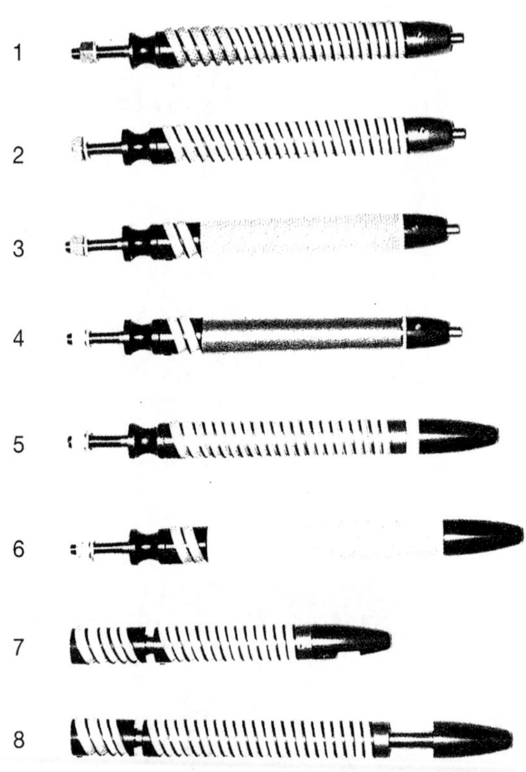

1. Breithalterzylinder mit 5 Kantenrädchen und 18 Rädchen differenzial, parallel oder Torsion
   Temple with 5 selvage burrs and 18 burrs progressive, parallel or torsion
2. Breithalterzylinder mit 2 Kantenrädchen und 22 Rädchen differenzial, parallel oder Torsion
   Temple with 2 selvage burrs and 22 burrs progressive, parallel or torsion
3. Breithalterzylinder mit 2 Kantenrädchen und gerillter Kunstgummiwalze
   Temple with 2 selvage burrs and threaded rubber barrel
4. Breithalterzylinder mit 2 Kantenrädchen und fein behauener Stahlwalze
   Temple with 2 selvage burrs and fine punched steel barrel
5. Breithalterzylinder mit 2 Kantenrädchen und 20 Kunstgummirädchen differenzial, parallel oder Torsion
   Temple with 2 selvage burrs and 20 rubber rings progressive, parallel or torsion
6. Breithalterzylinder mit 2 Kantenrädchen und 2-gängig gerillter Kunstgummiwalze
   Temple with 2 selvage burrs and rubber barrel threaded in two directions
7. Breithalterzylinder mit 2 Kantenrädchen und 17 Nadelrädchen speziell für Einleger
   Temple with 2 selvage burrs and 17 burrs – suited for tuck-in
8. Breithalterzylinder mit 2 Kantenrädchen und 25 Nadelrädchen für Einleger
   Temple with 2 selvage burrs and 25 burrs for tuck-in

Appendix 5

Units and Conversion Factors

1. U.S. CUSTOMARY UNITS

Length	1 foot = 12 inches
	1 yard = 3 feet
	1 mile = 1,760 yards = 5,280 feet
	1 inch = 1,000 mils
Area	1 square foot = 144 square inches
	1 square yard = 9 square feet
	1 acre = 43,560 square feet
	1 square mile = 640 acres
Volume	1 cubic yard = 27 cubic feet
	1 cubic foot = 1,728 cubic inches
	1 gallon = 231 cubic inches
liquid or fluid measures	
	1 pint = 4 gills = 16 ounces
	1 quart = 2 pints
	1 gallon = 4 quarts
	1 cubic foot = 7.4805 gallon
dry measures	
	1 quart = 2 pints
	1 peck = 8 quarts
	1 bushel = 4 pecks
Weights	1 ounce = 16 drams = 437.5 grains
	1 pound = 16 ounces = 7,000 grains
	1 ton (US) = 2,000 lb
Pressure	1 atmosphere = 14.7 psi

2. SI PREFIXES

multiplication factors	prefix	SI symbol
1 000 000 000 000 000 000 = 10^{18}	exa	E
1 000 000 000 000 000 = 10^{15}	pecta	P
1 000 000 000 000 = 10^{12}	tera	T
1 000 000 000 = 10^{9}	giga	G
1 000 000 = 10^{6}	mega	M
1 000 = 10^{3}	kilo	k
100 = 10^{2}	hecto	h
10 = 10^{1}	deka	da
0.1 = 10^{-1}	deci	d
0.01 = 10^{-2}	centi	c
0.001 = 10^{-3}	milli	m
0.000 001 = 10^{-6}	micro	μ
0.000 000 001 = 10^{-9}	nano	n
0.000 000 000 001 = 10^{-12}	pico	p
0.000 000 000 000 001 = 10^{-15}	femto	f
0.000 000 000 000 000 001 = 10^{-18}	atto	a

3. SI UNITS

quantity	unit	SI symbol	Formula
Base Units			
length	meter	m	
mass	kilogram	kg	
time	second	s	
electric current	ampere	A	
thermodynamic temperature	Kelvin	K	
amount of substance	mole	mol	
luminous intensity	candela	cd	
Derived Units			
acceleration	meter per second square		m/s^2
area	square meter		m^2
density	kilogram per cubic meter		kg/m^3
energy	joule	J	N.m
force	Newton	N	$kg.m/s^2$
frequency	hertz	Hz	1/s
power	Watt	W	J/s
pressure	Pascal	Pa	N/m^2
velocity	meter per second		m/s
voltage	volt	V	W/A
volume	cubic meter		m^3
work	joule	J	N.m

1 meter = 100 cm = 1,000 millimeter
1 angstrom = 10^{-8} centimeter

1 ton = 1,000 kg
1 kilogram = 1,000 gram

1 kilogram = 9.807 Newton
1 Newton = 100,000 dyn

1 liter = 100 centiliter = 1,000 milliliter = 1,000 cubic centimeter
1 cubic meter = 1,000,000 cubic centimeter

1 atmosphere = 1.01 × 10^5 Pascal
1 kg/cm^2 = 0.9678 atmosphere
1 bar = 10^6 dynes/cm^2 = 10^5 Newton/m^2
1 Pascal = 1 Newton/m^2
1 cm of mercury (0°C) = 13.60 cm of water (4°C)
1 kWh = 3.6 × 10^6 Joules
1 Watts = 1 Joules/second = 1.34 × 10^{-3} HP

4. CONVERSION BETWEEN SI AND U.S. CUSTOMARY UNITS

1 miles = 1.609 kilometer
1 yard = 0.9144 m
1 inch = 25.4 mm
1 m^2 = 1.196 square yard
1 acre = 4,047 m^2
1 cubic inch = 16.39 cubic centimeter
1 lb = 0.453 59 kg = 453.59 g
1 quart = 0.9464 liters
1 ounce = 28.35 grams
1 ounce (liquid) = 0.02957 liters
1 kg/m^2 = 9.807 Pascal = 0.001422 pounds/square inch
1 lb/square foot = 4.882 kg/m^2
1 ton (metric) = 2,205 lbs
1 BTU = 1,054 Joules
1 Joule = 0.7376 ft-lbs
1 HP = 550 ft-lb/second
1 Newton = 0.2248 lbs
1 m/s^2 = 3.281 feet/s^2

Metric to U.S. Customary Conversion Factors

to get	multiply	by
inches	centimeters	0.3937007874
feet	meters	3.280839895
yards	meters	1.093613298
miles	kilometers	0.6213711922
ounces	grams	0.03527396195
pounds	kilograms	2.204622622
gallons (liquid)	liters	0.2641720524
fluid ounces	milliliters (cc)	0.03381402270
square inches	centimeter squares	0.1550003100
square feet	meter squares	10.76391042
square yards	meter squares	1.195990046
cubic inches	milliliters (cc)	0.06102374409
cubic feet	cubic meters	35.31466672
cubic yards	cubic meters	1.307950619

U.S. Customary to Metric

to get	multiply	by
microns	mils	25.4
centimeters	inches	2.54
meters	feet	0.3048
meters	yards	0.9144
kilometers	miles	1.609344
grams	ounces	28.34952313
kilograms	pounds	0.45359237
liters	gallons (liquid)	3.785411784
milliliters (cc)	fluid ounces	29.57352956
square centimeters	square inches	6.4516
square meters	square feet	0.09290304
square meters	square yards	0.83612736
milliliters (cc)	cubic inches	16.387064
cubic meters	cubic feet	0.02831684659
cubic meters	cubic yards	0.764554858

5. TEMPERATURE CONVERSION

°F = 9/5 (°C) + 32

°C = 5/9 [(°F) − 32]

°K (Kelvin) = °C + 273

Sources

- ASTM
- Mark's Standard Handbook for Mechanical Engineers, McGraw-Hill Book Company, 1978.
- Standard Mathematical Tables, CRC Press, 1982.
- Handbook of Mathematical, Scientific and Engineering Formulas, Tables, Functions, Graphs, Transforms, Research and Education Association, 1994.

Index

X

Y

Z